IFMBE Proceedings Volume 34

Series Editor: R. Magjarevic

Previous Editions:

IFMBE Proceedings NBC 2011, "15th Nordic-Baltic Conference on Biomedical Engineering and Medical Physics"
Vol. 34, 2011, Aalborg, Denmark, CD

IFMBE Proceedings CLAIB 2011, "V Latin American Congress on Biomedical Engineering CLAIB 2011"
Vol. 33, 2011, Habana, Cuba, CD

IFMBE Proceedings SBEC 2010, "26th Southern Biomedical Engineering Conference SBEC 2010 April 30 – May 2, 2010 College Park, Maryland, USA",
Vol. 32, 2010, Maryland, USA, CD

IFMBE Proceedings WCB 2010, "6th World Congress of Biomechanics (WCB 2010)",
Vol. 31, 2010, Singapore, CD

IFMBE Proceedings BIOMAG2010, "17th International Conference on Biomagnetism Advances in Biomagnetism – Biomag2010",
Vol. 28, 2010, Dubrovnik, Croatia, CD

IFMBE Proceedings ICDBME 2010, "The Third International Conference on the Development of Biomedical Engineering in Vietnam", Vol. 27, 2010, Ho Chi Minh City, Vietnam, CD

IFMBE Proceedings MEDITECH 2009, "International Conference on Advancements of Medicine and Health Care through Technology", Vol. 26, 2009, Cluj-Napoca, Romania, CD

IFMBE Proceedings WC 2009, "World Congress on Medical Physics and Biomedical Engineering",
Vol. 25, 2009, Munich, Germany, CD

IFMBE Proceedings SBEC 2009, "25th Southern Biomedical Engineering Conference 2009",
Vol. 24, 2009, Miami, FL, USA, CD

IFMBE Proceedings ICBME 2008, "13th International Conference on Biomedical Engineering"
Vol. 23, 2008, Singapore, CD

IFMBE Proceedings ECIFMBE 2008 "4th European Conference of the International Federation for Medical and Biological Engineering", Vol. 22, 2008, Antwerp, Belgium, CD

IFMBE Proceedings BIOMED 2008 "4th Kuala Lumpur International Conference on Biomedical Engineering",
Vol. 21, 2008, Kuala Lumpur, Malaysia, CD

IFMBE Proceedings NBC 2008 "14th Nordic-Baltic Conference on Biomedical Engineering and Medical Physics",
Vol. 20, 2008, Riga, Latvia, CD

IFMBE Proceedings APCMBE 2008 "7th Asian-Pacific Conference on Medical and Biological Engineering",
Vol. 19, 2008, Beijing, China, CD

IFMBE Proceedings CLAIB 2007 "IV Latin American Congress on Biomedical Engineering 2007, Bioengineering Solution for Latin America Health", Vol. 18, 2007, Margarita Island, Venezuela, CD

IFMBE Proceedings ICEBI 2007 "13th International Conference on Electrical Bioimpedance and the 8th Conference on Electrical Impedance Tomography", Vol. 17, 2007, Graz, Austria, CD

IFMBE Proceedings Vol. 34

Kim Dremstrup, Steve Rees, and Morten Ølgaard Jensen
(Eds.)

15th Nordic-Baltic Conference on Biomedical Engineering and Medical Physics (NBC 2011)

14-17 June 2011, Aalborg, Denmark

Springer

Editors

Kim Dremstrup
Aalborg University
Department of Health Science and Technology
Fredrik Bajersvej 7D
9220 Aalborg
Denmark
Email: kdn@hst.aau.dk

Morten Ølgaard Jensen
Århus University
The Department of Thoracic and Cardiovascular Surgery
Brendstrupgårdsvej 100
8200 Aarhus
Denmark
Email: morten.jensen@ki.au.dk

Steve Rees
Aalborg University
Institute for Health Science and Technology
Fredrik Bajersvej 7D
9220 Aalborg
Denmark
Email: sr@hst.aau.dk

ISSN 1680-0737
ISBN 978-3-642-21682-4 e-ISBN 978-3-642-21683-1

DOI 10.1007/ 978-3-642-21683-1

Library of Congress Control Number: 2011930650

The IFMBE Proceedings is an Official Publication of the International Federation for Medical and Biological Engineering (IFMBE)

Typesetting & Cover Design: Scientific Publishing Services Pvt. Ltd., Chennai, India.

Printed on acid-free paper

9 8 7 6 5 4 3 2 1

springer.com

Foreword

Dear colleagues and friends,

It is our great pleasure to welcome you to the 15. Nordic – Baltic Conference on Biomedical Engineering and Medical Physics – NBC15 here in Aalborg, Denmark.

The motto of the conference is "To bring together science, education and business in Cooperation for health".

The Conference is held every third year in one of the Nordic – Baltic countries under the auspices of IFMBE - the International Federation for Medical and Biological Engineering.

In 2002 the conference was held in Reykjavik, Iceland, in 2005 in Umeå, Sweeden, in 2008 in Riga, Latvia and this year NBC15 take place on June 14-17, 2011 in Aalborg, Denmark.

The conference is organized together with the annual meeting for the Danish Society for Biomedical Engineering which will be held in parallel from Wednesday to Friday. Also on Thursday the event Windows of Opportunity is organized together with The Biomed Community organization. This event will gather entrepreneurs, inventors and investors. This time we also have incorporated a "Students Day" where our young future colleagues can participate in the scientific events and visit the exhibitions.

Also adding to the value of the conference is the 41 companies that have an exhibition at the conference site, showing their newest devices and equipment.

Conference papers is published in the IFMBE processing series and is available both in printed form and on this disk.

We are sure you will enjoy NBC2011 both scientifically and socially, and we will do our best to make NBC15 an outstanding event.

We welcome you in Aalborg!

On behalf of theNBC15 Organizing Committee:

Chairman Kim Dremstrup
President of the Danish Society for Biomedical Engineering
Head of Department
Dep. For Health Science and Technology
Aalborg University

Co-chair Steve Rees
Associate Professor
Center for Medical Model based Decision Support Systems
Aalborg University

Co-chair Morten Ølgaard Jensen
Associate Professor
The Department of Thoracic and Cardiovascular Surgery
Århus University

Preface to the IFMBE Proceeding for 15th Nordic – Baltic Conference on Biomedical Engineering and Medical Physics

Name

15th Nordic–Baltic Conference on Biomedical Engineering and Medical Physics
www.nbc15.dk

Short Name

NBC-2011

Venue

Aalborg, Denmark
June 14–17, 2011

Organized by

Danish Society for Biomedical Engineering
www.dmts.dk

Aalborg University
Department for Health Science and Technology
www.hst.aau.dk

In Co-operation with

IFMBE – International Federation for Medical and Biological Engineering
http//www.ifmbe.org

Proceedings Editors

Kim Dremstrup
Steve Rees
Morten Ølgaard Jensen

International Advisory Committee

Herbert F. Voigt	USA
Ratko Magjarević	Croatia
Metin Akay	USA
Shankar M. Krishnan	USA
Per Ask	Sweden
James Goh Cho Hong	Singapore
Thomas Sinkjær	Denmark

Scientific Committee

Herbert F. Voigt	USA
Ratko Magjarević	Croatia
Metin Akay	USA
Shankar M. Krishnan	USA
Per Ask	Sweden
James Goh Cho Hong	Singapore
MogensHørder	Denmark
Thomas Sinkjær	Denmark
Egon Toft	Denmark
Hans Nygaard	Denmark
Jørgen Arendt Jensen	Denmark
Kim Dremstrup	Denmark
Nils Fogh Andersen	Denmark
Morten Ølgaard Jensen	Denmark
Ole Kæseler	Denmark
Nico J.M. Rijkhoff	Denmark
Michel Dalstra	Denmark
Erik Morre Pedersen	Denmark
Peter Johansen	Denmark
Helge B. Sørensen	Denmark
Michael Hasenkam	Denmark
Johannes Struijk	Denmark
Pia Elberg	Denmark
Steen Andreassen	Denmark
Dario Farina	Germany
Natalie Kersting	Denmark
Hans Stødkilde-Jørgensen	Denmark
Stig Kjær Andersen	Denmark
Lars Mandrup	Denmark
Lars Hansson	Sverige
Jens VingeNygaard	Denmark
Thomas Sinkjær	Denmark
Winnie Jensen	Denmark
Carsten Dahl Mørk	Denmark
Pascal Madeleine	Denmark
Uwe Kersting	Denmark
Ole Hejlesen	Denmark
Birthe Dinesen	Denmark
Trine Fink	Denmark
Jeppe Emmersen	Denmark
Thomas Sangild Sørensen	Denmark
Lasse Riis Østergaard	Denmark

Local Organising Committee

Kim Dremstrup	Denmark
Per Overgaard Rasmussen	Denmark
Calle Thøgersen	Denmark
Benedikte Kruuse Lindvig	Denmark
Morten Ølgaard Jensen	Denmark
Steve Rees	Denmark
Daimi Frederiksen	Denmark
Hans Jørgen Clausen	Denmark
Henrik Kruckow	Denmark
Svend Erik Bodi	Denmark

Sponsors and Partners

Partners

IFMBE

Dansk Medicoteknisk Selskab - Danish Society for Biomedical Engineering -

AALBORG UNIVERSITY

BIO MED COMMUNITY

Sponsors

NMS MEDICO

DOF Det Obelske Familiefond

MAQUET

TRUMPF

Table of Contents

A Review of Telemedicine Services in Finland

Vikramajeet Khatri, Carrie B. Peterson, Sofoklis Kyriazokos, and Neeli R. Prasad

Center for TeleInfrastruktur (CTiF), Aalborg University, Aalborg, Denmark
{vkhatri,cbp,sk,np}@es.aau.dk

Abstract— **Telemedicine is gaining popularity due to the provision of ubiquitous health care services that is a fundamental need for every socialized society. In this paper, telemedicine services in Finland are discussed, as well as how they came into existence, how they are funded, evaluated, and what are their impacts on health care systems and society. Telemedicine services like teleradiology, telelaboratory, telepsychiatry and remote consultations, are being offered in all hospital districts. Primary health care centers in Finland are lacking telemedicine services, and are planning to have them. Electronic Patient Records (EPR), with e-referral and e-discharge letters, have prevented patients from unnecessary repeated laboratory examinations and treatments. The e-Archive (Finland's national EPR) is in the planning stage, making EPR on national level, to promote ease of access to patient records and ubiquitous care. The e-Prescription project is also in the planning stage, which aims to enhance drug safety, prevent forged prescription, and prevent threat to a patient's life.**

Keywords— **Telemedicine, Teleradiology, Finland, Ubiquitous Care.**

I. INTRODUCTION

Telemedicine results from the contribution of Information and Communication Technology (ICT) towards heath care, and the improving health and welfare of society. This is achieved by providing ubiquitous health care services to remote regions. Telemedicine has many advantages, such as serving people in remote areas due to unavailability or lack of health care professionals, and improving health care quality via consultations with specialists. The biggest considerable advantage of telemedicine is the savings of time (travel to appointments, requirements for both patient and professional to be available, administrative tasks, etc.), cost (organizational work load, administrative resources, reduced travel, utilization of consultation services at a distance, etc.), and effort for a patient.

Finland is a Nordic country in Northern Europe, with a population of 5.3 million people. Finland's northern areas cover about 30% of the total area, even more, these areas are sparsely populated. Even though citizens in these areas may have access to primary health care, they are consistently lacking specialized care. For patients in northern areas, it can be very difficult to visit Oulu district hospital for special care. Teleradiology was the first telemedicine application started in northern areas of Finland that improved the health care system and eventually benefited patients, heading towards Finland's goal for the completion of the ubiquitous health care dream.

Finland has good international relations and supports international research and development programs, particularly in the areas of ICT and health care services. Finland cooperates with its neighboring country Russia in many development programs and has bilateral agreements on education, health and economic co-operation. Finland is among the first three countries who established the first international teleradiology connection in the world, and it was established between university hospitals of Oulu (Finland), Reykjavik (Iceland), and Tromsoe (Norway) [1].

After an introduction to telemedicine highlights in Finland, the article is organized as follows: section 2 highlights the background of telemedicine in Finland; section 3 describes about the methodology involved in this paper, how the literature was collected and reviewed; section 4 discusses current applications of telemedicine in Finland, and factors associated with its implementation and evaluation; and section 5 summarizes the literature studied. The paper concludes with discussion and future implications for telemedicine systems in Finland.

II. BACKGROUND AND STATUS OF TELEMEDICINE SERVICES

This section discusses about the Finnish Health Care system, the background of telemedicine services in Finland, how they were evolved, evaluated, implemented and adopted in the Finnish Health Care system. Furthermore, it describes the current telemedicine development, and the status of telemedicine services according to a survey made in 2005.

Teleradiology refers to electronic transfer of radiological images such as x-ray, computed topography (CT) images and magnetic resonance images (MRI) from one clinical setting to another for diagnostic purposes. The first experiments took place in 1969, but did not enter the practical world until the

K. Dremstrup, S. Rees, M.Ø. Jensen (Eds.): 15th NBC on Biomedical Engineering & Medical Physics, IFMBE Proceedings 34, pp. 1–8, 2011.
www.springerlink.com

beginning of 1990's. Telemedicine services were a major interest in the sparsely populated northern areas, but services quickly spread around the country [2]. Finland has an extensive health care system, comprised of 21 hospital districts including five university hospital districts (Helsinki, Tampere, Kuopio, Oulu and Turku). One hospital district provides specialized health care to several primary health care centers in its area [3]. Private health care in Finland comprises of private clinics and private hospitals. The physicians working at private clinics are mostly specialists who work full time at a public hospital. By 1994, all five university hospital districts had teleradiology services implemented. Hospitals utilize teleradiology services to transmit radiological images to specialists such as neurosurgeons, and the neurosurgeons, after analyzing and studying the images used to report or consult, would contact the client's hospital via telephone earlier as all data networks were implemented simplex i.e. one-way, but later on they started reporting and consulting via videoconferencing.

Finland has implemented an electronic patient record (EPR) system as a primary patient database in its health care system; however, some records are kept and presented in traditional paper format. Oulu University Hospital has used multimedia medical records since 1995, and now they have merged e-referrals and e-discharge letter features to this. In 2005 [4], 16 out of 21 hospital district were providing e-referrals and e-discharge features to its subsidiary health care centers. These features allow health care professionals to view a patient's electronic record along with laboratory results and the imaging database, thus avoiding unnecessary examinations. Imaging databases include x-ray, and DICOM (the Digital Imaging and Communications in Medicine) format radiological images such as computed tomography (CT), ultrasound (US), and magnetic resonance imaging (MRI) images [1]. The EPR usage in Finnish health care system in 2005 [4] is shown in Table1.

Finland has also produced the first pocket-sized Nokia Communicator PDA (Personal Digital Assistant) device with integrated GSM phone, under the EU funded MEMODA project (Mobile Medical Data) during the years 1998-2000. These PDA terminals were utilizing GSM data pathways, helping physicians to view DICOM images on a secure connection and proved to be most effective for neurosurgery department. These PDA terminals were enhanced during the years 2002-2004 under the EU funded PROMODAS project (Professional Mobile Data Systems). The major enhancement was upgrading transport technology to GPRS (General Packet Radio Systems) that eventually reduced the system operating costs, and it is in clinical use these days [1].

The pharmacies in Finland are required to check every prescription by law. According to The Association of Finnish Pharmacies, pharmacies have to cope with over half a million unclear or inaccurate prescriptions for medicine every year, such as wrong dosage for a medicine or unavailability of drug in the market or prescribed medicine effects CNS (central nervous system). These checks have also revealed forged prescriptions and even fake physicians [5]. Therefore, Finland started a national e-Prescribing pilot-project in 2004-2006 [6], covering two hospital districts and a couple of primary health care centers involved with it. A doctor creates a prescription with a legacy system, signs it with electronic signature, and sends a SSL secure message to national prescription database referred to be as the Prescription Centre. When a patient goes to a pharmacy, pharmacist accesses the database, makes required changes, marks dispensing information on the electronic prescription, signs the markings with a personal smart card, and saves the markings to the prescription in the database. Then, the medicine is dispensed to the patient.

Table 1 EPR usage in Finnish health care system

Quantity	Hospital District	Primary Health Care	Private Health Care Providers	Status	Usage intensity
17	×			In use	> 90%
1	×			In use	50 – 60%
2	×			In use	25 – 49%
1	×			Planning	
229		×		In use	> 90%
3		×		Testing	
8		×		Planning	
11		×		Merging with neighbor	
25			×	In use	> 90%
3			×	---	

The Prescription Centre is accessible to health care professionals and pharmacists through a professional smart card, issued by Valvira (National Supervisory Authority for Welfare and Health). The Prescription Centre will contain other information along with medicine name such as pricing, interchangeable products, and clinical nutrients. The legislation for e-Prescription has been accepted in December 2006, and a national e-Prescription database has been created by the Social Insurance Institution (KELA). After a successful implementation, the patients will still have a right to choose prescription on paper [6]. It is aimed to be fully integrated with different EPRs to cover all pharmacies, and to reside continuously updated knowledge about all prescribed drugs of the patients, which will offer a platform for drug safety decisions. The prescription information is stored in the Prescription Centre for 30 months only, and then is archived in the Prescription Archive for 10 years, and then destroyed. It will help health care professionals to view, subject to a patient's oral consent, a patient's previous treatments, medication, avoid adverse drug interactions, and overlaps. The stored data can be used for supervision, drug safety operations, the payments for drug reimbursements, and research.

The health care system uses different systems for information management, which makes the distribution of patient's records complex, limiting use of systems, increasing costs, and paper archive preferred. Therefore, the Government of Finland decided to implement EPR on a national level rather than on a regional level, and store the records in a uniform technical format so that it can be distributed and accessed evenly. The National EPR project is expected to be finished by the end of 2011, and is maintained and handled by the Social Insurance Institution (KELA). The legislation for the National EPR was laid out in December, 2006 [4], and it will reside on a national public key infrastructure (PKI) for health care professional. The patients can refuse publishing of their records in the directory database, and their records can only be seen with an oral consent. The National EPR will offer citizens to view health information, such as reference and discharge letters, certificates, statements, results of examinations, and log data about visits to the personal health records, eventually making the system more secure to view without oral consent of a patient.

Other telemedicine applications include: sending laboratory or pathology results to physicians or specialists; telepsychiatry, teleophthalmology, teledentistry, distance teaching for other health care institutes and personnel via videoconferencing; and forwarding digital real-time reading parameters (pulse rate, oxygen saturation, blood pressure, ECG, etc.) of a patient in an ambulance heading towards the hospital.

III. Methodology

This section reveals the method of the study; namely, how the literature was obtained, the challenges and problems in accessing data, efforts to access and gain information, literature contents, and what information was of interest are explained in this section.

Initially, a search was made for various scientific articles regarding telemedicine focusing on the impact, progression, projects, and applications. The author hoped to find sufficient information through searching e-journals, e-databases, universities' publication databases, and organizations' published information. However, this proved to be a much more difficult task than was expected. One of the main hindrances to finding accurate and current information was language barriers. The official languages of Finland are Finnish and Swedish; Swedish being spoken and written in the metropolitan areas only. Because of this, it was very difficult to obtain literature and other information in English. The literature search started from e-journals, e-databases, search engines, and moved ahead to contact organizations, universities, library services of Aalborg University, Tampere University of Technology, Aalto University, individual professors, in addition to Pirkanmaa Hospital, and authors of different publications which were accessible only through direct exchange. After contacting individual authors, it was soon apparent that most of the journals are in the Finnish language and only the abstract is available in English, even though the language of the article may be listed differently in literature publication databases. After contacting library services in Finland, they suggested to look into TelMed – the leading database for medical publications in Finland, which eventually helped to access 3 more publications. While searching for pertinent information, it became painfully clear to the authors, that there is a serious gap in information regarding telemedicine in Finland. Further, we can understand from a European Union point of view that much more information could be disseminated regarding past and present telemedicine initiatives, particularly if it were made available in a common language, i.e. English.

The search for literature resulted in 30 papers and 4 research and review reports. Most of the papers were review articles – telemedicine pros and cons, project implementation phases, uses, and future aspects, but none of the papers revealed the technical aspects of interest: topology, operation principles and management. Out of the 30 papers obtained, 20 of the papers were dated 1991 -1999, while the rest were published in the year 2000 or later, no information or articles were found for the year 2010. Papers were accessed through e-journals, e-databases and universities research centers (Telemedicine Laboratory, Tampere University of Technology, and Finn TeleMedicum – Center of Excellence for TeleHealth, University of Oulu) while the reports were accessed from National Institute of Health & Welfare (THL) and its underlying centers: the National Research and Development Centre for Welfare and Health (Stakes) and Finnish Office for Health Technology Assessment (Finohta).

IV. Telemedicine Applications and Factors Affecting Implementation and Evaluation in Finland

In this section, qualities of telemedicine applications in Finland are described. There are many factors affecting the successful implementation and use of telemedicine systems, including funding and reimbursement, licensing and insurance barriers, and social acceptance. Some of these factors, funding, current applications and social acceptance are discussed here, to give an overview about the status and evaluation of telemedicine applications in Finland.

A. Funding

In this section, the various sources of funding for telemedicine projects in Finland are discussed, how the telemedicine projects are funded and run, which organizations are key players for it, and which specific area they are involved in. The organizational structure for funding is widely distributed, varying from public to private sector, all contributing towards the better health and welfare for the Finnish Society.

The Ministry of Social Affairs and Health in Finland [7] is the top level organization for administration, innovation, and management of health services in Finland. The National Institute for Health and Welfare (THL) [8] is expert in the research and development of health and welfare. THL runs many research centers under their umbrella, including the National Public Health Institute (KTL), the National Research and Development Centre for Welfare and Health (Stakes), and the Finnish Institute of Occupational Health (TTL). These research centers are involved in research and development for societal health and welfare, and funded by THL.

The Technical Research Centre of Finland (VTT) [9] is the biggest funding source for multi-technological applied research projects in Finland, and the biggest research organization in Northern Europe. VTT is an international scientific technology network that runs research projects and research programs associated with universities to develop, enhance, and innovate the technology to put the applied research to improve competencies into action. Along with other technologies, VTT provides high-end technology solutions and innovation services in Telemedicine as well.

The Academy of Finland [10] is the prime funding agency for basic research in Finland. The academy operates within the administrative sector of the Ministry of Education. It allocates funding of about 300 million Euros for the highest quality, and produces the most innovative, scientific research. Universities are the most important partner for the academy as research is involved, it supports and funds research projects, research programs, and Centers of Excellence. Centers of Excellence (CoE) offer excellent opportunities to carry out high quality research with six-year funding. The Academy of Finland also encourages the mobility of researchers (to and from Finland), such as FiDiPro (Finland Distinguished Professor), to extend and improve research collaboration, businesses, industry, and public administration internationally as well as nationally. Internationally, the academy cooperates with a number of other countries as well as with international funding organizations.

The Finnish Funding Agency for Technology and Innovation (Tekes) [11] is another main public funding organization for innovative research and development that works with the top innovative companies and research units in Finland. Tekes supports the projects that contribute towards the greatest benefits in the economical and social sectors in Finland. Along with other fields of innovative interest, Tekes funds many projects in Telemedicine as well.

Sitra, the Finnish Innovation Fund [12] is an independent public fund which promotes the welfare of Finnish society and has a mission to build a successful Finland for tomorrow under the supervision of the Finnish Parliament. Sitra co-operates closely with both the public and private sectors. Sitra chooses and changes programs themes aiming at the welfare of the society. Sitra enhances impact of its programs by various methods that include research, strategy process, innovative experiments, business development, and investment in internalization. Currently, Sitra does not focus actively on any health care program but, in the past, a health care program has been completed. This particular health care program was a research, training, and experimental program, having paper-free health care and seamless service as one of the key areas.

KanTa, the National Archive of Health Information, is a collective name for several national medical information systems, which are e-Prescription, e-Archive (national EPR), and online access by citizens to view their medical and prescription data. There lies a problem of funding in KanTa; the State will fund construction and operational costs to KanTa till 1st April 2011 only. Afterwards, the system will rely on funding obtained via user fees, which will be set at a level sufficient [6].

Finland also participates in a European Union (EU) Commission's Seventh Framework Programs (FP7) project, titled ISISEMD (Intelligent System for Independent living and SElfcare of seniors with cognitive problems or Mild Dementia). The ISISEMD project focuses on the elderly living people, having some problems or a mild loss of memory (dementia). In the past, the EU has funded three telemedicine projects under Fifth Framework Programs

(FP5) in Finland, titled RUBIS, PROMODAS (Professional Mobile Data Service) and MOMEDA (Mobile Medical Data) [1]. The remainder of funding comes from the private sector and giant companies of interest, such as Nokia and Remote Analysis, who want to innovate and develop their products for the welfare and health system in Finland. Most of the research projects in Finland today are funded by a co-operation of these funding agencies, e.g. a project funded by Nokia, Tekes, Intel, and Nvidia Graphics. These funding agencies start a research program or project and hire scientific staff or handover research to universities in order to evaluate the larger, "real" picture.

B. Current Applications

In this section, the current status of telemedicine applications and implementation are discussed. This section reveals more about the role that telemedicine applications have played in the Finnish health care system, how hospitals started utilizing telemedicine services and how it benefited both parties of the patients and the health care system.

Finland has many telemedicine applications currently in use: teleradiology, telelaboratory, telepsychiatry, teleopthalmology, teledermatology and teledentistry. Video conferencing is the key part of most used telemedicine service, where the physician is at one location while the patient (and the nurse) is at another location. It is used to consult a specialist of a hospital e.g. for patients with psychological or ophthalmological problems, and the services are known as telepsychiatry, teleopthalmology and teledentistry respectively [2]. In 2005 [4], 11 out of 21 were providing remote consultations and 21 out of 179 primary health care centers had purchased videoconferencing equipments, the growth is expected to develop as more health care centers are either planning or testing it. Video conferencing improves the quality of health care especially in case of telepsychiatry [13] – mental health care, expands the co-operations between primary and secondary health care units, it is currently used in all hospital districts, almost all primary and secondary care units, and is planned to expand further. The telemedicine applications mentioned above have been implemented in university hospital districts and other hospitals in Finland; meanwhile, the other applications are in the pipeline.

- Teleradiology

Finland has many telemedicine applications currently in use: Currently, 18 out of 21 (86%) hospital districts in Finland utilize this application [2]. In 1969, initial experiments took place when radiological images were transmitted between Helsinki hospital district and Oulu hospital district using the broadcasting network of Finnish national television (YLE). Some hospitals started teleradiology services utilizing existing copper telephone lines (POTS) to transmit X-ray images, but later upgraded to using Integrated Systems Digital Network (ISDN) lines as transport medium. The teleradiology and telemedicine applications network expanded widely along with the passage of time, Asynchronous Transfer Mode (ATM) dominated data transport technology, replacing ISDN lines and the YLE broadcast network there. While ATM was a dominant data transport technology, some hospitals also utilized Ethernet 10Mbps connections, because of compatibility issues - the equipment didn't support any ATM cards as an interface. The imaging database can be viewed in three ways: regional database, regional PACS (Picture Archiving and Communication System) or EPR having e-referral and e-discharge letters. In 2005, 52 out of 179 primary health care centers were utilizing some teleradiology services, where as it use in district hospitals is summarized in Table2 [4].

Table 2 Teleradiology services in Finnish health care system

Measure		No. of hospital districts
Teleradiology	Production Phase	16
	Pilot Phase	2
	Usage > 90%	5
Regional Archive	Production Phase	10
	Pilot Phase	3
	Usage > 90%	3
	Either teleradiology or regional archive	18

- Telelaboratory

Telelaboratory refers to electronic distribution of laboratory results from one location to another location and this application between hospitals is very common nowadays. According to a 2005 survey [1], 90% of the hospital districts were using electronic methods of distribution for laboratory results, where-as 27% of the primary care centres were receiving daily laboratory results electronically via a regional database and the rest were either at planning or testing stage. In earlier times, Integrated Services Digital Network (ISDN) lines were used for inter-laboratory communication, which was later on replaced by ATM connections along with data transport technology upgradation.

- Telepsychiatry

Telepsychiatry refers to interactive psychiatric consultations over distance that enables simultaneous sound and video connections between two or more interested parties (patient - psychiatrist or patient, nurse/physician - psychiatrist). The primary communication method used in telepsychiatric consultation is teleconferencing, because a psychiatrist tries to understand the problem through therapy and observing a patient, and these observations include physical movement, thoughts, reaction to certain actions, expressions and many other factors of a patient that are often difficult to quantify but are known indicators in Psychiatry. Initially, 3 pairs of ISDN lines (384 Kbps) were used for teleconferencing, as the technological revolution continued, Finland switched to using ATM connections.

- Teledentistry

Finland has a shortage of dental health care professionals, lacking odontology services in sparsely populated regions. Odontology is a branch of dentistry that deals with the teeth, their structure, development, and their diseases. Teledentistry refers to provision of dental services at remote end using videoconferencing, but it is mostly used for distance learning specialist education and clinical consultation purposes in dentistry. Turku university hospital odontological clinic hosts specialist training, which is distributed to various health care centers and hospitals in Western Finland. Videoconferencing was made possible through standard videoconference equipment and wireless intraoral camera technology, utilizing ISDN as well as TCP/IP network connections. Wireless intraoral camera is a tiny digital camera that fits comfortably in one's mouth and shows a clear real time view of one's smile and teeth to a dentist for analysis and diagnostic purposes [14]. In some health care centres, computer-aided consultation is also utilized for patient diagnosis and treatment planning. For a patient, some photographs using digital camera or digital images for x-ray films are obtained and sent via email for consultation, saving patient's time and effort [15]. Teledentistry is expected to develop further in near future.

The data transport technology used in telemedicine services is compared in Table 3.

Table 3 Data transport technology for telemedicine services

Application	Transport Technology
Teleradiology	YLE Network, POTS, ISDN, ATM
Telelaboratory	ISDN, ATM
Telepsychiatry	ISDN(3 pairs – 384 Kbps), ATM
Teledentistry	ISDN, TCP/IP

C. *Social Acceptance*

It seems as though the Finnish society has accepted telemedicine applications from a technical point of view, but there continue to be hindrances to social acceptance. The patients had to wait a long time for appointments, but recent system has improved, and resulted in less awaiting times for appointment. In near future, a hindrance will appear, when KELA will start charging fees about e-Prescription and e-Archive services from patients.

V. LITERATURE REVIEW

This section summarizes the papers studied and a review of literature as show in Table 4.

Table 4 Literature review

No.	Study	No. of participants	Methodology	Tool	Results
1.	Teledentistry (specialist education) [16]	26 specialists	Cost analysis	Videoconferencing	Costs saving per student €40,000. Attracted more students
2.	E-health development project 'ProViisikko' [17]	5 hospital districts	Process innovation	Analytical tool Interactive consulting centre	Enhance patient care, patients can book appointments, check laboratory test results online, and receive SMS acknowledgment.
3.	Child and adolescent psychiatry [18]	42 child and adolescent units in 21 hospital districts	Questionnaire (qualitative and quantitative analysis)	Videoconferencing	Savings of time & cost, availability of mental health services, can be improved by encouraging hospital staff to utilize videoconference on a proper schedule and improve technical support.

Table 4 *(continued)*

4.	EPR and general practitioner (GP) [19]	GPs in 8 health care centers	Use EPR system to look for specific data	3 types of EPR systems (2389 patient cases accessed)	Need to overcome shortage of qualified personnel to use and enter data correctly into EPRs.
5.	Extending EPR to e-referral and discharge letters [20]	12 university clinics and primary health care centers of 13 municipalities	Send XML message between EPR or on a secure server using VLAN or VPN.	EPR, imaging and laboratory database	Saves time in referral management, avoids unnecessary repeat imaging and laboratory examinations. Legalization for national electronic signatures and patient privacy are awaited.
6.	Development of working process [21]	--	Literature review Case study (interviewing and quantitative analysis of care process)	Technology (ICT)	A hypothesis – e-health services can be effective tool in improving and empowering patients in their own care, suggestion to start e-health pilot project for diabetes care.
7.	Fundus screening of type 2 diabetes patients [22]	Primary health care centers in South-Ostrobothnia, approx. 3000 patients screened each year	Take mobile unit to local centre, take fundus images and updates on central archive	Mobile digital fundus screening unit and central archive	Type 2 diabetes patients' fundus screening performed according to national health guidelines (one in 2.5 years) avoiding diabetic retinopathy and reducing university hospital workload.

VI. Conclusion

Finland is a pioneer in ICT services, and home to giant ICT company Nokia. Finland is sparsely populated, especially in northern areas, where telemedicine services can improve and provide specialized health care to the society. The Finnish Health Care system has been utilizing telemedicine services since 1994 to its community. The statistics presented in 2005 reveal that almost all of the district hospitals have teleradiology, telelaboratory, and remote consultations services to offer to primary health care centers. About 30% of primary health care centers have bought videoconferencing equipments to support remote consultations, and created links to district hospitals, the rest are planning to have them in near future. All hospitals utilize EPR, which is no longer a good measure for accounting telemedicine services. Therefore, the merging of e-referral and e-discharge letters with EPR have extended telemedicine services, and helped in avoiding repeated examinations and viewing patient history. In order to cope with incorrect prescription and drug safety, e-Prescription project is a good step forward, which will help in avoiding overlaps, incorrect dosages, and prevent threats to a patient's life. The e-Archive (national EPR) project will help to digitize all hospital records, creating a uniform technical format for documents, making ease of access of patient records. The national EPR will include e-discharge, e-referral letters as well; the patient can refuse publishing of his/her information, and can check through log files that who viewed his/her records. The other applications are the fundus screening of type 2 diabetes patients, which prevents diabetes retinopathy (sight blindness), and teledentistry that cops with shortage of dental care professionals in Finland.

Due to language barriers, it was difficult to find literature about telemedicine services in Finland in English. After searching through library databases, e-journals, e-databases, internet, and finally contacting some Finnish universities research centers and institutions, the information was gathered to study and review telemedicine services in Finland. Majority of the literature was dated 1991-1999, but no any recent information in the last two years about telemedicine services in Finland was found. The funding for the development, implementation and evaluation of telemedicine projects is supported by various funding bodies, but a barrier for e-Prescription project appears in April 2011. The Finnish society has an acceptance to telemedicine services, but hindrances are expected to be in near-future regarding charging fees for e-Archive and e-Prescription services. Telemedicine services have contributed towards the betterment and welfare of the Finnish society.

References

1. Jarmo Reponen. Radiology as a part of comprehensive telemedicine and ehealth network in northern Finland. International Journal of Circumpolar Health, 63:4 429-435, 2004
2. European Health Telematics Association: country report – Finland. Retrieved from http://www.i2health.org/forum/tasks-sources/task-force-sustainable-telemedicine-and-chronic-disease-management/annex-to-the-ehtel-briefing-paper/country-report-finland. Accessed 14 October 2010.
3. Hospital District of Helsinki and Uusima: health services in Finland. Retrieved from http://www.hus.fi/default.asp?path=59,404,4024,4023 . Accessed 14 October 2010

4. Jarmo Reponen, Ilkka Winblad, Päivi Hämäläinen. Current Status of National eHealth and Telemedicine Development in Finland. Studies in Health Technology and Informatics, 134: 199-208, 2008
5. Pharamcies Deal with Thousands of Incorrect Prescription. (2010, December 11). Retrieved December 13, 2010, from YLE - News: http://www.yle.fi/uutiset/news/2010/12/pharmacies_deal_with_thousands_of_incorrect_prescriptions_2212660.html
6. Funding – National Archive of Health Information (KanTa), Finland. Accessed 11 December 2010. Retrieved from https://www.kanta.fi/web/en/funding
7. Ministry of Social Affairs and Health, Finland. Retrieved from http://www.stm.fi . Accessed 14 October 2010.
8. National Institute for Health and Welfare, Finland. Retrieved from http://www.thl.fi/en_US/web/en/research;jsessionid=6C831396E60D4AEA474673F8F293BDA0. Accessed 14 October 2010.
9. Technical Research Centre of Finland (VTT). Retrieved from http://www.vtt.fi . Accessed 14 October 2010.
10. The Academy of Finland. Retrieved from http://www.aka.fi . Accessed 14 October 2010.
11. The Finnish Funding Agency for Technology and Innovation (Tekes), Finland. Retrieved from http://www.tekes.fi . Accessed 14 October 2010.
12. Sitra, the Finnish Innovation Fund, Finland. Retrieved from http://www.sitra.fi . Accessed 14 October 2010.
13. Marja-Leema Mielonen, Leena Väisänen, Juha Moring, Arto Ohinmaa, Matti Isohanni. Implementation of a Telepsychiatric Network in Northern Finland. Current Problems Dermatology, 2003, 32, 132-140.
14. Ignatius E, Mäkelä K, Happonen R-P, Hallikainen S, Perälä S. Videoconferencing as a teaching tool in specialist education in dentistry. Finnish Dental Journal, 17: 958- 962, 2004
15. Ignatius E, Mäkelä K, Perälä S. Computer aided dental consultation. Finnish Dental Journal, 16: 864-868, 2003
16. Eino Ignatius, Kari Mäkelä, Risto-Pekka Happonen, Sami Perälä. Teledentistry in dental specialist education in Finland. Journal of Telemedicine and Telecare, 12: 46-49, 2006
17. Teemu Paavola, Kari Mäkelä, Virpi Pyykkö, Sami Perälä. A national Finnish e-health development project "ProViisikko". Journal of Telemedicine and Telecare, 12: 67-69, 2006
18. Lilli Pesämaa, Hanna Ebeling, Marja-Leena Kuusimäki, Ilkka Winblad, Matti Isohanni, Irma Moilanen. Videoconferencing in child and adolescent psychiatry in Finland - an inadequately exploited resource. Journal of Telemedicine and Telecare, 13: 125-129, 2007
19. K Mäkelä, I Virjo, J Aho, P Kalliola, A-M Koivukoski, H Kurunmäki, M Kähärä, L Uusitalo, M Valli, V Voutair, S Ylinen. Electronic patient record systems and the general practitioner: an evaluation study. Journal of Telemedicine and Telecare, 11: 66-68, 2005
20. J Reponen, E Marttila, H Paajanen, A Turula. Extending a multimedia medical record to a regional service with electronic referral and discharge letters. Journal of Telemedicine and Telecare, 10: 81-83, 2004
21. Noora Ekroos, Kari Jalonen. E-health and diabetes care. Journal of Telemedicine and Telecare, 13: 22-23, 2007
22. Riku Lemmetty, Kari Mäkelä. Mobile digital fundus screening of type 2 diabetes patients in the Finnish county of South-Ostrobothnia. Journal of Telemedicine and Telecare, 15: 68-72, 2009

Repeatability of Pressure Oscillation Amplitudes during the Interrupter Measurement of Respiratory Resistance

J. Kivastik[1], J. Talts[1], K. Jagomägi[1], R. Raamat[1], and M. Vasar[2]

[1] Department of Physiology, University of Tartu, Tartu, Estonia
[2] Children's Clinic, University of Tartu, Tartu, Estonia

Abstract— **Interrupter resistance (Rint) technique for assessing respiratory mechanics requires minimal cooperation and can therefore be successfully performed in young children. Analysis of recorded oscillations of the mouth pressure (Pmo) has been suggested to provide additional indices of change in airway mechanics. The aim of this study was to establish the repeatability of pressure oscillation amplitudes.**

Children performed two sets of Rint measurements. Further analysis of Pmo tracings was performed using MATLAB software. Pmo data were normalized to the last recorded pressure and afterwards oscillation amplitudes (Amp) were found as the difference between the first Pmo maximum and minimum. Intra-measurement repeatability was assessed by the coefficient of variation (CV) and between-test repeatability - by the coefficient of repeatability (CR).

92 young children (aged 3 to 7 years) were studied (49 of them healthy, 18 wheezers and 25 coughers). Median CV values for both measurements were 14% and 15% for Rint, and 14% and 13% for Amp. Our between-test Rint repeatability was similar to that of previous studies (CR was 0.23 kPa·L^{-1}·s or 33.3% of baseline value). CR for Amp was 0.24 or 27.6% of baseline value. There was no significant difference between groups of children.

We measured short term repeatability for the most simple pressure oscillation amplitude and found that this is similar to Rint repeatability.

Keywords— **airway resistance, interrupter technique, pressure oscillations, amplitude analysis**

I. Introduction

The diagnosis and monitoring of airway disease in young children is more difficult than in older ages because children under 5-6 years of age can rarely perform repeated forced expirations needed for lung function measurements. A number of techniques applicable to lung function measurement in young children have been introduced several decades ago but these methods did not become popular in clinical physiology for a long time. Automated and portable commercial devices became available later, and because these techniques require passive co-operation only and measurements can be performed in children down to the age of 2 years, these methods have now become more widely used. One of the possibilities is to measure respiratory resistance by the interrupter technique.

The method involves a brief (100 ms) occlusion of the airflow during the tidal breathing while pressure measured at the airway opening equilibrates with alveolar pressure. The interrupter resistance (Rint) can be calculated when dividing the driving pressure by flow at the mouth immediately prior to occlusion. Measurement of Rint has been used to determine bronchodilator response (BDR) and also bronchial hyperresponsiveness (BHR) in young children, and detailed guidelines have been published [1, 2].

Immediately after occlusion, there is a very rapid change in pressure, followed by damped pressure oscillations and finally, there is a relatively slow rise in pressure. In addition to calculating just one Rint value, there have been attempts to pay more attention to the dynamic behavior of oscillations on the mouth pressure-time transient [Pmo(t)] to provide additional indices of change in airway mechanics [3-7]. Different algorithms to find the pressure oscillation amplitudes have been studied, however, the short-term repeatability of pressure oscillations has not been assessed.

The aim of our study was to establish the repeatability of pressure oscillation amplitudes using a commercial device in order to measure reliable baseline values and assess bronchodilator or bronchoconstrictor effects.

II. Materials and Methods

A. Subjects

Study subjects were children aged 3-7 years who attended the respiratory outpatients' clinic in Tartu Children's Clinic, healthy siblings or those who came after receiving the invitation sent to local kindergartens. All children had to be free of respiratory tract infection in the preceding 3 weeks and not wheezing at the time of testing.

When attending the study, a questionnaire was completed by parents. Questions about respiratory symptoms, diagnosed asthma, eczema and hayfever were asked. Subjects who reported wheeze during the previous year were classified as "wheezers". Children with recurrent or persistent

K. Dremstrup, S. Rees, M.Ø. Jensen (Eds.): 15th NBC on Biomedical Engineering & Medical Physics, IFMBE Proceedings 34, pp. 9–12, 2011.
www.springerlink.com

cough (i.e. with at least three coughing episodes in the last 6 months or every day for three consecutive weeks) and with no wheeze during the last 12 months were classified as "coughers".

The Ethics Review Committee on Human Research of the University of Tartu approved the study and written informed consent to take part was obtained from the parents on behalf of their children.

B. Equipment and measurements

Rint was assessed using a single MicroRint device (Micro Medical, UK) throughout the study. During normal and quiet breathing, children performed two sets of Rint measurements (15 min apart). One set consisted of up to 10 interruptions on the peak flow of the expiration.

The valve of such a device closes within 10 ms and remains closed for 100 ms. Sampling frequency of the pressure signal was 2000 Hz. Before the analysis all Pmo(t) graphs were checked using Rida software (Micro Medical, UK). Methodology of interrupter resistance measurement and graph rejection criteria we used have been described by others [8, 9]. For example, manual rejection was performed in case of tracings with a horizontal or declining pressure signal suggesting leakage at the mouth. The mean of 5-10 acceptable readings was taken as a measurement.

C. Data analysis

In addition to absolute Rint values obtained from a MicroRint device we also calculated z-scores [z = (measured value – reference value)/RSD, where RSD is the residual standard deviation in the reference population] using previously published reference data [10].

Further analysis of Pmo(t) tracings was performed using MATLAB (MathWorks Inc., USA). Prior to the oscillation analysis, Pmo(t) curves were normalized by dividing every pressure value by the last recorded pressure of that curve, in order to avoid the possible effect of interruptions occurring at different flows [3-4]. Oscillation amplitude (Amp) was found as the difference between the first mouth pressure maximum and minimum (Figure 1).

Intra-measurement repeatability was assessed by the coefficient of variation (CV) which was calculated for all parameters as the ratio of the standard deviation to the mean of the 5–10 individual readings (in %).

Between-test repeatability was assessed by the coefficient of repeatability (CR): twice the standard deviation of the mean difference between two sets of values. To compare CR for Rint and Amp we also expressed it as a percentage of the baseline value.

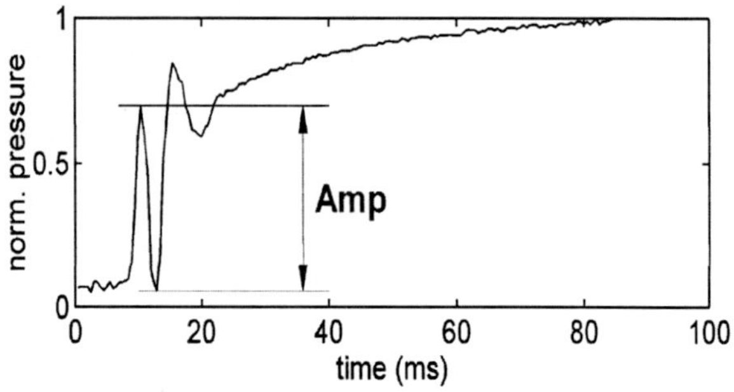

Fig.1 Finding the oscillation amplitude (Amp) as the difference between the first pressure maximum and minimum

To examine whether the variability was independent of the level of Amp, the differences between paired measurements were plotted against their means (Bland-Altman plot).

III. Results

92 young children (aged 3 to 7 years) were studied, their anthropometric data and Rint values in comparison with reference data are shown in Table 1.

Table 1 Anthropometric data and interrupter resistance (Rint) values by groups

	Healhty (n=49)	Coughers (n=25)	Wheezers (n=18)
Male/female	23/26	12/13	12/6
Age* yr	5.9 (0.8)	5.7 (1.2)	5.6 (1.1)
Height* cm	117.2 (4.6)	115.2 (5.3)	116.4 (4.2)
Rint* $kPa \cdot L^{-1} \cdot s$	0.68 (0.16)	0.68 (0.20)	0.73 (0.27)
Rint z-score*	−0.88 (1.50)	−1.07 (1.49)	−0.70 (2.19)

* mean (SD)

Intra-measurement repeatability: median coefficients of variation for both measurements were 14% and 15% for interrupter resistance (with a range from 5 to 48%), and 14% and 13% for oscillation amplitude (with a range from 3 to 36%). To visualize the repeatability of pressure oscillations, a set of normalized Pmo(t) graphs from one measurement can be seen in Figure 2.

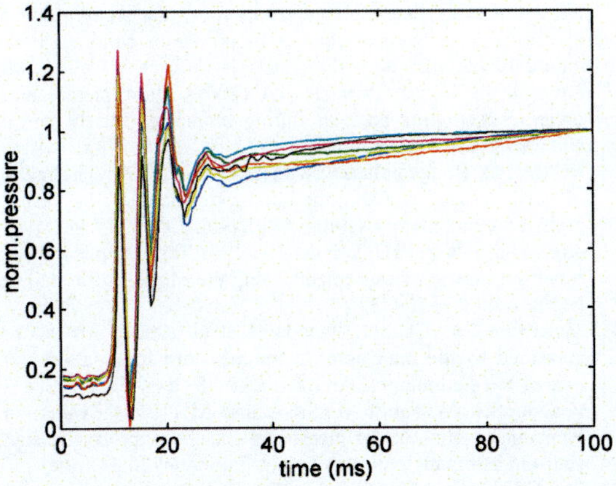

Fig.2 Normalized mouth pressure curves from one child (7.4 years old boy with Rint value of 0.5 kPa·L^{-1}·s, mean Amp value is 1.09, CoV=13%)

Our between-test repeatability for Rint was similar to that of previous studies (mean CR for the whole group was 0.23 kPa·L^{-1}·s or 33.3% of baseline value). Mean CR for Amp was 0.24 or 27.6% of baseline value. Data by different groups are given in Table 2.

Table 2 Measured values and coefficients of repeatability (CR) of Rint and Amp by groups

	Value 1*	Value 2*	CR§
Rint (kPa·L^{-1}·s)			
All	0.69 (0.20)	0.68 (0.17)	0.23 (33.3)
Healthy	0.68 (0.16)	0.68 (0.15)	0.23 (33.8)
Coughers	0.68 (0.20)	0.68 (0.19)	0.19 (27.9)
Wheezers	0.73 (0.27)	0.71 (0.20)	0.31 (42.5)
Amp			
All	0.87 (0.17)	0.86 (0.15)	0.24 (27.6)
Healthy	0.87 (0.18)	0.86 (0.15)	0.28 (32.2)
Coughers	0.86 (0.15)	0.87 (0.15)	0.22 (25.6)
Wheezers	0.90 (0.19)	0.88 (0.17)	0.18 (20.0)

* mean (SD), § CR as percentage of the value 1 is given in brackets

Mean difference between two measurements of oscillation amplitudes for the whole group was 0.006 and this difference did not depend on height (r=0.02, p=0.06). The differences in amplitudes are plotted against the mean amplitude from two measurements in Figure 3.

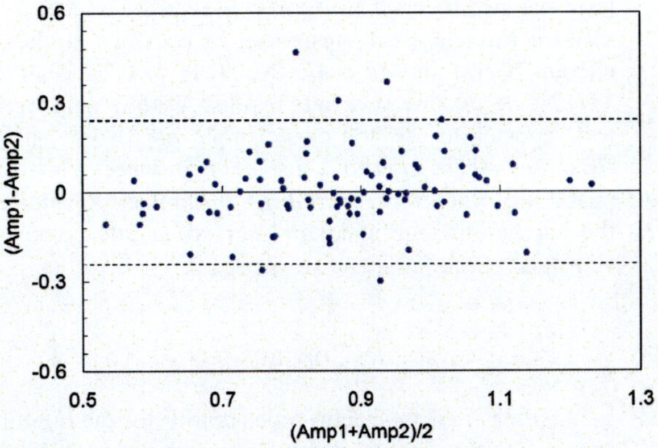

Fig.3 Bland-Altman plot of individual differences between two measurements (Ampl-Amp2) against mean amplitude values (Amp1+Amp2)/2. The dashed lines indicate 95% limits of agreement

IV. Discussion

The interrupter resistance measurement has been shown to be feasible for studying lung function in preschool children with mild bronchoconstriction. Most common method to calculate Rint is to use linear back-extrapolation of the mouth pressure-time transient. Mouth pressure curves from asthmatic children are considerably more concave with respect to the X-axis than in adults, and this can represent a problem when back-extrapolation is used. Therefore, different algorithms for deriving the change in pressure from Pmo(t) have been proposed, but there is still no concencus about the best method [2, 11-14].

One of the reasons to pay more attention to pressure oscillations is that amplitude analysis does not depend on equilibration of mouth and alveolar pressure, which may not occur in cases of severe bronchoconstriction [4]. If airway resistance increases (e.g. during BHR testing), oscillation amplitudes decrease, and the opposite happens during BDR assessment: airway resistance decreases and amplitudes increase. Several characteristics of postocclusional oscillations have been described, in our previous study we found that pressure amplitudes were more sensitive to detect changes in airway mechanics during BHR testing than frequency and damping factors [7].

Rint measurements are usually combined with BDR assessment, and because of that there is a need for a cut-off value to decide whether a change in Rint is caused by a pharmacological intervention or is that within the limits of short-term repeatability which reflects the variability of the measuring instrument and the biological variability of the disease. Most studies suggest using a decrease in Rint at least the size of the short-term repeatability to define a significant bronchodilator response. In previous studies, Rint CR has been found to range from 0.15 to 0.28 $kPa \cdot L^{-1} \cdot s$ [8, 15-17], i.e. in line with our results. Within- and between-test repeatability of the most simple oscillation amplitude from this study showed a good concordance with Rint repeatability, therefore, according to our data an increase in the oscillation amplitude for more than 30% would be a significant bronchodilator response.

V. Conclusions

We measured short term repeatability for the mouth pressure oscillation amplitude and found that this is similar to Rint repeatability. Therefore, we suggest that oscillation amplitude analysis could be implemented in the software of commercial devices so that it could be further evaluated for clinical use.

Acknowledgment

This study was supported by Estonian Science Foundation (grant 7322 and 7723) and Estonian Ministry of Education and Research (SF0180125s08).

References

1. Beydon N, Davis SD, Lombardi E, et al. (2007) American Thoracic Society/European Respiratory Society Working Group on Infant and Young Children Pulmonary Function Testing. An official American Thoracic Society/European Respiratory Society statement: pulmonary function testing in preschool children. Am J Respir Crit Care Med 175:1304–1345
2. Beydon N, Calogero C, Lombardi E (2010) Interrupter technique and passive respiratory mechanics. Eur Respir Mon 47:105–120
3. Frey U, Schibler A, Kraemer R (1995) Pressure oscillations after flow interruption in relation to lung mechanics. Respir Physiol 102:225–237
4. Frey U, Kraemer R (1995) Interrelationship between postocclusional oscillatory pressure transients and standard lung function in healthy and asthmatic children. Pediatr Pulmonol 19:379–388
5. Frey U, Kraemer R. (1997) Oscillatory pressure transients after flow interruption during bronchial challenge test in children. Eur Respir J 10:75–81
6. Bridge PD, Wertheim D, Jackson AC, McKenzie SA (2005) Pressure oscillation amplitude after interruption of tidal breathing as an index of change in airway mechanics in preschool children. Pediatr Pulmonol 40:420–425
7. Kivastik J, Talts J, Primhak RA (2009) Interrupter technique and pressure oscillation analysis during bronchoconstriction in children. Clin Physiol Funct Imaging 29:45–52
8. Bridge PD, Ranganathan S, McKenzie SA (1999) The measurement of airway resistance using the interrupter technique in preschool children in the ambulatory setting. Eur Respir J 13:792–796
9. Arets HG, Brackel HJ, van der Ent CK (2003) Applicability of interrupter resistance measurements using the MicroRint in daily practice. Respir Med 97:366–374
10. McKenzie SA, Chan E, Dundas I et al. (2002) Airway resistance measured by the interrupter technique: normative data for 2-10 year olds of three ethnicities. Arch Dis Child 87:248–251
11. Phagoo SB, Wilson NM, Silverman M (1995) Evaluation of the interrupter technique for measuring change in airway resistance in 5-year-old asthmatic children. Pediatr Pulmonol 20:387–395
12. Pao CS, Healy MJ, McKenzie SA (2004) Airway resistance by the interrupter technique: which algorithm for measuring pressure? Pediatr Pulmonol 37:31–36
13. Seddon P, Wertheim D, Bridge P, Bastian-Lee Y (2007) How should we estimate driving pressure to measure interrupter resistance in children? Pediatr Pulmonol 42:757–763
14. Oswald-Mammosser M, Charloux A, Enache I, Lonsdorfer-Wolf E, Geny B (2009) A comparison of four algorithms for the measurement of interrupter respiratory resistance in adults. Respir Med 103:729–735
15. Lombardi E, Sly PD, Concutelli G et al. (2001) Reference values of interrupter respiratory resistance in healthy preschool white children. Thorax 56:691-695
16. Beelen RM, Smit HA, van Strien RT, et al (2003) Short and long term variability of the interrupter technique under field and standardised conditions in 3-6 year old children. Thorax 58:761–764
17. Beydon N, M'buila C, Bados A, et al. (2007) Interrupter resistance short-term repeatability and bronchodilator response in preschool children. Respir Med 101:2482–2487

Author: Jana Kivastik
Institute: Department of Physiology, University of Tartu
Street: Ravila Str 19
City: Tartu
Country: Estonia
Email: jana.kivastik@ut.ee

Finite Element Implementation of a Structurally-Motivated Constitutive Relation for the Human Abdominal Aortic Wall with and without Aneurysms

M.S. Enevoldsen[1], K.-A. Henneberg[1], L. Lönn[2], and J.A. Jensen[1]

[1] Biomedical Engineering, Department of Electrical Engineering, Technical University of Denmark, Kgs. Lyngby, Denmark
[2] Department of Radiology and Department Vascular Surgery, Rigshospitalet, University of Copenhagen, Copenhagen, Denmark

Abstract— **The structural integrity of the abdominal aorta is maintained by elastin, collagen, and vascular smooth muscle cells. Changes with age in the structure can lead to development of aneurysms. This paper presents initial work to capture these changes in a finite element model (FEM) of a structurally-motivated anisotropic constitutive relation for the "four fiber family" arterial model. First a 2D implementation is used for benchmarking the FEM implementation to fitted biaxial stress-strain data obtained experimentally from four different groups of persons; 19-29 years, 30-60 years, 61-79 years and abdominal aortic aneurysm (AAA) patients. Next the constitutive model is implemented in an anisotropic 3D FEM formulation for future simulation of intact aortic geometries. The 2D simulations of the biaxial test experiment show good agreement with experimental data with a standard deviation below 0.5% in all cases. The maximum axial and hoop stress in the group of AAA patients was 94.9 kPa (±0.283 kPa) and 94.3 kPa (±0.224 kPa) at maximum stretch ratios of 1.043 and 1.037, respectively. In the 3D simulations, the maximum stress is also found to occur in the AAA patient group, with the highest stress in the circumferential direction (275 kPa). Comparison with an already published isotropic model indicates that the latter underestimates the peak stress significantly. Based on these results it is concluded that the four fiber family model has been successfully implemented into a 3D anisotropic finite element model and that this model can provide more accurate insight into the stress conditions in aortic aneurysms.**

Keywords— **Biomechanics, aortic aneurysms, four fiber family model, anisotropic finite element analysis.**

I. INTRODUCTION

The wall of the normal human aorta is a layered structure consisting of three layers; the intima, the media and the adventitia. The primary structural components of the aortic wall are the elastic fibers (elastin and associated microfibrils), collagen fibers and vascular smooth muscle cells (vSMC). With age the structure of the aortic wall changes, it becomes stiffer, and more vulnerable to damage leading to diseases like atherosclerosis and aneurysms. So, it is interesting to construct a simulation model to capture these structural changes and gain more insight into the pathology of these diseases from a biomechanical point of view. The current challenge is to determine the arterial wall stress and strength accurately. This presents some difficulty, because arterial tissue is anisotropic and nonlinear in the stress-stretch relationship, displays pseudo-elastic behavior, and changes material properties with age due to structural change and remodeling. The aim of this work is to implement the structurally-motivated phenomenological "four fiber family" model introduced by Baek et al. [1] for simulation of biomechanical properties in the human aorta with and without aneurysms. As a first step, a 2D finite element model (FEM) implementation is presented and used as a benchmark to numerically reproduce the stress-strain relations obtained in biaxial stress-strain experiments [2,3]. Next, the four fiber family model is implemented in a 3D anisotropic FEM and its ability to reveal detailed stress-strain information in arterial tissue is compared to that of an already published isotropic model [8].

II. MATERIALS AND METHODS

A. Constitutive framework

It is assumed that the aortic wall is a constrained mixture of four locally parallel families of collagen fibers (axial, circumferential, symmetric diagonal) embedded in an amorphous isotropic matrix dominated by elastic fibers. The biomechanical properties of a normal abdominal aorta and an aneurysm are described using the general formulation of the Cauchy stress (true stress) [4]

$$\boldsymbol{\sigma} = -p\mathbf{I} + 2\mathbf{F}\frac{\partial W}{\partial \mathbf{C}}\mathbf{F}^T , \tag{1}$$

where $\boldsymbol{\sigma}$ [Pa] is the Cauchy stress tensor, p is a Lagrange multiplier, $\mathbf{I}$ is the identity tensor, $\mathbf{F}$ is the deformation gradient tensor, W [Pa] is the SEF and $\mathbf{C}=\mathbf{F}^T\mathbf{F}$ is the right Cauchy-Green tensor. In order for the SEF to be as general as possible the model accounts for compressibility by splitting the SEF in a purely volumetric elastic response, $W_{vol}(J)$, and a purely isochoric elastic response, $W_{iso}(\mathbf{C},\mathrm{M}^{(k)})$ [5],

$$W\left(\mathbf{C},\mathrm{M}^{(k)}\right) = W_{vol}(J) + W_{iso}\left(\mathbf{C},\mathrm{M}^{(k)}\right), \tag{2}$$

where J=det(**F**) is the deformed-to-undeformed volume ratio and $M^{(k)}$ being a unit vector describing the direction of orientation of the collagen fiber families. Here the aortic tissue is assumed to be incompressible. To infer incompressibility the so-called penalty method is used in the finite element implementation. Here the tissue is modeled as slightly compressible by applying a very high bulk modulus in the volumetric elastic response, which has the simple form

$$W_{vol}(J) = \frac{\kappa}{2}(J-1)^2, \tag{3}$$

where κ [Pa] is the bulk modulus [4]. The isochoric response is modeled by the four fiber family constitutive relation [1]

$$W_{iso}\left(\mathbf{C}, \mathrm{M}^{(k)}\right) = \frac{c}{2}(I_C - 3) + \sum_{k=1}^{4} \frac{c_1^{(k)}}{4c_2^{(k)}} \left\{ \exp\left(c_2^{(k)} \left(IV_C^{(k)} - 1 \right)^2 \right) - 1 \right\} \tag{4}$$

where c, $c_1^{(k)}$, $c_2^{(k)}$ are material parameters, I_C is the first invariant of **C** and $IV_C^{(k)} = M^{(k)}\mathbf{C}M^{(k)}$ is the fourth invariant of **C**. This model has proven useful by providing increased insight into differences in the mechanical behavior due to structural abnormalities in the arterial wall [6]. For detailed information on the material parameters used in this study we refer to [6]. In brief, the determination of material properties is based on biaxial testing of tissue slabs from four different age/patient groups; 19-29 years, 30-60 years, 61-79 years, and AAA patients. Within each group the mean value of each material parameter is used.

B. Simulation of biaxial and inflation-extension test of arteries

Biaxial tension test of arteries is a well-known method for deducing the biomechanical properties of arteries [7]. Here we have simulated the biaxial testing of both normal abdominal aorta and pathological AAA tissue described by Vande Geest et al. [2,3] using COMSOL Multiphysics v4.1 (COMSOL AB, Stockholm, Sweden). In the simulation a tension value of 120 N/m is applied to the tissue corresponding to the circumferential tension per unit axial length in a thin-walled cylindrical tube pressurized to 113 mmHg, and the resulting stretch ratios and Cauchy stress components in the tissue sample are calculated. The inflation-extension test is also a commonly used experiment for determination of arterial properties, since the normal geometrical configuration of the artery is preserved [4]. Here a uniform internal pressure, P_i = 15 [kPa] is applied corresponding to 113 mmHg, which results in a radial force on the interior wall of a circular axis-symmetric cylinder. The cylinder has a radius of 1 cm and a length of 5 cm.

C. Analysis of simulated experiments

The implementation of the four fiber family model involves programming equations (1) – (4) into the finite element program. As the 2D variant of eq. (1) was used by Ferruzzi et al [6] to estimate the parameters of the constitutive model from the biaxial test data, the same equation can serve as a reference for benchmarking a 2D FEM implementation. In this perspective, stress-strain relations computed with a correct FEM implementation should be superimposed on the stress-strain relations calculated by hand using eq. (1). After benchmarking the implementation against biaxial test data the model is tested for predictability in the 3D case by comparing the anisotropic model to the isotropic model proposed by Raghavan and Vorp [8]. In this paper a negative Cauchy stress is interpreted as a compressive stress, and a positive stress is interpreted as a tensile stress. In addition, the unloaded configuration of the tissue is assumed to be stress free.

III. Results

A. Simulation of biaxial test

Comparison of the numerical simulation of the biaxial test and the analytical solution for the Cauchy stress components is shown in Fig. 3. The superimposition of the FEM results on the hand calculated curves confirms a correct FEM implementation of the model. The maximum stress values are seen in the circumferential direction (hoop stress) ranging from 85-175 kPa (638-1313 mmHg) compared to 85-165 kPa (638-1238 mmHg) for the axial direction. The maximum standard deviation was below 0.5% for both the hoop and axial stress; ±0.224 kPa and 0.283 kPa respectively for the AAA patient group, which has the highest standard deviation compared to the other groups (not shown). In general the aortic tissue becomes less compliant with age, and AAA tissue is significantly stiffer than normal abdominal aortic tissue. However, using the mean values of the material parameters indicate that the biomechanical properties of the normal AA for the groups 30-60 years and 61-79 years are similar. The tissue from the group 61-79 years is less compliant in the axial direction compared to the group of 30-60 year-olds. But in the circumferential direction the difference is minimal. This clearly shows that the anisotropy of the aortic tissue is captured by the constitutive relation, since the stretch ratios in the two directions are different from each other for all four groups of test subjects.

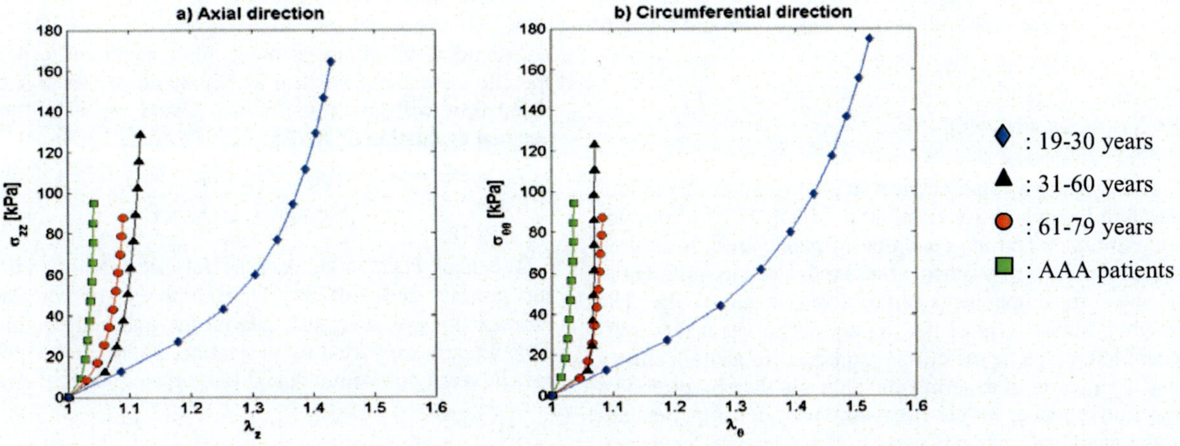

Fig. 1 Stress-stretch plot comparing the known analytical solution for biaxial loading. (a) shows the axial Cauchy stress (σ_z) as a function of axial stretch ratio (λ_z) for all four patient groups. (b) shows the Cauchy hoop stress ($\sigma_{\theta\theta}$) as a function of circumferential stretch ratio (λ_θ) for all four patient groups. The solid lines are the solutions of the experimental fit, and the symbols indicate the numerical solution.

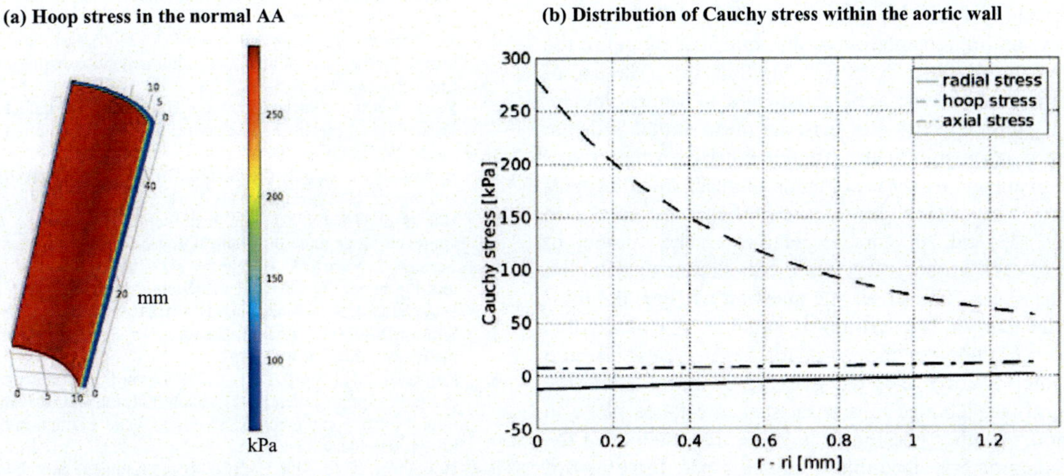

Fig. 2 (a) simulated inflation-extension experiment showing the amount of hoop stress within the aortic wall for the AAA patient group. (b) shows the stress distribution within the aneurismal wall for the AAA patient group. The dashed line is the hoop stress, the dash-dot line is the axial stress, and the solid line is the radial stress.

B. Simulation of inflation-extension test

The four fiber family constitutive relation was implemented in an anisotropic 3D FE model and applied to a circular, axis-symmetric cylinder. To exploit symmetry only one quarter of the cylinder is simulated. The simulation result for the hoop stress is shown in Fig 2a for the AAA patient group. A maximum hoop stress of 275 kPa is seen at the innermost part of the cylinder, and 60 kPa at the external part of the cylinder. With the new 3D model it is possible to investigate the anisotropic nature of the stress distribution within the aortic wall for the AAA patient group (see Fig 2b). The largest stress component is the hoop stress. The axial stress is almost constant varying from 8-10 kPa, and the radial (outward) stress is 15 kPa at the inner wall, corresponding to 113 mmHg, and zero at the external part of the wall. Comparing the results of the anisotropic model to the isotropic model for AAA tissue suggested by Raghavan and Vorp [8] (results not shown) the isotropic model underestimates the magnitude of the stress components within the wall (peak hoop stress is 70 kPa) and the change in stress distribution within the wall is more uniform within the aortic wall.

IV. Discussion and conclusion

In this paper a finite element implementation of the four fiber family constitutive relation in COMSOL Multiphysics is presented. When the condition of plane stress is secured in the biaxial test protocol, the axial and circumferential Cauchy stress components can be deduced analytically. The analytical model represented by eq. (1) serves a dual purpose. First, it can be used to fit parameters of a strain energy density function to experimental data and thereby provide a constitutive model for the stress-strain relationship. Secondly, the analytical model can serve as a reference for benchmarking numerical models such as finite element models. The former application was used by Ferruzzi et al [6] to develop a constitutive anisotropic model of arterial tissue from biaxial test data obtained by Vande Geest et al. [2]. The latter application was used successfully in this paper to benchmark an implementation of the four fiber model in the finite element program COMSOL Multiphysics. The use of mean values of the material parameters in finite element models is common [10]. But here the mean values indicate that there is not a significant difference between the groups of 31-60 years and 61-79 years in the biomechanical properties. This is surprising due to a significant difference in mean age (43 and 70 years respectively). The number of subjects in each group is the same with similar distribution among the sexes. This raises the question whether the division in the current age groups is suitable. An alternative could be to subdivide the group of 31-60 year-olds into smaller intervals of five or ten years, since it seems that the most significant change in arterial structure takes place in this period. Another possibility is to use the median of the material parameters, since this would eliminate the effect of outliers in the different patient groups. This exploration of the parameter space, together with extension of the mechanical tests to include inflation-extension tests of both normal abdominal aortic and aneurismal tissue, could improve the current model. In addition, with these improvements it might also be possible to obtain more complete knowledge about when the critical damage to the aortic tissue is most likely to occur. The model considered here is purely passive and does not account for the contribution from activation of vascular smooth muscle cells. The reasons for not including the active part are two-fold. There is lack information on the change in smooth muscle activity in normal AA. Secondly, AAA contains limited amounts of smooth muscle cells, [4,6].

Extending this finite element implementation to patient-specific model geometries with matching patient-specific blood flow will give the clinician a very powerful tool for detailed evaluation of AAAs.

Acknowledgment

We thank Prof. David A. Vorp for information on the tissue samples and Prof. Jay D. Humphrey and Mr. Jacopo Ferruzzi for providing the data on the material parameters. This work is supported by project no. 55562 at the Technical University of Denmark and Radiometer Medical Aps.

References

1. Baek S, Gleason RL, Rajagopal KR, Humphrey JD (2007) Theory of small on large: Potential utility in computations of fluid-solid interactions in arteries. Comput Method Appl M 196:3070-3078
2. Vande Geest JP, Sacks MS, Vorp DA (2004) Age dependency of the biaxial biomechanical behavior of human abdominal aorta. J Biomech Eng – T Asme 126:815-822
3. Vande Geest JP, Sacks MS, Vorp DA (2006) The effect of aneurysm on the biaxial mechanical behavior of human abdominal aorta. J Biomech 39:1324-1334
4. Humphrey JD (2002) Cardiovascular solid mechanics: cells, tissues and organs. NY:Springer, New York
5. Holzapfel GA (2000) Nonlinear solid mechanics – a continuum approach for engineering. John Wiley& Sons, Chichester
6. Ferruzzi J, Vorp DA, Humphrey JD (2010) On constitutive descriptors the biaxial mechanical behavior of human abdominal aorta and aneurysms. J R Soc Interface DOI:10.1098/rsif.2010.0299
7. Sacks MS (2000) Biaxial mechanical evaluation of planar biological materials. J Elasticity 61:199-246
8. Raghavan ML, Vorp DA (2000) Toward a biomechanical tool to evaluate rupture potential of abdominal aortic aneurysm: identification of a finite strain constitutive model and evaluation of its applicability. J Biomech 33:475-482
9. Humphrey JD, Taylor CA (2008) Intracranical and abdominal aneurysms: similarities, differences, and need for a new class of computational models. Annu Rev Biomed Eng 10:221-46
10. Vorp DA (2007) Biomechanics of abdominal aortic aneurysm. J Biomech 40:1887-1902

Author: Marie Sand Enevoldsen
Institute: Biomedical Engineering, Department of Electrical Engineering, Technical University of Denmark
Street: Oersteds Plads, Building 349
City: Kgs. Lyngby
Country: Denmark
Email: mse@elektro.dtu.dk

Assessment of the Optical Interference in a PPG-LDF System Used for Estimation of Tissue Blood Flow

J. Hagblad[1], M. Folke[1], L.-G. Lindberg[2], and M. Lindén[1]

[1]Mälardalen University, School of Innovation, Design and Engineering, Västerås, Sweden
[2]Linköping University, Department of Biomedical Engineering, Linköping, Sweden

Abstract—**The aim of this study is to assess the optical cross interference in a system including laser Doppler flowmetry (LDF) and photoplethysmography (PPG) with regard to the illuminating power of PPG-LEDs and distance between the light detector/s and light source/s.**

Reduced or missing blood perfusion can lead to pressure ulcers. Monitoring changes in blood flow in areas prone to pressure ulcer development would be a valuable tool for prevention of pressure ulcer development.

The probe, with one to two LDF-channel/s and two PPG-channels (PPG_G/560 nm and PPG_{IR}/810 nm), covers 10 cm x 10 cm. Influence from PPG-LEDs to the LDF-system and influence from the LDF-laser to the PPG-system was investigated. Three different light intensities were used for the PPG-LEDs.

Recordings were repeated using two different placements of the LDF-fibre, changing the distance between light source/s and light detector/s of the reciprocal technique.

The LDF did not show any influence from light from the PPG.

PPG_G is more affected by laser light than PPG_{IR}. Laser light influenced PPG_G, most at lowest intensity of the PPG-LEDs. The influence of the laser light to the PPG-channels is less in the outer position of the LDF-fibre.

Interference can be totally avoided by switching, only measuring by one technique at a time. Rapid flow changes are then not possible to monitor fully. When rapid blood flow variations at different vascular depths are of interest to monitor, placement of the LDF-fibre in the outer position and use of a higher light intensity of the PPG-LEDs might be an alternative. However, interference still can be present, and further, the measurement volume of LDF will be different from that covered by PPG-channels.

Keywords—**PPG, LDF, interference, peripheral flow, pressure ulcer**

I. INTRODUCTION

Local damage to skin and tissue in combination with reduced or missing blood perfusion can lead to pressure ulcers. The prevalence of pressure ulcers in hospitals in five European countries was estimated to 18 % 2006 [1]. A system giving the possibility to monitor both slow and fast blood flow changes in areas prone to pressure ulcers development would be a valuable tool for prevention. It is of interest to monitor both fast and slow blood flow changes at several vascular depths.

For none invasive tissue blood flow measurement we combined two methods: laser Doppler flowmetry (LDF) [2] and photoplethysmograpy (PPG) [3].

LDF is a technique utilizing monochromatic laser light for assessing the microcirculation of a small volume, less than 1 mm^3 according to Monte Carlo simulations [4]. The light is emitted through an optical fibre and scattered. Some of the light is reflected by moving red blood cells and the frequency is shifted. The frequency shift is used to estimate the total perfusion of the underlying tissue. The perfusion is presented in arbitrary units scaling linearly to the velocity and concentration of red blood cells.

PPG is based on absorption of light in tissue and blood. Monochromatic light illuminates the tissue and back reflected light is collected by a photo detector. Variations in the signal correlate to changes in several parameters, whereof pulsative changes in blood volume and blood flow are regarded as most important. Different wavelengths of the light and different distance between light source and detector can be utilized to monitor different tissue volumes (typical depths 5-20 mm) [5].

A probe combining PPG and LDF has previously been developed and evaluated regarding the ability to discriminate between blood flows at different tissue depths [6]. It was, however, fixed in a wooden frame making it stiff and having a linear sensor configuration. A new flexible probe has been developed with the PPG-LEDs arranged to cover a measurement area of 10 cm x 10 cm. Initial tests of this new probe has been performed [7], but also called for further investigations. The aim of the present study is to assess the optical cross interference between the PPG- and LDF-channel/s in the multi-technique system with regard to the illumination power of PPG-LEDs and distance between the light detector/s of and light source/s of the reciprocal technique.

II. MATERIALS AND METHODS

A. Optical probe

The layout of the probe is a matrix of LEDs of two different wavelengths surrounded by five photo detectors at

K. Dremstrup, S. Rees, M.Ø. Jensen (Eds.): 15th NBC on Biomedical Engineering & Medical Physics, IFMBE Proceedings 34, pp. 17–20, 2011.
www.springerlink.com

two different distances for the PPG recordings. The shallow blood flow is measured by light from green LEDs (placed 4 mm from the detectors) and the deeper tissue layers are reached by infrared LEDs (placed 25 mm from the detectors). All electrical and optical components are embedded in the flexible silicon sheet (10 cm x 10 cm and 5 mm thickness) with an extended tail to protect the connection cables. A sketch of the probe is presented in Figure 1. The two sets of LEDs are alternately switched, and the total light reaching the photo detectors is collected into two PPG-channels, PPG_G (λ= **560** nm) and PPG_{IR} (λ= **810** nm). Due to this internal switching algorithm of the PPG-system, no inference between the PPG-channels is present. The intensity of the PPG-LEDs can be adjusted and the system is calibrated to achieve suitable signal amplification.

Two flat probes containing a LDF-fibre can be inserted into the silicon sheet (Perimed 415-242 SPP, Perimed, Järfälla, Sweden), either in an inner (~20 mm from closest photo detector) or outer position (~30 mm from closest photo detector), Figure 1. In this study, one LDF channel was used, and the two positions were used sequentially.

A custom made program collects the LDF- and PPG-channel/s at a sampling rate of 75 Hz. As a way to address the optical interference, the system also includes an alternating control signal to switch between LDF and PPG measurements, with a switching time of ~20 s.

To assess the influence from PPG-LEDs on the LDF-system, LDF first was recorded with only the laser activated and then with the laser and PPG-LEDs activated.

The PPG-channels were recorded in three ways; using only laser from the LDF as light source, only the PPG-LEDs activated and with both systems activated.

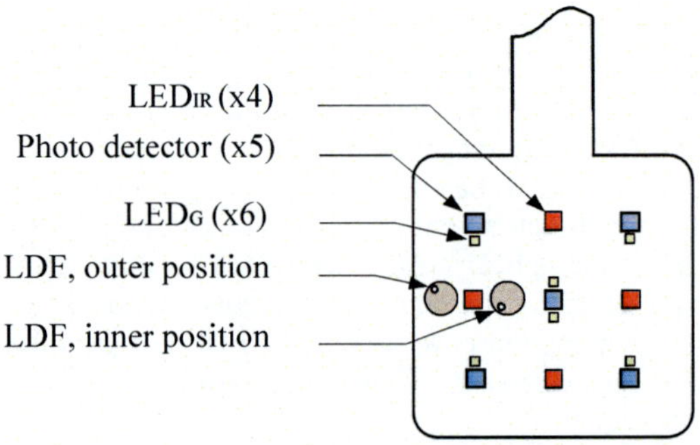

Fig. 1 Sketch of the probe, size 10 cm x 10 cm.

B. Subjects

For this study, three subjects considered healthy participated. The study was approved by the Research ethical committee at Linköping, Dnr M166-06.

C. Measuring procedure

To minimize stray light, the study was conducted with the curtains drawn and the lights out. Room temperature was 20°C-22°C.

To allow for blood flow stabilization, the measuring procedure begun with the subjects resting in a hospital bed in supine position for 15 min. The probe was then placed at the lower back, with maximum area in skin contact, while the subjects sat up in the bed. With the probe in position, prone position was resumed. The subjects remained silent and still during the measurement procedure, to minimise motion induced artefacts.

Three different illumination power levels of the PPG-LEDs were set at four minutes each. The switching algorithm was turned on for the first two minutes; each part is active for ~20 s at a time, thus ensuring no interference between LDF and PPG. The PPG recording was active during the LDF segment as well, resulting in recording of PPG with only laser active, with only LEDs active and with both laser and LEDs active. Further, LDF both with and without the presence of light from the PPG-LEDs could be recorded.

These series of recordings was first performed with the LDF-fibre placed in the inner position of the silicon sheet and then repeated using the outer placement of the LDF-fibre.

D. Data analysis

The mean value of the LDF-channel was calculated for each of the scenarios. Blood flow in the PPG-channels is represented as the peak-to-peak value of the AC-part of the signal. A custom made Matlab (Mathworks, Natick, MA, USA) program was used to extract the peak-to-peak value from the PPG-channels, and the mean for 15 s of each segment was calculated. To determine if any significant difference between the normal and influenced signal could be detected Students t-test for paired data was conducted.

III. Results

Representative recordings of the channels are presented in Figure 2, where the signals from the different combinations of light sources can be seen. Mean values and fractions of signal contributions from the different combinations of light sources of PPG and LDF, respectively, are presented in Table 1.

The LDF-channel did not show any changes depending on the presence of light from the PPG-LEDs or placement of the LDF-fibre, Figure 2. Table 1 shows that the fractional contribution in Laser/(Laser+LEDs) varies between 83 % - 119 %.

Table 1, at inner position of the LDF-fibre, shows that the laser light gives a smaller contribution to the PPG_G-channel when a higher intensity of the PPG-LEDs is used. Further, it can be seen that PPG_G is more affected than PPG_{IR}. Subjects B and C show most influence of the laser light on PPG_{IR} in the highest illumination power.

At outer position of the LDF-fibre, the laser light gives a smaller contribution to the PPG_G-channel when a higher illumination power of the PPG-LEDs is used, but the influence is not so pronounced as for PPG_G. PPG_G is more affected than PPG_{IR} also in the outer position.

When comparing inner and outer position of the LDF-fibre, it can be seen that the influence of the laser light on the PPG-channels (both PPG_{IR} and PPG_G) is less in the outer position.

T-tests for paired data was performed between each signal and influenced signal ($P<0.05$), e.g., the column LEDs and LEDs+Laser in the PPG_G in Table 1. Significant differences were achieved for the PPG_G in both inner and outer position of the LDF-fibre, but not in any other case.

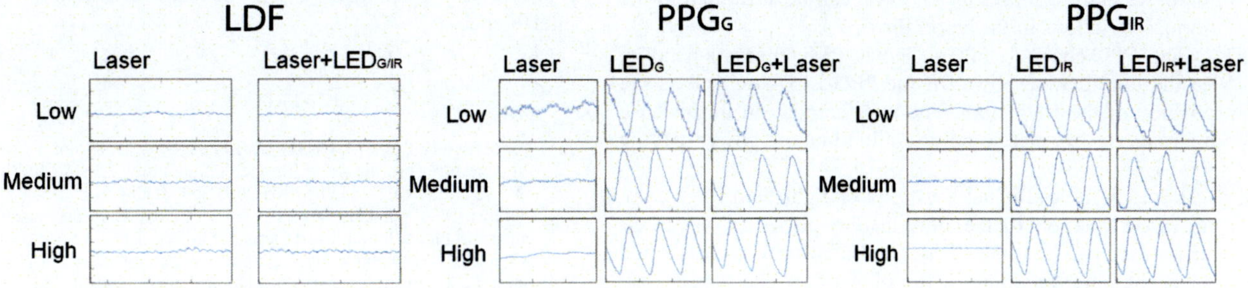

Fig. 2 Representative recordings of the channels: LDF, PPG_G and PPG_{IR} using the inner position of the LDF-fibre at low, medium and high illumination power of the PPG-LEDs. To the left the LDF-channel measured with only laser light as a light source (Laser) and both LED_{G+IR} and laser light (LED_{G+IR}+Laser). In middle and to the right: The two PPG-channels measured with only laser light as a light source (Laser), with only LEDs as a light source ($LED_{G/IR}$) and with both light sources ($LED_{G/IR}$+Laser).

Table 1: The mean values of the three channels PPG_G, PPG_{IR} and LDF are presented with different light sources, laser, IR or green LEDs. LDF is presented as a mean value of the signal. $PPG_{G/IR}$ is presented as a mean value of the amplitude of the AC-part of the signal.

Inner position of the LDF-fibre

Subject	Power of LEDs	LDF Laser	LDF Laser+LEDs	Laser+LEDs / Laser	PPG_G LEDs	PPG_G Laser	PPG_G LEDs+Laser	Laser / LEDs+Laser	PPG_{IR} LEDs	PPG_{IR} Laser	PPG_{IR} LEDs+Laser	Laser / LEDs+Laser
A	Low	0.051	*0.060*	118 %	1.690	0.444	1.950	23 %	1.430	0.085	1.360	6 %
	Medium	0.069	0.068	99 %	2.292	0.175	2.330	8 %	0.236	0.010	0.225	4 %
	High	0.106	0.101	95 %	1.400	0.075	1.499	5 %	0.525	0.007	0.564	1 %
B	Low	0.037	0.031	83 %	2.160	0.370	2.420	15 %	0.607	0.029	0.600	5 %
	Medium	0.044	0.046	105 %	1.925	0.146	2.110	7 %	0.170	0.011	0.202	5 %
	High	0.064	0.064	101 %	1.395	0.071	1.500	5 %	0.693	0.056	0.741	8 %
C	Low	0.060	0.060	100 %	1.850	0.405	2.040	20 %	1.260	0.067	1.150	6 %
	Medium	0.070	0.073	103 %	1.530	0.112	1.524	7 %	0.753	0.020	0.847	2 %
	High	0.088	0.081	92 %	0.992	0.052	1.169	4 %	0.462	0.075	0.542	14 %

Outer position of the LDF-fibre

Subject	Power of LEDs	LDF Laser	LDF Laser+LEDs	Laser+LEDs / Laser	PPG_G LEDs	PPG_G Laser	PPG_G LEDs+Laser	Laser / LEDs+Laser	PPG_{IR} LEDs	PPG_{IR} Laser	PPG_{IR} LEDs+Laser	Laser / LEDs+Laser
A	Low	0.153	0.160	104 %	1.904	0.080	1.946	4 %	0.565	0.009	0.583	2 %
	Medium	0.117	0.113	97 %	1.328	0.040	1.377	3 %	1.328	0.012	1.377	1 %
	High	0.094	0.096	101 %	1.636	0.023	1.711	1 %	1.538	0.027	1.549	2 %
B	Low	0.052	0.053	103 %	2.692	0.171	2.850	6 %	1.420	0.032	1.148	3 %
	Medium	0.058	0.062	106 %	1.745	0.105	1.728	6 %	1.336	0.027	1.318	2 %
	High	0.078	0.082	105 %	1.464	0.032	1.600	2 %	1.259	0.011	1.370	1 %
C	Low	0.075	0.089	119 %	1.950	0.105	2.140	5 %	1.160	0.037	1.230	3 %
	Medium	0.061	0.062	101 %	2.010	0.105	1.951	5 %	1.110	0.016	1.029	2 %
	High	0.076	0.084	110 %	1.460	Missing	1.626		1.162	0.038	1.253	3 %

IV. Discussion

In this study, methods to avoid optical interference in a system combining PPG and LDF have been investigated. The system has been developed to allow measurements of blood flow variations at various vascular depths, especially with the focus on investigation of the aetiology of pressure ulcer development [6][8].

No changes in the LDF signal level are found due to influence of light from the PPG-LEDs. The differences, increases and decreases up to 15 %, can be attributed to the normal variations in the blood flow.

The PPG channel, PPG_G, shows differences depending on the illumination power of the PPG-LEDs as well as the position of the LDF-fibre. The interference can in this case be reduced by increasing the illumination power of the PPG-LEDs.

PPG_{IR} is less affected by the laser light than PPG_G, but no direct relation to the illumination power of the PPG-LEDs used can be found. Though, the influence is less when using the outer position for the LDF-fibre. The reason why PPG_{IR} is less affected of the laser light is probably that the IR light has a larger penetration depth than both the green light and the laser light.

Using the outer position for the LDF-fibre and a higher light illumination power of the PPG-LEDs might decrease the interference between the techniques. However, some interference is still present in this investigation and an extended study is needed.

Using the outer position for the LDF-fibre further has the disadvantage that the blood flow measured by the LDF is slightly outside the tissue volume covered by the PPG-channels.

By switching between the different light sources, interference can be totally avoided. Then rapid flow changes however, such as reactive hyperaemia, are not possible to monitor fully in the whole tissue volume, due to the fact that just one of the techniques at a time is active (~20 s).

The switching mode is favourable when reasonable slow flow variations are under investigations such as response to prolonged times in one position. By switching, the LDF-fibre can be kept in the inner position. Thereby both systems evaluate the blood flow variations in the same tissue volume but at various depths.

Interesting physiological phenomena to be investigated by this probe is variations in blood flow over time with or without provocation on different support surfaces.

V. Conclusions

By switching between the different light sources, interference can be totally avoided. When rapid blood flow variations at several vascular depths are of interest to evaluate, placement of the LDF-fibre and illumination power of the PPG-LEDs might be used to minimize the interference.

References

1. Vanderwee K, Clark M, Dealey C, Gunningberg L, Defloor T (2006) Pressure ulcer prevalence in Europe: a pilot study. J Eval Clin Pract 13:2:227-235 DOI 10.1111/j.1365-2753.2006.00684.x
2. Holloway GA, Watkins DW (1977) Laser Doppler measurement of cutaneous blood flow. J Invest Dermatol 69:306-309
3. Allen J (2007) Photoplethysmography and its application in clinical physiological measurement. Physiological measurement 28:R1-R39 DOI 10.1088/0967-3334/28/3/R01
4. Fredriksson I, Larsson M, Strömberg T (2009) Measurement depth and volume in laser Doppler flowmetry. Microvasc Res 78:1:4-13DOI 10.1016/j.mvr.2009.02.008
5. Lindberg L G, Öberg P Å (1991) Photoplethysmography: Part 2. Influence of light source wavelength. Med Biol Eng Comput 29:48-54 DOI 10.1007/bf02446295
6. Hagblad J, Lindberg L-G, Kaisdotter Andersson A, Bergstrand S, Lindgren M, Ek A-C, Folke M, Lindén M (2010) A technique based on laser Doppler flowmetry and photoplethysmography for simultaneously monitoring blood flow at different tissue depths. Med Biol Eng Comput 48:5:415-422 DOI 10.1007/s11517-010-0577-2
7. Hagblad J, Folke M, Lindberg L-G, Lindén M (2010) A probe combining two optical techniques for monitoring of blood flow in different vascular depths – investigation of interference. 7th International Conference on Wearable Micro and Nano Technologies for Personalized Health, Berlin, Germany, 2010
8. Bergstrand s, Länne T, Ek A-C, Lindberg L-G, Lindén M, Lindgren M (2010) Existence of Tissue Blood Flow in Response to External Pressure in the Sacral Region of Elderly Individuals – Using an Optical Probe Prototype. Microcirculation 17(4):311-319 DOI 10.1111/j.1549-8719.2010.00027.x

Author: Jimmie Hagblad
Institute: Mälardalen University, IDT
Street: Box 883
City: Västerås
Country: Sweden
Email: jimmie.hagblad@mdh.se

A Flexible Sensor System Using Resonance Technology for Soft Tissue Stiffness Measurements –Evaluation on Silicone

Anders P. Åstrand[1,3], Ville Jalkanen[1,3], Britt M. Andersson[1,3], and Olof A. Lindahl[2,3]

[1] Department of Applied Physics and Electronics, Umeå University, Umeå, Sweden
[2] Department of Computer Science, Electrical and Space Engineering, Luleå University of Technology, Luleå, Sweden
[3] Centre for Biomedical Engineering and Physics, Umeå University, Umeå, Sweden

Abstract— **One of the most common forms of cancer among men in Europe and the United States is prostate cancer. The cancerous tissue is less soft, and has different biomechanical properties compared to healthy tissue. It has been shown that tactile sensors can be used to distinguish this difference. If a piezoelectric sensor is set to oscillate at its resonance frequency through a feed back circuit, a frequency shift is observed when the sensor comes in contact with a surface. This shift can be correlated to the stiffness of the tissue. A flexible instrument has been developed, with which it is possible to scan both flat and spherical bodies and where the sensor can be tilted to have different contact angles. Measurements performed in this study on flat silicone discs of different stiffness showed a relationship between both the frequency shift and the impression depth for the different silicone discs, when a constant force was applied. The results are promising for future studies on silicone with different geometries and finally on prostate tissue to complete the evaluation.**

Keywords— **Resonance sensor, Piezoelectric, Prostate cancer, Detection, Frequency shift.**

I. INTRODUCTION

The possibility to use piezoelectric resonance sensors in medical research in order to detect differences in stiffness in tissue has been shown by Omata and Terunuma [1]. The stiffness of tissue, or areas of tissue, can vary due to different pathological conditions and diseases and thereby a change in the biomechanical properties. This has been shown in previous studies and indicates the potential for several biomechanical applications [2]. Piezoelectric resonance technology has been used to measure the stiffness of the liver to indicate liver fibrosis [3] to detect lymph nodes containing metastases [4] and to measure the differences in stiffness and elasticity of the skin in order to detect oedema and lesions [5, 6]. Studies of prostate tissue show the possibility to differentiate between benign and malignant prostate glands [7]. In a study by Lindberg et al [8] a resonance sensor probe was attached to an instrument with a counter balance arm. Measurement of the frequency shift with constant applied force was done successfully on silicone discs and prostate tissue. A theoretical model relates the measured stiffness trough the frequency shift at constant force to the Young´s elastic stiffness modulus of the measured object [9].

Prostate cancer is one of the most common forms of cancer. In the US, the American Cancer Society estimates that about 218.000 new cases will be diagnosed in 2010 [10]. In Sweden, about 10.000 men were diagnosed with prostate cancer in year 2009 which was an increase by 18% compared to year 2008 [11].

To be able to study the presence of cancerous tissue on the capsule of the whole prostate after radical prostatectomy, to scan for cancer in the surface layer, it is necessary to be able to have a fixture that can rotate. Due to the spherical shape of the prostate it is also important to have the possibility to tilt the vertical translation stage so its movement is normal to the curved surface during measurements. For measurements on slices of the prostate it is necessary to be able to fixate such specimens in a horizontal position. In order to further evaluate and optimize the possibilities to distinguish cancerous from normal tissue, detailed studies with modified sensor probe constructions are needed. For example the size of the sensor might affect the resolution of stiffness variations in the tissue. For all these reasons an instrument with high flexibility is needed.

The aim of this study was to develop a flexible instrument and to verify the basic functionality of the instrument. It was also to show the possibility to distinguish soft and hard areas on tissue by using flat silicone discs of different stiffness cast in Petri dishes.

II. MATERIALS AND METHODS

A. *Theory*

The piezoelectric material used in this study was a ceramic made of lead zirconate titanate (PZT). Previous

K. Dremstrup, S. Rees, M.Ø. Jensen (Eds.): 15th NBC on Biomedical Engineering & Medical Physics, IFMBE Proceedings 34, pp. 21 – 24, 2011.
www.springerlink.com

studies by Omata and Terunuma in 1992 [1] found that two PZT elements combined, one used to generate oscillations and one for detecting the vibrations, which was connected to an electronic circuit worked as a resonance transducer.

Oscillation was generated by an electronic feedback circuit consisting of an amplifier, a band pass filter and a phase-shift circuit. The signal from the pickup was constantly transferred to the feedback circuit where the phase-shift circuit ensured that a zero phase condition between the pickup and drive signals was established. This kept the whole sensor oscillating at its resonance frequency. When the tip of the sensor came into contact with a material there will be a change in the frequency Δf, (Eq. 1). This is shown in the schematic picture in Fig. 1.

$$\Delta f = f - f_0 \quad (1)$$

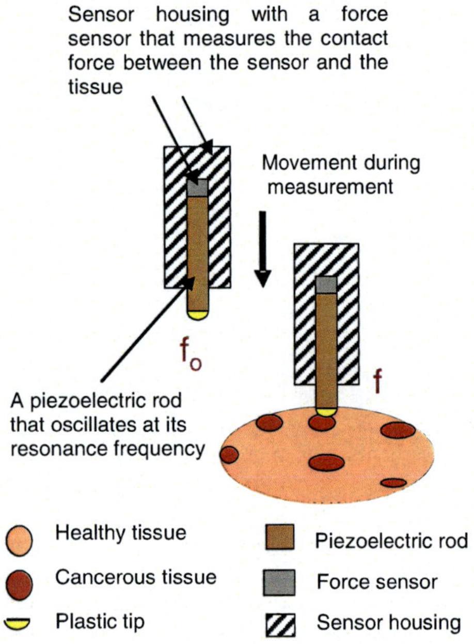

Fig. 1 A schematic picture of the sensor probe, before and during contact with a specimen. The resonance frequency f_0 changes to f during contact with the specimen

B. Experimental Setup

The exchangeable sensor probe used for the measurements in this study consisted of a piezoelectric resonance sensor in the shape of a hollow rod with the diameter of 5mm (Morgan electro ceramics, Bedford, Ohio, USA) and a strain gauge used as a force sensor (PS-05KC; Kyowa, Tokyo, Japan) that was mounted in a probe with an aluminium casing as explained in previous studies by Lindberg et al [8]. In this instrument, the probe and the vertical translation stage was attached to a manual rotary stage (NT55-030; Edmund Optics, York, UK). This made it possible to approach the measured object with different angles.

Movement of the probe was controlled via a LabView® (National Instruments, Austin, Texas, USA) program by two translation stages in the horizontal directions (M4424; Parker Hannifin Corporation, Daedal Division, Irwin, PA, USA) modified with stepper motors (17HD1008; Moons', Shanghai, P.R China) attached to the micro meter screws. Vertical movement was controlled by a motorized stage controller with a resolution of 2,5μm (Pollux Box; Micos, Irwin, PA, USA) running a compact translation stage (VT-21; Micos, Irwin, PA, USA).

The signals from the piezoelectric sensor f, the force transducer and the translation stages were collected at a sampling rate of 1kHz trough a shielded I/O connector block (NI SCB-68; National Instruments, Austin, Texas, USA) and into a computer with a data acquisition card (NI6036E; National Instruments, Austin, Texas, USA).

Flat or plane specimens could easily be attached with brackets to a platform with threaded holes (Fig. 2). For measurements on spherical bodies the instrument had a special holder that could rotate the specimen around its horizontal axis (Fig. 3).

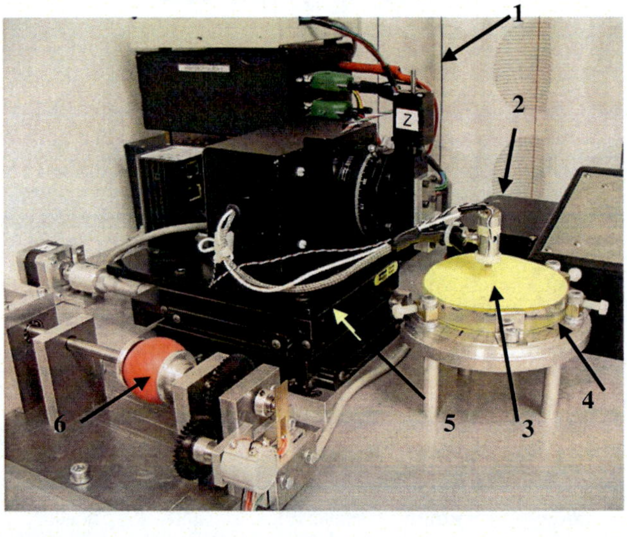

Fig. 2 The complete instrument in position to measure on a flat specimen

1 Vertical translation stage, 2 Sensor probe, 3 Silicone model, 4 Platform, 5 Horizontal stages, 6 Holder for spherical specimens. The flat silicone model in the Petri dish on the platform has temporarily been covered by a yellow paper for better viewing

The instrument shown in Fig. 2 exemplifies a position to measure on a flat silicone model in a Petri dish that is attached to the mounting platform. In Fig. 3, the instrument has been detached from the horizontal translation stages and turned 90 degrees clockwise and reattached. The position of the sensor head has been tilted so that the movement of the vertical translation stage is normal to the surface plane of the spherical silicone model.

Fig. 3 The instrument in exemplified position to measure at an angle on to the surface of a spherical body. 1 Rotation stage, 2 Vertical translation stage, 3 Sensor tip, 4 silicone model. The spherical silicone model has temporarily been covered by a red balloon for better viewing

C. Tissue Models

It has been found that for preliminary research and evaluation of new sensor techniques regarding human soft tissue characterization, silicone models are very suitable [12, 13]. They can be cast with different stiffness and it is possible to enclose small pieces of harder or softer objects inside the silicone models when casting them.

To simulate human tissue in this study, measurements were done on a two-component silicone (Wacker SilGel 612; Wacker-Chemie GmbH, Germany). Homogeneous silicone models were cast in Petri dishes (diameter 87mm, height 13mm).

This type of silicone has been used earlier by Eklund et al [10]. By mixing the two components A and B at different ratios in accordance with the manufacturer, the silicone could be made to get the different degrees of stiffness as shown in table 1.

Table 1 Two-component silicone mixing ratio according to manufacturers instructions. Stiffness measured by penetration depth according to DIN ISO 2137 using a 150g hollow cone [14]

Mixing ratio A : B, by weight		Penetration value (mm x 10^{-1})
4 : 2,5	Hardest	49
4 : 3,0		126
4 : 3,2		163
4 : 3,5		221
4 : 3,75	Softest	251

D. Measurements

In this study, the five different mixing ratios covered in table 1 were used. Flat silicone models were cast and cured in room temperature and at the given mixing ratios by weight, using a laboratory scale (BL 310; Sartorius GmbH, Göttingen, Germany).

Measurements were made with 10 repetitions on each silicone model, with a period of two minutes in between each measurement. In this study each measurement lasted

4 s and the total penetration depth was set to 1 mm at a constant penetration speed of 5 mm/s with a sampling rate of 1 kHz. The position of the probe was held at maximum penetration depth for rest of the 4 s before it was lifted.

The frequency shift, Δf, when a force of 40mN was applied was used to verify the instruments capability to distinguish the difference in stiffness between the five different silicone discs. The penetration depth I_p of the sensor probe during the applied force of 40mN was also studied as this also reflects the stiffness of the different silicone mixtures [8].

III. Results

The mean and the standard deviation for 10 measurements of the Δf and I_p was plotted against the cone penetration values (Fig. 4) from Table 1. Measurements were made on flat silicone models.

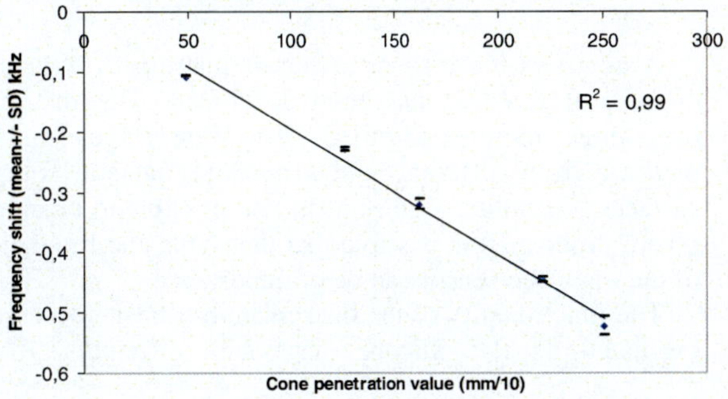

(a)

Fig. 4 Measurements at 40mN force applied on five flat silicone discs with different stiffness plotted as a function of cone penetration values. a) The measured frequency shift. b) The measured penetration depth

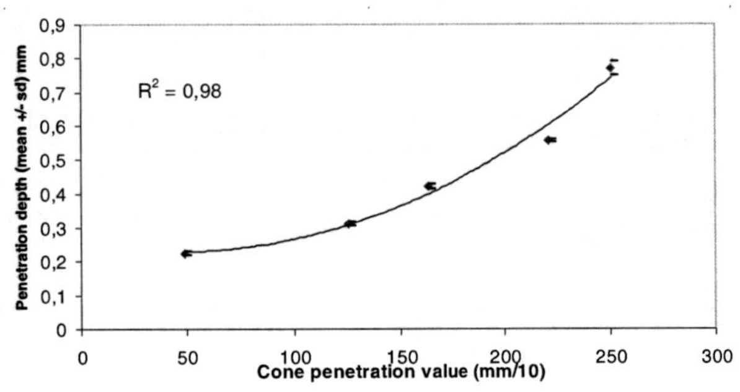

(b)

Fig. 4 *(continued)*

IV. Discussion

The study shows that the Δf of the resonance sensor could measure the stiffness of different silicone mixtures in accordance with the DIN ISO standard 2137. The frequency shift was linearly correlated to the cone penetration depth (R^2=0.99, $p<0.05$). The impression depth I_p was also shown to be correlated with a second order polynomial (R^2=0.98, $p<0.05$) to the cone penetration depth. The main technical novelty that is important for this study compared to the study made by Lindberg et al [8] is the motor controlled vertical translation stage and that the sensor can be tilted to approach the measured object with a chosen angle. The possibility to exchange the sensor probe is another important novelty. The impression speed is higher in this study, 5mm/s, compared to 1mm/s used in the system by Jalkanen et al [13]. The significance of different impression speeds may be evaluated in future studies and also the possibility to evaluate different sensor tip configurations.

V. Conclusions

A new flexible resonance sensor instrument with many degrees of freedom has been developed. The instrument uses three motor controlled translation stages and can measure from different angles on both flat and spherical surfaces. For future studies, it is also possible to change the sensor probe as it is possible that the shape and the material of the tip of the sensor can be of importance.

The functionality of the instrument has been successfully evaluated on flat silicone discs with different stiffness. Further studies on silicone models with different geometries and finally on prostate tissue with both cancerous and healthy areas must be done to evaluate this new instrument in its fullest.

Acknowledgement

The study was supported by The Industrial graduate School at Umeå University, BioResonator AB and by grants from Objective 2 North Sweden-EU Structural Fund.

References

1. Omata S, Terunuma Y (1992) New tactile sensor like the human hand and its applications. Sens Actuat. 35:9-15
2. Lindahl OA, Constantinou CE, Eklund A et al. (2009) Tactile resonance sensors in medicine. J. Med. Eng. Tech. 33:263-273
3. Kusaka K, Harihara Y, Torzilli G et al. (2000) Objective evaluation of liver consistency to estimate hepatic fibrosis and functional reserve for hepatectomy. J. Am. Coll. Surg. 191:47-53
4. Miyaji K, Furuse A, Nakajima J et al. (1997) The stiffness of lymph nodes containing lung carcinoma metastases – a new diagnostic parameter measured by a tactile sensor. Cancer 80:1920-1925
5. Lindahl OA, Ängquist K-A, Ödman S (1991) Impression technique for the assessment of oedema – Technical improvement and methodological evaluation of a new technique. Med. & Biol. Eng. & Comput. 29:591-597
6. Sasai S, Zhen Y-X, Suetake T et al. (1999) Palpation of the skin with a robot finger: an attempt to measure skin stiffness with a probe loaded with a a newly developed tactile vibration sensor and displacement sensor. Skin Res. Tech. 5:237-246
7. Jalkanen V, Andersson BM, Bergh A et al. (2006) Resonance sensor measurements of stiffness variations in prostate tissue in vitro – a weighted tissue proportion model. Physiol. Meas. 17:1373-1386
8. Lindberg P, Andersson B, Bergh, A et al. (2006) Prostate cancer detection with an improved resonance sensor system: parameter evaluation in a silicone model and on human prostate tissue *in vitro*. Med Bio Eng Comput. 44:1053-1059
9. Jalkanen V, Andersson BM, Bergh, A et al. (2008) Explanatory models for a tactile resonance sensor system – elastic and density-related variations of prostate tissue in vitro. Physiol. Meas. 29:729-745
10. American Cancer Society (11/22/2010) at http://www.cancer.org
11. Socialstyrelsen (12/17/2010) Cancer Incidence in Sweden 2009 at http://www.socialstyrelsen.se
12. Eklund A, Bergh A, Lindahl OA (1999) A catheter tactile sensor for measuring hardness of soft tissue: measurement in a silicone model and in an *in vitro* human prostate model. Med. Biol. Eng. Comput. 37:618-624
13. Jalkanen V, Andersson BM, Bergh A et al. (2006) Prostate tissue stiffness as measured with a resonance sensor system: a study on silicone and human prostate tissue *in vitro*. Med. Biol. Eng. Comput. 44:593-603
14. Wacker-Chemie (01/19/2011) Wacker-Chemie, GmbH, Germany at http://www.wacker.com

Possibility to Use Finapres Signal for Augmentation Index Estimation

K. Pilt, K. Meigas, M. Viigimaa, and K. Temitski

Tallinn University of Technology/Department of Biomedical Engineering, Tallinn, Estonia

Abstract— **The possible method for augmentation index estimation from a Finapres signal is described. The experiments were carried out on volunteers. The augmentation indices were calculated from the Finapres signal by using the proposed method and compared with the Sphygmocor reference device. As result the correlation between Sphygmocor and Finapres augmentation indices was found to be r=0.82 (p<0.001) and the regression model was constructed.**

Keywords— **Augmentation index, pressure wave, derivatives, Sphygmocor, Finapres.**

I. Introduction

Atherosclerosis is pathology, which causes the most of the early deaths in Europe. In case the atherosclerosis is diagnosed in early stadium, the pathogenesis can be retarded with effective treatment. Before the fats are buildup on walls of artery, the first changes are taking place in the walls of blood vessels. The walls become stiffer and thicker. In this stadium the development of atherosclerosis can be delayed.

The stiffness of the arteries can be estimated by calculating the augmentation index, which is measured from the registered pressure wave. Previous research results prove that the risen augmentation index value and the classical cardio vascular risk factors have a strong relationship [1].

The aortic augmentation index is the ratio between pulse pressure and the amount of additional pressure in systole, which is caused by the back reflected pressure wave from the bifurcation [2]. Additional pressure is also called as augmentation pressure. In case the higher stiffness of the aorta, the back reflected pressure wave arrives earlier to the heart and causes the higher augmentation pressure. In addition the augmentation index depends on the amplitude of the back reflected pressure wave, which is primarily related to the resistance of arterioles.

Aortic augmentation index can be measured noninvasively with recognized device as Sphygmocor. Firstly the pressure wave is registered from radial artery by using the applanation tonometry. As the next step the pressure wave in the aorta is calculated from radial artery waveform by using the universal transfer function. The augmentation index is calculated from the aorta waveform.

The Sphygmocor measurement technique is operator dependent and complex to carry out as the radial artery has to be precisely flattened between tonometry and bone during the waveform registration [3]. In addition the measured waveform has to meet the quality standards, which is determined by the characteristics of the waveform. Dependent on patient physiology the waveform, which meets the standards, can be measured after several trials. As result, from the 10 second long measurement, the average augmentation index is given.

Finapres is used to measure the beat-to-beat peripheral blood pressure by using the finger cuff. The cuff is attached around the finger and following measurement protocol does not need additional operator support. The measurement technique is relatively simple and the blood pressure waveform is registered almost continuously. In this research we are interested to estimate the augmentation index from Finapres signal, which can be related to aortic augmentation index. In this article we propose a method to calculate the augmentation index for Finapres measured blood pressure waveform. In addition we compared the calculated augmentation index values with Sphygmocor results and constructed linear regression model.

II. Methods

Aortic augmentation index is determined on the basis of pressure waveform. Aorta pressure waveform (Figure 1a) consists of primary pressure wave, which starts from left ventricle, and late back reflected pressure wave. Pressure wave reflects back from aorta bifurication and arrives back to left ventricle in the systole of heart cycle. Reflected wave influence on systole peak of pressure wave is characterized by aortic augmentation index. Augmentation index determines the additional pressure in left ventricle, which is caused by reflected wave. Augmentation index is calculated by the following formula [4]:

$$AIx\% = \frac{(P_1 - P_2)}{PP} \cdot 100\%, \qquad (1)$$

where PP is pulse pressure (systolic pressure – diastolic pressure), P_1 is reflected wave caused pressure and P_2 is

K. Dremstrup, S. Rees, M.Ø. Jensen (Eds.): 15th NBC on Biomedical Engineering & Medical Physics, IFMBE Proceedings 34, pp. 25–28, 2011.
www.springerlink.com

pressure of initial wave. P_1 - P_2 is called augmentation pressure.

The pressure wave travels along the blood vessels towards the periphery. The shape of the pressure pulse changes and shows an increase in amplitude, a steepening of the front, and a moderate fall of the mean pressure. This wave phenomenon is a direct consequence of the distensibility of the arterial wall [5]. On Figure 1b is given pressure waveform, which is registered from radial artery with Sphygmocor. On Figure 1a is also given the calculated pressure waveform in the aorta by using the transfer function. From figure 1b it is visible that the initial wave amplitude has increased compared to aortic waveform. The reflected wave is visible on radial artery waveform between initial wave maximal point and dicrotic notch. The reflected wave increase or decrease in the aortic waveform causes also the similar changes in radial artery waveform. The pressure of initial wave P_2 and reflected wave caused pressure P_1 can be similarly detected from radial artery pressure waveform (Figure 1b). In this way the augmentation index can be calculated for the radial artery pressure wave. Augmentation indices for radial artery and aortic waveform are different due to the shape of pressure waves. Still the augmentation indices are assumed to be related to each other.

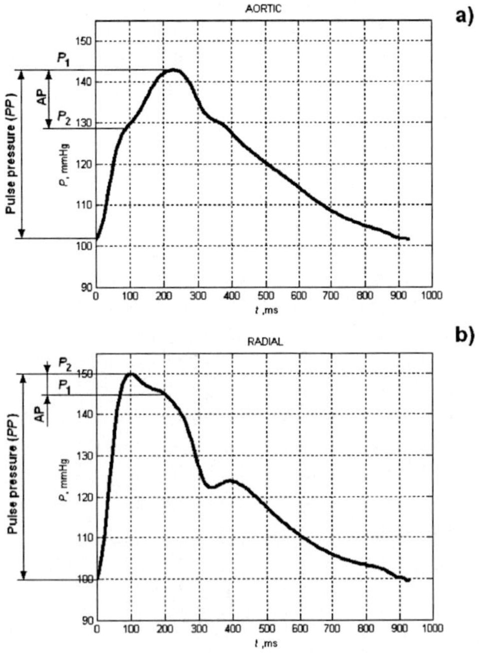

Fig. 1 a) Aortic pressure waveform, which is calculated from radial artery waveform by using Sphygmocor transfer function. b) Radial pressure waveform, which is obtained by using Sphygmocor.

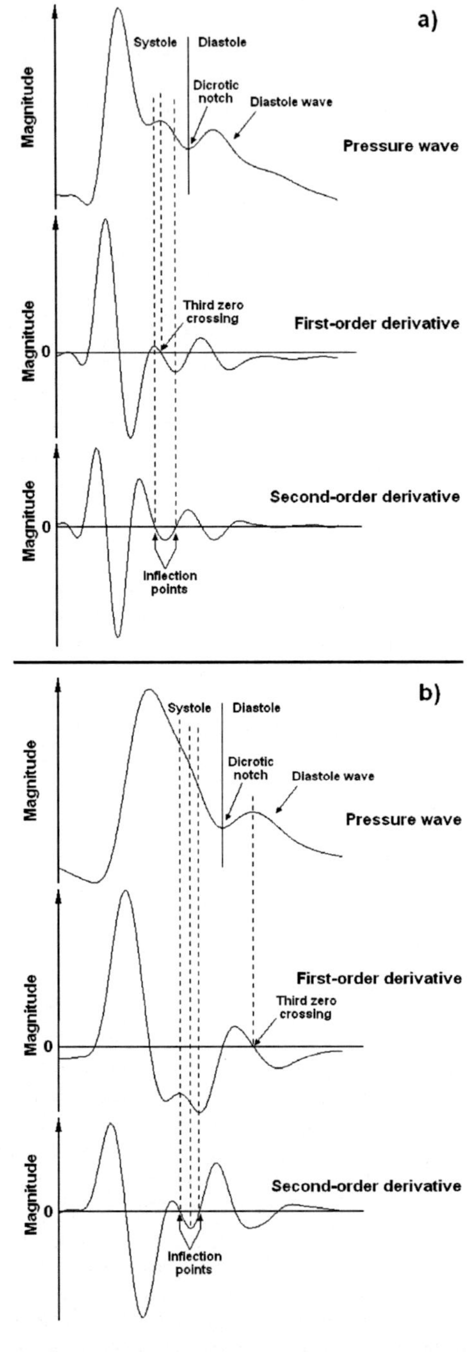

Fig. 2 Augmentation index calculation from Finapres signal by using first- and second-order derivatives. a) The reflected wave is visible on 26 years old subject's pressure waveform. b) The reflected wave is mixed with initial wave on 64 years old subject's pressure waveform.

The peripheral pressure wave, which is measured with Finapres, has similar waveform with radial artery (Figure 2a, b). The signal shape characteristics depend on the subject age and the mechanical properties of the blood vessels. For the younger subjects the initial wave is followed by the reflected wave, which is clearly visible. The dicrotic notch following the reflected wave separates the waveform into systole and diastole part. In cases of older subjects the waveform is smoother than in the younger subjects. The reflected wave is moved towards the initial wave and the two waveforms are visually harder to separate. For the augmentation index calculation the reflected wave has to be detected from the Finapres signal. The signal shape changes can be determined by analyzing the derivatives of the waveform.

The maximal points of the first derivative determine the fastest changing parts of the pressure wave. In every heart cycle the maximal point of the first derivative corresponds to the raising front of the pressure wave. The following local maximal points of the first derivative within one heart cycle can be related to the fronts of reflected or diastole wave.

The first derivative crosses zero at the places, where the pressure waveform starts to decrease or increase and it is related to the local maximal and minimal points of the pressure wave. In cases of the younger subjects the reflected wave can be easily detected from the first derivative signal by using the third zero crossing after the maximal point within the heart cycle. The same approach can not be applied for the older subject as the signal is smoother, and as a result the third zero crossing can be at the location of a diastole wave.

In the places where the second derivative of the pressure wave changes sign, the signal switches from being concave to convex, or vice versa. The point where this occurs is called an inflection point. It is noticeable that in younger subjects the reflected wave maximum falls between two inflection points where the second derivative is negative. Similarly the inflection points are visible in the signals of older subjects. For the older subjects the reflected wave can be detected at the place where the second derivative has a minimal value between two inflection points.

For the augmentation index calculation the algorithm firstly finds within one heart cycle the pressure wave maximal and minimal points. According to equation (1) the pulse pressure, *PP*, is calculated by subtracting the maximal value from minimal. Also the maximal value of pressure wave is taken as P_2 in equation (1) as it is the pressure of initial wave. In the second step, the two inflection points previously described from the second derivative are detected. Following this, the zero crossing point of first derivative is found in the range between two inflection points. In cases where the zero crossing is found, the pressure wave value at this point is taken as P_1 in equation (1). In cases where it is not found, the P_1 value from pressure wave is taken at the place where second derivative has minimal value between two inflection points. Because of the peripheral pressure waveform the augmentation index is negative.

III. RESULTS

The signal measurements with Finapres and Sphygmocor were carried out on group of volunteers. The group consisted of 28 subjects (19 male, 9 female). The age varied between 20 and 64 years (mean: 37 years). The measurements were carried out while the subject was in resting position. The subject was in resting position at least 20 minutes before the measurements. The room temperature was monitored constantly around 23 degrees during the experiments.

Firstly the augmentation index measurement was carried out with Sphygmocor. In all measurements the operator index was above 85. The operator index varies from 0 to 100 and depends on registered pressure wave average pulse height, pulse height variation, diastolic variation and shape variation. The calculated augmentation index is considered reliable in case the operator index is higher than 85.

As follows the pressure waveform was registered from subject's right hand middle finger during one minute by using Finapres device. The return-to-flow calibration was not applied as we are not interested in absolute values of pressure wave. The analogue signal was digitalized with National Instruments PCI MIO-16-E1 data acquisition card and registered with LabView environment.

In addition for the further signal processing the electrocardiographic (ECG) signal was obtained and digitalized

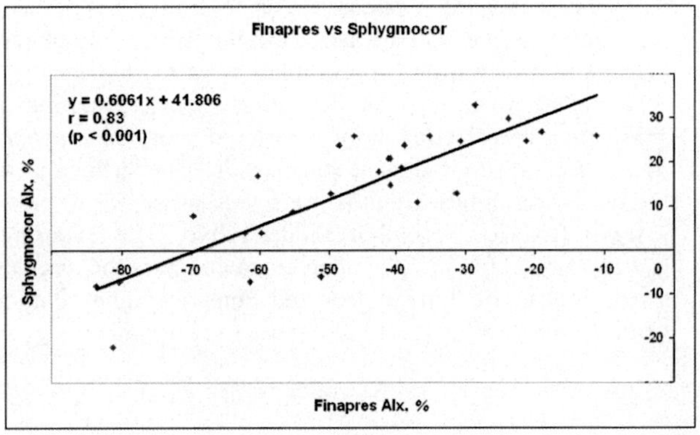

Fig. 3 Augmentation index of Finapres and Sphygmocor data set plot with regression line.

with ADInstruments PowerLab 4/20T device and registered with ADInstruments Chart software. It was ensured that the two systems were recording signals synchronously [6]. The signals were digitized with 1kHz sampling frequency. R-peaks of the ECG signal were detected in post processing by using Hamilton-Tompkins algorithm [7]. The heart cycle starting points in Finapres signal were determined by the R-peaks of the ECG signal.

Post processing of the Finapres signals was carried out in MATLAB. The signal was filtered with high- and lowpass FIR filters. The cut-off frequencies were 0.1Hz and 30Hz respectively.

The augmentation index changes with breathing. It is considered that the maximal length of one breath period is 10 seconds [8]. For the augmentation index calculation the 10 seconds long segment of the Finapres signal was chosen, where there are no distortions due to device calibration. For every heart cycle, within chosen segment, the augmentation index was calculated for Finapres signal as was described before. Based on augmentation index of every heart cycle the average was calculated.

The correlation coefficient between Sphygmocor results and augmentation indices, which are calculated from Finpres signal, is r=0.83. The probability of getting a correlation by random chance is less than 0.001. On Figure 3 is given the dataset with linear regression model. Based on these results it can be assumed that there can be linear relationship between Sphygmocor and Finapres augmentation indices. Augmentation index for Finapres signal can be estimated by using previously described method. Aortic augmentation index can be calculated by entering the augmentation index of Finapres signal into the constructed regression model.

IV. Conclusions

The idea of this research was to find out whether the Finapres signal can be used for the aortic augmentation index estimation and to compare it with Sphygmocor results. The method for augmentation index calculation from Finapres signal consists of registered pressure wave first- and second-order derivative analysis. The reflected wave from aorta bifurication is detected either from pressure wave first- or second-order derivative. The experiments were carried out on 28 volunteers to compare the augmentation indices of Sphygmocor and Finapres signal. The augmentation indices were calculated from Finapres signal by using proposed method and compared with Sphygmocor results. The correlation between Sphygmocor and Finapres augmentation indices is r=0.82 (p<0.001). From presented results it can be assumed that there is linear relationship between augmentation indices. Based on results the regression model was constructed. By using the regression model the aortic augmentation index can be estimated from augmentation index of Finapres. To improve the linear model more experiments has to be carried out. As future study the constructed linear model has to be evaluated by using Bland-Altman plot.

Acknowledgment

This work was supported by the Estonian Science Foundation Grant no. 7506, by the Estonian targeted financing project SF0140027s07, and by the European Union through the European Regional Development Fund.

References

1. Nürnberger J et al. (2002) Augmentation index is associated with cardiovascular risk. J Hypertens 20:2407–2414
2. Nichols W.W, O'Rourke M.F (2005) McDonald's blood flow in arteries. Hodder-Arnold, London
3. Kelly R, Hayward C, Ganis J, Daley J, Avolio A, O'Rourke M (1989) Noninvasive registration of the arterial pressure pulse wave form using high-fidelity applanation tonometry. J Vasc Med Biol 1:142-149
4. Kelly R, Hayward C, Avolio A and O'Rourke M (1989) Noninvasive determination of age-related changes in the human arterial pulse. Circulation 80:1652-1659
5. van de Vosse F N and Stergiopulos N (2011) Pulse Wave Propagation in the Arterial Tree. Annu Rev Fluid Mech 43:467-499
6. Pilt K, Meigas K, Viigimaa M, Temitski K, Kaik J (2010) An experimental measurement complex for probable estimation of arterial stiffness, Conf. Proc. IEEE Eng. Med. Biol. Soc. vol.. 1, EMBC, Buenos Aires, Argentina, 2010, pp 194–197
7. Hamilton P.S, Tompkins W.J (1986) Quantitative investigation of QRS detection rules using the MIT/BIH arrhythmia database. IEEE Trans Biomed Eng 12:1157-1165
8. Asmar R (1999) Arterial Stiffness and Pulse Wave Velocity. Clinical applications. Elsevier, Paris, pp 25-55

Author:	Kristjan Pilt
Institute:	Department of Biomedical Engineering
Street:	Ehitajate tee 5
City:	Tallinn, 19086
Country:	Estonia
Email:	Kristjan.pilt@gmail.com

Onto-oncology: A Mathematical Physics Unifying the Proliferation, Differentiation, Apoptosis, and Homeostasis in Normal and Abnormal Morphogenesis and Neural System

K. Naitoh

Waseda University, Faculty of Science and Engineering, Tokyo, Japan

Abstract— **The macroscopic theory describing the rhythmical morphogenetic process and the standard circuit in brain (Naitoh, Proc. of ICBME, 2008 & JJIAM 2011) also reveals the essential features underlying cancer expanding infinitely, which is with proliferation, differentiation, apoptosis, and homeostasis. Extremely large number of molecules in healthy and cancerous processes can be classified into six macroscopic groups. Then, the macroscopic theory, which is based on an ordinary differential equation system with only six variables, will bring a new fundamental platform for eradicating cancer.**

Keywords— **Sevenfold beat, ontogeny, apoptosis, cancer, brain, neural network.**

I. Introduction

A cycle of about seven cell divisions is the basic morphogenetic cycle determining the bifurcation points of organs in human beings. [1, 2] For example, about seven cell divisions of a fertile egg produce blast cysts and the next seven divisions lead to another new stage with the complex structures of ectodermal and endoderm cells. In the later part of the morphogenesis of human beings, the first heart beat (beating cardiomyocytes) occurs at about seven divisions after mesoderm formation and hands and legs also emerge at seven divisions after the first heart beat. [1, 2]

It is also known that, when suprachiasmatic nuclei (SCN) in brain are removed, the clock shows a cycle of about 3-4 hours, although the normal circadian clock in the brain has a cycle of 24-25 hours. [3] This means that the sevenfold-beat cycle is also present in the brain.

The number of beats for the morphogenetic process and the brain system is about between four and nine in many cases with the averaged value being about six or seven.

What is the bio-standard circuit controlling the various processes from the origin of life to human beings? Researches on the complex biochemical reaction networks will be important for finding the standard circuit. However, the network is too complex to find the essence.

Our previous model [4-7] reveals various temporal aspects underlying the morphogenetic processes and the brain including the circadian clock and fundamental neural network circuit.

In the present report, we will show that the model also clarify the fundamental mechanism underlying cancer, which was mysterious, because the process of cancer is complicatedly related to various aspects such as differentiation, proliferation, apoptosis, gene change, immune system, and homeostasis.

II. Four-cylinder bioengine: as the standard amplifier

Living beings employ huge types of genes and proteins. Even the primitive microorganisms such as archaebacteria need many types of proteins in order to survive. (Fig. 1a) A lot of databases on the genes and proteins have been developed in the world. However, we may miss many genes and proteins unknown in the database and also the imperfect database may lack some important bio-molecules for maintaining living beings. Thus, it will be important to find the standard pattern of bio-molecular network system, by classifying all of the molecules into some macroscopic groups. (Fig. 1b)

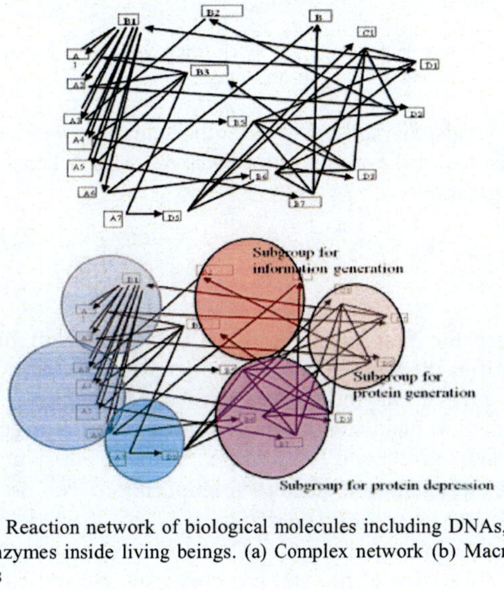

(a)

(b)

Fig. 1 Reaction network of biological molecules including DNAs, RNAs, and enzymes inside living beings. (a) Complex network (b) Macroscopic groups

K. Dremstrup, S. Rees, M.Ø. Jensen (Eds.): 15th NBC on Biomedical Engineering & Medical Physics, IFMBE Proceedings 34, pp. 29–32, 2011.
www.springerlink.com

In biological systems, at least two types of enzyme categories x21 and x22, which are for replicating gene groups and enzyme systems respectively, are necessary for achieving a closed reaction cycle. Examples of the former and latter are DNA replicase and ribosome protein, respectively. This leads to the conclusion that two types of gene groups x11 and x12 for generating the two types of enzyme systems are inevitable and serve to code the two enzyme systems (Fig. 2). The core cycle for self-replication can be modeled by these four molecular categories, i.e., two gene groups and two enzyme systems, which are denoted as categories x11, x12, x21, and x22 shown in Fig. 2. [4, 5, 7] This four-cylinder system of the four categories works as a closed loop, if elements such as nTP and amino acids are input from the outside with energy. This is the minimum 4-cylinder engine (minimum hypercycle), which produces molecules exponentially in cases where there are no degradations. It is stressed that only two groups cannot amplify both groups, because the PCR system with only two molecular groups cannot induce enzymes.

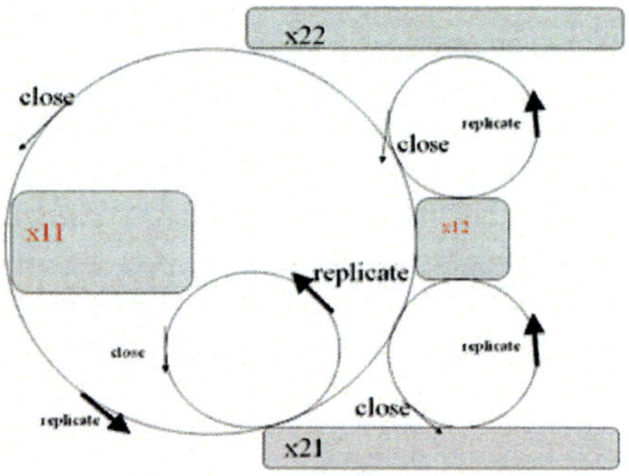

Fig. 2 Four-cylinder engine for living beings (Minimum hyper-cycle: two gene groups and two enzyme systems).

It yields the following essential equation system for the densities of the four categories $x_{ij}(N)$, with N denoting time (or generation).

$$x_{ij}^{N+1} - x_{ij}^{N} = \alpha_{ij} * x_{1j}^{N} * x_{2i}^{N}, \quad x_{ij} \ge 0, \ (i = 1,2, j = 1,2) \tag{1}$$

where the first subscript i in $x_{ij}(N)$ on the left hand side signifies the molecular type: $i = 1$ for the gene group and $i = 2$ for the enzyme system; the second subscript j in $x_{ij}(N)$ on the left hand side indicates function: $j = 1$ for generating the gene group and $j = 2$ for generating the enzyme system; and the operator * denotes multiplication. The notation α_{ij} indicates an arbitrary constant.

The simplest hypercycle of x11, x12, x21, and x22 works as "the engine of life" for promoting more complex reaction networks.

Then, single cell systems such as bacteria and archaebacteria will employ the present 4-cylinder system as proliferation, because the number of cells will continue to increase, if energy is supplied in continuous cultivation systems.

III. FOUR-CYLINDER BIOENGINE WITH NEW CANDIDATES

If a small amount of the other gene group x1n (n>2), unrelated to x11 and x12, comes around the minimum hyper-cycle in Fig. 2, the production of protein x2n (n>2) coded by x1n will be generated automatically. There are many possibilities for x1n. When many types of the gene groups for x1n (n>2) come to the pool of x11, x12, x21, and x22, many types of proteins corresponding to these gene groups can be generated. Some of these proteins will be for muscle. This is an important step toward a cell system.

IV. SIX-CYLINDER BIOENGINE: AS THE STANDARD CLOCK CIRCUIT FOR THE MORPHOGENETIC PROCESSES

Next, let us think that a gene group which codes negative enzyme system for depressesing gene productions such as Oct-4 and SOX-2 [11, 12], x13, is incorporated into the foregoing core cycle of x11, x12, x21, and x22. (The morphogenetic process of multi-cellar systems in mammals must include negative controllers such as Oct-4 and SOX2 for producing tissues and organs.) This depression gene group x13 attaching DNA would be one among a large number of possible gene groups mentioned in the above section. When this x13 accidentally met the four-cylinder bio engine, the depression protein groups x23 automatically increases. Thus, the six-cylinder bioengine in Fig. 3 works as a standard clock for various purposes.

It yields the following essential equation system for the densities of the six categories $x_{ij}(N)$, with N denoting time (or generation).

$$x_{ij}^{N+1} - x_{ij}^{N} = \alpha_{ij} * (x_{1j}^{N} - \beta_{ij} * x_{23}^{N}) * x_{2i}^{N},$$
$$x_{ij} \ge 0, \quad x_{1j}^{N} - \beta_{ij} * x_{23}^{N} \ge 0, \quad (i = 1-2, j = 1,-3) \tag{2}$$

where the first subscript i in $x_{ij}(N)$ on the left hand side signifies the molecular type: $i = 1$ for the gene group and $i = 2$ for the enzyme system; the second subscript j in $x_{ij}(N)$ on the left hand side indicates function: $j = 1$ for generating the gene group, $j = 2$ for generating the enzyme system, and $j = 3$ for depressing the system; and the operator * denotes multiplication. The notation a_{ij} indicates an arbitrary constant. [4-7]

Computational results obtained with Eq. 2 clearly show that the antagonism between the negative controller x23 and the positive replication factors x21 and x22 induces bifurcation events at rhythmic intervals constituting about seven divisions, although the intervals are slightly chaotic and the vibrational amplitude is attenuated. (Fig. 4a) [The density ratio of x12 and x23 signifies the amount of DNA uncovered by protein group x23. The condition of x12/x23 > 1.0 means that a part of x12 is not covered by x23. The condition of x12/x23 < 1.0 means that x12 is completely blocked by redundant x23. When x23 is dense in stem or induced pluripotent stem (iPS) cells, it means that these cells can be reprogrammed by the presence of much Oct-4. This oscillation of x12/x23 will lead to changes in the gene combination for expressions, i.e., the emergence of new organs.] The present six-cylinder bioengine can be used as normal differentiation in the morphogenetic process.

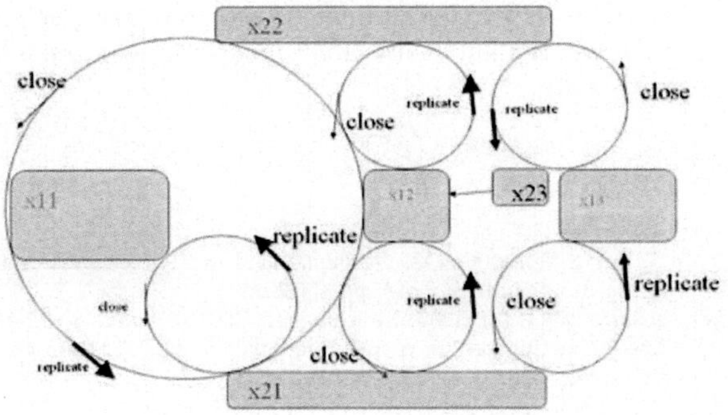

Fig.3 Six-cylinder engine for living beings.

V. SIX-CYLINDER BIOENGINE FOR THE HOMEOSTASIS WITH HEALTHY APOPTOSIS

Apoptosis should also work while synchronizing with the above six-cylinder engine for the morphogenetic process of several organs and bio-units, because fade-away process such as that of cells between fingers during the morphogenetic process is controlled by apoptosis. This implies that the apoptosis process will also work with the six-cylinder engine of the form in Eq. 2.

It is also known that, while proliferation occurs, normal apoptosis controls the sizes of organs at the staeady state of adult related to homeostasis. Thus, homeostasis is in a fusion of the prolifeation based on 4-cylinder engine and apoptosis due to 6-cylinder engine.

VI. CANCER

Healthy cells proliferate and differentiate with the mechanism described by Eq. 2 and Fig. 3, while maintaining the rhythmic intervals constituting about seven divisions. However, the control system based on x13 and x23 does not work well for aggregation of cancer cells. Thus, cancer cells dominated by only the four-cylinder engine with four variable system of x11, x12, x21, and x22 may essentially lead to their monotone increases. [Analysis based on Eq. 1 shows exponential increases of the four molecular groups. It is stressed that cancer spreads at a high speed incessantly.]

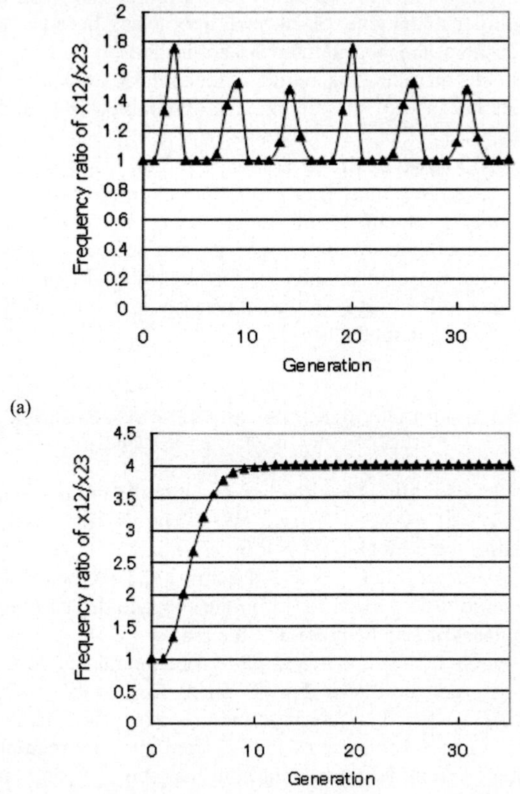

Fig. 4 Density oscillations of bio-molecule groups: Time-evolutions of x12/x23 while parameter a_{ij} is varied.

(a) $a_{11}=1, a_{12}=1, a_{13}=1, a_{21}=1, a_{22}=10, a_{23}=1$

(b) $a_{11}=1, a_{12}=1, a_{13}=1, a_{21}=10, a_{22}=1, a_{23}=1$

It is stressed in Fig. 4b that excessive activation of α_{21} assisting replication of genes brings no oscillation. Actually, cancer does not degrade telomere in DNA, which means excessive activation of genes. Thus, the 6-cylinder bioengine having large replication of genes is close to the 4-cylinder one.

VII. Regenerative Organs

Some organs inside the human beings such as liver are naturally regenerated when a part is eliminated, although ones such as heart are not regenerative. It is well known that gecko can regenerate the tail.

Regeneration can basically be possible by the six-cylinder bioengine, which controls the morphogenetic process, because regeneration and morphogenesis are like "two sides of the same coin". Moreover, regeneration is also with the six-cylinder bioengine of normal apoptosis, because the sizes of the organs regenerated are controlled.

Thus, the artificial regeneration technology will also be difficult because of the complexity of double six-cylinder engines.

It is well known that regeneration based on iPS cells often bring cancer. We must find a narrow passage between the four-cylinder bioengine of cancer and the double six-cylinder bioengines of morphogenesis and normal apoptosis.

Determination of parameter α_{ij} for healthy and cancerous conditions will bring us a guideline principle for realizing safe regenerative medicines.

VIII. Six-cylinder bioengine: as the standard clock circuit for the brain system

Next, let us remind the circadian clock in the brain, having a cycle of 24-25 hours. We know that, when suprachiasmatic nuclei (SCN) in brain are removed, the clock shows a cycle of about 3-4 hours. [3] This means that the sevenfold-beat cycle based on the six-cylinder engine of molecules will also be present in the brain.

The fundamental topological pattern of neural network in brain will also be that in Fig. 3. Then, six variables of Di and xij (i=1-2, j=1-3) are redefined as activation level of neurons. The other systems inside brain such as memory, thinking, pyramidal area, emotion, language, visual and auditory senses may also be governed by this seven-beat cycle of Fig. 3.

IX. Conclusion and Discussion

A standard circuit common to embryo, apoptosis, cancer, and neural network is shown in this report. We solved Eq. (2) for various values of the parameters. Computational results obtained by varying the parameters in Eq. (2) show that the time cycles are between four and nine beats, while varying α_{i2} between 0.0 and 5.0. The cycle is about sevenfold beat on the average. These values are close to those in various living systems. [4] However, we should search the other evidence more. [15]

We should also examine the spatial feature underlying the normal and abnormal processes in living beings.

We revealed the inevitability of the specific size ratios of various biological molecules such as five nitrogenous bases, twenty amino acids, nucleic acids, and proteins. [7-10, 13] In our previous report [14], we also simulated the three-dimensional morphogenetic process of the complex cerebral shape in the brain, including its main blood vessels, as well as that of asymmetric inner organs such as the liver and symmetric outer organs like the arms and legs. Thus, the fusion of this report on temporal aspect and our previous models on space [7-14] may outline the whole process.

Acknowledgment

This article is part of the outcome of research performed under a Waseda university Grant for special research project (2009B-206).

References

1. Gilbert S F (2006) Developmental biology. 8th edition. Sinauer Associates, Sunderland, MA, pp 211-251
2. Nishikawa S (Ed.) (2008) Ultimate Stem Cell, Newton 6. pp 12-55
3. Bear M F, Connors, B W and Paradiso M A (2007) Neuroscience, Lippincott Williams & Wiklins Inc. USE.
4. Naitoh K (2011) Morphogenic economics. Japan Journal of Industrial and Applied Mathematics, Vol. 28, No.1.
5. Naitoh K (2008) Proceedings of 13th Int. Conf. on Biomedical Engineering, Springer-Verlag.
6. Naitoh K (2009) Proceedings of the 2009 Conference on Chemical, Biological, and Environmental Engineering (CBEE), Singapore.
7. Naitoh K (2008) Artificial Life and Robotics, 13, pp 10-17.
8. Naitoh K (2010) Artificial Life and Robotics, 15, pp 117-127.
9. Naitoh K (2010) J. of Cosmology, 5, pp 999-1008.
10. Naitoh K (2001) Japan Journal of Industrial and Applied Mathematics, 18-1, pp 75-105.
11. Takahashi K and Yamanaka S (2006) Cell 126, pp 663-676.
12. Jin Y (2008) Proceedings of international symposium on induced pluripotent stem cell research frontier and future, Kyoto. pp 14-15.
13. Naitoh K, Inoue H, Hashimoto K (2010) Topo-embryology. WCB 2010. IFMBE Proceedings, Springer-Verlag pp 1163-1166.
14. Naitoh K (2008) Artificial Life Robotics, Springer, 13 pp 18.
15. Tannock I F et al. (editors) (2005) The Basic Science of Oncology. (The McGraw-Hill Companies, Inc.)

Author: Ken Naitoh
Institute: Waseda University, Faculty of Science and Engineering
Street: 3-4-1 Ookubo
City: Shinjuku, Tokyo
Country: Japan
Email: k-naito@waseda.jp

Supervised Neuro-fuzzy Biofeedback for Computer Users

A. Samani[1], A. Kawczyński[2], and P. Madeleine[1]

[1] Center for Sensory-Motor Interaction (SMI)/ Department of Health Science and Technology, Aalborg, Denmark
[2] Department of Athletes Motor Skills, Sport Institute, University School of Physical Education, Wroclaw, Poland

Abstract— **The design of an advanced biofeedback system was introduced using neuro-fuzzy concept. Eleven healthy volunteers took part in six sessions over two weeks in which computer work was performed for 10 min. The six sessions were divided into two identical parts where each part was taken place in two consecutive days. After the first session, the subjects underwent excessive eccentric exercises of shoulder elevation to induce muscle soreness. The second session was performed immediately after the exercises and third session 24 hours after the exercises. The second part was performed exactly one week after the first part. Surface electromyography (EMG) of descending and ascending trapezius, deltoideus anterior and serratus anterior was recorded. Linear and non-linear indices of muscular load were calculated from EMG signals. The first session was utilized as the benchmark of normal muscle condition during computer work and all the rest as mal-functioning/altered condition. A neuro-fuzzy system was trained and tested to discriminate between the first session and all the rest. Using a greedy forward search strategy most discriminative features were found. A high sensitivity ~90% but a low specificity ~60% was observed. It was concluded that apart from the trapezius, the deltoideus and serratus anterior should also benefit from a biofeedback design. Combining of such system with timing constraint on biofeedback alarming can render a viable biofeedback system aiming at preventing musculoskeletal disorders.**

Keywords— **delayed onset muscle soreness, sample entropy, permutation entropy, extended permutation entropy, work related musculoskeletal disorders.**

I. Introduction

Biofeedback approaches aim at preventing the occurrence of work related musculoskeletal disorders (WMSD). Using surface electromyography (EMG) as input signal, biofeedback has been shown effective to decrease the muscle activation level or change the activation pattern in the descending part of the trapezius [1, 2]. WMSD are prevalent among computer users and especially around the shoulder region.

Although biofeedback has been successful to decrease the load on muscles, as a side effect, the muscular load might be simply redistributed to other synergistic muscles. In particular, an increase of muscular load has been observed in the ascending trapezius, anterior deltoid and serratus anterior when descending trapezius rendered the source of biofeedback [3]. Additionally, serratus anterior and ascending parts of trapezius work synergistically to provide the required force for lower scapula rotation [4]. This indicates that the design of an innovative biofeedback system should be capable of merging the information from a series of muscles.

Most biofeedback designs have used simple measures (e.g. EMG amplitude and gaps) as source of information for generating feedback [1, 2, 5, 6]. These types of information need long time-series (longer than one minute) to provide consistent results [2]. Additionally, the definition of "safe working", condition is highly arbitrary [1, 2].

To simulate a malfunctioning condition of musculoskeletal system, modalities of experimental muscle pain may be of relevance because they cause transitory well-controlled muscle pain that can mimic clinical conditions [7]. In particular, repetitive eccentric exercises at maximal or supra-maximal levels have been used to induce delayed onset muscle soreness (DOMS) as an endogenous muscle pain model [8].

In this study, we used DOMS to induce a situation reminiscent of "not safe" working condition. The biofeedback system role was to discriminate between normal and "not safe" condition.

II. Method

A. Subjects

Eleven right handed males subjects (aged 25,3±5,0 years; height 177,9±5,8 cm; weight 69,8±6,4 kg) participated in the present study. All participants reported no pain in the shoulder region prior to the experiment and had no history of neck-shoulder disorders. The study was approved by the local ethics committee (No. N-20070004MCH) and conducted in accordance with the Declaration of Helsinki.

K. Dremstrup, S. Rees, M.Ø. Jensen (Eds.): 15th NBC on Biomedical Engineering & Medical Physics, IFMBE Proceedings 34, pp. 33–36, 2011.
www.springerlink.com

B. *Experimental Protocol*

The experiment was composed of six sessions divided in two identical parts. The second part was performed exactly seven days after first part. Each part was performed in two consecutive days. The procedure for each part composed of following items during first part: 1) recording of 30 second instructed rest to compute the resting level of EMG 2) to perform reference contractions consisting 3 different task bilateral shoulder elevation (shrug), abduction and flexion for five seconds in upright sitting position used for normalization of SEMG. Root mean square (RMS) of EMG activity was calculated over 250 ms epochs moving in steps of 100 ms and finally taken average. Reference voluntary electrical activation (RVE) was chosen the maximum of calculated RMS values for three reference contractions 3) 10 minutes of standardized computer work. These four steps were repeated before (session one), immediately after (session two) and 24 hours after (session three) eccentric exercises. Session four to six were performed similarly one week later. The detailed information regarding the eccentric exercises and computer worked were performed as explained below:

a) Eccentric exercises

The eccentric exercise protocol used to induce DOMS and consisted of 50 eccentric contractions of right shoulder which was divided into 5 bouts including 10 contractions at maximum voluntary contraction level and the bouts were separate by 2 min resting period. During each eccentric contraction subject had to counteract the vertical force exerted by a dynamometer as much as possible over previously defined shoulder range of motion. The subject raised both shoulders and wearing a corset during the exercise to avoid lateral bending.

b) EMG recordings and features

EMG was collected from four sites, namely, descending and ascending parts of trapezius muscle, deltoideus anterior and serratus anterior. Bipolar surface electrodes were aligned (inter-electrodes distance of 2 cm) along the direction of the muscle fibers. Electrodes were placed (i) ~20% lateral to the midpoint between the acromion and the C7 vertebra for the second upper (descending) part, (ii) ~33% medial to the midpoint between the medial scapular and the T8 vertebra for the lowest (ascending) part (iii) within the elongated oval area 2 cm below the lateral end of the clavicle (deltoid anterior) (iv) on the fourth rib from the top on the midaxillary line (serratus anterior). The reference electrode was placed over the C7 spinal process. The EMG signals were amplified 2000 times, and band pass filtered [5-1000 Hz]. SEMG signals were sampled at 2048 Hz, converted in digital form by a 12 bit A/D converter and stored on disk. SEMG signals were digitally band-pass filtered (Butterworth, 4th order, 30-500 Hz). Furthermore, a notch filter (4th order Butterworth band stop with rejection width one Hz centered at two first harmonics of the power line frequency (50 Hz)) was used in case of line interference. Normalized root mean square (RMS), sample entropy (SaEn), permutation entropy (Pe) and its extended version (ExPe) values (see the appendix for detailed information) were estimated for 0.5 s non overlapping epochs and averaged over 20 second window as EMG features. Relative rest time (RRT) over 20 s non-overlapping epochs was computed [9]. In total 20 features of EMG time-series were extracted over 20 second windows.

c) Neuro-fuzzy classifier

A neuro-fuzzy network was constructed to model the input-output relation. This means that given the extracted the aforementioned EMG features, it could discriminate between the first session of computer work and all the other sessions. The importance of this discrimination is that the firs session implies a normal working and other sessions imply some sort of malfunctioning/ altered condition. The network was based on a Sugeno-type fuzzy inference system. The training algorithm was composed of a combination of the least-squares method and the back-propagation gradient descent method. The training algorithm adjusts the membership function parameters and emulates the given training data set [10].

The classifier was tested through a 5-fold cross validation and a greedy forward search algorithm was utilized to point out the most discriminative feature [11] in cost of no significant loss of sensitivity and specificity.

III. Results

Twelve features out of 20 calculated features which are shown in Table 1 were selected by the search algorithm.

The classifier showed a high sensitivity ~90% but low specificity ~60% using the selected features. Including the removed features based on search algorithm could not improve the sensitivity and specificity more than one percent.

IV. Conclusions

A proper biofeedback should probably be based on several synergistic muscle and both linear and nonlinear indices of EMG variability. Particularly, in this study the importance of deltoideus and serratus anterior beside trapezius was highlighted.

Furthermore, the use of non-linear indices highlighted that these features may provide useful information about the true structure of motor variability. This information could not be achieved by means of linear indices [12, 13].

For example, nonlinear approaches can provide some information on the occurrence of recurrent patterns throughout the same time-series [14]. Such methods have been used in ergonomics studies recently [15], but their discriminative importance to classify the working condition had not been investigated before.

Table 1 Calculated feature from four sites of EMG recordings, the hashed cells in front of each feature (normalized RMS (NRMS), relative rest time (RRT), sample entropy (SaEn), permutation entropy (Pe) and extended permutation entropy (ExPe)) indicates that the feature has been selected by the search algorithm

Site of recording	feature	
Descending trapezius	NRMS	
	RRT	▒
	SaEn	▒
	Pe	▒
	ExPe	▒
Ascending trapezius	NRMS	▒
	RRT	
	SaEn	
	Pe	
	ExPe	

Site of recording	feature	
Deltoideus anterior	NRMS	▒
	RRT	
	SaEn	▒
	Pe	▒
	ExPe	▒
Serratus anterior	NRMS	▒
	RRT	
	SaEn	
	Pe	▒
	ExPe	▒

The current design is quite conservative (60% specificity), however, applying a time constraint on the interval of successive alarms may mitigate this issue. Additionally, adding some new features to the initial feature vector may improve the performance of the design. For example, including recurrence quantification analysis could probably improve the performance of the design.

Appendix

Sample entropy (SaEn): is calculated through the steps listed below:

Step 1. Form a time series of data $u(1),\ldots u(N)$, given N raw data values from measurements equally spaced in time. Normalize the time series to its standard deviation.

Step 2. Fix the embedding dimension m, an integer, and the tolerance distance r, a positive real number. The value of m represents the length of compared runs (a window), and r effectively represents a filter.

Step 3. Form a sequence of vectors $X(1),\ldots X(N-m+1)$ in $\Re^m$ real m-dimensional space, as $X(i)=[u(i),\cdots,u(i+m-1)]$.

Step 4. Use the sequence $X(1),\ldots X(N-m+1)$ to construct, for each i, $C_i^m(r)=\{$ number of $X(j)$ such that $d[X(i),X(j)]\leq r\}/(N-m)$.

$d[X(i),X(j)]$ for vectors $X(i)$ and $X(j)$ is defined as $d[X(i),X(j)]=\max|u(i+k-1)-u(j+k-1)|$ for $k=1,\cdots m$ and $i\neq j$.

Step 5. Next, define $\Phi^m(r)=\dfrac{\sum_{i=1}^{N-m} C_i^m(r)}{N-m}$

Step 6. SaEn= $-\ln\left(\dfrac{\Phi^{m+1}(r)}{\Phi^m(r)}\right)$.

Permutation entropy (Pe) and its extended version (ExPe): First, the input time-series is fragment into a sequence of motifs composed of "n "consecutive samples of time series (in this case n=3), second, identify each motif as belonging to a particular category of relative position of the "n" samples with respect to each other (in this case 6 categories). These categories could be formed like peaks, troughs and slopes as local morphological structure of a time-series. Third, count the number of motifs of each of the six categories, to obtain the probability of occurrence of each motif in the time-series (p_i) and finally, calculate the Pe of the resultant normalized probability distribution of the motifs, using the standard Shannon uncertainty formula.

$$Pe=\frac{\sum_{i=1}^{Number\ of\ motifs} p_i}{\ln(Number\ \ of\ \ motifs)}.$$

If the aforementioned probability is extended into two dimensions having the distribution of distance of motifs in the same category with respect to each other on the other dimension, extended permutation entropy is calculated as the Shannon entropy of the two dimensional probability distribution.

Acknowledgment

This work was financially supported by Gigtforeningen and Det Frie Forskningsråd_Teknologi og Produktion.

References

1. P. Madeleine, P. Vedsted, A. Blangsted, G. Sjøgaard and K. Søgaard, "Effects of electromyographic and mechanomyographic biofeedback on upper trapezius muscle activity during standardized computer work," Ergonomics, vol. 49, pp. 921-933, 2006.
2. H. J. Hermens and M. M. R. Hutten, "Muscle activation in chronic pain: its treatment using a new approach of myofeedback," Int. J. Ind. Ergonomics, vol. 30, pp. 325-336, 2002.
3. G. Palmerud, H. Sporrong, P. Herberts and R. Kadefors, "Consequences of trapezius relaxation on the distribution of shoulder muscle forces: an electromyographic study," J. Electromyogr. Kinesiol., vol. 8, pp. 185-193, Jun, 1998.
4. A. Holtermann, P. J. Mork, L. L. Andersen, H. B. Olsen and K. Søgaard, "The use of EMG biofeedback for learning of selective activation of intra-muscular parts within the serratus anterior muscle:: A novel approach for rehabilitation of scapular muscle imbalance," J. Electromyogr. Kinesiol., vol. 20, pp. 359-365, 2010.
5. A. Holtermann, K. Søgaard, H. Christensen, B. Dahl and A. K. Blangsted, "The influence of biofeedback training on trapezius activity and rest during occupational computer work: a randomized controlled trial," Eur. J. Appl. Physiol., vol. 104, pp. 983-989, 2008.
6. J. V. Basmajian, Biofeedback: Principles and Practice for Clinicians. Williams & Wilkins, 1989.
7. T. Graven-Nielsen, "Fundamentals of muscle pain, referred pain, and deep tissue hyperalgesia," Scand. J. Rheumatol., vol. 35, pp. 1-43, 2006.
8. E. A. Dannecker, K. F. Koltyn, J. L. Riley 3rd and M. E. Robinson, "The influence of endurance exercise on delayed onset muscle soreness," J. Sports Med. Phys. Fitness, vol. 42, pp. 458-465, Dec, 2002.
9. A. Samani, A. Holtermann, K. Søgaard and P. Madeleine, "Active pauses induces more variable electromyographic pattern of the trapezius muscle activity during computer work," J Electromyogr Kinesiol., vol. 19, pp. e430-437, 2009.
10. J. S. R. Jang, "ANFIS: Adaptive-network-based fuzzy inference system," IEEE Transactions on Systems, Man and Cybernetics, vol. 23, pp. 665-685, 1993.
11. R. Kohavi and G. H. John, "Wrappers for feature subset selection," Artif. Intell., vol. 97, pp. 273-324, 1997.
12. U. H. Buzzi, N. Stergiou, M. J. Kurz, P. A. Hageman and J. Heidel, "Nonlinear dynamics indicates aging affects variability during gait," Clin. Biomech., vol. 18, pp. 435-443, 2003.
13. A. B. Slifkin and K. M. Newell, "Noise, information transmission, and force variability," Journal of Experimental Psychology Human Perception and Performance, vol. 25, pp. 837-851, 1999.
14. S. M. Pincus, "Approximate entropy as a measure of system complexity," Proceedings of the National Academy of Sciences, vol. 88, pp. 2297-2301, 1991.
15. P. Madeleine and T. M. T. Madsen, "Changes in the amount and structure of motor variability during a deboning process are associated with work experience and neck–shoulder discomfort," Appl. Ergon., vol. 40, pp. 887-894, 2009.

Author: Afshin Samani
Institute: Center for sensory motor interaction (SMI)
Street: Fredrik bajers vej, 7 E1, 104b
City: Aalborg east
Country: Denmark
Email: afsamani@hst.aau.dk

Manipulation of Grating Lobes by Changing Element Shape

Svetoslav Ivanov Nikolov and Henrik Jensen

BK Medical Aps, R&D Applications and Technology, Herlev, Denmark

Abstract— **In this paper we present two approaches to manipulate the positions and levels of grating lobes in linear arrays by modifying the element shape. The first approach is to push the grating lobes outside the imaging plane in elevation direction by using skewed elements. The second approach is to use interwoven element (cut in a zig-zag fashion) to move the zero in the radiation pattern of the individual elements closer to the main lobe relative to the position of the first grating lobe. The performance of the suggestied designs is evaluated using simulations in Field II. The suppression of grating lobes is 20 dB for the interwoven elements.**

Keywords— **grating lobes, arrays, ultrasound, radiation pattern.**

I. RADIATION PATERN OF ARRAYS

Imaging in modern ultrasound scanners is performed using linear or curved arrays. The aperture is thus sampled. This spatial sampling results in aliasing of spatial frequencies. The artifacts created by this aliasing are generally referred to as "grating lobes". Different approaches to minimize the level of grating lobes have been investigated over the years. One approach is to remove the grating lobes by removing the periodicity in the array design [1]. Another approach is to select sets of elements so that the position of grating lobes in the transmit radiation pattern are cancelled by the zeros in the receive radiation pattern [2].

The goal of this study is to investigate methods to reduce the grating lobes by manipulating the shape of the elements without increasing the channel count.

The notation used in the paper is shown in Fig. 1: pitch is the distance between the centers of two transducer elements; width is the physical width of the element in the azimuth direction; height is the width of the physical element in the elevation direction; and kerf is the distance between two elements.

The relation between the pitch d_x, the width w and the kerf d_e is:

$$d_x = w + d_e \qquad (1)$$

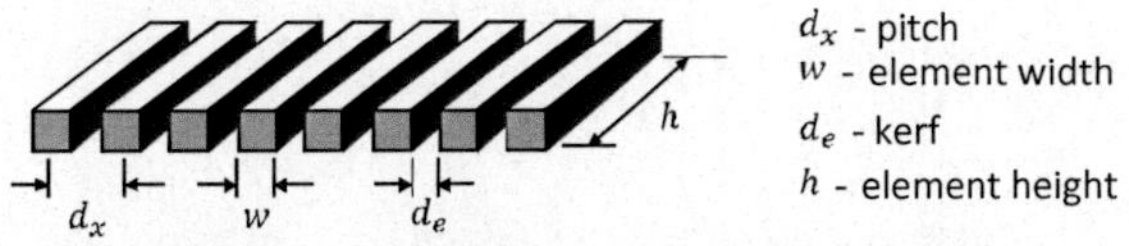

Fig. 1 Symbols and terms describing transducer geometry.

The pulse-echo radiation pattern of a linear array, which elements are omnidirectional can be approximated for the continuous-wave case as:

$$A_\delta(\theta) = \frac{\sin^2\left(\frac{k}{2} N d_x \sin\theta\right)}{\sin^2\left(\frac{k}{2} d_x \sin\theta\right)} \qquad (2)$$

where $\theta$is the off-axis angle and k is the wavenumber. The wavenumber k represents a spatial frequency defined as

$$k = \frac{2\pi}{\lambda} = \frac{2\pi f_0}{c}$$

where f_0 is the center frequency, c is the speed of sound and λ is the wavelength.

The radiation pattern $A_\delta(\theta)$ in (2) is derived using Fraunhoffer approximation which is valid for the far field and for the focal region in the near field. It represents essentially the Fourier transform of the apodization function of the array. It has a periodic nature with maxima located at those angles θ for which the denominator is equal to zero, i.e.

$$\frac{k}{2} d_x \sin\theta = n\pi$$

where n is an integer number. When $d_x = \lambda$, the grating lobes appear at $\theta = \pm 90^\circ$ off the main beam. Fore example, if the beem is steered at 30 degrees, then a grating lobe will appear at – 60 degrees.

The pulse-echo radiation pattern of an array with elements with finite width w is the product of the radiation pattern of the array $A_\delta(\theta)$ and of the individual element $A_{ele}(\theta)$:

$$A(\theta) = A_\delta(\theta) \times A_{ele}(\theta) \qquad (3)$$

The pulse-echo radiation pattern of the individual element can be approximated as

$$A_{ele}(\theta) = \left[\frac{\sin\left(\frac{k}{2} w \sin\theta\right)}{\frac{k}{2} w \sin\theta}\right]^2 \qquad (4)$$

Unlike arrays, elements do not have grating lobes, but a number of side lobes.

K. Dremstrup, S. Rees, M.Ø. Jensen (Eds.): 15th NBC on Biomedical Engineering & Medical Physics, IFMBE Proceedings 34, pp. 37–40, 2011.
www.springerlink.com

The pulse echo radiation pattern of an array consisting of elements with width w is shown in Fig. 2.

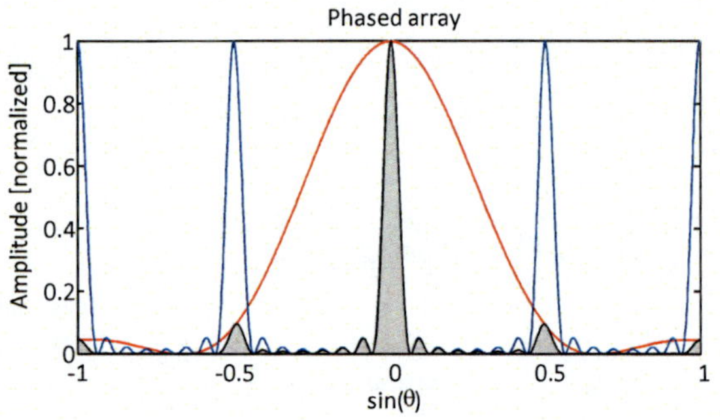

Fig. 2 Pulse-echo radiation pattern of a phased/linear array transducer $A(\theta)$. The periodic function in blue is the pattern of an array consisting on omnidirectional point elements. The red graph shows the pulse-echo radiation pattern of a single element. The product of the two gives the two-way pattern of the array, which has been shown with a grey filled graph.

The position of the zeros in the radiation pattern of a single element is inversely proportional to the width of the element w, and can be derived from:

$$\frac{k}{2} w \sin\theta = n\pi$$

The position of the peaks of the grating lobes is inversely proportional to the pitch of the array d_x. Usually the element width w is smaller than the pitch and the first grating lobes are placed within the main lobe of the radiation pattern of the elements.

A number of applications require beam steering. Two examples are spatial compounding and color flow mapping. The effect of steering on the radiation pattern is illustrated in Fig. 3. It shows the radiation pattern of an 8-element transducer with a pitch of 300 μm and element width of 275 μm. The radiation pattern of the single element does not change its position and steering the beam leads to a reduced level of the main beam. Simultaneously, the grating lobe moves towards the center of the radiation pattern of the single element and the ratio between the two is decreased from 38 dB to about 17 dB.

Notice that both the main lobe and the grating lobes are evaluated in the imaging plane. In the rest of the paper we will consider two approaches to decrease the level of the grating lobe: (1) move the grating lobe outside the imaging plane, and (2) move the grating lobe beyond the first zero in the radiation pattern of the single element.

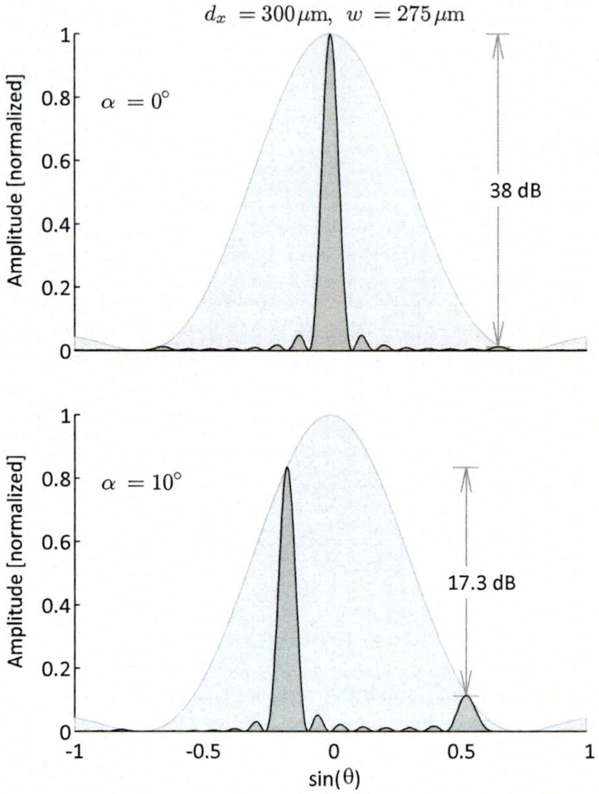

Fig. 3 Pulse echo radiation pattern of an array with $N = 8$ elements, pitch of 300 μm, element width of 275 μm, and a center frequency of 7.5 MHz.

II. Investigated Approaches

Rotate the transducer elements: A method for moving the grating lobes outside of the imaging plane is provided to us from the approximation of the radiation pattern, which can be loosely approximated by the Fourer transform of the function describing the aperture of the array. The properties of the 2D Fourier transform state that if the input function is rotated, then the spectrum is rotated too. An example is shown in Fig. 4

The top plot of Fig. 4 shows the position of the main lobe (red circle), the grating lobes (red-pink ellipses), and the imaging plane, which is illustrated by an orange rectangle.

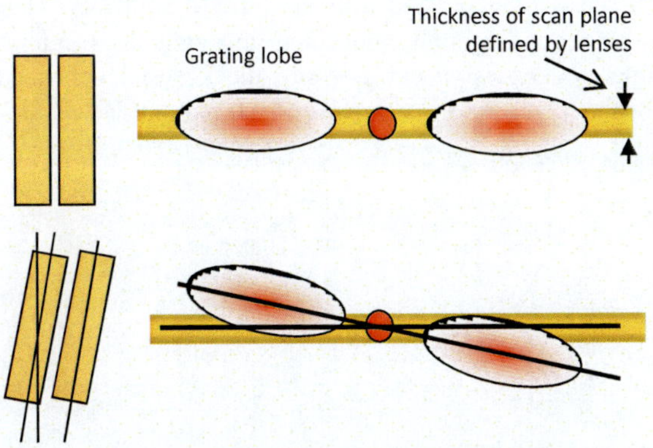

Fig. 4 A rotation of the direction of the elements results in a rotation of the position of grating lobes.

The bottom plot of Fig. 4 shows that by dicing the elements at an angle, we expect the grating lobes to rotate by the same amount, while the focusing in the elevation direction will remain largely unaffected, because of the effect of the acoustic lens. Since the focusing in elevation direction remains unmodified, we expect that the echoes coming from the grating lobes will be suppressed.

Interweaved element: Another method is suggested by Fig. 3. The reason why the difference in the levels of the main and the grating lobes decreases, when the beam is steered, is that the grating lobe of the array is already inside the main lobe of the individual element. What is needed is to manipulate the radiation pattern of the individual elements in such a way, that the first zero in the radiation pattern is placed closer to the main lobe than the grating lobe. In this way, when the beam is steered, the grating lobe will coincide with the zero and will be suppressed. To achieve this, every element must "collect" energy from an area that crosses the boundaries between two elements, which gives "interwoven elements" as shown in Fig. 5.

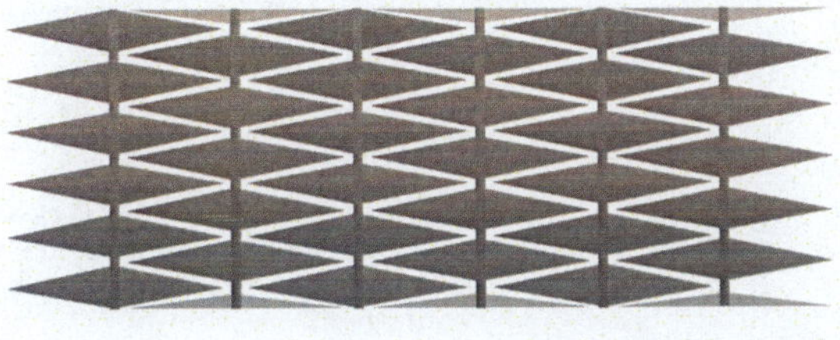

Fig. 5 Interweaved elements. The elements are cut in a zig-zag pattern.

III. PRACTICAL DESIGNS

A widespread process for manufacturing arrays is to subdice the transducer elements into square ceramic posts; fill-in the space between the ceramic posts with kerf material; depose electrodes as metallization on top and on the bottom of the elements; and finally, cut the metallization using laser to form the final elements. The process structure is shown in Fig. 6.

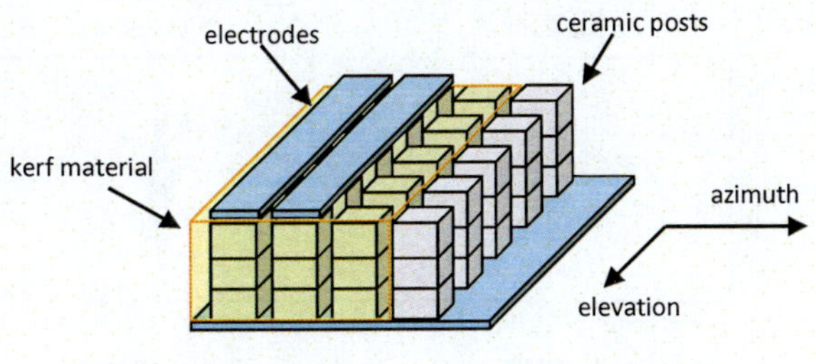

Fig. 6 Illustration of the internal structure of a typical array

A constraint on the design has been imposed, such that the cutting procedure is executed only over areas filled with kerf material.

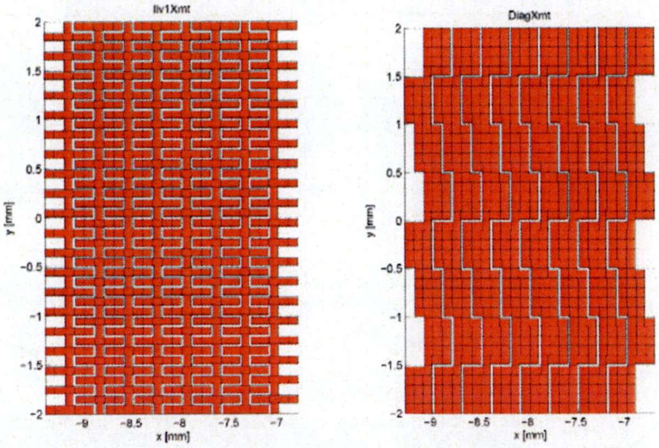

Fig. 7 Investigated element designs: interweaved elements on the left and diagonally cut elements on the right.

The resulting arrays are shown in Fig. 7. The design parameters are: center frequency f_0=8 MHz; ceramic post size 100 μm; element height h=4 mm; pitch d_x=300 μm. Assuming speed of sound c = 1540 m/s, the wavelength at the center frequency is 192 μm and the pitch is:

$$d_x = 1.56\,\lambda$$

IV. SIMULATION RESULTS

Simulations were done using Field II. The "physical" elements were sub-divided into 40x3 mathematical elements and their delays and apodization functions were set such as to simulate the element geometries shown in Fig. 7. The elevation focus was set at 20 mm. The sampling frequency used in the simulation was set to 100 MHz.

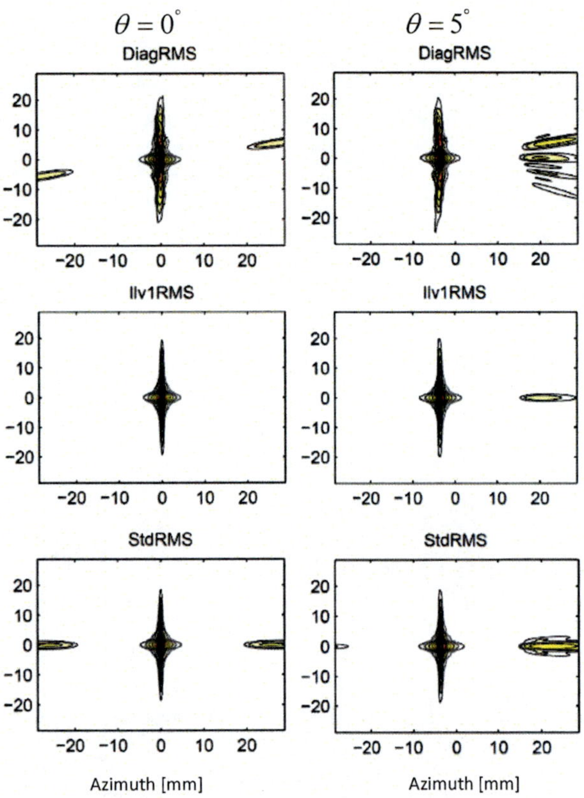

Fig. 8 The PSF at depth of 40 mm. The RMS values are shown. The contour lines are drawn at levels from -72 dB to 0 dB at a step of 6 dB.

Figure 8 shows the point spread function (PSF) at a depth of 40 mm. The contour plots are of the root-mean-square (RMS) values of the point spread function. The plots are shown for two steering angles – 0 and 5 degrees. The top row shows the PSF for the diagonally cut elements, the middle for the interwoven elements and the bottom for the standard array. Looking at the plots for θ=0, it can be seen that the grating lobes are rotated for the diagonal design and strongly suppressed for the interwoven elements, which is in unison with the expectations. When the beam is tilted, a plurality of grating lobes appear for the diagonal design. This can be explained in part by the fact that the elements are not cut diagonally, but rather in a staircase fashion. The interwoven design exhibits still the lowest level of grating lobes.

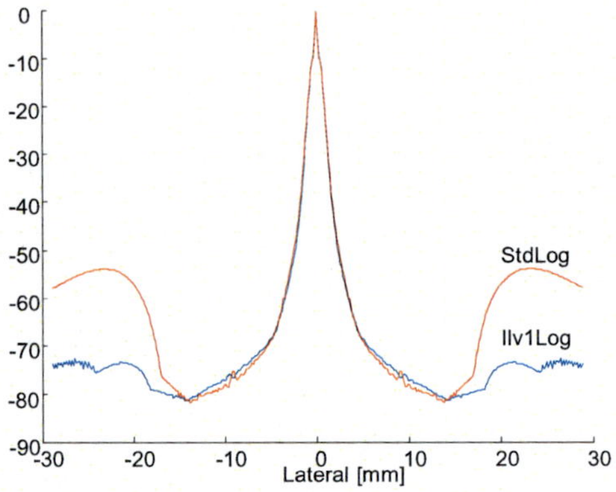

Fig. 9 The PSF of the interwoven design plotted against the PSF of atandard array at a depth of 40 mm and a center frequency of 8 MHz.

Figure 9 shows the maximum projection of the point spread function from Fig.8 for the interwoven and standard arrays. The level of the grating lobes for the interwoven design is about 20 dB below the level of the grating lobes of the standard array.

V. CONCLUSIONS

It is possible to manipulate the position and the level of the grating lobes of an ultrasonic array by changing the shape of the transducer elements. Cutting the elements at an angle pushes the grating lobes outside of the scan plane. Creating interwoven elements moves the first zeros of the radiation pattern of the individual elements closer to the main lobe and suppresses the grating lobes with more than 20 dB, for an array with a pitch larger than 1.5 wavelengths.

REFERENCES

1. Steinberg, B. (1973). Comparison between the peak sidelobe of the random array and algorithmically designed aperiodic arrays. *IEEE Transactions on Antennas and Propagation* 21, no. 3 (May): 366-370. doi:10.1109/TAP.1973.1140493.
2. Lookwood, G.R., and F.S. Foster. (1996). Optimizing the radiation pattern of sparse periodic two-dimensional arrays. *IEEE Transactions on Ultrasonics, Ferroelectrics and Frequency Control* 43, no. 1 (May): 15-19. doi:10.1109/58.484458.
3. Jensen, J A. (2004). Simulation of advanced ultrasound systems using Field II. In *IEEE International Symposium on Biomedical imaging from nano to macro*, 636-639.

Characterization of Pathological Tremor from Motor Unit Spike Trains

J.L. Dideriksen[1], J.A. Gallego[2], and D. Farina[1,3]

[1] Center for Sensory-Motor Interaction, Department of Health Science and Technology, Aalborg University, Aalborg, Denmark
[2] Bioengineering Group, Consejo Superior de Investigaciones Cientificas, Madrid, Spain
[3] Department of Neurorehabilitation Engineering, Bernstein Center for Computational Neuroscience, University Medical Center Göttingen, Georg-August University, Göttingen, Germany

Abstract— **Pathological tremor is the involuntary oscillation of a limb and is related to a diversity of diseases. Using a newly developed computational model for simulating surface EMG and complete discharge patterns of the motor unit population under the influence of neural oscillations in the central nervous system, the use of motor unit spike trains to characterize pathological tremor was investigated. Simulations were performed in 12 conditions, including different types of pathological tremor and varying levels of voluntary activity. The results showed that the cumulative spike trains of at least 5 motor units provided a better estimate of the central oscillations that it could be obtained from surface EMG in approximately 50% of the simulated conditions. In the remaining conditions there was no significant difference. The results indicate that motor unit spike trains constitute a useful signal for characterizing pathological tremor in vivo.**

Keywords— **Tremor, motor unit, electromyography, neuromuscular model.**

I. Introduction

Pathological tremor (in the following referred simply as tremor) is involuntary oscillatory movement of a limb and is among the consequences of a variety of neurodegenerative disorders, such as the Parkinson's disease [1, 2]. As these involuntary movements interfere with voluntary movement they are a source of great discomfort for patients.

While current treatments for tremor mainly rely on medication [3] or deep brain stimulation [4], the use of external devices for automatic tremor suppression, such as inertial loading [5] or functional electrical stimulation [6], has been given some attention. A crucial element in such devices is an effective methodology for tremor characterization so it can be suppressed in the optimal way without influencing the voluntary movement. Effective tremor characterization implies estimating the oscillatory element in the neural drive to the muscle.

Several physiological signals can be used for tremor characterization, however many applicable measureable signals reflect rather a series of non-linear transformations of the neural drive than the drive itself, thereby possibly masking the oscillatory element. For example, limb acceleration reflects not only muscle activation patterns but also highly non-linear relations, such as muscle force-length and force-velocity properties [7] and the force frequency relation of muscle force [8]. Although surface electromyography (sEMG) is often used as an indicator of the neural drive to the muscle (e.g. [9]), this measure is affected by several mechanisms potentially blurring this relation [10]. Instead, the current study investigates the use of motor unit (MU) discharge patterns for tremor characterization, since it has been shown that these reflects the neural drive in a linear fashion in healthy subjects [11]. In the analysis we compare the use of MU discharge patterns for tremor characterization with that of sEMG.

In the study we use a recently developed computational model [12] to generate data, rather than using experimentally obtained data, since this holds a number of advantages with respect to the aim of the study. First, the model simulates the behavior of the entire motor neuron pool which allows for a more complete assessment of how MUs reflect tremor than can be achieved with the limited number of MUs that can be identified experimentally. Furthermore, unlike in experimental conditions, in the model the central oscillations are known and can thus be used as a reference signal when assessing the quality of the MU-based tremor characterization.

II. Methods

A. Computational model

The applied computational model was a refined version of a model that comprised of and extended a number of previously published models [12]. In the model, the voluntary drive required to maintain a predefined angular trajectory was continuously estimated using a PID-control algorithm. Oscillations (band-passed filtered white noise) were imposed on this drive to simulate the central origin of tremor. Along with this drive, simulations of afferent feedback were used as the input to the motor neuron pool estimating the complete discharge pattern of all motor neurons. Based on the discharge pattern of the motor neuron pool, the sEMG, muscle force, and limb inertia were estimated.

K. Dremstrup, S. Rees, M.Ø. Jensen (Eds.): 15th NBC on Biomedical Engineering & Medical Physics, IFMBE Proceedings 34, pp. 41–44, 2011.
www.springerlink.com

By adjusting the model input, the tremor type (frequency, intensity) and degree of voluntary activation could be varied. The model proved capable of simulating experimentally observed temporal and spectral characteristics of the limb inertia and sEMG as well as the general trends of the MU population discharge patterns [12].

As the current application of the model requires a high level of precision of MU discharge patterns with regards to real experimental conditions the model was modified in two ways to ensure the necessary accuracy. First, the equation that is used to estimate the time of the next discharge for each MU (Eq. 4 in [13]) was modified. As discussed in [14], this equation was adopted from a neuromuscular model of static, isometric conditions [13], where the time of discharge $n+1$ of a MU was estimated based on the excitation level at the discharge n. In spite proving to be able to simulate experimentally general trends in discharge patterns [14], this model was not optimally equipped for predicting valid inter-spike intervals (ISI) in conditions with rapid changes in the excitation level, which is the case for tremor. Therefore, the ISI between discharge $n+1$ and n predicted at the time of discharge n was adjusted according to the equation

$$T_m(n+1) = T_m(n) + \widetilde{ISI}_m(T_m(n)) - (\widetilde{ISI}_m(t) - \widetilde{ISI}_m(t-1)) \quad (1)$$

Where T denotes discharge time, m denotes MU number, n denotes discharge number, $\widetilde{ISI}$ denotes the predicted ISI, according to the excitation level at a given point in time, and t denotes the time index. In this way the time of the next discharge was continuously adjusted according to changes in the excitation level.

Secondly, the filter used for simulating the central oscillations were fine-tuned in order for the model output to exhibit highly similar temporal variability in peak tremor frequency as those experimentally observed in 20 patients with Parkinson's disease [14]. In this study two parameters were derived from the sEMG in 0.5 s windows: Tremor consistency (percentage of time spent at the most commonly occurring tremor peak-frequency) and tremor bandwidth (the range on continuous non-empty bins in the tremor peak-frequency histogram around the most commonly occurring tremor peak-frequency). By adopting this methodology, the filter settings were optimized in order for these parameters for the simulations were within the reported range.

B. Simulation paradigms

In the simulations two tremor frequencies (5 and 10 Hz), representing low- and high frequency tremor [1]), and three tremor intensities, representing low-, medium, and high intensity tremor, were applied. The simulations were performed with a static, neutral target angle with two levels of inertial loading (5 and 15% of the maximum voluntary contraction MVC level; MVC) to represent the range of contraction levels most often used in daily activities. This yielded a total of 12 combinations of simulation settings. For each setting the simulated contraction duration was 40 s in which the target angular trajectory was set to a static, neutral (0 degrees) target angle. Furthermore, the sEMG signals were simulated 10 times for each condition by randomly assigning the locations of the muscle fibers within the muscle to replicate the variability observed for measurements on different subjects.

C. Data analysis

MU spike trains and sEMG were analyzed in 1 s non-overlapping windows. 50 groups of 10 MUs were randomly selected from the active MU population. First, the analysis was performed with the spike train of only the first MU of each group; then with the cumulative spike train of the first and second MUs and so on until all 10 MUs of each group were included in the cumulative spike train. In this way, 10 cumulative spike trains for each of the 50 groups and 10 sEMG signals were analyzed in each window.

The analysis of performance was performed in two steps. First, the peak-frequency error (difference in the peak spectral frequency in the 3-14 Hz band (range of tremor frequencies (reference)) between the central oscillator and the cumulative spike trains / sEMG) was calculated. If this error exceeded ±2 Hz this data was excluded from the next step of the analysis. The percentage of exclusions indicated the ability of the signals to roughly reflect the central oscillations.

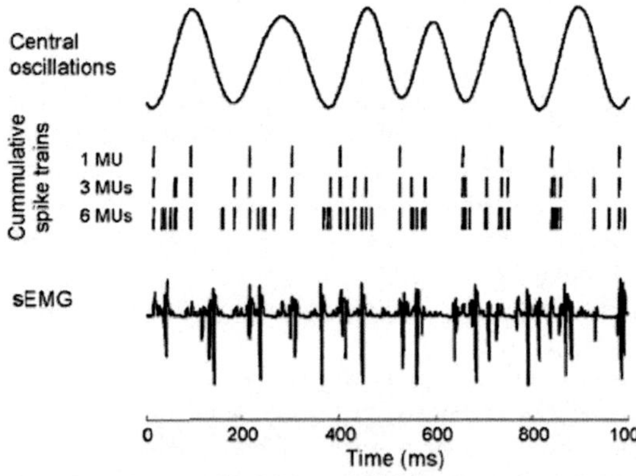

Fig. 1 The central oscillations, cumulative spike trains and sEMG in 1 s of a contraction with a neutral, static target angle, an inertial load of 5% MVC with 6 Hz low intensity tremor

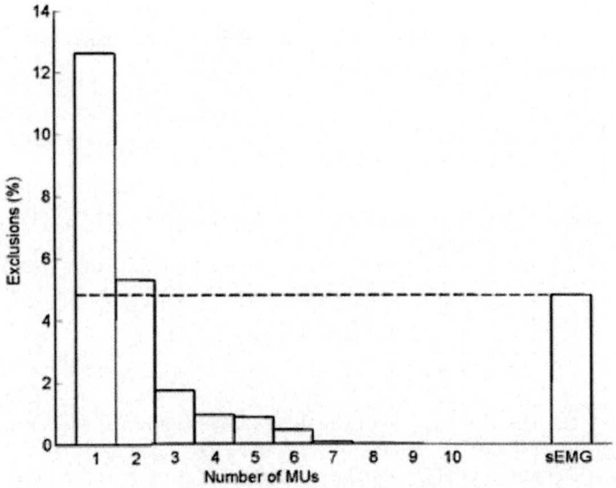

Fig. 2 The number of excluded data-windows (du to a peak frequency error > 2 Hz) from cumulative spike trains consisting of 1-10 MUs and sEMG. The simulation settings were 10 Hz medium-intensity tremor with an inertial load of 15% MVC.

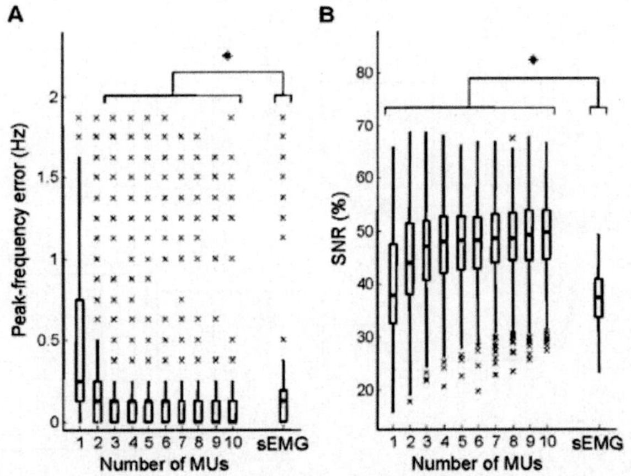

Fig.3 The peak-frequency error (A), and SNR (B) based on cumulative spike trains consisting of 1-10 MUs and sEMG from all non-excluded data-windows. * indicates significant difference ($p<0.05$) between cumulative spike trains and sEMG.

In the next step of the analysis one additional performance variables were derived: Signal-to-noise-ratio (SNR), estimated as the percentage of spectral power in the 3-12 Hz band (the usual spectral band for tremors [1]) of the cumulative spike trains / sEMG located within ±1 Hz of the central oscillator frequency.

The spectral analysis was performed using the fast Fourier transform using zero-padding to increase the spectral resolution. The peak frequency error and SNR was compared between cumulative spike trains and sEMG using the Wilcoxon Signed Rank Test with a significance level of 0.05.

III. Results

Figure 1 depicts an example of the imposed central oscillations and how these are reflected in the MU pool and in the sEMG signal. Cumulative spike trains consisting of 1, 3, and 6 randomly selected MUs are presented. Including more MUs in the cumulative spike train increased the correlation with the central oscillations. In this example this correlation exceeds that of sEMG and the central oscillations when having included more than two MUs.

Figure 2 and 3 show the representative results from one simulation setting: 10 Hz medium-intensity tremor with an inertial loading of 15% MVC. Figure 2 illustrates the number of exclusions (data-windows with a peak-frequency error > 2 Hz). At this simulation setting approximately 5% of the estimated peak-frequencies from sEMG exceeded the threshold, whereas when including 3 MUs or more less than 2% of the estimated peak-frequencies from cumulative spike trains were excluded. When including 8 or more MUs no estimates were excluded. Figure 3 indicates the peak-frequency error and SNR for the cumulative spike trains and sEMG. In this example, 3 MUs a required in the cumulative spike train to achieve a significantly better estimate of the peak-frequency than the sEMG-based estimate, whereas the spike train of a single MU is sufficient to exceed the performance of sEMG with regards to SNR.

Figure 4 depicts histograms of the number of simulations where the estimation based on cumulative spike trains were superior to the estimation based on sEMG with respect to the three performance variables (exclusions, peak-frequency error, and SNR) when varying the number of MUs included in the cumulative spike trains. These results indicate that when including 4-6 MUs a significantly better estimation across all performance variables is obtained using cumulative spike trains. In other cases there was no difference in the performance, except in one case where sEMG-based estimation yielded a lower number of exclusions even when 10 MUs were included in the cumulative spike train.

IV. Discussion

This study compares the use of MUs and sEMG with regards to providing an accurate characterization of the central oscillations causing tremor. Using computational model for simulations of the discharge pattern of the motor neuron pool and sEMG in tremor, the analysis revealed that across most conditions spanning a wide range of contraction conditions, the use of cumulative spike trains consisting of a limited number of randomly selected MUs provides an equally good or better characterization of these central oscillations.

Currently MU spike trains may be identified using invasive or non-invasive techniques. Non-invasive techniques

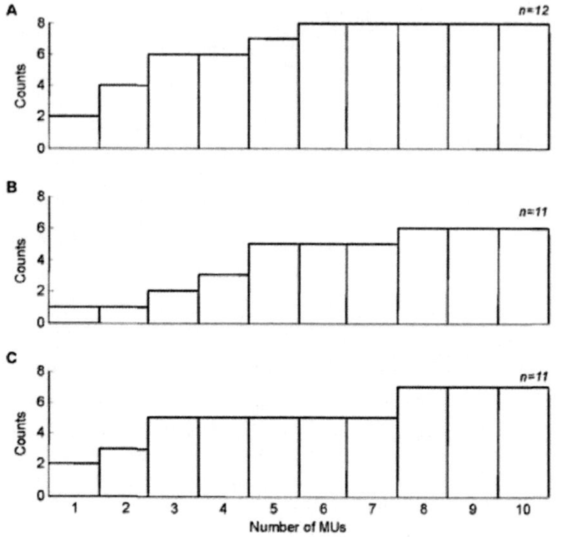

Fig. 4 Histograms indicating the number of simulation settings for which the estimation based on cumulative spike trains of different lengths are superior to that based on sEMG for number of excluded time-windows (A), peak-frequency error (B) and SNR (C). In the other cases, except for one in the number of exclusions (A) where sEMG consistenly provided the best performance, there was no significant difference between the performances. One simulation setting could not be included in the estimation of peak-frequency error and SNR since in this case all sEMG data-windows were excluded (thus n=11 for B and C).

based on high-density sEMG recordings has been shown to be able to provide the discharge pattern from on average more than 10 MUs from one contraction [15]. Therefore, the identification of the cumulative spike train of a number of MUs sufficient for an accurate estimation of tremor, as shown in this study, is feasible in practical conditions and with non-invasive approaches. A major limitation with the current sEMG decomposition techniques, however, is that they can only be used in an off-line setting. Therefore, using cumulative spike trains to characterize tremor to control an on-line tremor suppression system still requires advances in decomposition methods. Another application of the proposed methodology, however, could be within the field of diagnosis of the pathology causing tremor. Since the central oscillators in different types of tremor are believed to be located in different brain centers [2] it is possible that the precise estimation of the oscillator provided by the current technique may aid the process of making the correct diagnosis. This is particularly relevant since misdiagnosis of the pathology underlying tremor is common [16].

In conclusion, this study shows that a superior approximation of the central oscillations in tremor can be obtained by using cumulative spike trains from as little as 5 MUs compared to estimating it from sEMG.

Acknowledgment

This work was supported by the EU project "TREMOR" (project ref 224051).

References

1. Deuschl G et al. (1998) Consensus statement of the Movement Disorder Society on tremor. Mov Disord 13(suppl 3): 2-23
2. Deuschl G et al (2001) The pathophysiology of tremor. Muscle Nerve 24(6): 716-735
3. Obeso J.A. et al (2010) Missing pieces in the Parkinson's disease puzzle. Nat Med 16(6): 653-661
4. Vaillancourt D.E. (2003) Deep brain stimulation of the VUM thalamic nucleus modifies several features of essential tremor. Neurology 61(7): 919-925
5. Rocon E. et al (2007) Design and validation of a rehabilitation robotic exoskeleton for tremor assessment and suppression. IEEE Trans Neur Sys Rehab Eng 15(3): 367-378
6. Prochazka A. et al (1992) Attenuation of pathological tremors by functional electrical stimulation I: Method. Ann Biomed Eng 20(2): 205-224
7. Zajac F.E. (1989) Muscle and tendon: properties, models, scaling, and application to biomechanics and motor control. Crit Rev Biomed Eng 17(4): 359-411
8. Bawa P. et al (1976) Frequency response of human soleus muscle. J Neurophysiol 39(4): 788-793
9. Kooistra R.D. et al (2007) Conventionally assessed voluntary activation does not represent relative voluntary torque production. Eur J Appl Physiol 100(3): 309-320
10. Farina d. et al (2004) The extraction of neural strategies from the surface EMG. J Appl Physiol 96(4): 1486-1495
11. Negro F. et al (2011) Linear transmission of cortical oscillations to the neural drive to muscles is mediated by common projections to populations of motor neurons in humans. J Physiol, *In press*
12. Dideriksen J.L. et al (2010) An integrative model of the surface EMG in pathological tremor. Proc IEEE EMBC
13. Fuglevand, A.J. et al (1993) Models of recruitment and rate coding organization in motor-unit pools. J Neurophysiol 70(6): 2470-2488
14. O'Suilleabhain P.E. et al (1998) Time-frequency analysis of tremors. Brain 121(11): 2127-2134
15. Holobar A. et al (2009) Estimating motor unit discharge patterns from high-density surface electromyogram. Clin Neurophysiol 120(3): 551-562
16. Louis E.D. (2009) Essential tremors: A family of neurodegenerative disordes? Arch Neurol 66(10): 1202-1208

Corresponding author:

Author: Jakob Lund Dideriksen
Institute: Department of Health Science and Technology
Street: Fredrik Bajersvej 7
City: Aalborg
Country: Denmark
Email: jldi@hst.aau.dk

Quantification of Indoxyl Sulphate in the Spent Dialysate Using Fluorescence Spectra

J. Holmar[1], J. Arund[1], F. Uhlin[1,2], R. Tanner[1], and I. Fridolin[1]

[1] Department of Biomedical Engineering, Tallinn University of Technology, EST-19086 Tallinn, Estonia
[2] Department of Nephrology, University Hospital, Linköping, S-581 85 Linköping, Sweden

Abstract— **The aim of this study was to investigate the possibility to determine the amount of Indoxyl Sulphate (IS) in the spent dialysate using fluorescence spectra.**

Eight uremic patients from Linköping were studied during their three dialysis treatments in one week at the Department of Dialysis and Nephrology at Linköping University Hospital. Dialysate samples were taken during each treatment and analyzed, IS concentration was estimated using HPLC method, and fluorescence spectra was measured with spectrofluorophotometer. The fluorescence spectral values were transformed into IS concentration using regression model from total material noted as fluorescence method (F). Achieved results were compared regarding mean values and SD.

Mean value of IS estimated by HPLC was 1.21±0.77 mg/l and by F 1.22±0.72 mg/l. Concentrations were not significantly different (p≤0,05).

This study indicates, that it is possible to estimate the concentration of IS using only fluorescence values of the spent dialysate.

Keywords— **Indoxyl Sulphate, dialysis, fluorescence, spectra, uremic toxins.**

I. Introduction

Indoxyl sulphate (IS) (MW 251 D) is metabolized by the liver from indole, which is produced by the intestinal flora as a metabolite of tryptophan. The production of indole in the gut may be greater in uremic patients than in normal subjects because of the effect the uremic milieu has on the composition of intestinal flora. IS is one of the well known substances of a group of protein-bound uremic retention solutes which increases the rate of progression of renal failure. IS impairs osteoblast function and induces abnormalities of bone turnover and strongly decreases the levels of glutathione, one of the most active antioxidant systems of the cell [1].

It is found that IS is the most abundant compound in uremia and has been linked to endothelial damage, inhibition of endothelial regeneration and repair and endothelial free-radical production. IS is one of uremic retention solute with cardiovascular damaging potential and it belongs to the list of uremic toxins [2, 3].

Early studies by HPLC have shown that the uremic retention solute IS can be observed by fluorescence measurements in the plasma as well as in ultrafiltrate [4, 5].

A good possibility to estimate concentrations of different compounds in the spent dialysate using ultra violet absorbance spectra and processed UV absorbance spectra has shown in earlier studies[6-9]. It is found by our group that some compounds absorbing UV radiation are not fluorescent and other way around; this means that fluorescence spectra may add selectivity to optical method for determination of the constituents in the spent dialysate.

The aim of this study was to investigate the possibility of using fluorescence and fluorescence spectral data for determination of IS in the spent dialysate.

II. Materials and Methods

This study was performed after approval of the protocol by the the Regional Ethical Review Board, Linköping, Sweden. An informed consent was obtained from all participating patients.

Eight uremic patients, one female and seven males, mean age 77±7 years were included in the study. All patients were on chronic three -weekly on-line HemDiaFiltration (ol-HDF) at the Department of Nephrology, University Hospital of Linköping, Sweden. The dialysis machine used was a Fresenius 5008 (Fresenius Medical Care, Germany). The dialyzers used were in all treatments FX 800 (Fresenius Medical Care, Germany), with an effective membrane area of 1.8 m^2, with an ultra filtration coefficient of 63 ml/h mmHg. The duration of the ol-HDF treatments varied between 180 to 270 minutes, the dialysate flow was 500 mL/min, the blood flow varied between 280-350 mL/min. All patients were dialyzed via artery-venous fistulas using a "two-needle" system. The auto sub system mode for calculation of the on-line prepared substitution volume varied between 12.2 to 29.7 litres per session.

During each dialysis following samples from the drain tube of the dialysis machine were taken: 9-25 minutes after the start of the session and at the end of ol-HDF session(180-268 min.). All spent dialysate/ultrafiltrate was also collected in a tank where the last dialysate sample was taken, when the treatment was completed and after careful stirring was performed. In addition the treatments were monitored on-line with UV-absorbance. If a self-test of the dialysis machine occurred during the planned sampling time, the sample was taken when the UV-absorbance curve reached baseline level again which occurred within 2-3 minutes. Pure dialysate was collected before the start of a dialysis session, used as the reference solution, when the dialysis machine was prepared for starting and the conductivity was stable. pH of the collected samples was neutral. Samples were freezed and transported to Tallinn.

Concentration of IS was determined by fluorescence signal during HPLC analysis, in Tallinn Technical University. UltiMate 3000 HPLC instrument (Dionex) was used[10]. Concentration determination methodology used in Tallinn was similar to one described in [5].

Spectrofluorophotometer (SHIMADZU RF-5301) was used for the fluorescence measurements. Fluorescence analysis was performed over an excitation (EX) wavelength range of 220 - 500 nm, emission (EM) wavelength range of 220-800 nm and with excitation increment 10 nm. An optical cuvette with an optical path length of 4 mm was used. Measurements were performed at the room temperature (ca. 22° C). The obtained fluorescence values were processed and presented by software Panorama fluorescence and the final data processing was performed in EXCEL (Microsoft Office Excel 2003).

Linear correlation coefficients (R) and the R-squared values (R^2) for full data matrix were determined on the basis of the fluorescence spectral values and IS concentration values. The best EX/EM wavelength pair for estimating the concentration of IS was found and regression model was determined.

The obtained relationship was used for generating a concentration calculation algorithm to estimate IS concentration.

The accuracy (BIAS) and precision (SE) were calculated as follows using concentrations from the HPLC analysis as reference:

$$BIAS = \frac{\sum_{i=1}^{N} e_i}{N} \qquad (1)$$

where e_i is the i-th residual(difference between laboratory and optically determined concentration values for the i-th measurement) and N is the number of observations.

$$SE = \sqrt{\frac{\sum_{i=1}^{N} (e_i - BIAS)^2}{N}} \qquad (2)$$

III. Results

Figure 1 illustrates examples of 3D fluorescence spectra obtained over the excitation wavelength range of 240-500 nm and emission wavelength range of 250-800 nm on the pure dialysate sample and on the spent dialysate samples taken at 10 and 207 min after the start of a dialysis session.

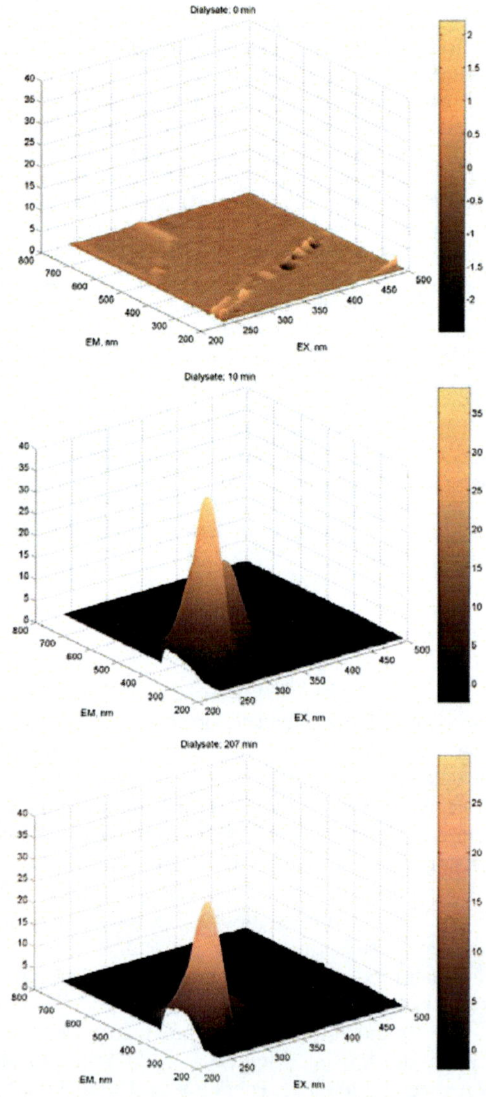

Fig. 1 Examples of fluorescence spectra's of spent dialysate taken before, at the start and at the end of the dialysis procedure. EX=240-500nm, EM=250-800nm

Linear correlation coefficients were determined for full data matrix on the basis of the fluorescence spectral values and IS concentration values of all samples (Figure 2).

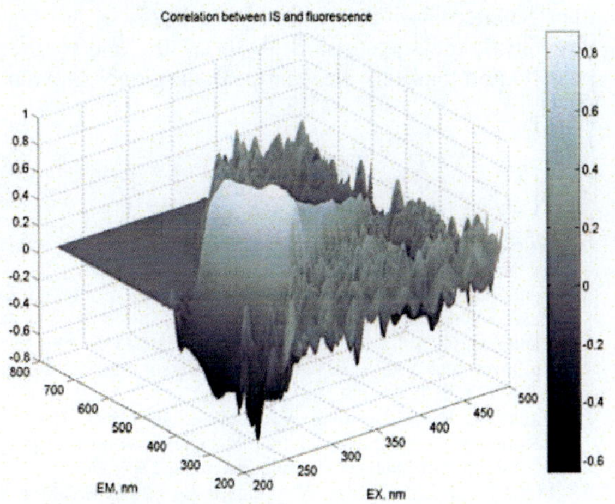

Fig. 2 Matrix of correlation coefficients between IS concentrations and fluorescence spectral values over EX wavelengths 240-500 nm and EM wavelengths 250-800 nm

It was found that the best EX/EM wavelength pair for estimating the concentration of IS is 300/358(EX/EM) (Figure 3).

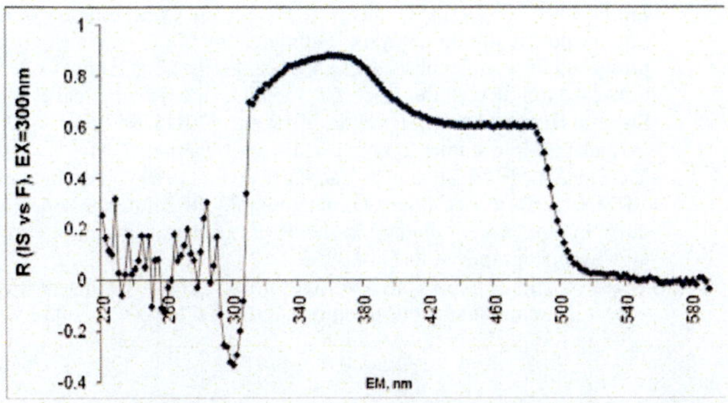

Fig. 3 Value of correlation coefficient r between IS concentration and fluorescence spectral data at EX wavelength 300 nm and EM wavelengths 220-590 nm

A linear regression model was built using spectral values from those wavelengths. Concentration values were calculated on the basis of the regression model for IS, using fluorescence spectral values at 300/358 nm (EX/EM).

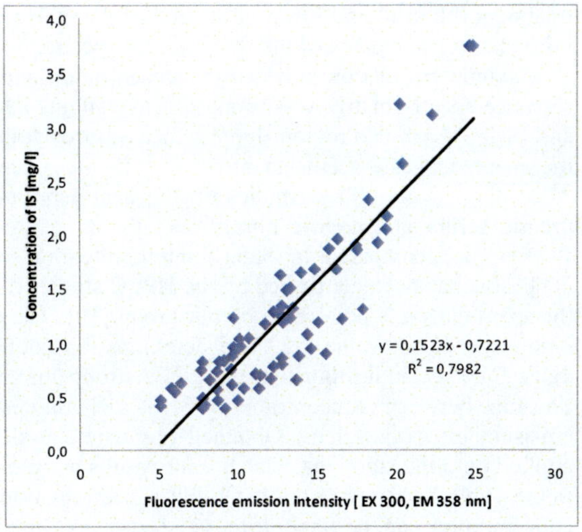

Fig. 4 A regression line between IS concentration in dialysate and of fluorescence emission intensity (EX=300 nm, EM=358 nm)

Table 1 shows a summary of the results regarding the IS concentration in mean and standard deviation values (Mean +/- SD) from the standardized methods (Lab) and new optical method (F), linear correlation coefficient (R) and the R-squared value (R^2) between the IS concentration from F and concentration measured at the laboratory (Lab), the accuracy (BIAS) and precision (SE) for the new method to measure concentration of IS.

Table 1 concentration of IS estimated by different methods

	IS mg/L
N	69
Lab (Mean +/- SD)	1.21 +/- 0.77
F (Mean +/- SD)	1.22 +/- 0.72
R	0.89
R^2	0.80
BIAS [mg/L]	0.01
SE [mg/L]	0.34

IV. Discussion

As seen from the figure 1, some of distinctive fluorescence maxima at specific regions are clearly seen. Moreover, the fluorescence amplitude is proportional to the content of eliminated uremic retention solutes in the spent dialysate being higher in the beginning of the dialysis treatment (10 min) and lower at the end of the dialysis (207 min) at specific regions of the fluorescence spectra.

Maximal correlation between IS concentration and fluorescence spectra matrix was found at wavelengths EX=300 and EM=358. In this region fluorescence of other solutes in the spent dialysate seems to affect less IS measurements. According to the HPLC studies on the heat-deproteinized uremic serum and uremic ultrafiltrate, IS has a prevalent fluorescence compared to other uremic retention solutes [11]. This has been confirmed by the HPLC studies of IS in the spent dialysate performed by our group [10]. The correlation coefficient value 0.89 and determination coefficient value 0.80 are indicating that there is a strong linear relationship between concentration of IS and fluorescence emission intensity values. Obtained regression model was applied onto the study material and the results are presented in the Table 1. As seen from the Table 1 determination of IS concentration can be done with satisfactory accuracy and precision applying the novel fluorescence method, F.

The clinical aim in the future is to develop an on-line monitoring system that could offer an estimation of the removal of several clinical important solutes during haemodialysis. The present technical approach may help to confirm the previous knowledge and broaden the coming information about the uremic toxin, IS removal during dialysis and a positive impact to the patients according to needs in chronic renal failure therapy [12]. The optical technique for measuring concentration of different uremic toxins may give a useful, rapid and cost-effective tool for clinicians to estimate the effectiveness of dialysis procedure.

V. CONCLUSIONS

This study investigated whether fluorescence spectral data can be used for determination of the amount of indoxyl sulphate in the spent dialysate. It was found that fluorescence spectral data can be used for estimating this uremic toxin in the spent dialysate. Thereby it is possible in the future to use fluorescence spectral data in the optical system for estimating the dialysis procedure.

New clinical trials giving access to a larger amount of data for analyzing fluorescence spectra's of the spent dialysate will be issue of the next studies.

ACKNOWLEDGMENT

The authors wish to thank all dialysis patients who participated in the experiments. The work is supported in part by the County Council of Östergötland, Sweden, the Estonian Science Foundation Grant No 8621, by the Estonian targeted financing project SF0140027s07, and by the European Union through the European Regional Development Fund.

REFERENCES

1. Wishart D., C. Knox, A. Guo et al., (2009) Hmdb: A knowledgebase for the human metabolome. Nucleic acids research 37: D603.
2. Vanholder R. and G. Glorieux, (2003) An overview of uremic toxicity. Hemodialysis International 7: 156-161.
3. Vanholder R., S. Laecke, F. Verbeke et al., (2008) Uraemic toxins and cardiovascular disease: In vitro research versus clinical outcome studies. NDT plus 1: 2.
4. Vanholder R., R. De Smet, V. Jacobs et al., (1994) Uraemic toxic retention solutes depress polymorphonuclear response to phagocytosis. Nephrology Dialysis Transplantation 9: 1271-1278.
5. Meert N., S. Eloot, M. A. Waterloos et al., (2009) Effective removal of protein-bound uraemic solutes by different convective strategies: A prospective trial. 24: 562-570.
6. Jerotskaja J., I. Fridolin, K. Lauri et al., (2009) An enhanced optical method for measuring concentration of uric acid removed during dialysis, Springer, 11th International Congress of the Medical Physics and Biomedical Engineering, Munich, Germany, 2009, pp. 9-12.
7. Jerotskaja J., F. Uhlin and I. Fridolin, (2008) A multicenter study of removed uric acid estimated by ultra violet absorbance in the spent dialysate, Springer, 14th Nordic-Baltic Conference on Biomedical Engineering and Medical Physics, Riga, Latvia, 2008, pp. 252-256.
8. Jerotskaja J., F. Uhlin, I. Fridolin et al., (2009) Optical online monitoring of uric acid removal during dialysis. Blood purification 29: 69-74.
9. Jerotskaja J., F. Uhlin, K. Lauri et al., (2010) Concentration of uric acid removed during dialysis. Estimated by multi wavelength and processed ultra violet absorbance spectra, IEEE, 32nd Annual International Conference of the IEEE Engineering in Medicine and Biology Society, Buenos Aires, Argentina, 2010, pp. 5791-5794.
10. Arund J., "Hplc studies, unpublished work," Tallinn, 2011.
11. De Smet R., P. Vogeleere, J. Van Kaer et al., (1999) Study by means of high-performance liquid chromatography of solutes that decrease theophylline/protein binding in the serum of uremic patients. Journal of Chromatography A 847: 141-153.
12. Meert N., E. Schepers, R. De Smet et al., (2007) Inconsistency of reported uremic toxin concentrations. Artificial organs 31: 600-611.

Author: Jana Holmar
Institute: Department of Biomedical Engineering, Technomedicum, Tallinn University of Technology
Street: Ehitajate tee 5
City: 19086 Tallinn
Country: Estonia
Email: jana@cb.ttu.ee

Pressure Algometry and Tissue Characteristics: Improved Stimulation Efficacy by a New Probe Design

S. Finocchietti, L. Arendt-Nielsen, and T. Graven-Nielsen

Laboratory for Musculoskeletal Pain and Motor Control, Center for Sensory-Motor Interaction (SMI),
Department of Health Science and Technology, Aalborg University, Denmark

Abstract— **Pressure algometry is broadly utilized to assess deep tissue sensitivity. In this study the relation between pressure-induced pain in humans and stress/strain distribution within the deep tissue was evaluated. A 3-dimensional finite-element computer model was utilized to describe stress/strain distribution in tissues of the lower leg during pressure stimulation. The computer model was validated based on data recorded by computer-controlled pressure-induced muscle pain in 6 subjects. An indentation of 7 mm was sore for all subjects and at this level data were extracted from each simulation. Simulations were performed with two probe designs (cylindrical and semispherical). The principal stress peaked in the skin and was decreased to about 10% in the underlying muscle tissue. The principal strain peaked in adipose tissue and was reduced in muscle tissue to 80%. The probe evoked a strain peak in adipose tissue at 0.12 (cylindrical) and 0.24 (semispherical); in muscle tissue 0.10 and 0.20 respectively. The shear strains were also reduced using the semispherical tip. The human pressure pain thresholds with the semispherical tip were significantly smaller compared with the flat probe (P<0.05). The results suggest that pressure-induced muscle pain is most effectively induced by semi spherical probes, while flat ones activate superficial structures. The probe design is considered an important factor during pressure pain assessments and should take into account when performing clinical studies.**

Keywords— **pressure algometry, muscle pain, finite element.**

I. Introduction

Pressure algometry is a widely used method for assessment deep tissue sensitivity [1]. So far there is limited knowledge about how pressure is distributed in the deep tissues and it is unknown which probes are better accounting for nociceptors excitation in deep tissue during pressure stimulation.

Probe shape and diameter are crucial factors in pressure pain assessment. Traditionally, the classical probes used in clinical studies have a contact area of 1 cm^2 and are flat, usually covered by a rubber disk in order to minimize the sharp contact edges, in line with the Fischer algesiometer [2], which has been recommended for measurement of the muscle pain threshold. Anyway no scientific explanation was ever provided for the use of this kind of probe. Rather, a sensation of sharpness is usually felt on the interface between the probe and the skin and was shown to be mediated by superficial tissue's nociceptors and influence the deep pressure pain assessment [3]. A good way to reduce this sharpness is to reduce the shear strains that are created with the contact between the skin and the probe and so, new probe endings have to be designed.

In the present study the perception of muscle pain is related to the stress and strain distribution in the anterior tibialis muscle undergoing pressure stimulation with probes of different shapes. Parts of data or methods have been presented elsewhere [4].

II. Methods

Pressure algometry measurements on the tibialis anterior muscle with 2 different probe shapes were performed and a 3 dimension finite-element model (FE) was developed to describe the stress and strain distribution in the deep tissues.

Specifically, the right lower leg of a healthy volunteer (male, 25 years old, 4.4 mm subcutaneous adipose thickness) was imaged using a 1.5 T Signa MRI scanner (GE Medical System, Milwaukee, WI, USA). The imaging protocol was based on a matrix of 512 x 512 pixels, 60 slices, 3 mm slice thickness, echo time (TE: 13.664 ms) and repetition time (TR: 660 ms) providing a clear anatomical delineation.

The transverse MRI slice comparable to the site of experimental assessment sites was used to construct a FE model of the calf within the COMSOL Multiphysics software (Comsol AB, Sweden). The anatomical structure were segmented, imported in COMSOL and extruded in the 3rd direction. The transverse MRI slices were meshed with triangular elements. The mesh was enhanced in the contact area with the probe, with an element size of 1 mm in the skin and subcutaneous adipose tissues, and 5 mm in the muscle tissue. The 2-dimentional mesh was swept in the longitudinal direction creating a 3-dimentional tetrahedral mesh. Convergence tests show that different mesh densities do not improve significantly the geometry, loading and constrain parameters. All nodes of the bones were fixed, while the other ones were left free resulting in non-constricted boundary conditions. In this way the bone was considered as a rigid and uncompressible element. The probe was fixed in 2 directions and allowed to move in the direction orthogonal to the skin surface. A contact surface between the tip of the probe and the skin was defined and a contact penalty factor was set as the ratio between a given stiffness (1 GPa), and the maximum element size in the contact surface (1.6 mm).

The stimulation probe was applied to the skin perpendicularly. Rubber-like, non-linear Neo Hookean material models are used (table 1). Two simulations were performed in relation to the probe used: cylindrical tip and semi-spherical tip (figure 1A).

Table 1 Neo-hookean materials moduli of the three tissues constituent the FE model, based from data [5].

Tissue	Bulk modulus (k) [kPa]	Shear modulus (m) [kPa]
Skin tissue	30000	2000
Subcutaneous adipose tissue	36	1
Muscle tissue	92.8	5.952

In parallel, six healthy subjects participated in the experimental study where pressure pain thresholds (PPTs) and tolerance (PPTOs) are recorded using a computer-controlled pressure algometer with a mechanical footplate (Aalborg University, Denmark) (figure 1B). PPT was defined as the point at which a sensation of pressure changed into a sensation of pain. PPTO was defined as the maximum pain tolerated.

All statistical analyses were conducted using SPSS (IBM SPSS, USA). A two way mixed intra-class correlation coefficient (ICC) with absolute agreement measure is used to compare the relation between tissue indentation and pressure stimulation intensity between the data obtained with the experimental pressure algometry session and the FE model. ICC values above 0.75 indicate good reliability. The data are presented as means and standard deviation (SD).

The experimental data were analysed with a two 1-way analysis of variance (ANOVA) with probe design (cylindrical, semispherical) as factor. $P < 0.05$ was considered significant.

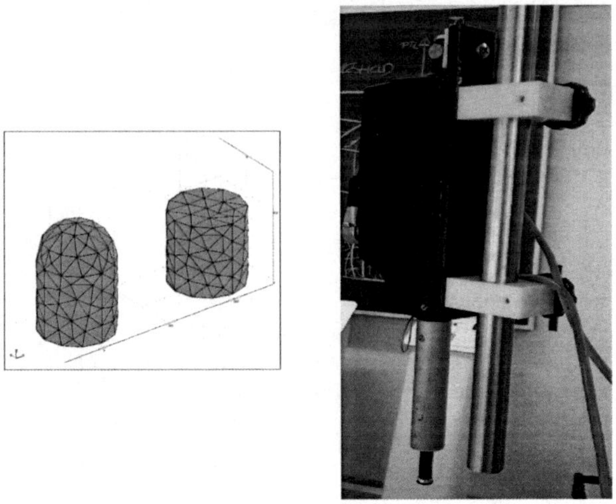

A B

Fig. 1 A: The 2 different probe used. Semi spherical tip on the left and cylindrical, flat tip with straight edges on the right. The section of both probes is 1 cm^2. B. The computer-controlled pressure algometer. The shin below the probe was stabilized in a vacuum cushion

III. Results

The reliability between the experimental session data and the model data was good, ICC coefficient was 0.956. This showed that the two dataset are stable over time and that the FE model can be considered as a valid approximation of the real experiment session.

The pressure values measured for PPT and PPTO were significantly affected by probe design (table 2). Two one-way ANOVA with main factor probe design were performed on PPT and PPTO, respectively. The PPT and PPTO were significantly smaller with the semispherical tip compared to the cylindrical one ($P<0.05$). An indentation of 7 mm was painful for all subjects and at this level data were extracted from each simulation. The VAS values at 7 mm indentation were significantly higher with cylindrical probe compared with the semispherical probe ($P<0.03$).

Table 2 Mean (SD, n=6) pressure pain threshold (PPT), pressure pain tolerance (PPTO), and VAS score values at an indentation depth equal to 7 mm

Probe	PPT [kPa]	PPTO [kPa]	VAS score at 7 mm indentation
cylindrical	461 ± 35	967 ± 63	5.2 ± 0.2
semispherical	395 ± 46	855 ± 51	6.5 ± 0.3

The data from all the 3 directions of principal stress and strain were extracted, but the data from the 1st and 3rd are not presented, because their profile was not significantly different from results in the 2nd direction. The strain profile usually had a clear maximum whereas the stress profile seemed more flat and uniform. Furthermore, the principal stress peaked always in skin layer at maximum of 650 kPa (figure 2); the highest values were always around the sharp edges of the probe. There was a general high stress distribution under the edges of the cylindrical tip. With the semispherical tip, the principal stress was wider distributed under the probe.

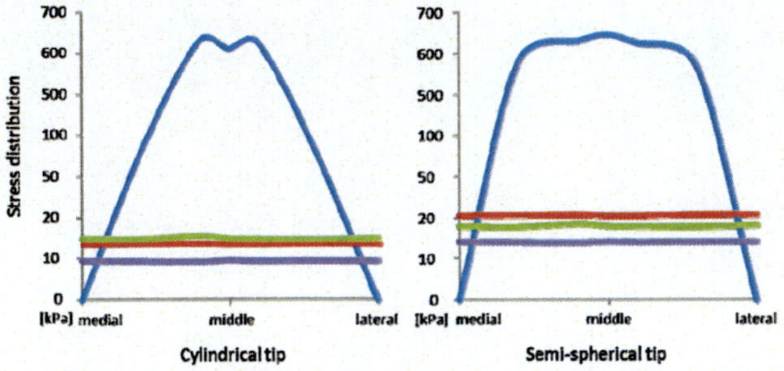

Fig. 2 Stress distribution below the different tissues: Skin (in blue), subcutaneous adipose (in red), superficial muscle (in green), and deep muscle (1 cm deep, in violet) in the medial-lateral plane for a cylindrical (left) and semi-spherical (right) probe at 7 mm of indentations

Examining the deeper layers, in absolute values, the principal stress was reduced to about 10% in the subcutaneous adipose tissue and muscle tissue. The principal strain under the contact with the cylindrical tip peaked at 0.12 in the adipose tissue and was reduced in the muscle tissue to about 66%. The skin layer showed a general low principal strain, always lower than 0.015. The strain value at the deep muscle was 0.04. The principal strain under the semi spherical tip was higher and peaked around 0.24 in the subcutaneous tissue. At the muscle site the mean value in the superficial muscle was 0.13, while for the deep muscle it was 0.10 (figure 3).

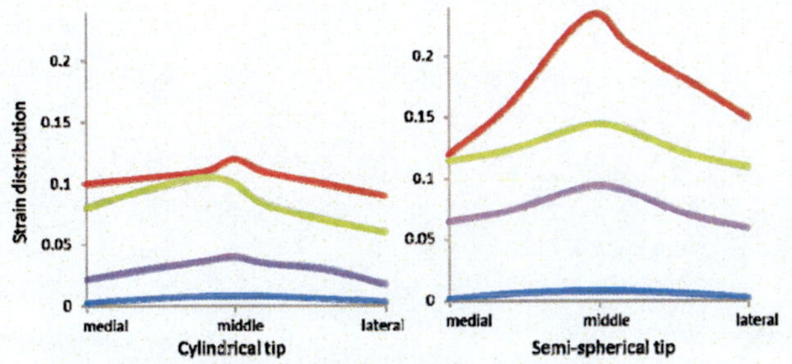

Fig. 3 Strain distribution below the different tissues: Skin (in blue), subcutaneous adipose (in red), superficial muscle (in green), and deep muscle (1 cm deep, in violet) in the medial-lateral plane for a cylindrical (left) and semi spherical (right) probe at 7 mm of indentations

The shear strain due to the contact between the probe and the skin was reduced by approximately 62% when using the semi-spherical compared with the cylindrical tip (figure 4).

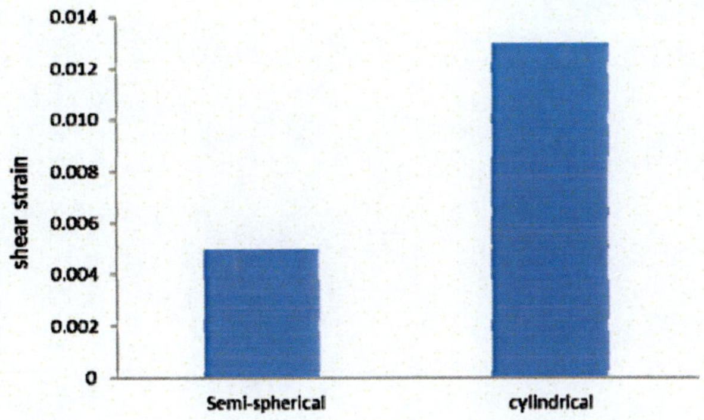

Fig. 4 Shear strain on the contact area between probe and skin. The semispherical probe show smaller shear strain in comparison to the flat one

IV. Conclusions

Pressure algometry data showed that the probe shape is highly important for the assessment of PPT and PPTO in the anterior tibialis muscle. Probably, the complete adhesion between probe and skin tissue favors the muscle strain, reduce the share strain. Moreover, the stress and strain contribution to the other layers is changing with probe's design. Semispherical tips may minimize the deformation in the epidermis, and enable a preferential activation of deep afferents. Semi-spherical probes are suggested to be applied in clinical evaluations and pharmacologic studies.

Acknowledgment

The study was supported by Svend Andersen Fonden (Aalborg, Denmark). The authors have no conflicts of interest to report.

References

1. Graven-Nielsen T. Fundamentals of muscle pain, referred pain, and deep tissue hyperalgesia. Scand J Rheumatol Suppl 2006; 35:1-43.
2. Fischer AA. Pressure algometry over normal muscles. Standard values, validity and reproducibility of pressure threshold. Pain 1987; 30: 115-126.

3. Greenspan JD, McGillis SLB. Stimulus Features Relevant to the Perception of Sharpness and Mechanically Evoked Cutaneous Pain. Somatsens Mot Res 1991; 8:137-147.
4. Finocchietti S, Nielsen M, Mørch CD, Arendt-Nielsen L, Graven-Nielsen T. Pressure-induced muscle pain and tissue biomechanics: A computational and experimental study. Eur J Pain 2011; 15(1):36-44.
5. Tran HV, Charleux F, Rachik M, Ehrlacher A, Tho MC. In vivo characterization of the mechanical properties of human skin derived from MRI and indentation techniques. Comput Methods Biomech Biomed Engin 2007; 10:401-407.

Preliminary Experimental Verification of Synthetic Aperture Flow Imaging Using a Dual Stage Beamformer Approach

Ye Li and Jørgen Arendt Jensen

Center for Fast Ultrasound Imaging, Dept. of Electrical Engineering, Technical University of Denmark, Lyngby, Denmark

Abstract— **A dual stage beamformer method for synthetic aperture flow imaging has been developed. The motivation is to increase the frame rate and still maintain a beamforming quality sufficient for flow estimation that is possible to implement in a commercial scanner. With the new method high resolution images can be obtained continuously, which will highly increase the frame rate. The flow velocity is estimated by using a time-domain cross-correlation technique. The approach is investigated through experiments with the SARUS scanner (Synthetic Aperture Real-time Ultrasound System). A flow rig generates a parabolic laminar flow, and the SARUS scanner is used for acquiring the data from individual channels of the transducer. The experimental results showed that increasing the number of imaging lines used for the estimation form 4 to 24 reduces the standard deviation from 21% to 7.6%. The parameter study showed that the number of cross-correlation functions for averaging and length of the search range influence the performance.**

Keywords— **Synthetic aperture flow imaging, synthetic aperture sequential beamforming, SARUS.**

I. Introduction

Synthetic aperture (SA) ultrasound imaging has many benefits compared to conventional imaging methods. Due to the complete data set, it is possible to have both dynamic transmit and receive focusing to improve contrast and resolution, and the data can be beamformed in any direction at any position. Therefore, SA flow imaging can achieve flow velocity mapping with a high frame rate and high resolution. However, a full SA method needs to acquire individual element data, which consumes huge amounts of memory, if the number of emissions is high. The high number of calculations is a challenge and makes real time implementation expensive. A synthetic aperture sequential beamforming (SASB) approach has been suggested without the need for storing RF-data and with a reduced system complexity [1]. Compared to a full SA setup only a single RF-line is beamformed and stored for each emission. That reduces the number of calculations and complexity significantly.

The approach was developed and tested through simulations [2]. The results indicated that it was possible to employ a dual stage beamformer approach for synthetic aperture flow imaging [2]. The standard deviation on the estimates was found to be less than 3% (compared to the peak velocity).

In this paper, a preliminary experiment was made to verify the method. A flow rig system with a gear pump generates a parabolic laminar flow profile inside the tube. A commercial linear array scans the tube at a fixed angle. The SARUS (Synthetic Aperture Real-time Ultrasound System) [3] scanner system amplifies and digitizes the receiving signals and stores the data in the local memory. The stored data sets are processed off-line to estimate the flow velocity profile.

II. Method for SASB Flow Imaging

This section describes how the dual stage beamformer approach and synthetic aperture flow imaging work.

A. Dual stage beamformer approach

At every emission, a focused beam is transmitted and received by a subaperture. The transmit elements are Hamming weighted to reduce edge effects. In the first stage the data is beamformed at the same fixed focus in both transmit and receive. The constructed B-mode RF lines are then input to a second beamformer, which is fully dynamically focused. In the second stage, the emit focus points are considered as virtual sources, since a spherical wave emanates from this point in a limited angular region.

In receiving beamforming, the time of flight (TOF) is calculated to find the correct signal samples. The round trip TOF is calculated by tracing the path of the ultrasound wave, which is

$$t_{tof} = \frac{|\vec{r}_{ve} - \vec{r}_e| \pm |\vec{r}_{ip} - \vec{r}_{ve}| \pm |\vec{r}_{ve} - \vec{r}_{ip}| \pm |\vec{r}_r - \vec{r}_{ve}|}{c}, \quad (1)$$

where $\vec{r}_{ve}$, $\vec{r}_e$, $\vec{r}_{ip}$, and $\vec{r}_r$ are the positions of the virtual element, the emitting element, the imaging point, and the receive element, respectively. The virtual sources are at the same position as the focal points. The wave starts from the transmit origin to the imaging point through the transmit focal point and propagates back to the receive element through the receive focal point.

K. Dremstrup, S. Rees, M.Ø. Jensen (Eds.): 15th NBC on Biomedical Engineering & Medical Physics, IFMBE Proceedings 34, pp. 53–56, 2011.
www.springerlink.com

The beamforming method is illustrated in Fig. 1. The image from the second stage is constructed by selecting samples corresponding to the same position from the first stage B-mode lines. Therefore, each high resolution image point is obtained by summing the corresponding low resolution image point in the first stage.

The wave beam pattern in Fig. 1 is determined by the opening angle α, which can be expressed as

$$\alpha = 2\arctan\frac{1}{2F^{\#}} , \tag{2}$$

where $F^{\#}$ is the ratio of the focal point depth and the size of the active sub-aperture. The choice of focal point depth and the sub-aperture length can influence the transmitted wave field and the point spread function (PSF).

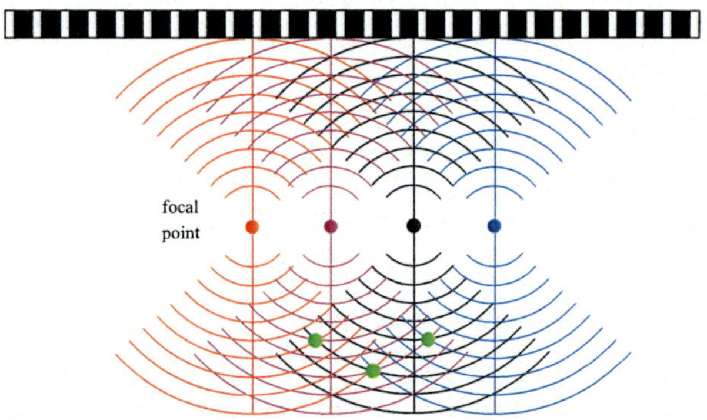

Fig. 1. Illustration of the second beamformer. Four image lines from the first stage are the inputs to the second stage here. They only carry information within the transmit pattern, which is defined by the opening angle. The high resolution points indicated as green dots are obtained by adding all low resolution points overlapping at that position.

B. Velocity estimator

At every emission the low resolution image is replaced by the latest one, thus, high resolution images can be obtained continuously. Due to the scatterer moving in a certain direction, the two consecutive signals are time shifted versions of each other, which can be expressed as [4]

$$H^{n}(t) = H^{n-N}(t - 2\frac{v_z}{c}NT_{prf}) , \tag{3}$$

where t is the propagation time from pulse emission, n is the global emission index, N is the number of emissions in one sequence, v_z is the velocity along the axial direction and T_{prf} is the pulse repetition time.

The cross-correlation function is calculated between two high resolution images from the same emission sequence, which can be formulated as

$$\begin{aligned} R_{n-N,n}(\tau) &= \frac{1}{2T}\int_T H^{(n-N)}(t)H^{(n)}(t+\tau)dt \\ &= \frac{1}{2T}\int_T H^{(n-N)}(t)H^{(n-N)}(t-t_s+\tau)dt \\ &= R_{n-N,n-N}(\tau - t_s), \end{aligned} \tag{4}$$

where t_s gives the times shift of the peak of the cross-correlation function. The axial velocity can then be expressed as

$$v_z = \frac{ct_s}{2NT_{prf}} \tag{5}$$

The summation of several cross-correlation functions lowers the variance and improves the estimate. The velocity profile can be achieved by finding the cross-correlation function as a function of depths.

III. Experimental setup

The flow rig system is illustrated in Fig. 2. The fluid flows through the flow circulation system, which includes a gear pump drive, (Cole-Parmer, Model: 75211-15) that can vary the speed and a Danfoss MAGFLO Flowmeter (Type MAG 3000) for reference. The mean flow volume is measured by the flowmeter. An air-trap device removes air bubbles from the closed circuit. A 1.2 m long metal tube is used to develop the laminar flow, and the tube inside the water tank is made of rubber.

The concentrated Doppler fluid, (Dansk Fantom Service) consists of Glycerol, Orgasol, Triton X 100 (2%), Nabenzoate, Na_2EDTA diluted 1:9 with demineralized water. Dextran is added to the solution to increase the viscosity as the flow phantom is intended to mimic the blood properties. The final kinematic viscosity of the flow phantom is 3.6 mm^2/s which is within the acceptable range.

The measurement parameters can be found in Table 1. The angle between the ultrasound beam and the laminar flow is 70 degrees. The center of the tube is 28 mm from the surface transducer. The transmitting pulse and the data acquisition are performed by the SARUS scanner [7]. Arbitrary waveforms can be emitted and all channel data from the transducer are acquired simultaneously.

Each transmit sequence contains four emissions, which are active at different positions. The centers of the sub-apertures are several λ away from each other. At each emissions, 64 elements are active and transmit a focused beam. The scattering signals are received by the same elements. The acquisition was made for 256 emissions using this transmit sequence, thus 252 high resolution lines can be obtained. The high resolution signal was convolved with a match filter to remove noise outside the pulse frequency range. Stationary echo cancelling was done by calculating the mean value of several high resolution signals and then subtracting this from the signals.

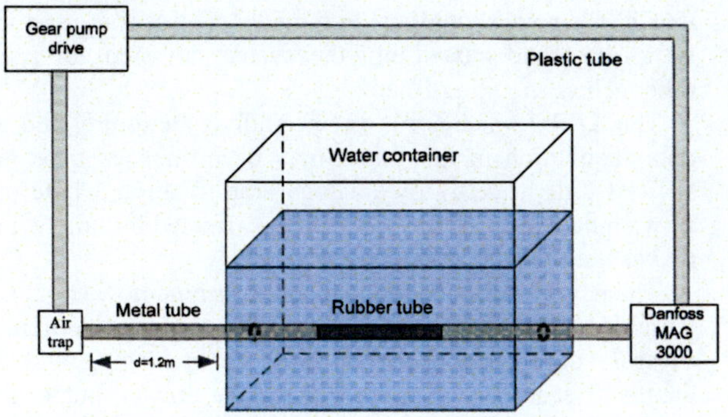

Fig. 2. Flow rig system. The flow is pumped out to the air trap to remove air bubbles, then moves through a 1.2 m long metal tube to develop a laminar flow. A flow meter measures the volume flow as a reference.

Table 1 Parameters for the measurement

Parameters	Value	Unit
Transducer center frequency	7	MHz
Sampling frequency	70	MHz
Transducer pitch	0.208	mm
Transducer height	4.5	mm
Pulse repetition frequency	4	KHz
Radius of rubber tube	6.5	mm
Number of transmit element	64	
Number of receive element	64	
Focal depth (virtual source)	10	mm
Volume flow speed	20	L/h

After applying the match filter and stationary echo cancelling, the signals were cross-correlated with the signals from the same sequence and added to the other cross correlation functions to obtain the velocity. Furthermore, the velocity profile is obtained as the cross-correlation was found as a function of depths.

The resulting image is shown in Fig. 3. There are four emissions with seven λ spacing between each emission. 24 lines were used for each velocity estimate. The standard deviation varies between 1.8% and 20%, with an average over the profile of 6.4% relative to the axial peak velocity of 0.034 m/s and the average bias is 7.6% over the profile.

Fig. 4 shows the velocity profiles obtained from the same emission sequence, but with different number of averaging number. Averaging 48 cross-correlation functions yield better velocity estimates. The standard deviation and bias are 4.4% and 7.6% relative to the peak velocity.

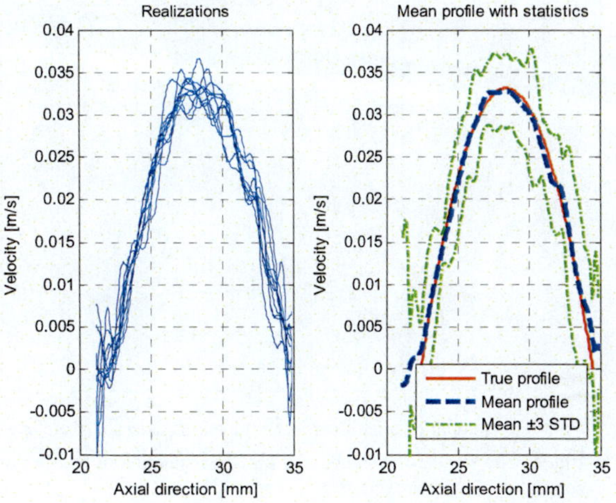

Fig. 3. Velocity profiles are obtained by four emissions with seven λ spacing. 24 cross correlation functions are averaged to get one profile. The left graph shows 10 estimated profiles and the right shows the mean profile with ±3 standard deviations.

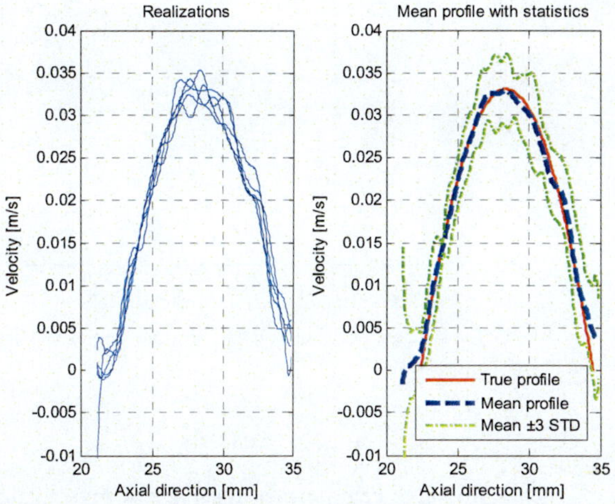

Fig. 4. Velocity profiles are obtained by four emissions with seven λ spacing. 48 cross correlation functions are averaged to get one profile. The left graph shows 5 estimated profiles and the right shows the mean profile with ±3 standard deviations.

The results of the parameter study are shown in Fig. 5 and Fig. 6. It is obvious that both the averaging number and number of λ in the searching range have an influence on the estimation performance. Using more cross correlation functions for one estimate gives better performance and a longer searching range gives more reliable results.

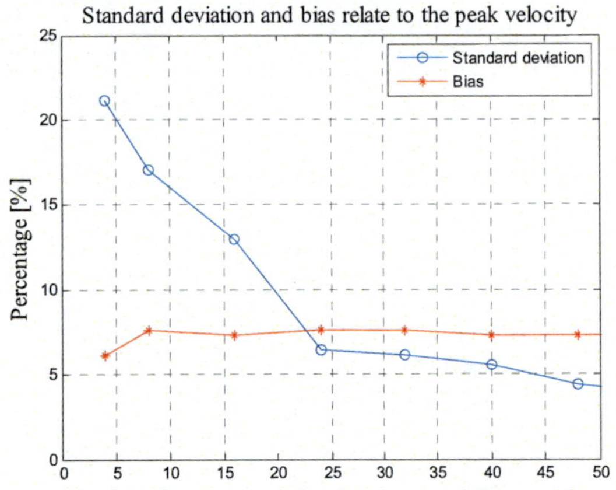

Fig. 5. The standard deviation and bias as a function of number of cross-correlation functions averaged.

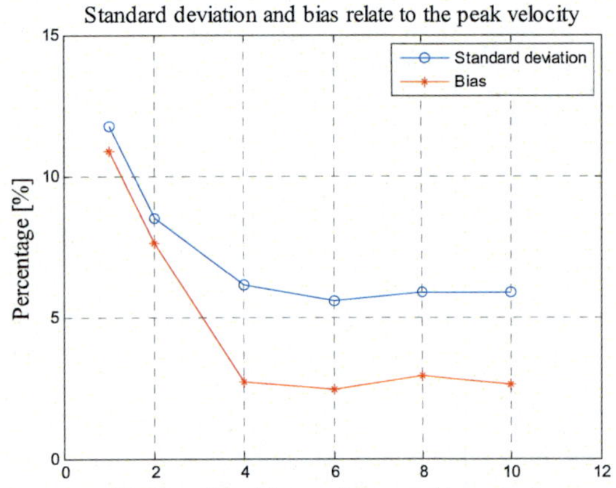

Fig. 6. The standard deviation and bias as a function of number of λ in the search range.

IV. CONCLUSIONS

The two stage beamformer approach is investigated experimentally for synthetic aperture flow imaging. Full synthetic aperture can acquire data continuously yielding high resolution images available for processing, and the data can be focused both in transmit and receive in any direction and position. However, this approach needs more allocated memory to store all channel data and demands more calculations. The new method has to store beamformed image lines from the first stage only instead of all the individual channel data. This reduces the number of calculations significantly and lowers the complexity of the system. Although the number of stored image lines is reduced, the beamforming quality is sufficient for flow imaging.

The preliminary experiment in this paper proved that it is possible to use the new method for synthetic aperture flow imaging, and that a fast flow imaging system can be built with the continuous data acquisition. Although a simple match filter was applied and mean subtraction used for stationary echo cancelling, the system is open for using other echo cancelling filters.

The SARUS scanner is not yet fully functioning and the data acquisition in this paper may be influenced by a low SNR (1.7 dB) due to low analogy amplification. Therefore, flow signals may at times have been affected by this, which makes the real signal difficult to detect.

The parameter study was done to examine how some key variables influence the estimation performance. The number of cross correlation functions used for averaging and the length of searching range influence the results, but in this preliminary experiment also the low SNR caused a decreased performance, when fewer emissions were averaged. Thus, the suggested method makes it possible to build a synthetic aperture flow imaging system with fewer calculations and a lower complexity of the system.

ACKNOWLEDGMENT

This work is supported by grant 027-2008-3 from the Dasnish Advanced Technology Foundation.

REFERENCES

1. J. Kortbek and J. A. Jensen, "Synthetic aperture sequential beamforming," in *Proc. IEEE Ultrason*. Symp., pp. 966-969, 2008.
2. Y. Li and J. A. Jensen, "Synthetic aperture flow imaging using a dual beamformer approach" in *Proc. IEEE Ultrason*. Symp., 2010.
3. J. A. Jensen, H. H. Lund, R. T. Nielson, B. G. Tomov, M. B. Stuart, S. I. Nikolov, M. Hansen and U. D. Larsen, "Performance of SARUS: A synthetic aperture real-time ultrasound system", in *Proc. IEEE Ultrason*. Symp., 2010.
4. S. I. Nikolov and J. A. Jensen, "In-vivo synthetic aperture flow imaging in medical ultrasound," IEEE Trans. Ultrason., Ferroelec., Freq. Contr., vol. 39, pp. 366-380, 1992.
5. O. Bonnefous and P. Pesque, "Time domain formulation of pulse Doppler ultrasound and blood velocity estimation by cross correlation," Ultrason. Imaging, vol. 8, pp 73-85, 1986.
6. P. M. Embree and W. D. O'Brien, "Volumetric blood flow via time-domain correlation: Experimental verification," IEEE Trans. Ultrason., Ferrolelec., Freq. Contr., vol. 37, pp. 176-189, 1990a.

An fMRI Investigation of Auditory Pathway Using Different Paradigms and Analysis Procedures

M. Ryn, E. Charyasz, M. Erb, and U. Klose

Diagnostic and Interventional Neurardiology, University Hospital Tuebingen, Germany

Abstract— **The purpose of the study was to improve the visualization of the auditory pathway using different paradigms and analysis procedure of functional magnetic resonance imaging (fMRI). In particular, the signal of white matter (WM) was used in statistical analysis to remove distortions and improve the quality of correlation maps. Ten healthy volunteers were examined on a 3T scanner with music stimulation and three different paradigms: block design, continuous sound and resting state. The auditory pathway could successfully be visualized. Development in post-processing using the repetition time and WM as regressors and also correlation analysis provided an improvement in visualization of cortical as well as subcortical structures. The results demonstrate a tight functional relation between auditory cortex and brainstem areas in the human brain.**

Keywords— **Functional magnetic resonance imaging, brainstem, auditory cortex, white matter, correlation.**

I. Introduction

Functional magnetic resonance imaging (fMRI) is an appropriate tool to study cortical [1] and even subcortical activation [2- 4] if auditory stimuli are applied. The auditory pathway consists of the auditory cortex (AC) and several subcortical structures such as: medial geniculate body (MGB), inferior colliculus (IC), nuclei of lateral lemniscus (nLL), superior olivary complex (SOC) and cochlear nucleus (CN). The visualization of the auditory pathway still needs some improvement due to the small size of the nuclei, effect of pulsatile brainstem motion and parallel brain activity which can interfere with the detection of localized BOLD signal.

The motion in the brain arises from several factors, such as heart beat, vasculature, cerebrospinal fluid movement and tissue deformation. Temporal signal fluctuations due to cardiac pulsations have shown to be significant in fMRI. Cardiac gated measurements can reduce the effect of the brainstem motion [5-7].

The calculated signal changes in a white matter reference region or the whole white matter may allow removing signal distortions which may interrupt the real acoustic activations in the brain [8]. The evaluation from an area obtained by segmentation of the white matter was already suggested by other investigators [9].

The additional analysis of functional connectivity can be also useful tool to improve the detection of auditory pathway. It can describe the connections between functionally related regions. The synchronized changes in signal intensity, metabolic rate and blood flow suggest neural connections between different functional parts of the brain [10] and can offer a useful characterization of cortical and subcortical interactions.

The aim of this study was to improve the visualization of auditory pathway using different technical parameters, cardiac gated measurements, using signal from the whole white matter as a regressor in the statistical analysis and correlation analysis to improve the detection of brainstem nuclei. Furthermore different paradigms as block design, continuous sound and resting state were used.

II. Materials and Mathods

A. Subjects

Ten healthy and right handed volunteers (age range 26.8±6.7 eight females, two males) participated in fMRI experiments. All subjects had normal hearing abilities and gave their informed consent to participate in the study.

B. Sound stimuli

Auditory stimuli consisted of music (rock and classical) presented binaurally. Music was cut into parts of 2sec in *WavePad Sound Editor (NCH Swift Sound)* and these parts were stored as standard 16-bit uncompressed WAVE files with a sampling rate of 22050 Hz. Loudness of each stimulus was adapted individually for each subject by presenting stimuli with different intensities to the subject during scanner operation.

C. Paradigms

Each subject was examined by using three experimental paradigms: block design, continuous sound and resting state.

Block design paradigm consisted of 7 blocks of 'ON' condition and 8 blocks of 'OFF' condition. During the 'ON'

K. Dremstrup, S. Rees, M.Ø. Jensen (Eds.): 15th NBC on Biomedical Engineering & Medical Physics, IFMBE Proceedings 34, pp. 57–60, 2011.
www.springerlink.com

epoch the continuous music track was applied to the subjects' ears. During the 'OFF' block there was a rest, without delivering any stimuli. Each block consisted of 10 scans which corresponded to approximately 20s.The length of the played music depended of the cardiac cycle.

In the *continuous sound* paradigm there was just one 'ON' condition without any rest conditions. The classical music was applied continuously.

Resting state paradigm consisted only of scanner noise. Volunteers were instructed to remain relaxed with their eyes closed but not fall asleep.

D. Data acquisition

Imaging was conducted on a *3 T* scanner *(Tim Trio Siemens Erlangen, Germany).*

Neural activation was detected by BOLD differences using *T2*-* weighted EPI. *fMRI* data were collected using an echo time (TE) of 30 ms or 34ms (conditional on slice thickness), the repetition time (TR) depended on the cardiac cycle. Start of acquisition was timed to start every second heart beat which was measured with a finger tip pulse sensor. A total of 153 images per slice were acquired. The volumes consisted of 10 slices in first experiment slice thickness was 2.5mm, in plane resolution 2.9x2.9 mm, in second one slice thickness 1.7 mm and in plane resolution 2.5x2.5mm. Anatomical images were acquired using a MPRAGE sequence (TR= 2300ms, TE= 3.03ms, TI= 1100ms, flip angle= 8 degrees, slice thickness= 1mm, FoV= 256mm, voxel size 1.0x1.0x1.0mm).

E. Data analysis

Pre-processing of the data was conducted on SPM5 (http.//www.fil.ion.ucl.ac.uk). Statistical analysis was performed according to GLM. In this step measured variable repetition times were used in *fMRI* model specification as a regressor (first analysis) to remove the effect of the variable length of the heartbeat on the signal intensity.

To improve the results further a second analysis was carried out. The average signal from the whole white matter was used as a second regressor. Regions of auditory cortex and brainstem were identified on the basis of the anatomical atlases using threshold P<0.001 uncorrected.

Afterwards an additional correlation analysis was performed. Coordinates of the seed regions were chosen in the conventional t-test analysis from the highest t-value in the auditory cortex on each side. The signal intensities in all measured volumes at the position of the seed voxel built the reference time course which was used to calculate the correlation coefficients from the whole brain. The correlation maps were created which showed functionally region, related auditory cortex and brainstem.

III. RESULTS

A. Conventional SPM analysis (with TR and WM regressors)

In total of 20 sessions (10 for first experiment, 10 for second) activation in response to music stimuli was detected in all individuals in the right and left auditory cortex (AC). The T values (block design) for local maxima in AC were higher for analysis with two regressors (TR and WM). Mean T values for analysis only with TR regressor were 14.5 and 12.40 (respectively for right and left side) and for both regressors 18.4 (right side) and 16.3 (left side). Fig. 1 presents the T maps for both analyses.

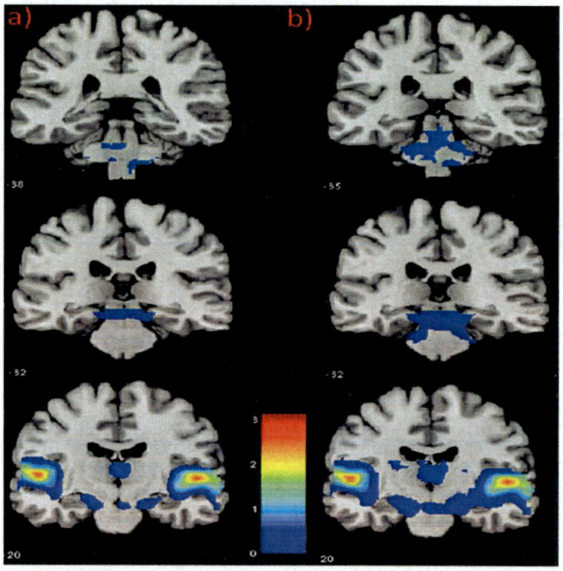

Fig. 2 Example of mean T maps for the first experiment: analysis using only TR as a regressor (a) and TR and WM as regressors (b).

The highest increase of parameter T is visible in auditory cortex; however the change of T-value is also visible in brainstem structures using the signal from white matter as a second regressor. The average percentage of detectability for medial geniculate body (MGB) was 22.5% (first analysis) and 57.5% (second analysis), and 17.5% and 80% for inferioir colliculus (IC). The percentage of detectability for nuclei of lateral lemniscus (nLL) was 12.5% (first experiment) and 30% (second experiment), for cochlear nucleus

(CN) 45% and 90% respectively and the superior olivary complex (SOC) was detected with 15% and 30%. Numbers of activated subcortical structures are presented in the Table 1 (block design paradigm). Because of the observed increase of detection rate in brainstem both regressors were used in the continuous sound and resting state paradigm analysis.

Table 1 Average numbers of activated localizations which were found in conventional analysis for block design paradigm: with TR as a regressor (first analysis) and in the second one with two regressors (TR and WM) for both experiments.

Structures	TR regressor R	L	TR and WM regressors R	L
IC	3	4	16	16
nLL	3	2	5	7
SOC	4	2	5	7
CN	8	10	17	19
MGB	5	4	11	12

B. Correlation analysis

Reference time courses form both analyses were used to obtain the correlation maps. The quality of the maps using two regressors was improved due to elimination of the signal interference. With the threshold of correlation coefficients 0.5, localizations in the brainstem were found in all of experimental paradigms. The signal curves at these positions were strongly correlated with the auditory cortex.

In the *block design paradigm* (10 sessions for first experiment and 10 sessions for second experiment) the average percentage of detectability for IC and MGB was 80% and 82.5%. Nuclei of LL were detected with 77.5%, SOC with 72.5% and CN with 75%. The results of numbers of positions found in correlation analysis are demonstrated in Table 2. Figure 2a illustrates example of correlation map for block design.

For *continuous sound paradigm,* data were obtained from nine sessions with slice thickness 2.5mm and from eight sessions for 1.7mm. Results from first experiment are presented in the Fig. 2b. The average percentage of detectability was: 50% for MGB, 68% for IC, 47 and 38% (nLL and SOC respectively) and for CN 44%. The numbers of the structures are presented in the Table 2.

Resting state paradigm was carried out in nine sessions for first experiment and in eight sessions for the second one. Fig. 2c illustrates the correlation maps. The numbers of brainstem structures are presented in the Table 2. MGB was detected with 38%, SOC with 41%. The detection for CN, IC and nLL was 26, 59 and 35% respectively.

Table 2 Average numbers of activated localizations (correlation analysis with both TR and WM regressors) for all three paradigms for both experiments.

Structures	Block Design R	L	Continuous sound R	L	Resting State R	L
IC	16	16	11	12	12	8
nLL	17	14	8	8	5	7
SOC	15	14	6	7	8	6
CN	15	15	7	8	5	4
MGB	17	16	9	8	7	6

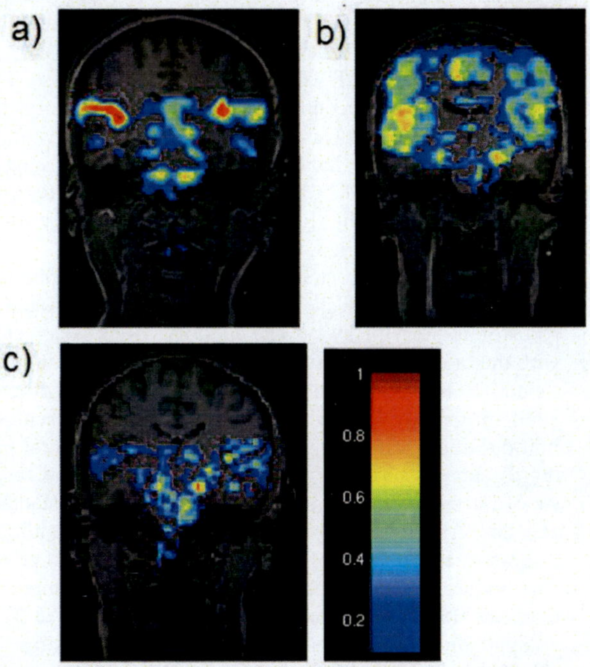

Fig. 2 Example of correlation maps for first experiment (slice thickness 2.5mm), analysis with both regressors for different subjects and different experimental paradigms: a) block design, b) continuous sound and c) resting state.

IV. Discussion

The present study was designed to assess the visualization of cortical (auditory cortex) and subcortical structures (medial geniculate body, inferior colliculi, nuclei of the lateral lemniscus, super olivary complex and cochlear nucleus) of auditory pathway. We successfully visualized fMRI activation in areas of the auditory pathway using a block paradigm with application of music as stimuli. Cardiac gated method was effectively used to minimize the pulsative movement artifacts resulting in higher detection rate in regions of auditory cortex and brainstem what improve the reliability of measurements. The same effect was observed also in other auditory fMRI studies [5- 7].

Activation in auditory cortex, inferior colliculus and cochlear nucleus was seen in most subjects, while a BOLD response within others subcortical structures such as nuclei of the LL, SOC and MGB were seen less recurrently. Most likely IC (6x6x4mm^3) and CN (3x3x7mm^3) were seen more commonly because of its larger size compared with the other brainstem nuclei (nucleus of LL and SOC 2x2x5mm^3). In previous reports, the auditory cortex and subcortical structures such as cochlear nucleus, inferior colliculi was also detected more often than the rest of the structures and they were repeatable in several experiments [4].

Acquired maps of T-values with only one regressor (TR) resulted in lower detection rate for brainstem structures than analysis with two regressors (TR and WM). Activation in the auditory pathway increased on average 60%. Using the signal from the whole white matter as a second regressor helped to minimize the signal changes which disturb acoustic activation in the brain.

Correlation maps created using additional functional connectivity analysis showed us the relation between regions of auditory pathway. The auditory cortex is correlated with the brainstem structures with the high correlation coefficients (0.8 to 0.97). To improve selective visualization of auditory cortex or structures of brainstem it was essential to introduce additional analysis [7]. For all experimental paradigms such as block design, continuous sound and resting state the signal from white matter was used in statistical analysis. After using signal from white matter as additional regressor, the quality of correlation maps was improved. The highest detection rate for all brainstem structures was observed for standard block design paradigm (77.5 %) in comparison to continuous sound (40%) and resting state (39.8%).

V. Conclusion

This approach shows that functional MR imaging reveals functional connectivity between cortical and subcortical regions. The calculation of the averaged white matter signal is easily implemented and gave better results within the brainstem using the applied analysis techniques. The investigation of correlation analysis in this area combined with the cognitive tasks such as auditory stimuli can be an important tool for evaluating auditory network in the human brain and improve the detection. The results provide improvement in visualization of subcortical structures in auditory pathway comparison with the other studies.

Acknowledgment

This study was supported by Landesgraduiertenförderung (LGFG) scholarship for young scientist.

References

1. Buckner RL, Bandettini PA, O'Craven KM et al. (1997) Detection of cortical activation during averaged single trials of a cognitive task using functional magnetic resonance imaging. Proc. Natl. Acad. Sci. USA 93:14878–14883
2. Langers DRM, Pim van Dijk, Backes WH (2005) Lateralization connectivity and plasticity in the human central auditory system. NeuroImage 28:490-499
3. Heβelmann V, Wedekind Ch, Kugel H et al. (2001) Functional magnetic resonance imaging of human pontine auditory pathway. Hear Res 158:160-164
4. Hawley M, Melcher JR, Fullerton BC (2005) Effects of sound bandwidth on fMRI activation in human auditory brainstem nuclei. Hear Res 204(1-2):101-110
5. Guimaraes AR, Melcher JR, Talavage TM et al. (1998) Imaging subcortical auditory activation in humans. Humans Brain Mapp. 6:33-41
6. Zhang WT, Mainero C, Kumar A et al. (2006) Strategies for improving the detection of fMRI activation in trigeminal pathways with cardiac gating. NeuroImage 31:1506-15012
7. Ryn M, Erb M, Klose U (2009) Detection of fMRI activations after acoustic stimulation by correlation analysis. IFMBE Proceedings 25:193-196
8. Desjardins AE, Kiehl KA, Liddle PF (2001) Removal of confounding effects of global signal in functional MRI analyses. NeuroImage 13:751-758
9. Spence JS, Carmack PS, Gunst RF et al. (2006) Using a white matter reference to remove the dependency of global signal on experimental conditions in SPECT analyses. NeuroImage 32:49-53
10. Stein T, Moritz CH, Quigley M et al. (2000) Functional connectivity in the thalamus and hippocampus studied with functional MR imaging. Am J Neuroradiol 21:1397–1401

Use macro [author address] to enter the address of the corresponding author:

Author: Michalina Ryn
Institute: University Hospital Tuebingen, Neuroradiology
Street: Hoppe-Seyler Str.3
City: 72076 Tuebingen
Country: Germany
Email: michalina.ryn@googlemail.com

Spatiotemporal QRST Cancellation Method for 3-Lead ECGs

C. Klamor, K. Grimmel, N. Lentz, and A. Bolz

Karlsruhe Institute of Technology (KIT), Institute for Biomedical Engineering (IBT), Karlsruhe, Germany

Abstract— **The analysis and characterization of atrial tachyarrhythmias like atrial fibrillation (AF) require an extraction of the atrial activity (AA) from present ECG recordings by eliminating all ventricular activity (VA). This contribution develops a new QRST cancellation approach with an independent component analysis (ICA) for 3-lead ECG recordings. The small set of leads leaves residual parts of the QRS complex in the extracted AA. Therefore a post-processing with a discrete wavelet transformation (DWT) is used to detect position and duration of all remaining QRS complexes for further elimination. The result is a VA-free signal sequence describing only the AA which offers the requirement for a safe AF frequency determination needed for further characterizations.**

Keywords— **ECG, QRST cancellation, blind source separation, wavelet transformation.**

I. INTRODUCTION

Atrial fibrillation (AF) is one of the most common cardiac disorders, effecting mainly people in their sixties or older and among other things leading to a 5-times higher risk to suffer a stroke. Most accepted therapy is electrical or pharmaceutical cardioversion. Because of the high recurrence rate of AF after cardioversion, it is important to predict the chance of success for this treatment. Several studies show that it is possible to evaluate the prospects by determining the AF frequency [1], [2]. For that reason it is essential to eliminate every ventricular influence in the electrocardiogram ECG, consisting of QRS complex and T-wave. This process is known as QRST cancellation.

Already existing cancellation approaches use average beat subtraction [3], [4], spatiotemporal alignment [5] or blind source separation methods (BSS) [6]. Especially a recently developed BSS technique using independent component analysis (ICA) is able to obtain useful results in performing AF extraction on a 12-lead ECG [7].

The aim of this work is to achieve an equally dependable outcome for ECG recordings utilizing only three leads by applying an ICA method followed by a post-processing using a discrete wavelet transformation.

II. USED ECG DATA

The ECG recordings used to test the developed algorithm originate exclusively from the Universtitätsklinikum Hamburg-Eppendorf. All of them are long-term recordings (up to 48 hours) consisting of paroxysmal AF. With the help of existing annotations, it was possible to extract several AF episodes of 60 s duration.

For evaluation reasons, an artificial AF episode of the same length was generated by applying a saw-tooth function (Fig.1(a)). This distribution is combined with a sinus-rhythm ECG without P-wave to create an artificial AF-ECG (Fig.1(b)). Because the shape of the artificial AF is known, it is possible to review the morphology of the QRST-cancelled AF-ECG to estimate the extraction quality.

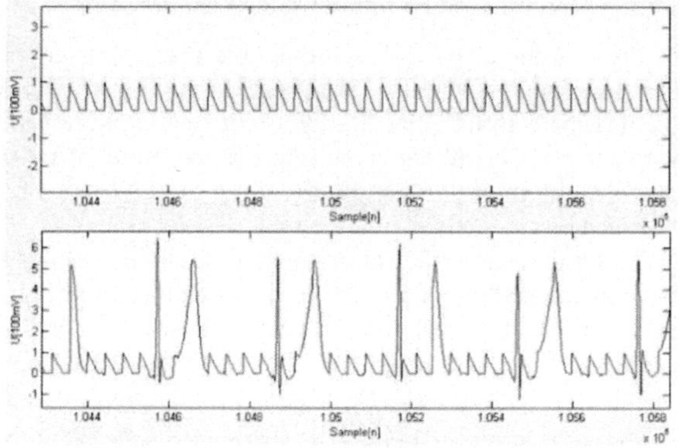

Fig. 1 Artificial AF (top) and resulting artificial AF-ECG (bottom)

III. PRINCIPLES OF BLIND SOURCE SEPARATION (BSS)

BSS methods are widespread procedures for separating components out of one mixed signal. These methods are often implemented if the source signals are unknown and at the same time only multiple recordings of one mixed signal exist.

In this case the recording of one lead of a surface ECG represents one observation of linear combinations of the sources, which can be divided into atrial activity (AA), ventricular activity (VA) and noise.

A condition for applying all BSS techniques is the lack of correlation between the source signals. For the here used independent component analysis (ICA) the condition is even stricter, because the sources need to be statistically independent [8]. That AA and VA occur independently, especially in phases of AF, was proven in former studies [3], [9].

Additionally it is important to realize that a combined signal can only be separated in as many sources as counts of observations are available. This means, for separating AA and VA at least two leads are needed. For this is a theoretical limit not considering interferences, 3-lead ECG recordings are utilized.

The mixed signal that should be examined emerges from all present sources. This can be described mathematically by a vectorial representation using a mixing matrix A.

$$\vec{x} = A\vec{s} \qquad (1)$$

K. Dremstrup, S. Rees, M.Ø. Jensen (Eds.): 15th NBC on Biomedical Engineering & Medical Physics, IFMBE Proceedings 34, pp. 61–64, 2011.
www.springerlink.com

In equation (1) the vector $\vec{s}$ represents the source signals and $\vec{x}$ all observations of the mixed signal. The task of the ICA is to solve the inverse problem.

$$\vec{s} = A^{-1}\vec{x} \quad (2)$$

In general, the ICA solves this equation by an optimization model using a gradient descent. The used FastICA algorithm employs negentropy as measurement for the maximal independence between the requested source signals [8].

IV. Pre-processing

Prior to the actual calculations take place, it is crucial to perform an accurate pre-processing. Most common disturbances in ECG recordings occur by thorax movements because of breathing resulting in baseline wandering, artifacts by muscular contractions and the $50/60\ Hz$ power frequency.

To remove baseline wandering and possible offsets a two-step median filter is used. In the first step the whole signal is scanned with a window of $200\ ms$ length sorting all samples by ascending order. Adjacent the sample originally located in the middle of the window is replaced by the now rearranged median value. This leads to the removal of the QRS complex and f-waves. For the second step the window length is set to $600\ ms$ to exclude all T-waves from the signal. The result is a signal sequence only representing the baseline wandering [10]. If this sequence is subtracted from the original ECG a drift- and offset-free ECG is achieved. Figure 2 illustrates the difference between the original ECG (top) and the filtered ECG (bottom). Additionally the signal sequences after first filtering step (green) and second step (red) are shown.

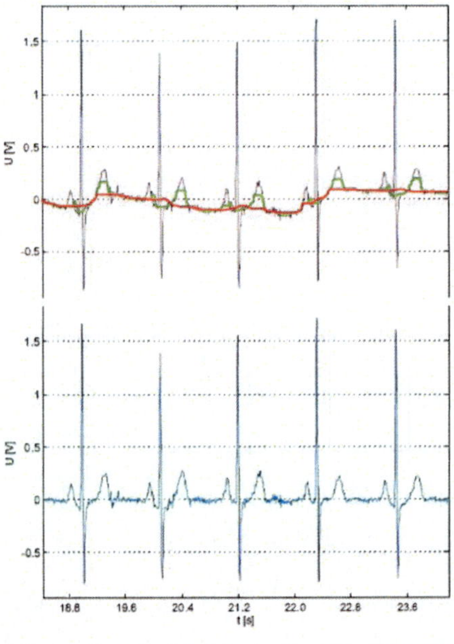

Fig. 2 Comparison of original (top) and median-filtered ECG (bottom)[10]

To remove the power frequency and noise originated by muscular activities a simple FIR low-pass filter with a cut-off frequency of $49\ Hz$ is applied. This is feasible, because in the end only the frequencies of the f-waves are analyzed. Their frequency band is situated in the range of $4.5 - 10\ Hz$, thus no necessary information will be lost.

V. AF extraction

For AA-extraction the already existing FastICA algorithm is used [11]. It includes a whitening procedure as a further pre-processing step. Whitening is required to harmonize all considered leads and to avoid an emphasizing of single recordings.

The use of FastICA produces three single signal components as illustrated in Fig.3. Every component in Fig.3 contains an augmented part of the ECG. One component enhances the QRS complex, another sequence amplifies the T-waves and the most important one emphasizes the AA and thus the f-waves. Because the output order is not constant further calculations have to be implemented.

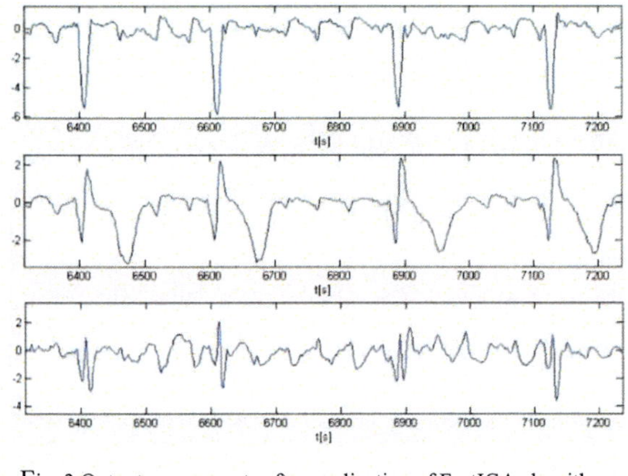

Fig. 3 Output components after application of FastICA algorithm

To identify the component which represents AA the most the kurtosis of every component can be calculated. The kurtosis is able to define, if a signal sequence has sub- or super-gaussian behavior [2]. Representations of the QRS complex or the T-wave are expected to have super-gaussion behavior. In contrast to the f-waves of a fibrillatory AA which develop sub-gaussian characteristics.

For these reasons the component with the lowest kurtosis value represents the AA. Table 1 contains the kurtosis values with variance for each component.

Table 1 Kurtosis values of ICA components

Augmented ECG part	Kurtosis
f-wave	5,67 +/-1,62
QRS complex	17,02 +/-5,89
T-wave	25,77 +/-3,14

Comparisons between the original recordings and the separated AA show, that the T-wave was cancelled successfully. But the signal still contains parts of the QRS complex. Because of these parts the kurtosis reaches no sub-gaussian values and it is hardly possible to determine the exact AF frequency.

In this case a post-processing is necessary to eliminate the remaining parts of the QRS complex. Therefore the discrete wavelet transformation is introduced.

VI. QRS ELIMINATION WITH A DISCRETE WAVELET TRANSFORMATION (DWT)

Wavelet transformations are often used in signal processing if additionally to a frequency analysis further temporal information is needed. Thereby the location of occurring frequencies can be defined.

For computational reasons a discrete wavelet transformation (DWT) is applied. The functionality of a DWT can be described with the use of filter banks [12]. In the process the original signal is successively low-pass filtered on several levels. The level count can be varied by scaling factors. For each level the separated high-frequency parts of the signal sequence are kept as detail signal. In the end the DWT represents the signal through an approximation signal and several detail signals according to the scaling factors. Fig. 4 illustrates the principle of operation of a DWT for a signal f.

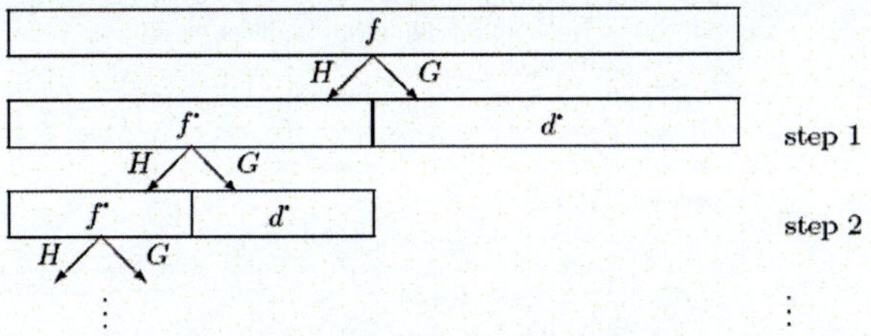

Fig. 4 DWT principle with filter banks

Because the QRS complex is displayed by frequencies in higher ranges than f-waves, the results of the DWT can be used to detect position and duration of each complex. In this case the remaining ECG signal is decomposed with 12 levels by applying a Daubechies-3-Wavelet as mother wavelet.

All parts of the QRS complex are located in the lower detail levels. For each respective level an adapted threshold is defined. If the detail signal on one level reaches the particular threshold these signal parts are marked as piece of a QRS complex. For a reliable recognition the results for several scaling factors are combined.

Once positions and durations of every QRS complex are known these sections can be eliminated from the residual ECG, illustrated in Fig. 5. The sequence in the top shows the original ECG in comparison to the QRST-cancelled ECG after AA extraction and QRS elimination on the bottom.

The QRS detection works dependably for different ECG morphologies and can also be utilized for usual ECG recordings without previous AA extraction.

The combination of multiple scaling factors stabilizes the detection and makes sure that possible interferences do not disturb the elimination process.

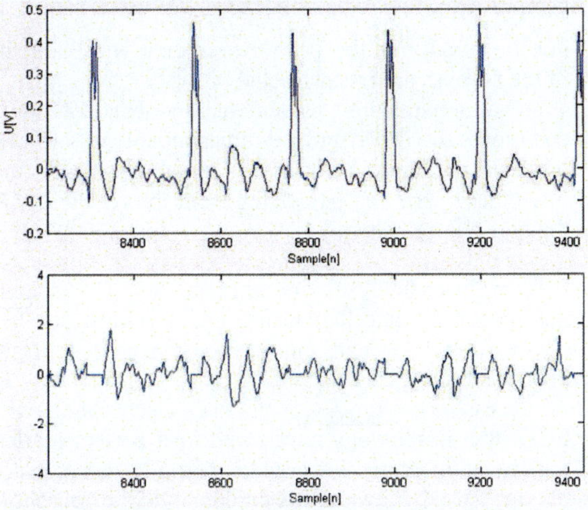

Fig. 5 Comparison of original ECG (top) and QRST-cancelled result (bottom)

VII. AF FREQUENCY DETERMINATION

The achieved cancellation performance allows a simple AF frequency determination by using a Fast-Fourier-Transformation (FFT). The outputs of the FFT are applied to calculate a power spectral density (PSD) which returns the fibrillation frequency.

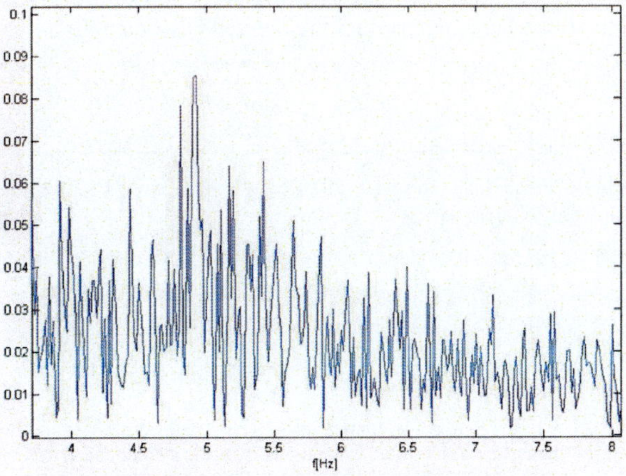

Fig. 6 Power spectral density with peak frequency at 4,92 Hz

Fig. 6 shows the plotted PSD according to the current frequencies. The chosen example has a density peak at $4{,}92\ Hz$. This peak describes the most common frequency in the considered sequence and represents the wanted frequency of the f-waves.

VIII. Results

The introduced method for QRST-cancellation using a two-step approach is a reliable tool for 3-lead ECGs. The ICA allows an AA extraction with almost perfect T-wave compression.

For the duration of the T-wave even the basic characteristics of the f-waves are reconstructed.

To eliminate the residual components of the QRS complexes the DWT provides a depandable way to calculate position and duration of every remaining QRS complex. Thus these parts can be deleted from the residual ECG and the pure AA remains.

To evaluate the result, the artificial AF-ECG is applied to the developed cancellation method. A comparison between the artificial AF and the extracted signal confirms the previous findings.

In areas where the AA is the only electrical signal in the ECG, the morphology is restored in a perfect matter. The regions with former T-waves contain a basic reconstruction of the used saw-tooth function with morphological irregularities but keep track of the overall AF frequency. All QRS positions are set to zero and contain no further information. However these intervals have a much shorter dimension than the remaining fibrillation episodes and no significant influence on the frequency determination.

Fig. 7 illustrates the evaluation results. The plot compares the original AF constructed by a saw-tooth function (blue) with the extracted artificial AF (red). The previously mentioned different areas of reconstruction are marked by different colors. The green region marks the best restored regions. Yellow represents the dilatation of the T-wave and the red area determines the deleted QRS complex.

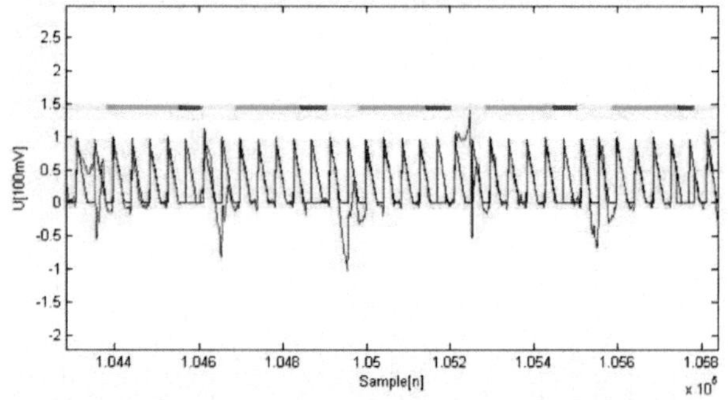

Fig. 7 Evaluation of artificial AF (blue) and extracted signal (red); reconstruction is rated by their quality using a color coded ranking list

IX. Conclusion

The developed cancellation method reduces the needed count of ECG leads for an ICA approach to three by introducing an additional computational effort to eliminate residual QRS complexes.

The ICA leads to negligible morphological changes to the reconstructed f-waves which however do not prevent a stable AF frequency determination.

References

1. Holmqiust, Frederik und Stridh, Martin. Atrial Fibrillatory Rate and Sinus Rhythem maintenance in Patients Undergoing Cardioversion of Persistent Atrial Fibrillation. s.l. : The European Society of Cardiology, 2006.
2. Bollmann, Andreas und Kanuru, Narendra K. Frequency Analysis of Human Atrial Fibrillation using the Surface Eletrocardiogram and Its Response to Ibutilide. s.l. : American Journal of Cardiology, 1998.
3. Shkurovich, Sergio und Sahakian, Alan. Detection of Atrial Activity from High-Voltage Leads of Implantable Ventricular Defibrillators using a Cancellation Technique. s.l. : IEEE Transactions on Biomedical Engineering, 1998.
4. Holm, Magnus und Pehrson, Steen. Non-invasive Assessment of the Atrial Cycle Length during Atrial Fibrillation in Man: Introducing, validating and Illustrating a new ECG Method. Lund : Elsevier, 1998.
5. Stridh, Martin und Sörnmo, Leif. Spatiotemporal QRST Cancellation Techniques for Analysis of Atrial Fibrillation. s.l. : IEEE Transactions on Biomedical Engineering, 2001.
6. Castells, F. und Rieta, JJ. Estimation of Atrial Fibrillatory Wave from Single-Lead Atrial Fibrillation Electrocardiograms using Principal Component Analysis Concepts. s.l. : Medical & Biological Engineering & Computing, 2005.
7. Castells, F. und JJ, Rieta. Spatiotemporal Blind Source Separation Approach to Atrial Activity Estimation in Atrial Tachyarrhythmias. s.l. : IEEE Transactions on Biomedical Engineering, 2005.
8. 8. Hyvärinen, Aapo, Karhunen, Juha und Oja, Erkki. Independent Component Analysis. New York : John Wiley & Sons, 2001. ISBN 0-471-40540-X.
9. Rieta, JJ und Castells, Francisco. Atrial Activity Extraction for Atrial Fibrillation Analysis Using Blind Source Separation. s.l. : IEEE Transactions on Biomedical Engeneering, 2004.
10. Klein, Kersten. Extraktion formsensitiver Merkmale aus dem EKG zur Erkennung von paroxysmalem Vorhofflimmern. 2010.
11. Gävert, H. und Hyvärinen, A. The FastICA package for MATLAB. [Online] 2005. [Citation date: 20. 09 2010.] http://www.cis.hut.fi/projects/ica/fastica/.
12. Stark, Hans-Georg. Wavelets And Signal Processing. Heidelberg : Springer, 2005. ISBN 3-540-23433-0.

Christian Klamor
Karlsruhe Institute of Technology (KIT),
Institute for Biomedical Engineering (IBT)
Kaiserstr. 12
76131 Karlsruhe
Germany
christian.klamor@gmail.com

Telerehabilitation for COPD Patients across Sectors: Using Technology to Promote Cooperation among Healthcare Professionals

B. Dinesen[1], O.K. Hejlesen[1,2], S.K. Andersen[1], and Egon Toft[1]

[1] Department of Health Science and Technology, Aalborg University, Aalborg, Denmark
[2] Department of Health and Nursing Science, University of Agder, Norway

Abstract— **A telerehabilitation programme for patients with chronic obstructive pulmonary lung disease has been developed. Aim: The aim of the paper is to explore how telerehabilitation technology influences cooperation and coordination between healthcare professionals in the Telekat network. Theory: Network theory has been applied. Methods: A triangulation of data collection techniques such as documentary materials, participant observations (n=163 hours), and qualitative interviews (n=24) has been used in a case study of the Telekat project. Findings: Telerehabilitation technologies influences cooperation and coordination between healthcare professionals in themes like: interdisciplinary decision making, mutual learning processes and new professional roles. Conclusion: A combination of telerehabilitation technologies is a facilitator for cooperation and coordination between healthcare professionals in a network.**

Keywords— **Co-operation, coordination, tele-rehabilitation, healthcare professionals.**

I. Introduction

Today's healthcare system appears to be fragmented in several aspects of the cross-sector coordination of patient care, treatment and rehabilitation. Applying telerehabilitation technologies in a cross-sector cohesive patient-care process brings about change in healthcare organizations and opens up new possibilities for service delivery and supporting integrated care working [1-2]. The term 'telerehabilitation'denotes care, treatment and rehabilitation cutting across sector lines by means of information or communication technology. Prior research on the implementation of telehomecare and telerehabilitation technologies and clinical task solving across sectors on the operational level is limited. Most studies have been conducted in rural and remote areas. A longitudinal study of the implementation process of telehomecare in rural and remote areas of Canada identified that telehomecare would help create networks between hospital and primary care providers but widespread use of telehomecare has not been reached [3]. The integration of telerehabilitation in work processes improved coordination of care [4]. Cooperation aided by telerehabilitation seems to work well and may be influenced by factors such as the personality of the healthcare professionals involved, their personal relations, preparation and experience [5]. It has been shown that task solving within telehomecare or telerehabiltation depends on the applications; however, further research is needed to clarify this [6-7]. An English study shows that integration of telehealth services has been difficult and clinicians find it hard to integrate the services into their deeply embedded local working structures. Additionally, it has been reported that, problems arise due to the complexity of the operational changes that results after the implementation of telerehabilitation technology [8].

This paper report on a part of a research project, namely the TELEKAT project ("Telehomecare, chronic patients and the integrated healthcare system"). The aim of the project is to develop a preventive home monitoring concept across sectors, enabling patients with server and very server chronic obstructive pulmonary disease (COPD) to avoid readmission to hospital, to perform self-monitoring of their condition, and to maintain rehabilitation activities in their homes.

The research question in this study was: *How will telerehabilitation technology influence cooperation and coordination between healthcare professionals in the Telekat network?*

II. Introduction of the Programme of Telerehabilitation for COPD Patients

Patients with (COPD) are a serious public health problem. It is estimated that 210 million people have COPD worldwide and more than 3 million people died of COPD in 2005 and this is equal to 5% of all deaths globally that year [9]. Reviews of the disease management programs for patients with COPD show that the programs are heterogeneous in terms of interventions, outcome measures and study design. However, quality of life is improved, and triple intervention programs have resulted in lower probability of at least one hospital admission compared to usual care. The reviews also conclude that there is a need for more research on

K. Dremstrup, S. Rees, M.Ø. Jensen (Eds.): 15th NBC on Biomedical Engineering & Medical Physics, IFMBE Proceedings 34, pp. 65–68, 2011.
www.springerlink.com

chronic disease management programs in patients with COPD across primary and secondary care [10-11].

The Telekat project has focused on developing a programme which will take place in the patients own homes and in collaboration with various healthcare professionals such as doctors and nurses at hospital and healthcare centre, district nursing and GP. The programme is based on a telehealth monitor box installed in the patient's home for four months. Using wireless technology, the telehealth monitor can collect and transmit data about the patient's blood pressure, pulse, weight, oxygen level, lung function, etc. via the Internet network, transmitting the data to a web-based portal or to the electronic health care record. Healthcare professionals such as GP, district nurses, nurses, doctors and physiotherapists at the health care centre or hospital can assess the patient's data, monitor the patient's disease and training inputs and provide advice to the patient. The patients and relatives can also view the data on the web portal and decide whom they want to share their data with (see figure 1). The patients receive an individual training program by a physiotherapist and may carry out home-based exercises.

A telerehabilitation team consisting of health care professionals from primary and secondary care meet virtually once a month to coordinate and discuss the individual rehabilitation programme for the COPD patients. An illustration of the telerehabilitation programme in the TELEKAT project is shown in figure 1 below.

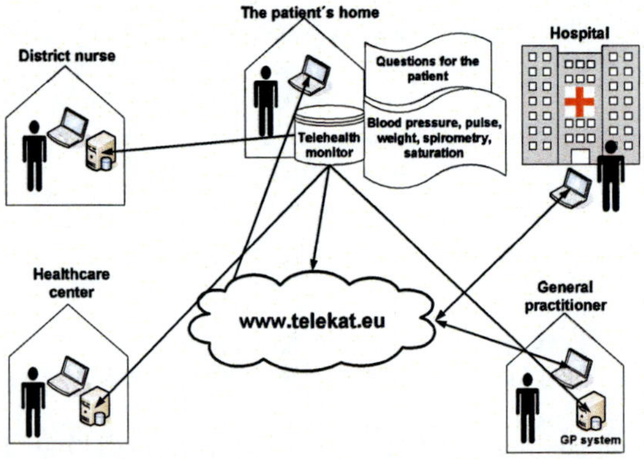

Fig. 1 The Telerehabilitation Programme

III. Theoretical framework

Network theory [12-13] has been applied, as it opens up the boundaries of the organizations and helps explain network dynamics and processes between the healthcare professionals in the Telekat network. A network is defined as: *''the basic social form that permits inter-organizational interactions of exchange, converted action, and joint production. Networks are unbounded or bounded clusters of organizations that, by definition, are non-hierarchical collectives of legally separate units''* [12, p. 46].

In the literature [12,13], there are different models of networks, and the Telekat network can be characterized as a systemic network that consists of five elements: parties, processes, vision, and architecture and culture. The parties are the resources of the network. A crucial element in relations between the network parties is trust. Network processes are centered on exchange of coordination, information and joint problem solving between the organizations. A vision for the network is a joint vision, in this case the telerehabilitation programme for COPD patients. The network architecture shapes the structural framework for cooperation and coordination. Formal and informal culture in the network constitutes the norms and values for interaction between the healthcare professionals.

Cooperation is a voluntarily arrangement in which two or more parties engage in a mutually beneficial exchange instead of competing. There are three types of *coordination* in a network [14]:

- Mutual adjustments: Parties in a network retain their autonomy and coordination is based on spontaneous interaction and is based on informal rules.
- Alliances: There exist no authorities and coordination is accomplished by negotiated rules.
- Corporate coordination means that parties develop a joint authority structure and they give up some of their authority.

Accordring to Denis et al. (1999) introducing new technologies as telerehabilitation technologies in a network can be seen as driver for change. The technology may affect the form of coordination between different individual groups of professionals and mobilize their interactions in somewhat different ways [15]. Professionals groups that have worked together for a time have developed mutual trust and tacit rules and an outside intervention like introducing new technologies alter boundaries between the professionals and produce new local and negotiated orders.

IV. Methods

The case study approach [16] is chosen as the overall research strategy for this study. The purpose is to make an explorative and in-depth study on how telerehabilitation technology influence cooperation and coordination between healthcare professionals in the Telekat network.

A randomised study (pilot phase: n=5; randomised study: n=111) has been conducted in order to create quantitative data on clinical and economical data as well (not reported in this paper). 60 patients participated in the intervention group who did home monitoring by the use of telerehabilitation technology. The control group consisted of 51 patients who followed traditional rehabilitation programme.

A triangulation of data collection techniques has been used including: documentary materials, 163 hours of participant-observation [17], qualitative interviews [18] with healthcare professionals: GP (n= 6), nurses and doctors at hospital (n= 6), nurses at the healthcare centre (n= 6) and district nurses (n=8).

All the transcribed interviews were coded with Nvivo 8.0 software and analyzed in steps inspired by Kvale (2009). The data were analyzed using a combination of deductive and inductive strategy. The code tree was formed on the basis of central definitions and concepts (in vitro nodes) from the theoretical framework and from interviews (in vivo nodes). Based on the data analysis themes that characterise how telerehabilitation technologies influences cooperation and coordination between healthcare professionals in the Telekat network was identified. The process was done in dialogue with research colleagues.

In relation to conducting a cases study one of the recurring discussions is on its generalizability. In order to optimize generalization of case studies works of reference [19] on the case study recommend strategic case selection or analytical generalization.

Ethical approval was obtained from the local Ethics Committees (August 27 2008/ N-20080049). The study was performed according to the Declaration of Helsinki. The project was reported to the Danish Data Protection Agency (August 7 2008).

V. Findings

Below are listed themes that characterise how telerehabilitation technologies influences cooperation and coordination between healthcare professionals in the Telekat network:

Interdisciplinary decision making: The healthcare professionals express that they benefit in their clinical decision making by being able to access and share clinical rehabilitation data across sectors from the web based portal. A doctor at the hospital express: *"I can share data on saturation, blood pressure, pulse and weight with the patients GP. This makes the clinical decision making more evidence based for the benefit of the patient".*

Mutual learning process on rehabilitation issues: Doctors and nurses at hospital and healthcare centre gained a better understanding of the context for counselling on rehabilitation activities in the homes of the COPD patients. The healthcare professionals expressed that the video meetings gave them the opportunity to create mutual learning and knowledge sharing on specific rehabilitation issues and patient cases. A district nurses express: *"When the video technology is running it is an easy way to counselling the lung specialist at the hospital at the video meetings and get new knowledge on rehabilitation issues. I learn from this interdisciplinary dialogue and it makes me be more reflective in clinical practice".*

New professional roles: The healthcare professionals expressed that they during the process of creating and testing the programme for telerehabilitation have created new roles in the cooperation between the healthcare professionals such as district nurses, GP´s, nurses and doctors at hospital and healthcare centre. The roles have changed concerning couseling on rehabilitation issues. The healthcare professionals at hospital and healthcare centre can via the new programme follow patients rehabilitation activities in their homes. District nurses and GP´s have gained more focus on their preventive role. A district nurses express: *"We have got a new dimension on preventive work and rehabilitation".*

Field notes from observations shows that the healthcare professionals have through mutual adjustments and "unformal" negotiation reached consensus on the new roles.

VI. Discussion

We have identified three themes that influence where telerehabilitation technologies influences cooperation and coordination between healthcare professionals in the Telekat network. The themes are: interdisciplinary decision making, mutual learning processes on rehabilitation issues and new professional roles. Through the process of developing the programme of telerehabilitation mutual trust has emerged between the healthcare professionals for the benefit of cooperation and coordination of activities related to telerehabilitation of COPD patients across sectors. The technology promotes interdisciplinary decision making between healthcare professionals across sectors. Prior research has showed the same tendency by the use of telehomecare [5, 20]. The technology seems to "glue" together the fragmentation of the cross-sector coordination of patients' rehabilitation processes. The healthcare professionals expressed the view that the video meetings facilitated mutual knowledge-sharing with specialists at the hospital and an interdisciplinary learning process. A research question to be asked is it the technology that facilitates the mutual learning process or it is synergy between the possibilities of the technology and development of an informal learning culture in the Telekat network. Observations from the case study indicate that it is a combination of technology and new informal interaction patterns between the healthcare professionals. Lamothe et al (2006) have also showed that telehomecare technologies create synergy and facilitate cooperation between healthcare professionals. In the telerehabilitation programme new professional roles has emerged between the healthcare professionals through a process of informal process of negotiations. Denis et al. (1999) highlights that new technology can restructure the professional's organization in a network and mobilize interactions in different ways. The case study of the Telekat project indicate that restructuring of the roles can be for the benefit of the patients and towards a more seamless healthcare system. A study [21] indicate that the interplay between new technology and existing professional practices and relationships go beyond simple issues of training and demands particular attention on the implementation process in order to get full benefit from the technology.

The case study call for limitations of generalization at one hand but gives us a view of the dynamics in the case of introducing new technologies into healthcare practice that is valuable for implementation of the technologies.

VII. Conclusions

A combination of telerehabilitation technologies such as telehealtmonitor, a joint webportal and video among healthcare professionals is a facilitator for cooperation and coordination between healthcare professionals in a network.

Acknowledgements

We wish to thank the COPD patients and relatives participating in the project as well as clinical and industrial partners (for details, see www.Telekat.eu). The Telekat project is funded by the Program for User-driven Innovation, the Danish Enterprise and Construction Authority, Center for Healthcare Technology, Aalborg University, and by various clinical and industrial partners in Denmark.

References

1. World Health Organization. Chronic obstructive pulmonary disease (COPD). Fact sheet N°315; 2009 Nov. [Cited 2010 8 December]. Available from: http://www.who.int/mediacentre/factsheets/fs315/en/
2. Lemmens KMM, Nieboer AP, Huijsman R (2009) A systematic review of integrated care use of disease management interventions in astma and COPD. Respir. Med. 103: 670-691
3. Peytremann-Bridevaux I, Staeger P, Brideaux PO, Ghali WA, Burnand B (2008) Effectiveness of Chronic Obstructive Pulmonary Disease- Management Programs: Systematic Review and Metaanalysis. Am J Med 121 (5): 433-443
4. Wensing M, Wollersheim H, Grol R (2006) Organizational interventions to implement improvements in patient care: a structured review of reviews. Imple Scie 22; 1:2
5. Dinesen B, Gustafsson J, Nøhr C, Andersen SK, Sejersen H, Toft E (2007) Implementation of the concept of home hospitalization for heart patients by means of telehomecare technology: integration of clinical tasks. Int J Integr Care May 30;7:e17
6. Lamothe L, Fortin JP, Labbe F, Gagnon MP, Messikh D (2006). Impacts of telehomecare on patients, providers, and organizations. Telemed J E Health 12(3):363-9
7. Aas IH (2001). A qualitative study of the organizational consequences of telemedicine. J Telemed Telecare 7(1):18-26.
8. Dinesen B, Gustafsson J, Nøhr C, Andersen SK, Sejersen H, Toft E (2007). Telehomecare technology across sectors: claims of jurisdiction and emerging controversies. Int J Integr Care Oct–Dec; 7: e43
9. Aas IH (2001). Telemedical work and cooperation J Telemed Telecare 7(4): 212-8.
10. Aas IH (2002). Changes in the job situation due to telemedicine J Telemed Telecare 8(1):41-7
11. May C, Harrison R, Finch T, MacFarlane A, Mair F, Wallace P (2003) Understanding the normalization of telemedicine services through qualitative evaluation. J Am Med Inform Assoc 10(6):596-604
12. Alter C, Hage J (1993) Organizations Working Together. Sage Publications, Newbury Park, California, pp. 46-191
13. Gustafsson J (2007) Ledarskap i interorganisatoriska nätverk för folkhälsa och välfärd. In: Axelsson R, Axelsson SB, editors. Folkhälsa i samverkan mellan professioner och samhällsektorer. [Public health in inter-professional relations and social sectors]. Pozkal (Poland): Studentlitteratur, 2007, pp. 61–86
14. Kickert WJM, Klijn EH, Koppenjan JFM (1999) Managing Complex Networks. Sage Publications, London, pp-19-23
15. Denis JL, Lamothe l, Langley A, Valette A (1999) The struggle to redefine boundaries in health care systems. In Restructuring the Professional Organization edited by David M. Brock, M. Powell and CR Hinnings. Routledge, 1999
16. Yin R (2009) Case Study Research Design and Methods. Sage Publications, London
17. Kristiansen S, Krogstrup HK (2004). Deltagende observation: introduktion til en forsknings metodik. [Participant observation: Introduction to research methodic]. Hans Reitzels Forlag, Copenhagen
18. Kvale S, Brinkmann S (2009) Interviews: Learning the Craft of Qualitative Research Interviewing. Sage Publications, Los Angeles; 2009. pp. 97-218.
19. Flyvberg B (200/) Five misunderstandings about case-study research. In: Seale C., Gobo G., Gubrium JF, Silverman D., editors. Qualitative Research Practice. Sage Publications, London, 2007, pp 390-404
20. Gaikwad R, Warren J (2009) The role of homebased information and communications technology interventions in chronic disease management: a systematic literature review. Health Inform J 15: 122
21. Hibbert D, Mair FS, May CR, Boland A, O`Connor J, Capewell S, Angus M (2004) Health professionals`response to the introduction of a home telehealth service. J Telemed Telecare 10:226-23

Author: Birthe Dinesen
Institute: Department of Health Science and Technology, Aalborg University
Street: Frederik Bajers Vej 7 D1
City: DK-9220 Aalborg
Country: Denmark
Email: bid@hst.aau.dk

The Properties of the Missing Fundamental of Complex Tones

T. Matsuoka[1,2] and Y. Iitomi[2]

[1] MIT, Cambridge, U.S.A.
[2] Faculty of Engineering, Utsunomiya University, Utsunomiya, Japan

Abstract— **The existence of the missing fundamental phenomenon is known, but its mechanism is unknown. We showed how the information of the missing fundamental f_0 explicitly appeared on the aggregated autocorrelogram of the output pulse train for input signal f_1 to one cochlear model and the output pulse train for input signal f_2 to another cochlear model (where f_1=nf_0 and f_2=(n+k)f_0). In practice, we listen to a complex tone of f_1 and f_2 with each ear. In this paper, we try to investigate the influence of the pitch and the number of harmonic components of complex tones on perceiving the missing fundamental.**

Keywords— **the missing fundamental, complex tone, perception, cochlear models experiment.**

I. Introduction

The frequency band of a telephone line channel is from 300 Hz to 3400 Hz. Although pitch frequency of speech is not in the frequency band, we can perceive the pitch over the telephone. The mechanism is unknown. It is considered that the missing fundamental is produced in the auditory system when we listen to a complex tone of f_1=nf_0 and f_2=(n+k)f_0 [1]. f_0 is known as the missing fundamental.

We confirmed that subjects were able to perceive the missing fundamental by the significant difference test (level 5 %) of the psycho-acoustic experimental results. In the psycho-acoustic experiments, subjects simultaneously listened to a pure tone of f_1 with his/her one ear and a pure tone of f_2 with his/her another ear [2]. In practice, we listen to a complex tone of f_1 and f_2 with each ear. .Subjects perceived the missing fundamental f_0 more easily in the psycho-acoustic experiment for two complex tones than for two pure tones. We made clear the mechanism and presented it at BIOSIGNAL2010 [3]. We try to investigate the influence of the pitch and the number of harmonic components on perceiving the missing fundamental.

II. Methods

We made cochlear models shown in Fig. 1 (The figure is the example in the case of two cochlear models. In this paper, we use two cochlear models, three cochlear models,

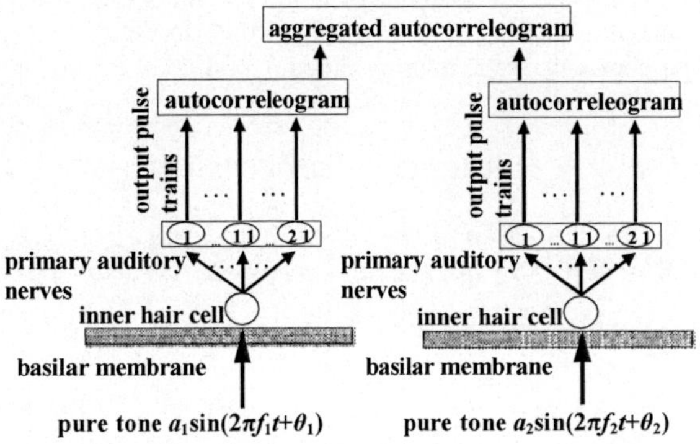

Fig. 1 Cochlea models, autocorrelogram and aggregated autocorrel ogram

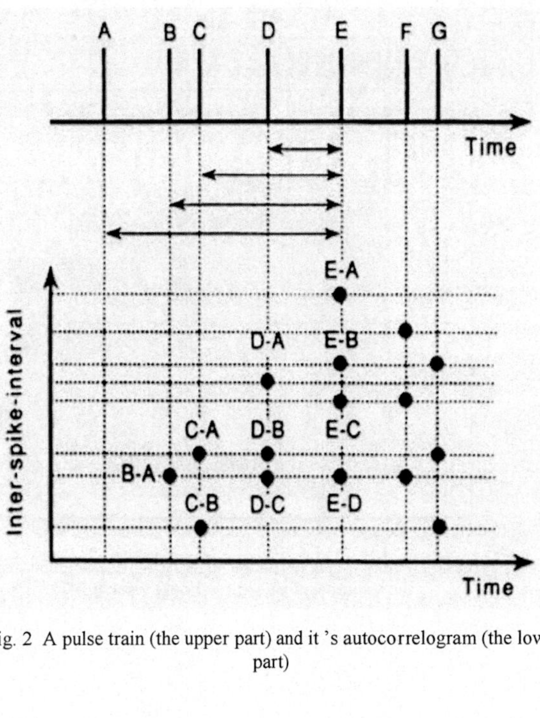

Fig. 2 A pulse train (the upper part) and it 's autocorrelogram (the lower part)

K. Dremstrup, S. Rees, M.Ø. Jensen (Eds.): 15th NBC on Biomedical Engineering & Medical Physics, IFMBE Proceedings 34, pp. 69–72, 2011.
www.springerlink.com

etc.). We showed how the information of the missing fundamental f_0 explicitly appeared on the aggregated autocorrelogram of the output pulse train for input signal f_1 to one cochlear model and the output pulse train for input signal f_2 to another cochlear model [2]. An example of autocorrelogram is shown in Fig. 2. In this paper, we have investigated the properties of the missing fundamental of complex tones having more than two components on cochlear models. We can't carry out the psycho-acoustic experiments for more than two tones, because we have only two ears.

We use the maximum peak except for the missing fundamental peaks (a.b.MPexceptforMFP) in Fig. 3 to see the perceptivity of the missing fundamental.

III. Results and Discussion

A. The experiment of the influence of the pitch on the extraction of the missing fundamental (Experiment 1)

We have carried out the following two experiments.

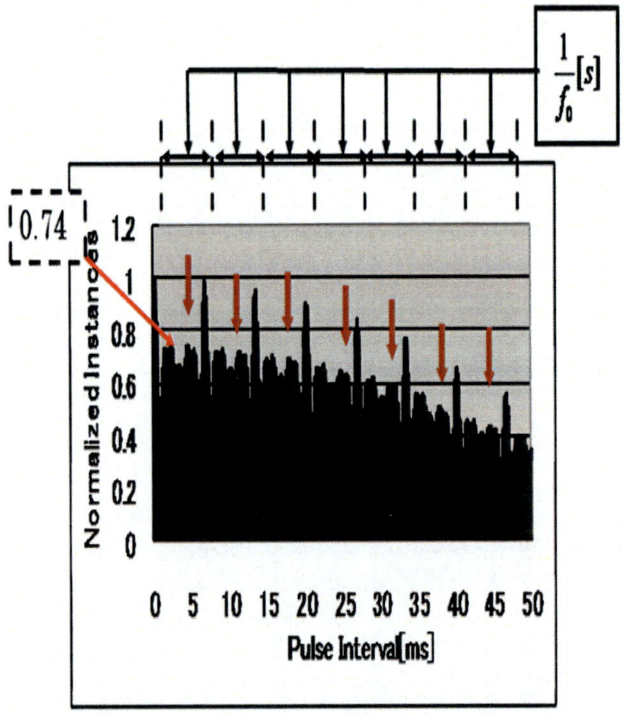

Fig. .3 The maximum peak except for the missing fundamental peaks (In this example, it is the peak of the value 0.74)

Table 1 Component frequencies sets for the experiment of the influence of the pitch (for the experiment 1)

	Fundamental Frequency[Hz]	Component Frequencies [Hz]			
①	f_0=150	450	600	750	900
②	f_0=225	450	675	900	
③	f_0=500	1500	2000	2500	3000
④	f_0=750	1500	2250	3000	
⑤	f_0=700	2100	2800	3500	4200
⑥	f_0=1050	2100	3150	4200	
⑦	f_0=1000	3000	4000	5000	6000
⑧	f_0=1500	3000	4500	6000	

Table 2 The value of the maximum peak except for the missing fundamental peaks in the experiment 1

	Fundamental Frequency [Hz]	Method(a)	Method(b)
①	f_0=150	0.74	0.72
②	f_0=225	0.77	0.77
③	f_0=500	0.99	0.964
④	f_0=750	0.995	0.989
⑤	f_0=700	0.998	0.984
⑥	f_0=1050	0.995	0.993
⑦	f_0=1000	×	0.993
⑧	f_0=1500	×	×

Experiment 1 is for investigating the influence of the pitch on the extraction of the missing fundamental and experiment 2 is for investigating the influence of the number of harmonic components on the extraction of the missing fundamental.

We have compared the experimental results in (a) with in (b) about each experiment.

(a) The case that each cochlear model is fed pure tones.

(b) The case that each cochlear model is fed complex tones.

For experiment 1, the component frequencies sets are shown in Table 1.

Here, the set of ① has the same lowest frequency and the same highest frequency to those of the set of ②. The ratio of the highest frequency to the lowest frequency is 1.5. ③ and ④, ⑤ and ⑥, ⑦ and ⑧ are in the same manner.

The experimental results are shown in Table 2. x means that the extraction of the missing fundamental is impossible. For every set, the value of MPexceptforMFP {the maximum peak except for the missing fundamental peaks} is smaller in (b), on having no concern with the height of the pitch, than in (a). It means that the extraction of the missing fundamental is more easy in (b)(complex tones inputs) than in (a)(pure tones inputs).

B. The experiment of the influence of the number of harmonic components on the extraction of the missing fundamental (Experiment 2)

The component frequencies sets for experiment 2 are shown in Table 3. The experimental results are shown in Table 4 for (a) and Table 5 for (b).

In the case that f_0 is in the range of low frequency (less than 4 components in ① and ②), the perception of the missing fundamental becomes easy not only in (a) but also in (b) according as increasing the number of harmonic components. In the case that f_0 is in the range of high frequency (less than 4 components in ③,④,⑤, and ⑥), the perception of the missing fundamental becomes easy only in (b) according as increasing the number of harmonic components.

Table 3 Component frequencies sets for the experiments of the influence of the number of harmonic components (for the experiment 2)

No.	* / f_0 [Hz]		2 components	3 components	4 components	5 components	6 components
①	150	450	600	750	900	1050	1200
②	225	450	675	900	1125	1350	1575
③	500	1500	2000	2500	3000	3500	4000
④	750	1500	2250	3000	3750	4500	5250
⑤	700	2100	2800	3500	4200	4900	5600
⑥	1050	2100	3150	4200	5250	6300	7350
⑦	1000	3000	4000	5000	6000	7000	8000
⑧	1500	3000	4500	6000	7500	9000	10500

* harmonic components
[Hz]

The reason has been made clear by the consideration using cochlear models. i.e. complex tones give the auditory system the information of the missing fundamental and the information of combination tones $\{(f_2 - f_1), (2f_1 - f_2), etc.\}$, where $(f_2 - f_1)$ equals f_0, and $(2f_1 - f_2)$ equals f_0, etc. The extraction of the missing fundamental f_0 becomes more easily for complex tones, by the information from combination tones, than for pure tones (the figure of the information is abbreviated).

The investigation of the influence of multiples of f_0 (for example, comparing one complex tone (in $f_1=2f_0$, $f_2=3f_0$., and $f_3=5f_0$.) with another complex tone (in $f_1=2f_0$, $f_2=4f_0$., and $f_3=5f_0$.)) is future works.

IV. CONCLUSIONS

The missing fundamental f_0 has been extracted more easily for complex tones in cochlear model experiments than for pure tones. The reason has been made clear by the consideration using cochlear models i.e. the extraction of the missing fundamental f_0 becomes more easily for complex tones by the information from combination tones than for pure tones.

The research results of the mechanism generating the missing fundamental can be applied to the followings. They are the research of forming neural networks for perception in the auditory system, improving cochlear implant, realizing electronic watermark, etc.

ACKNOWLEDGMENT

I am pleased to acknowledge the considerable assistance of Ph.D. Joseph S. Perkell (RLE at MIT).

REFERENCES

1. Greenberg S, Rhode W S (1987) Auditory processing of complex sound. In:Yost W A, Watson C S, editors. Lawrence Erblaum Associates pp225-236.
2. Matsuoka T, Ono Y (1998) Phase-locking by integral pulse frequency modulation and information of missing fundamental in pulse trains. 20th Annual Int Conf IEEE in MBS Proc.,Vol. 20, No. 6, Hong Kong, 1998, pp 3184-3187
3. Matsuoka T, Iitomi Y (2010) The perception of the missing fundamental of complex tones, BIOSIGNAL2010 Proc., Brno, Czech Republic, 2010, p15

Use macro [author address] to enter the address of the corresponding author:

Institute: MIT RLE rome 36-591
Street: 77 Massachusetts Avenue
City: Cambridge
Country: USA
Email: matsuokat@live.jp

Table 4 The value of the maximum peak except for the missing fundamental peaks in the experiment 2 (a)

	f_0 [Hz]	Number of Harmonic components				
		2	3	4	5	6
①	150	0.84	0.77	0.74	0.75	0.76
②	225	0.74	0.77	0.76	0.79	0.81
③	500	0.986	0.986	0.99	0.991	0.992
④	750	0.992	0.995	0.996	0.996	0.996
⑤	700	0.997	0.998	0.998	0.998	0.998
⑥	1050	0.994	0.995	0.996	0.997	0.997
⑦	1000	×	×	×	×	×
⑧	1500	×	×	×	×	×

Table 5 The value of the maximum peak except for the missing fundamental peaks in the experiment 2 (b)

No	f_0 [Hz]	Number of Harmonic Components				
		2	3	4	5	6
①	150	0.85	0.76	0.72	0.72	0.73
②	225	0.77	0.77	0.73	0.75	0.75
③	500	0.986	0.971	0.964	0.958	0.96
④	750	0.986	0.989	0.979	0.981	0.983
⑤	700	0.991	0.987	0.984	0.988	0.987
⑥	1050	0.997	0.993	0.991	0.991	0.991
⑦	1000	0.997	0.995	0.993	0.993	0.993
⑧	1500	0.997	×	×	×	×

An Influence of Multiple Affecting Factors on Characteristic Ratios of Oscillometric Blood Pressure Measurement

J. Talts, R. Raamat, K. Jagomägi, and J. Kivastik

Department of Physiology, University of Tartu, Tartu, Estonia

Abstract— **We studied how interactions between the arterial pressure pulse and mechanical characteristics of the arterial wall and occluding cuff can modulate the characteristic ratios used for oscillometric estimation of systolic and diastolic blood pressures (k_{syst} and k_{diast}, respectively). Using an integrated artery–cuff pressure/volume model with different arterial pressure pulses as input signals we obtained the oscillation envelopes and calculated characteristic ratios. For the tested range of affecting factors, k_{syst} varied from 0.41 to 0.81 and k_{diast} from 0.56 to 0.90. This gives evidence that oscillometric estimation may lead to substantial inaccuracies if fixed characteristic ratios are used. Errors can be reduced by considering changes in the pulse pressure amplitude and in the symmetry index of the artery-cuff pressure/volume relationship.**

Keywords— **Oscillometric blood pressure, characteristic ratio, accuracy of measurement, modelling.**

I. INTRODUCTION

Latest practice guidelines of the European Society of Hypertension [1] recommend oscillometric monitoring as a useful adjunct to conventional office blood pressure (BP) measurement. The accuracy of measurements and ability of physicians to interpret results have been noted as the key issues of this methodology.

Oscillometric BP is typically determined from the envelope of successive oscillometric pulse amplitudes obtained from the occlusive cuff during its inflation or deflation. The highest point of the envelope curve is generally regarded as the mean arterial pressure (MAP) [2]. A basic criterion which has been used to estimate systolic and diastolic blood pressures (SBP and DBP, respectively) is the height-based methodology [3]. In the height-based approach the systolic and diastolic pressures are determined using special fractions of the maximum oscillation amplitude [4]. These fractions are known as characteristic ratios (k_{syst} and k_{diast}, respectively).

A large variability in the results of oscillometric measurements is quite typical and can be explained by a number of different factors able to affect oscillometric estimation. Theoretical and experimental work has demonstrated that among the factors which can influence the oscillometric BP measurement are pulse pressure (PP), shape of the arterial pressure/volume (P/V) relationship, heart rate, cuff size and compliance [5–8]. Changes in these factors can have significant effects on the characteristic ratios and, subsequently, on the accuracy of estimation of SBP and DBP on the basis of fixed ratios.

The aim of this paper is to investigate by means of computer simulation how the interaction of the arterial pressure pulse with mechanical characteristics of the arterial wall and occluding cuff, can affect the values of characteristic ratios used for oscillometric estimation of SBP and DBP.

II. METHODS

In the present study, we used an asymmetric arctangent model of the arterial P/V relationship, which was initially used by us to model the finger arterial wall mechanics. This model was identified by photoplethysmographic recording of pulsations from finger arteries [9–10]. Considering the fact that in conventional oscillometric monitors, volume oscillations are picked up from the occluding cuff rather than directly from the tissue, we modified the model to consider also cuff-related parameters (artery-to-cuff signal transfer and cuff mechanics). Subsequently, we applied an integrated artery–cuff P/V model with arterial pressure pulses as an input signal and cuff volume oscillations (like those recorded by a conventional oscillometric device) as an output signal. There was a possibility to modify the shape of the P/V relationship. Thereafter, the oscillation envelopes were drawn and characteristic ratios k_{syst} and k_{diast} calculated to fit the oscillometrically estimated SBP and DBP values with corresponding values of the input pressure pulses.

The amplitude of the arterial pressure pulse is known as pulse pressure (PP), defined as PP=SBP–DBP. Transmural pressure (TP) can be calculated from intra-arterial pressure (IP) and cuff pressure (CP) as TP=IP–CP.

The artery–cuff P/V relationship determines how arterial pressure pulses are converted into cuff volume pulses. We characterised the shape of the used P/V relationship by two indices. Definitions of these indices are given in Fig.1.

First, similarly to arctangent steepness in Langewouters model [11] we used the steepness index $P_{0.5}$ (known as a half-maximum width of the pressure/compliance curve), indicating the rate of decrease of the maximum compliance C_{max} to its half value. It is generally known that a stiff artery

K. Dremstrup, S. Rees, M.Ø. Jensen (Eds.): 15th NBC on Biomedical Engineering & Medical Physics, IFMBE Proceedings 34, pp. 73–76, 2011.
www.springerlink.com

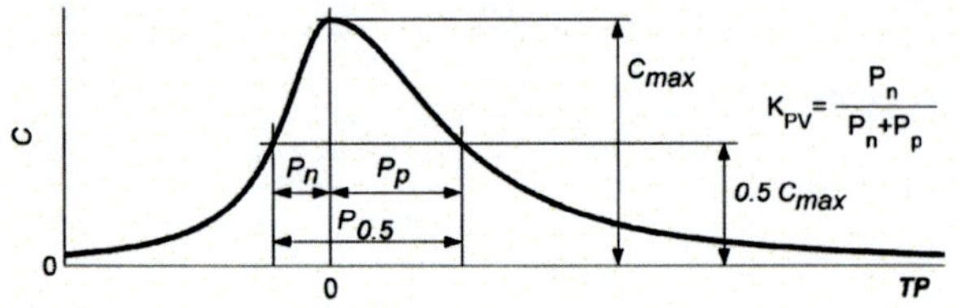

Fig. 1 Pressure/compliance curve, derivative of the asymmetric arctangent model of the artery–cuff P/V relationship

shows a lower steepness of the P/V relationship and, subsequently, a broader compliance curve than a flexible artery.

Second, we used the symmetry index K_{PV} to assess the asymmetry of the arterial wall relationship. This index characterises the relation between negative and positive transmural pressure areas (left and right sides) of the pressure–compliance curve. It has been noticed that the symmetry index K_{PV} tends to increase when stiffness of the artery increases.

The artery–cuff P/V model was identified from in vivo measurements in normal subjects by simultaneously recording volume pulsations from the appropriately sized upper arm cuff and pressure pulsations from the radial artery using the Millar tonometer.

We set the following values of pulse pressure: PP=30, 60 and 90 mmHg. To simulate alterations in the artery–cuff P/V relationship, the following values of model parameters were chosen: K_{PV}=0.35 and 0.50; $P_{0.5}$=30 and 50 mmHg. We considered a brachial artery with K_{PV}=0.35 and $P_{0.5}$=30 mmHg to represent a normal artery; that with K_{PV}=0.5 and/or $P_{0.5}$=50 mmHg was assumed to refer to an arterial wall stiffening.

III. Results

Induced modulations in k_{syst} and k_{diast} are listed in Table 1. For the tested range of affecting factors, k_{syst} ranged from 0.41 to 0.81 and k_{diast} from 0.56 to 0.90. Fig.2 and Fig.3 illustrate two example combinations of pulse pressure and artery-cuff properties leading to shown shapes of volume oscillations.

Table 1 Modulation in k_{syst} and k_{diast} induced by affecting factors

PP mmHg	K_{PV}	$P_{0.5}$ mmHg	K_{syst}	K_{diast}
30	0.35	30	0.55	0.82
30	0.35	50	0.68	0.90
30	0.50	30	0.70	0.70
30	0.50	50	0.81	0.81
60	0.35	30	0.44	0.74
60	0.35	50	0.51	0.80
60	0.50	30	0.60	0.60
60	0.50	50	0.67	0.67
90	0.35	30	0.41	0.71
90	0.35	50	0.45	0.75
90	0.50	30	0.56	0.56
90	0.50	50	0.61	0.61
		min	0.41	0.56
		max	0.81	0.90

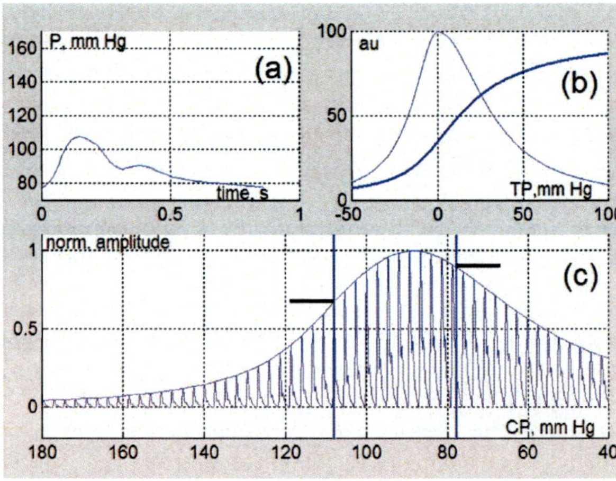

Fig. 2 Arterial pressure P (a), P-V (bold) and P-C (thin) relationships (b) and volume oscillations (c) in the case of factors PP=30 mm Hg, KPV=0.35, P05=50 mm Hg; row #2 in Table 1, leading to greater values of ksyst and k_{diast}

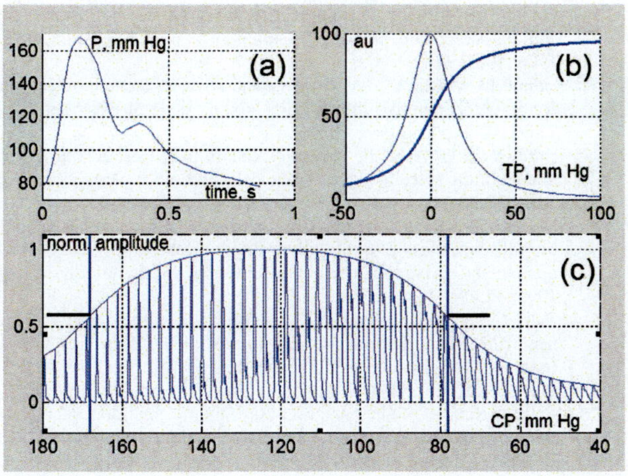

Fig. 3 Arterial pressure P (a), P-V (bold) and P-C (thin) relationships (b) and volume oscillations (c) in the case of factors PP=90 mm Hg, K_{PV}=0.50, P_{05}=30 mm Hg; row #11 in Table 1, leading to smaller values of k_{syst} and k_{diast}

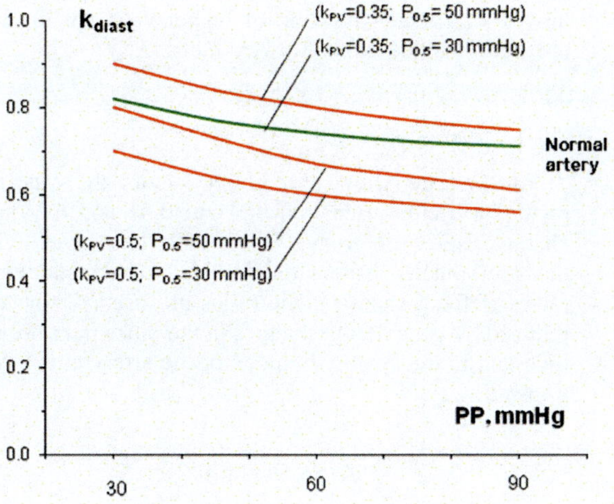

Fig. 5 k_{diast} as a function of pulse pressure PP when several combinations of affecting factors K_{PV} and $P_{0.5}$ are applied

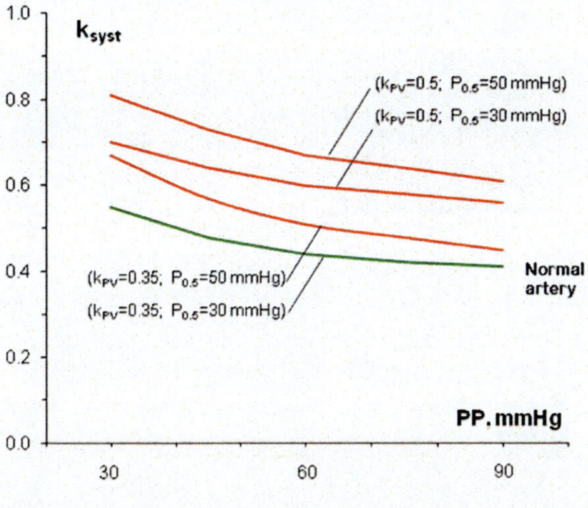

Fig. 4 k_{syst} as a function of pulse pressure PP when several combinations of affecting factors K_{PV} and $P_{0.5}$ are applied

Fig.4 demonstrates changes of k_{syst} to the pulse pressure increase from 30 to 90 mmHg when different combinations of affecting factors are applied. Fig.5 is an analogous illustration of changes of k_{diast}.

IV. Discussion

Our simulation clearly demonstrated that both characteristic ratios k_{syst} and k_{diast} were significantly influenced by the affecting factors examined in this study.

One can see that there was a general trend for k_{syst} as well as for k_{diast} to decrease when PP increased (Fig.4 and Fig.5). However, the rate and character of the fall depend on the interaction of factors K_{PV} and $P_{0.5}$. For a normal artery with K_{PV}=0.35 and $P_{0.5}$=30 mmHg, k_{syst} varies from 0.55 to 0.41 and k_{diast} from 0.82 to 0.71 if PP increases from 30 to 90 mmHg. For a stiff artery with K_{PV}=0.50 and $P_{0.5}$=50 mmHg, corresponding changes are larger (from 0.81 to 0.61 for k_{syst} and also for k_{diast}). The condition K_{PV}=0.50 means a symmetrical P/C curve and in this case constants k_{syst} and k_{diast} will be equal to each other. The difference between constants is a specific measure of the symmetry. In general, obtained changes in k_{syst} and k_{diast} are comparable with results of other simulation studies [5–6, 12].

As pointed out in the Methods section, the symmetry index K_{PV} and the steepness index $P_{0.5}$ can contain also information on the cuff-related parameters (artery-to-cuff transfer and cuff mechanics). This means that cuff fitting and cuff compliance can modify the values of the abovementioned indices. Cuff compliance has not proved to significantly affect the results of BP measurement [6–7]. In contrast, inappropriate cuff dimensions may lead to significant measurement errors between subjects [13]. Undercuffing gives rise to K_{PV}, while overcuffing brings K_{PV} to decrease. As it can be seen in Fig.2 and 3, an increase in K_{PV} moves k_{syst} in the direction of higher values while k_{diast} is moved in the direction of lower values.

V. CONCLUSIONS

For the tested range of affecting factors, the systolic characteristic ratio varied from 0.41 to 0.81 and the diastolic characteristic ratio from 0.56 to 0.90. This gives evidence that oscillometric estimation may lead to substantial inaccuracies if fixed characteristic ratios are used. Errors can be reduced by considering changes in the pulse pressure amplitude and in the symmetry index of the artery-cuff P/V relationship.

ACKNOWLEDGMENT

This study was supported by Estonian Science Foundation (grants 6947 and 7723) and Estonian Ministry of Education and Research (SF0180148s08).

REFERENCES

1. Parati G, Stergiou GS, Asmar R, Bilo G, de Leeuw P, Imai Y et al. European Society of Hypertension Practice Guidelines for home blood pressure monitoring. J Hum Hypertens 2010; 24:779–785
2. Mauck G, Smith C, Geddes L, Bourland J. The meaning of the point of maximum oscillations in cuff pressure in the indirect measurement of blood pressure – Part II. J Biomech Eng 1980;102:28–33
3. Ng K, Small C. Survey of automated non-invasive blood pressure monitors. J Clin Eng 1994;19:452–475
4. Geddes L, Voelz M, Combs C, Reiner D, Babbs C. Characterization of the oscillometric method for measuring indirect blood pressure. Ann Biomed Eng 1982;10:271–280
5. Forster FK, Turney D. Oscillometric determination of diastolic, mean and systolic blood pressure - a numerical model. J Biomech Eng 1986; 108:359–364
6. Ursino M, Cristalli C. A mathematical study of some biomechanical factors affecting the oscillometric blood pressure measurement. IEEE Trans Biomed Eng 1996; 43:761–778
7. Drzewiecki G, Hood R, Apple H. Theory of the oscillometric maximum and the systolic and diastolic detection ratios. Ann Biomed Eng 1994; 22:88–96
8. Raamat R, Talts J, Jagomägi K, Länsimies E. Mathematical modelling of non-invasive oscillometric finger mean blood pressure measurement by maximum oscillation criterion. Med Biol Eng Comput 1999; 37:784–788
9. Talts J, Raamat R, Jagomägi K. Identification of the dynamic component of the finger arterial pressure-volume relationship. In: Ursino, M (editor). Modelling in medicine and biology VI. Southampton: WIT Press; 2005. pp. 153–160
10. Talts J, Raamat R, Jagomägi K. Asymmetric time-dependent model for the dynamic finger arterial pressure-volume relationship. Med Biol Eng Comput 2006; 44:829–834
11. Langewouters GJ, Wesseling KH, Goedhard WJ. The static elastic properties of 45 human thoracic and 20 abdominal aortas in vitro and the parameters of a new model. J Biomech 1984; 17:425–435
12. Zheng D, Amoore JN, Mieke S, Smith FE, King ST, Murray A. Automated blood pressure measurement: Reasons for measurement variability uncovered. Comput Cardiol 2009; 36:21–24
13. O'Brien E. Review: a century of confusion; which bladder for accurate blood pressure measurement? J Hum Hypertens 1996; 10:565–572

Author: Jaak Talts
Institute: Department of Physiology
Street: Ravila 19
City: Tartu
Country: Estonia
Email: jaak.talts@ut.ee

Examples of Vector Velocity Imaging

Peter M. Hansen[1], Mads M. Pedersen[1], Kristoffer L. Hansen[1], Michael B. Nielsen[1], and Jørgen A. Jensen[2]

[1] Department of Radiology, Section of Ultrasound, Rigshospitalet, Copenhagen, Denmark
[2] Center for Fast Ultrasound Imaging, Dept. of Elec. Eng., Technical University of Denmark, Lyngby, Denmark

Abstract— **To measure blood flow velocity in vessels with conventional ultrasound, the velocity is estimated along the direction of the emitted ultrasound wave. It is therefore impossible to obtain accurate information on blood flow velocity and direction, when the angle between blood flow and ultrasound wave approaches 90°. The majority of the vessels in the human body is parallel to the surface and therefore positioned in a way that prevents proper placement and angulation of the transducer, when the velocity and direction of blood flow is to be estimated. Different techniques to circumvent this problem have been tried including Transverse Oscillation. This method has been tested in computer simulations, on flow phantoms and in-vivo, and subsequently validated against MRI angiography. Transverse Oscillation is now implemented in a commercial ultrasound scanner from BK Medical (UltraView). In this article UltraView is demonstrated on the carotid artery, jugular vein and femoral vein that all runs almost parallel to the skin and thus is angled near 90° to the ultrasound waves. Arterial and venous simple and complex flow with formation of vortices is demonstrated by scanning on the longitudinal axis with a 90° angle on the vessel. Moreover secondary flow in the abdominal aorta is illustrated by scanning on the transversal axis.**

Keywords— **Transverse oscillation, vector velocity, blood flow, velocity estimation.**

I. INTRODUCTION

Using conventional ultrasound the blood velocity is measured using the shift in position between pulse emissions [1]. This results in a change in frequency that occurs when either the source of the ultrasound wave or the reflector moves. In this case, the reflector is blood cells flowing through the vessel. The relation between the blood flow velocity in the direction of the ultrasound wave v_z, i.e. the axial velocity, and the true velocity v of the blood flow through the vessel is given by

$$v = \frac{v_z}{\cos\theta} = \frac{f_d c}{2 f_t \cos\theta} \tag{1}$$

where f_t is the frequency of the emitted ultrasound wave, f_d is the received frequency, c is the ultrasound velocity and θ is the angle between the ultrasound wave and the direction of the blood flow. From the equation it can be determined that the validity of the result deteriorates as θ approaches 90°, because cos θ approaches zero. A small error in measurement will therefore result in a larger error of estimate, and one should, for that reason, keep the angle at less than 60-70° [1] To achieve a valid result it is, thus, necessary in most cases to either angle the transducer manually or angle the emitted ultrasound waves electronically in relation to the targeted blood vessel. With ordinary color flow imaging it is possible to get an indication of whether the blood moves toward or away from the transducer, without any indication of velocity, and thus it is only a qualitative measurement. By using Spectral velocity estimation, where the operator corrects the angle based on the presumed direction of the blood flow from the B-mode image, it is possible to obtain a quantitative estimate of the velocity. However, it is only possible to obtain an estimate of the velocity in a small area of the vessel at a time, and the measurements are based on an assumption that the flow is laminar. It is therefore extremely difficult to apply the technique on blood vessels with complex flow, such as at bifurcations, junctions and where the blood passes valves or stenoses of any kind, as this leads to an unpredictable complex flow [2]; and it is in these areas as much information as possible about blood flow is needed, since they have predilection for arteriosclerosis [3-5]. The angulation problem has been tried solved in several ways, e.g. using speckle tracking [6], two ultrasound beams [7], plane wave excitation [8,9] and Transverse Oscillation [10]. The latter is the method presented in this article.

II. MATERIALS AND METHODS

The scans were performed on a voluntary, healthy, male subject at age 34 by an experienced radiologist. The study was performed after approval by The Danish National Committee on Biomedical Research Ethics.

Transverse Oscillation is a technique proposed by Jensen and Munk [10] and a similar method has been suggested by Anderson [11]. The conventional estimator is used to find the velocity in the axial plane v_z. By subsequently manipulating the receive apodization of the

K. Dremstrup, S. Rees, M.Ø. Jensen (Eds.): 15th NBC on Biomedical Engineering & Medical Physics, IFMBE Proceedings 34, pp. 77–80, 2011.
www.springerlink.com

transducer the transverse velocity component v_x is found using the same echo that gave v_z. From these two values the correct velocity and direction is calculated as a vector. The technique is the first of its kind to be implemented in a conventional scanner (Pro Focus 2202 UltraView, BK Medical), using a conventional Linear Array Transducer (8670, BK Medical). The only change to conventional ultrasound is the manipulation of the apodization when the echo is received. The technique has previously been tested in computer simulations and with flow-phantoms by Jensen and Udesen [12,13] and in vivo by Udesen et al [14]. In addition, Hansen et al., on the experimental scanner RASMUS, conducted in-vivo validation of Transverse Oscillation against MRI angiography and showed a strong correlation (R = 0.91) between the two modalities [15]. On UltraView, blood flow direction and velocity is illustrated with color-coded pixels. It is then possible to indicate the direction of flow with arrows, which is superimposed in real-time on these color-coded pixels.

To illustrate simple, complex, and secondary flow, scans of the common carotid artery, jugular vein, carotid bulb, femoral vein, and the abdominal aorta have been conducted. The secondary flow is the phenomenon that blood flowing antegrade through an artery, simultaneously revolves on an axis centrally in the vessel. The exact cause and reason for this is unknown. This phenomenon is known from MRI scans [16,17] and experimental off line ultrasound scanning using Plane Wave Excitation [18], but has not previously been shown in real time using a conventional ultrasound scanner.

III. Results

During one session a series of scans were performed on the volunteer using UltraView. The following five figures were selected from the recorded sequences to illustrate some of the possibilities using Transverse Oscillation. The scans targeted the carotid artery, the jugular and femoral vein and the abdominal aorta. All figures appear as they do on the scanner and there have been no offline processing. Please note the angle of insonation.

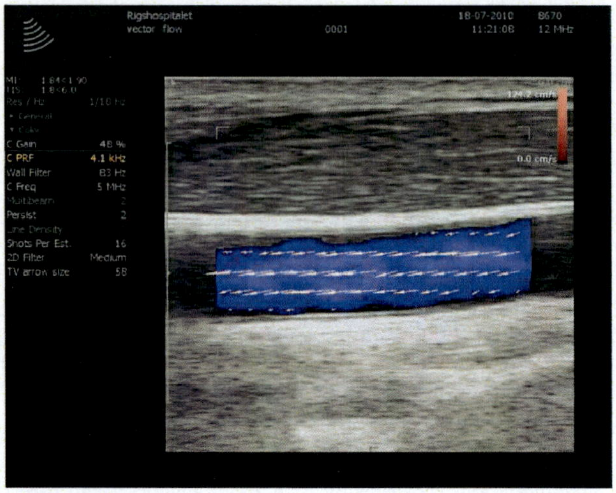

Fig. 1. Scan on the longitudinal axis of the common carotid artery illustrating simple blood flow.

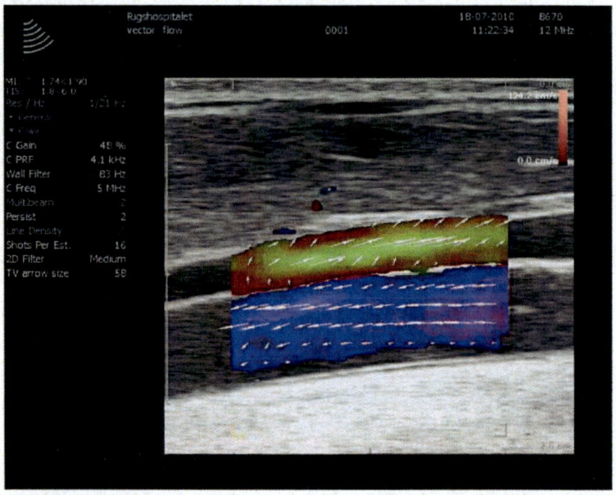

Fig. 2. Scan on the longitudinal axis of the common carotid artery and the jugular vein illustrating simple bidirectional blood flow.

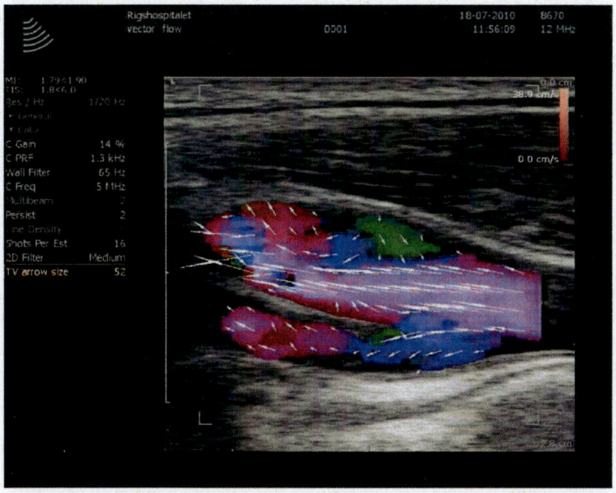

Fig. 3. Scan on the longitudinal axis of the carotid bifurcation showing the carotid bulb. The carotid bulb is a dilation of the vessel containing baroreceptors involved in regulating blood pressure. Because of the changing geometry in the vessel a vortex is formed in the carotid bulb. It appears at the top of the figure, where the green pixels illustrate retrograde flow.

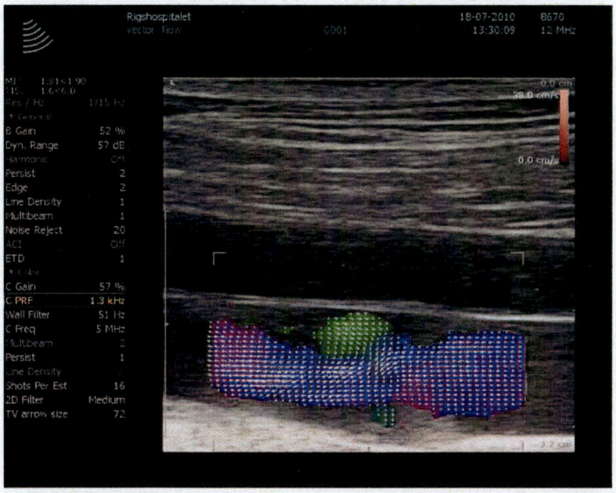

Fig. 4. Scan on the longitudinal axis of the femoral vein with disturbed flow at the passage of a venous valve. A vortex is formed in the pocket behind the valve.

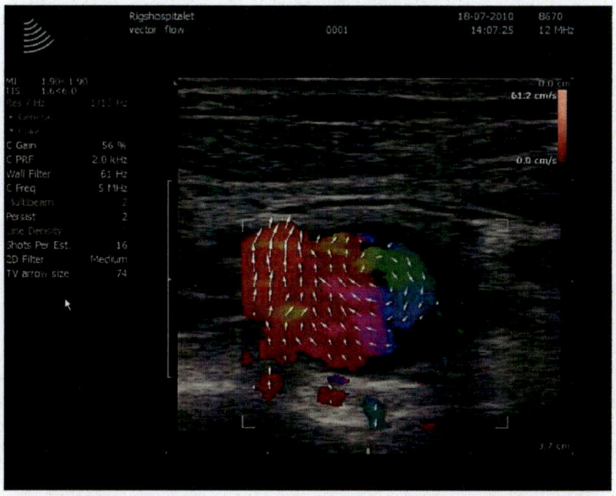

Fig. 5. Scan on the transversal axis of the abdominal aorta with illustration of the secondary flow.

IV. DISCUSSION

With UltraView it is for the first time possible to obtain a real-time angle independent estimate of the blood flow direction as a directly readable parameter. With Transverse Oscillation estimation of blood flow velocity becomes operator independent, since there is no need to do angle correction. Studies with Spectral velocity estimation have revealed that e.g. the manual angle correction causes significant error in estimating the blood flow at maximum velocity. Hoskins found 10-100% error, assuming only laminar flow is present [19]. With knowledge of the secondary flow, we know that laminar flow, if it exists, only presents itself in short sections of the pulse cycle. This further undermines the validity of the spectral method to estimate velocity. Transverse Oscillation can provide new information about the movement of the blood that will improve the diagnostic and prognostic studies of e.g. arteriosclerotic blood vessels, stenotic or insufficient heart valves and allow for assessment of perfusion grade of organs. This could be used to assess whether a transplanted organ will be accepted, or to assess the efficacy of tumor therapy. Since the angle independent estimate with Transverse Oscillation is calculated within the entire scan area, it is easy to simultaneously gather data from multiple points around an interesting area in the vessel, e.g. a stenosis, and show the flow velocity as a spectrogram. The technique has, as previously mentioned, been validated against MRI angiography, but several studies have pointed out limitations in the accuracy of velocity estimates obtained with MRI. An additional problem is that the obtained estimate is an average of several heartbeats, which masks inconsistencies linked to the individual heart beat [17,20]. The Transverse Oscillation technique, as applied in 2D in UltraView, does not take the secondary flow during scanning on the longitudinal axis into account. However, at Technical University of Denmark, Center for Fast Ultrasound Imaging a 3D vector flow technique is currently under development implemented on the experimental scanner SARUS [21].

V. Conclusion

An angle independent ultrasound technique to estimate the direction and velocity of blood flow is now implemented in a conventional scanner, making it suitable for use on patients. With UltraView, it is possible to illustrate the same flow phenomena real-time in-vivo, which was previously only found offline with the experimental scanner RASMUS. The scanner is the first of its kind, which naturally leads to some limitations that surely will be resolved in the near future. The earlier studies and this demonstration give reason to hope that Transverse Oscillation can provide new, clinically useful information about blood flow in both healthy and sick individuals. However, further development of the technique as well as further studies with different transducers, scan depths, flow types and with different angles of insonation are needed.

References

1. Jensen JA. (1996) Estimation of blood velocities using ultrasound: A signal processing approach. New York: Cambridge University Press
2. Hoskins PR (1997) Peak velocity estimation in arterial stenosis models using colour vector Doppler. Ultrasound Med Biol 23(6):889-897
3. Birchall D, Zaman A, Hacker J, Davies G, Mendelow D (2006) Analysis of haemodynamic disturbance in the atherosclerotic carotid artery using computational fluid dynamics. Eur Radiol 16(5):1074-1083
4. Cheng C, Tempel D, van Haperen R, van der Baan A, Grosveld F, Daemen MJ, Krams R, de Crom R (2006) Atherosclerotic lesion size and vulnerability are determined by patterns of fluid shear stress. Circulation 113(23):2744-2753
5. Richter Y, Edelman ER (2006) Cardiology is flow. Circulation 113(23):2679-2682
6. Trahey GE, Allison JW, von Ramm OT (1987) Angle independent ultrasonic detection of blood flow. IEEE Trans Biomed Eng 34(12):965-967
7. Fox MD (1978) Multiple crossed-beam ultrasound Doppler velocimetry. IEEE Trans Son Ultrason 25(5):281-286
8. Udesen J, Gran F, Hansen KL, Jensen JA, Thomsen C, Nielsen MB (2008) High frame-rate blood vector velocity imaging using plane waves: simulations and preliminary experiments. IEEE Trans Ultrason Ferroelec Freq Contr 55(8):1729-1743
9. Udesen J, Gran F, Hansen KL, Jensen JA, Nielsen MB (2007) Fast blood vector velocity imaging: Simulations and preliminary in-vivo results. IEEE Ultrasonics Symp:1005-1008
10. Jensen JA, Munk P (1998) A new method for estimation of velocity vectors. IEEE Trans Ultrason Ferroelec Freq Contr 45:837-851
11. Anderson ME (1998) Multi-Dimensional Velocity Estimation with Ultrasound using Spatial Quadrature. IEEE Trans Ultrason Ferroelec Freq Contr 45:852-861
12. Udesen J, Jensen JA (2003) Experimental investigation of transverse flow estimation using transverse oscillation. Proc IEEE Ultrasonics Symp:1586-1589
13. Udesen J, Jensen JA (2006) Investigation of transverse oscillation method. IEEE Trans Ultrason. Ferroelec Freq Contr 53:959-971
14. Udesen J, Nielsen MB, Nielsen KR, Jensen JA (2007) Examples of In-vivo Blood Vector Velocity Estimation. Ultrasound Med Biol 33(4):541-548
15. Hansen KL, Udesen J, Thomsen C, Jensen JA, Nielsen MB (2009) In-vivo Validation of a Blood Vector Velocity Estimator with MR Angiography. IEEE Trans Ultrason Ferroelec Freq Contr 56(1):91-100
16. Lee KL, Doorly DJ, Firmin DN (2006) Numerical simulations of phase contrast velocity mapping of complex flows in an anatomically realistic bypass graft geometry. Med phys 33(7):2621-2631
17. Steinman DA, Thomas JB, Ladak HM, Milner JS, Rutt BK, Spence JD (2002) Reconstruction of carotid bifurcation hemodynamics and wall thickness using computational fluid dynamics and MRI. Magnet Reson Med 47(1):149-159
18. Hansen KL, Udesen J, Gran F, Jensen JA, Nielsen MB (2009) In-vivo Examples of Flow Patterns With The Fast Vector Velocity Ultrasound Method..Ultraschall in Med 30:471-477
19. Hoskins PR (1999) A review of the measurement of blood velocity and related quantities using Doppler ultrasound. Proc Inst Mech Eng 213(5):391-400
20. Marshall I, Papathanasopoulou P, Wartolowska K (2004) Carotid flow rates and flow division at the bifurcation in healthy volunteers. Physiol Meas 25(3):691-697
21. Jensen JA, Hansen M, Tomov BG, Nikolov S, Holten-Lund H (2007) System architecture of an experimental synthetic aperture Real-time ultrasound system. IEEE Ultrason Symp:636-640

Author: Peter Møller Hansen
Institute: Department of Radiology, Rigshospitalet
Street: Blegdamsvej 9
City: 2100 Copenhagen Ø
Country: Denmark
Email: pdmhansen@gmail.com

Analysis of the Auditory Perception of Ultrasound Doppler Signals to Improve Pregnancy Risk Assessment

M. Ewerlöf[1], A. Thuring[2], K. Maršál[2], and T. Jansson[1]

[1] Department of Electrical Measurements, Lund University, Lund, Sweden
[2] Department of Obstetrics and Gynecology, Lund University, Lund, Sweden

Abstract— **In high-risk pregnancies, the transport of oxygen and nutrients from maternal to fetal blood via the placenta is often impaired. To assess the risk, pulsed Doppler ultrasound (US) is used to evaluate the flow velocity waveform in the umbilical artery with the pulsatility index (PI), which is derived from the velocity envelope of the Doppler power spectrum. However, simply listening to the Doppler signal can indicate to an experienced sonographer that the type of the blood flow is worse than the PI suggests. This is however dependent on the operator's experience and it may be difficult to estimate what influences the subjective judgement. Motivated by the description of the Doppler sounds by an experienced operator (AT) as having a "timbre", this study describes an analysis of Doppler sounds in search for an index or method with capacity to better evaluate the blood flow in the umbilical artery in high-risk pregnancies.**

A test was designed, where synthetically produced Doppler sounds with various spectral contents were played together with a variable sinusoidal sound signal. The task for the five test persons was to match the frequency of the sinusoidal signal to the Doppler sounds.

The tests indicated that the human ear is most sensitive to the lower frequencies of Doppler sounds. An analysis of pre-recorded sounds showed a difference in the lower frequencies of a sound considered to emanate from the umbilical blood flow of healthy fetuses with normally functioning placenta as compared to a pathological one. This might explain the difference between the sounds experienced by an operator.

As a suggestion to extract more information than the maximum envelope, the minimum frequency envelope was extracted from pre-recorded clinical sounds. Based on the pilot tests presented here, this shows to be a promising strategy.

Keywords— **synthetic Doppler signal, pulsatility index, flow velocity waveform, umbilical artery blood flow**

I. INTRODUCTION

For small for gestational age (SGA) fetuses, Doppler ultrasound is commonly used to evaluate the fetal status. Typically, the flow velocity acquired from the umbilical artery is analyzed, and the so-called pulsatility index is calculated to characterize the blood velocity waveform. It is derived from the maximum envelope of the Doppler spectrum as the difference between the maximum (systolic) and minimum (diastolic) velocity of the envelope during one heart cycle divided by the mean envelope velocity [1].

The pattern of the blood flow in the umbilical artery that is related to pathological changes in the placenta and, consequently, to the status of the fetus, can be characterized using five blood flow classes (BFC). PI within +/- 2 standard deviations (SD) from the population mean defines BFC Normal. When the resistance increases, the velocities in diastole decrease, thus increasing the PI. A PI increase to between +2 to +3 SD defines BFC I, and PI above +3 SD, BFC II. When the resistance to flow in the umbilical cord becomes even higher, the flow velocity during diastole reaches zero and the flow is classified as BFC IIIA. BFC IIIB represents a reversed flow during a part of diastole, which is associated with severe fetal hypoxia, often a fatal condition for the fetus [2]. The principal waveforms of BFC I–BFC IIIB are shown in Fig. 1.

When the PI is calculated using the maximum envelope, only a small part of the information of the Doppler signal is used. An experienced sonographer can however recognize differences between the Doppler audio signals recorded from uncomplicated and pathological cases having identical PI values.

The aim of this study is to investigate whether additional information can be extracted from the Doppler sound.

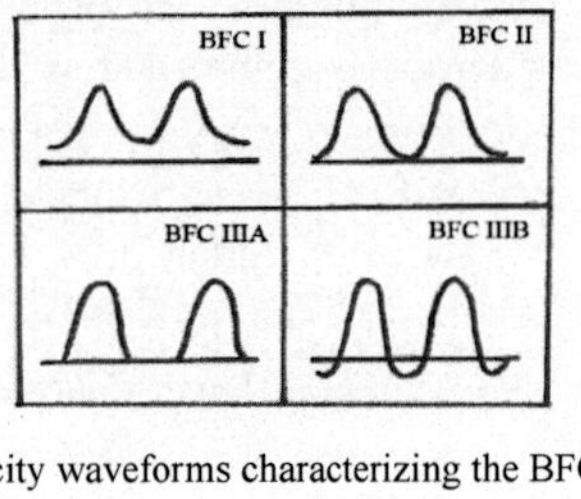

Fig. 1 Flow velocity waveforms characterizing the BFCs.

K. Dremstrup, S. Rees, M.Ø. Jensen (Eds.): 15th NBC on Biomedical Engineering & Medical Physics, IFMBE Proceedings 34, pp. 81–84, 2011.
www.springerlink.com

II. MATERIALS AND METHODS

The investigation is motivated by the experience of the sonographer that it is possible to make a more detailed evaluation of the Doppler sound, even when the PI is virtually the same. Two recordings of Doppler waveforms in BFC I with similar PI (1.41 and 1.53), but perceived as normal and borderline pathological, respectively, were chosen for closer analysis.

A typical Doppler spectrum was identified from each case. The one characterized as more pathological, had a velocity distribution that appeared to be more uniform (as for a parabolic flow). For the normal case there seemed to be a tendency towards a more plugged flow (Fig. 2).

To find an alternative summary measure to the PI, which instead is based on the auditory perception of the Doppler signal, a test was designed where a test person was asked to match a sinusoidal tone to variants of these representative spectra. The idea was that the sonographer would match the single tone to the one frequency perceived as the most dominant in the complex Doppler spectrum.

This study reports the results from tests where synthetic sounds were generated with the two typical spectra as models. The spectral properties of the Doppler sounds were varied to investigate what properties mainly influenced the sonographer's judgement. Based on the results, alternative signal processing strategies were tested to extract additional diagnostic information from the two clinical cases.

A. Tests with synthetic sounds

An algorithm was implemented in MATLAB, to generate synthetic sounds imitating the properties of ultrasound Doppler signals. These were played together with a sinusoidal tone of variable frequency, which could be changed by a test person via a graphical user interface, GUI. The test person was asked to match the sinusoidal signal to the Doppler sound in order to see what frequencies of the sound are considered most important, or dominant.

The synthetic sounds were created by filtering white Gaussian noise. A filter bank, which corresponds to the velocity distribution of the blood flow, is convoluted to the noise [3, 4]. The order and type of the filters needed to construct the synthetic signal were chosen by matching the synthetic spectra to spectra of recorded true sounds.

To imitate the wall-filter, a high-pass FIR filter with order 80 and cut-off frequency 150 Hz was used. To obtain the upper cut-off frequency of the spectrum corresponding to a parabolic flow, a 110th order low-pass FIR filter was applied.

For the plugged flow spectrum, it was necessary to start with the high amplitude peak. To produce the lower frequency slope of the peak, a high-pass IIR Butterworth filter with order 10 was applied to the noise. Then noise with lower amplitude, A, was added to imitate the plateau for the lower frequencies. A low-pass FIR filter with order 50 was used to create the upper frequency slope of the peak. Examples of both synthetic spectra are displayed in Fig. 2.

After the shapes of the spectra had been matched to the ones extracted from clinical recordings, the cut-off frequency, f_0, of the low-pass filters could be changed to simulate different maximum velocities of the blood flow for both types of spectra. For the spectrum corresponding to plugged flow, the width of the peak was changed by varying the order, N, of the Butterworth filter. The amplitude of the plateau with low-frequency components is changed with A.

A test session containing two parabolic (f_0 = 500 and f_0 = 1000) and two plugged (f_0 = 1000 Hz, A = 0.3 for both, and N = 2 and N = 10 respectively) spectra were created. The sounds were played through headphones for five test persons; one experienced sonographer and four test persons with limited or no experience in listening to Doppler sounds from blood flow recordings.

Each sound was played six times in a randomized order with the starting frequency of the sinusoidal set to zero for three of the repetitions, and $F_s/2$ (F_s is the sampling frequency = 6400 Hz) for the remaining three.

To examine the importance of low frequency components, a test session was created where the spectrum of a plugged flow with f_0 = 1000 Hz and N = 10 was used. The peak amplitude of the spectrum was left stationary, while the amplitude of the lower frequency band, A, was varied.

B. Analysis of pre-recorded sounds

To process the data of the recorded sounds in MATLAB, the power density spectrum, PDS, was calculated. Welch's method was used to reduce the amplitude variability of the spectrum: The signal was divided into smaller segments of 2048 samples each, which had an overlap of 50 %. Each segment was windowed by a Hanning window, and then

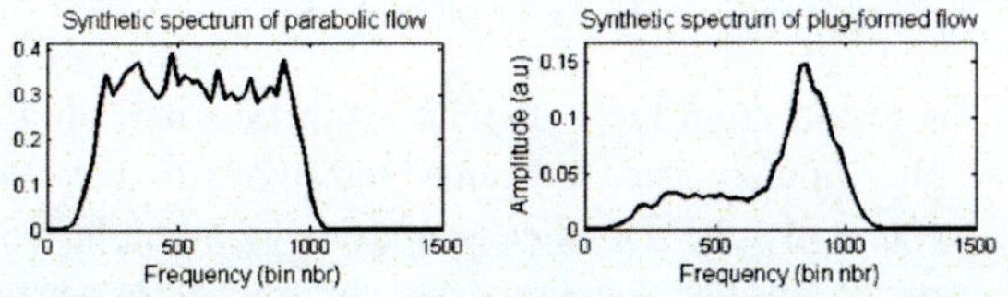

Fig. 2 Synthetic spectra of parabolic (left) and plugged flow (right).

Fourier transformed. Successive spectra were displayed as a sonogram.

The peak velocity, i.e. the maximum detectable frequency from the Fourier transform of each time interval, was found with the modified geometric method, MGM [5].

To appreciate how well variations in the low frequency components can be tracked in clinical recordings, the minimum frequency curve of the recorded sound was extracted. Here, the MGM was also used, but in a slightly different way. The cumulative sum was calculated from the maximum frequency of each segment towards zero. Then the same algorithm as used when calculating the maximum frequency was applied.

III. RESULTS

A. Tests with synthetic sounds

The results from the tests with synthetic sounds showed that the chosen sinusoidal frequencies were gathered in the lower part of the spectra for both high and low cut-off frequencies, Fig. 3.

It appears that for the spectra with a pronounced peak ("plugged flow"), the test persons did not perceive the maximum amplitude frequency as the most dominant (Fig. 4). The sinusoidal signal matched to the filtered noise was chosen to a lower or higher frequency, but not to the frequency with maximum amplitude in the spectrum. Furthermore, when the lower band of the spectrum was flat, the test person tended to more precisely find the lower cut-off frequency of the spectrum.

The test where the amplitude was varied in the lower frequency band showed a large influence also for very small amplitudes (Fig. 5). When the lower frequencies in the spectrum were 40 dB below the main peak, no sinusoidal frequencies were chosen to this frequency interval. At 20 dB, the test person managed to pick out these lower tones beside the main peak, which indicates a large influence from low frequencies on how the sounds are perceived.

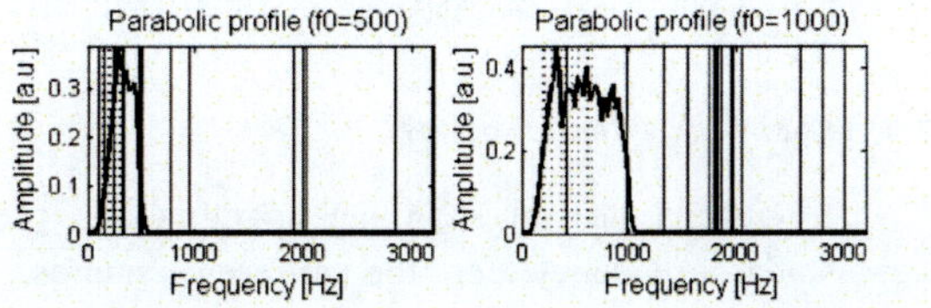

Fig. 3 Results from the test session showing the chosen frequency of the sinusoidal (vertical lines) for all test persons for two sounds with the spectrum of a parabolic flow (bold black line). The type of the lines show whether the start frequency is zero (dotted) or $F_s/2$ (solid).

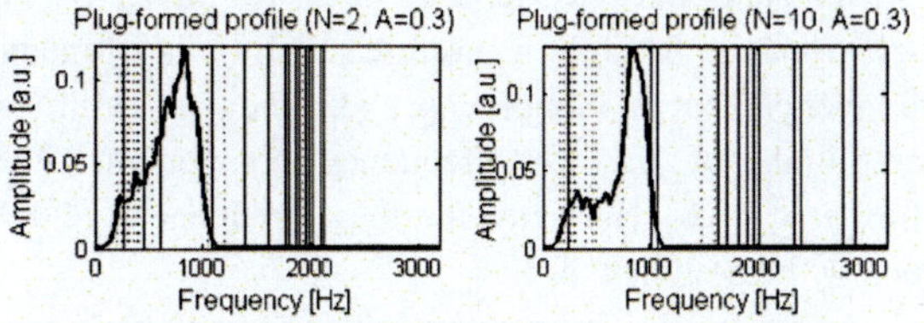

Fig. 4 Results from the test session showing the chosen frequency of the sinusoidal (vertical lines) for all test persons for two sounds with the spectrum of a plugged flow (bold black line). The type of the lines show whether the start frequency is zero (dotted) or $F_s/2$ (solid).

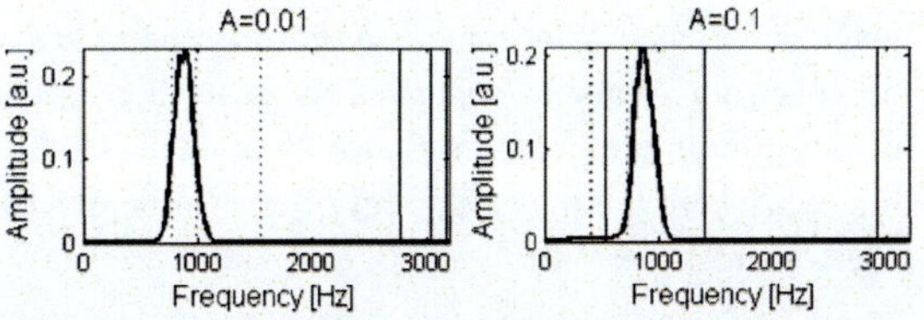

Fig. 5 Results from the test where the low-frequency amplitude, A, was varied. The chosen frequency of the sinusoidal is displayed as vertical lines and the spectrum of each sound as a bold black line. The type of the vertical lines show whether the start frequency is zero (dotted) or $F_s/2$ (solid).

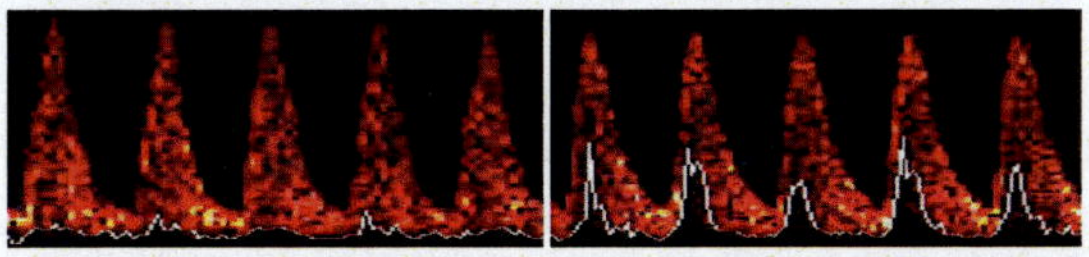

Fig. 6 Minimum frequency curve (white solid line) displayed together with the corresponding PDS for the two sounds; the borderline pathological sound to the left and the more normal to the right.

B. Analysis of pre-recorded sounds

Calculation of the minimum frequency curve for the two sounds showed a distinct difference between the sounds (Fig. 6). The sound graded as more abnormal by the sonographer lacks low-frequency components compared to the normal sound.

IV. DISCUSSION

A. Tests with synthetic sounds

The results do not show an unambiguous preferable sinusoidal frequency for the different sounds, and a characterization of the "timbre", described by the experienced sonographer, is hard to accomplish with this kind of test. Yet, some conclusions can be drawn regarding the sensitivity of the human ear. For instance, the chosen sinusoidal frequencies tend to lie in the lower part of the spectrum, indicating a preference for the fundamental frequencies of a complex sound. The test with varying amplitude of the low frequency components indicates further a high sensitivity for these frequencies, 20-40 dB below the dominant peak.

According to the experienced sonographer, it was hard to find a matching tone to some of the sounds in the test sessions, while in other sessions there were more matching sounds than one. The latter finding may indicate that a harmonic of a fundamental tone is considered a match, which can explain why some chosen frequencies are outside the spectrum frequency range. This is further emphasized by the grouping of chosen sinusoids at 1800-1900 Hz for the tests where the maximum cut-off is just below 1000 Hz (Figs. 3 and 4).

B. Analysis of pre-recorded sounds

The minimum frequency curve showed a considerable difference between the two selected cases. The more normal of the sounds had a distinct "window", while the other showed only a slight variation of the minimum frequency. Analysis of the lower frequencies of a sound could thus provide additional information that, in combination with the PI (based on the maximum envelope) could improve the evaluation of the Doppler signals. Whether this holds true in a larger population falls outside the scope of this study. It might also be that other signal processing strategies are more optimal to track the minimum frequency, such as for instance fundamental frequency analysis [6], but this is also outside the scope of this study.

V. CONCLUSIONS

The results from the tests with synthetic sounds indicate that the human ear is more sensitive to the lower frequencies of the Doppler sounds, which might be an explanation of the sonographer's different experience of Doppler sounds recorded from pathological umbilical artery blood flow as compared to those recorded from pregnancies with normal resistance to flow in the placenta.

Sounds considered more normal have a distinct "window" lacking the lower frequency components. This points on a drawback of the commonly used PI, which is calculated on the maximum frequencies during the heart cycle. An index that takes the entire frequency content of the Doppler sound into account would likely increase the diagnostic value of clinical obstetric Doppler examinations.

ACKNOWLEDGEMENTS

We thank Karin Johansson, Hans Hellsten and Kent Olofsson at Malmö Academy of Music for helpful discussions and valuable input.

REFERENCES

1. Evans D. H., McDicken W. N. (2000) Doppler ultrasound. Wiley
2. P. Malcus, J. Andersson, K. Maršál, P. Å. Olofsson (1991) Waveform pattern recognition—A new semiquantitative method for analysis of fetal aortic and umbilical artery blood flow velocity recorded by doppler ultrasound. Ultrasound Med. Biol. 17:453-460.
3. Kristoffersen K., Angelsen B. A. J. (1988) A time-shared ultrasound Doppler measurement and 2-D imaging system. Trans Biomed Eng. 35:285-295
4. Jensen J. A. (2006) Spectral velocity estimation in ultrasound using sparse data sets. J Acoust Soc Am 120:211-220
5. Moraes R., Aydin N., Evans D. H. (1995) The performance of three maximum frequency envelope detection algorithms for Doppler signals. J Vasc Invest, 1:126-134
6. Doval B., Rodet X. (1993) Fundamental frequency estimation and tracking using maximum likelihood harmonic matching and HMMs. ICASSP-93, IEEE International Conference on Acoustics, Speech, and Signal Processing, 1993, 1:221-224

Author: Tomas Jansson
Institute: Lund University
Street: PO Box 118
City: SE-221 00 LUND
Country: Sweden
Email: tomas.jansson@elmat.lth.se

Phonocardiographic Recordings of First and Second Heart Sound in Determining the Systole/Diastole-Ratio during Exercise Test

S.M.M. Rønved, I. Gjerløv, A. Brokjær, and S.E. Schmidt

Department of Health and Science technology, Aalborg University, Aalborg, Denmark

Abstract— **The objective was to examine whether a microphone can be used to detect first and second heart sound during an exercise test and if these recordings can be used to calculate S/D-ratios in healthy subjects. Furthermore the objective was to describe the changes in systolic and diastolic duration under cardiac stress.**

Nine healthy subjects (5M, 4F) completed a standardized exercise test while wearing a Panasonic microphone incorporated in a specially designed coupler. Recordings of heart sounds were made with Acarix Data Acquisition System at the end of each workload level.

As heart rate increased, the recordings became more difficult to interpret as did detection of first and second heart sounds. Recordings from female subjects were easier to interpret than those of male subjects. As heart rate increased, the S/D-ratio increased accordingly. The development of systole and diastole duration was very similar in male and female subjects.

First and second heart sound can be detected using a microphone, but noise at higher load levels necessitates the development of a noise-reducing filter. Furthermore the duration of systole and diastole was found altered during stress. The systole duration decreased minimally whilst the diastole duration decreased markedly as a function of higher heart rate.

Keywords— **Ischemic heart disease, heart sounds, exercise test, S/D-ratio, microphone.**

I. Introduction

Ischemic heart disease (IHD) is the most commonly occurring heart disease worldwide. The most common cause of IHD is coronary arterio sclerosis [1]. One of the diagnostic methods is an exercise test. Compared to myocardial scintigraphy, heart CT-scanning and Coronary Angiography, exercise test has the lowest cost and amount of side-effects [2]. However it is also the test with the lowest sensitivity [3-5]. Research suggests that an involvement of the ratio of systolic and diastolic duration (S/D-ratio) can increase the diagnostic value of the exercise test [6-8]. S/D-ratios during rest show no significant difference between healthy people and people with IHD, but during pulse elevation differences in S/D-ratios occur as a result of longer systole length in people with IHD than in people without IHD [6]. The extended systole is a result of slower contraction of the ischemic ventricles during systole. In healthy subjects the coronary perfusion increases while the diastole shortens at higher heart rates, owing to vasodilatation of the coronary blood vessels. The ability to vasodilate is compromised in people with severe coronary stenosis, which is often seen in people with IHD. This results in an insufficient blood supply to the myocardium resulting in a decreased contractibility of the ventricles and therefore a slower contraction [6,7].

Recording heart sounds implicate some difficulties due to the patient's movement during the test, the respiration sounds and background noise. These difficulties can influence the precision of the heart sound detection [9,10]. Both accelerometers and microphones have been used for recording heart sounds [8,11,12]. The frequency area of heart sounds is between 20 and 175 Hz, and both an accelerometer and a microphone are able to detect these frequencies [13,14]. The microphones advantage is the ability to detect higher frequencies than the accelerometer. This may be relevant if there is suspicion of coronary stenosis, where the turbulence in the coronary arteries create high frequency murmurs when the subject is lying still [12].

The objective of this study is to examine whether a microphone can be used to detect first (S1) and second (S2) heart sound during an exercise test, and if these recordings can be used to calculate S/D-ratios in healthy subjects. Furthermore this is an exploratory study which describes the changes in systolic and diastolic duration under cardiac stress.

II. Methods

Nine healthy subjects were enrolled in the study (M = 5, F = 4). Median age 32 (24-36). Informed consent was retrieved from all subjects prior to the exercise test. A Panasonic microphone was incorporated in a coupler, specially designed by the Department of Acoustics at Aalborg University, Denmark. The microphone detects the mechanical pressure differences in the coupler, caused by alterations of the sound pressure. The microphone records with a frequency of 48000 Hz. The heart sound recordings are synchronized with a 3-lead ECG in order to distinguish between the first and second heart sound. The microphone was fitted to the 3rd left intercostal space with a specially designed double adhesive plaster. Subsequently the subject cycled on a Monark Ergometric 894E ergometer bicycle. The workload was increased by 25 watt every two minutes with an initial workload of 25 watt. The subject cycled until subjective maximum endurance was reached. Afterward subjects that did not reach 80 % of maximum heart rate defined as 220 minus age ± 12 were excluded from the study [15]. Recordings of heart sounds were made for 10 seconds at the end of

K. Dremstrup, S. Rees, M.Ø. Jensen (Eds.): 15th NBC on Biomedical Engineering & Medical Physics, IFMBE Proceedings 34, pp. 85–88, 2011.
www.springerlink.com

each workload level. Acarix Data Acquisition System was used for recording the heart sounds and ECG.

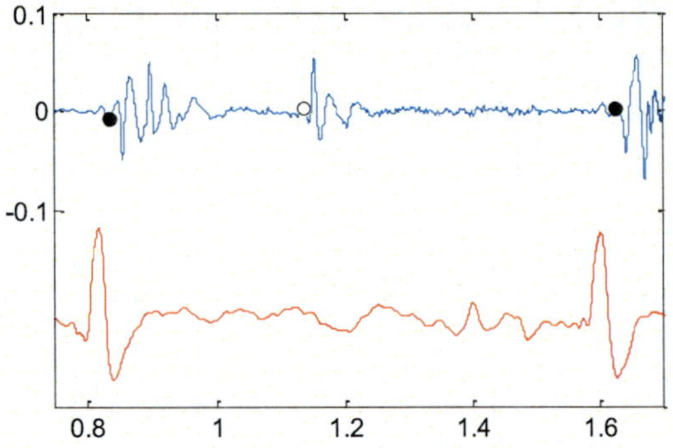

Figure 1 From HSPoints. Illustrates the definition of the onset of the systole and diastole. Black marker refers to the systolic onset and the white marker refers to the diastolic onset. The bottom line represents the ECG.

A. Localization of the S1 and S2

Data analysis was performed in MATLAB R2009b, where the program HSPoints was used for manual detection of heart sounds, based on the point where the variation from the baseline began, se figure 1. Subsequently the S/D-ratios were calculated. In order to make the data easier to analyze, a Butterworth filter was designed and used. A fourth order Butterworth band-pass filter with cut of frequencies at 25 Hz and 200 Hz was used

B. Statistics

For every subject at every work load level the mean was calculated for both systole and diastole duration. From these means the mean S/D-ratio and standard deviation was calculated in order to plot the data. Furthermore the heart rate was calculated.

III. Results

One subject did not reach maximum heart rate and was therefore excluded from the study. In 8 subjects usable recordings were made at each work level. However as the heart rate increased, the recordings became more difficult to interpret and therefore the marking of first and second heart sounds became more difficult. Figure 2 shows baseline recordings and recordings from the middle and the last load level of a male and a female subject. At higher heart rate, the heart sound signals contain more noise, as figure 2 shows, and therefore the detection of the onset of first and second heart sound is more difficult. In these situations the ECG gives a guideline for where the first heart sound has onset. When analyzing sequential recordings from the same subject the pattern of the first and second heart sound becomes recognizable, and therefore the detection the precise onset of the heart sounds becomes less difficult. Despite this fact, it is not easy to precisely define the onset point of the heart sound, as it is seen in figure 2. The interpreters noticed

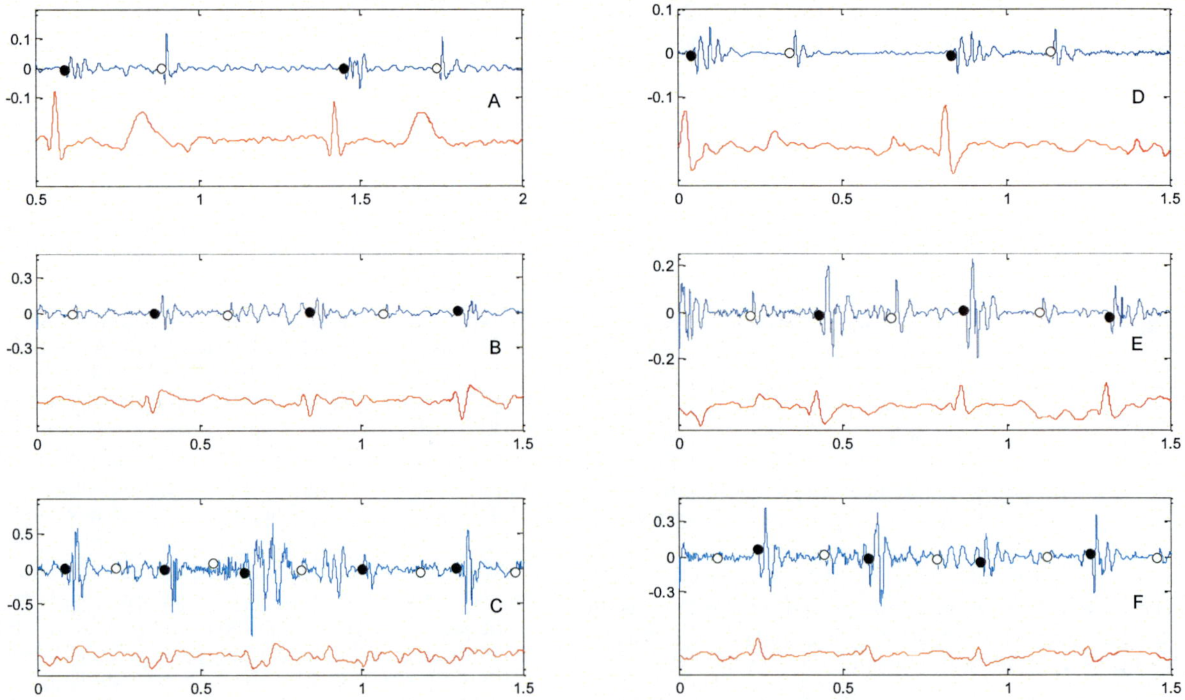

Figure 2 Measurement from baseline (A), middle load level (B) and maximal load level (C) for a male subject and measurement from baseline (D), middle load level (E) and maximal load level (F) a female subject. Black marker refers to the systolic onset and the white marker refers to the diastolic onset. The bottom lines represent the ECG

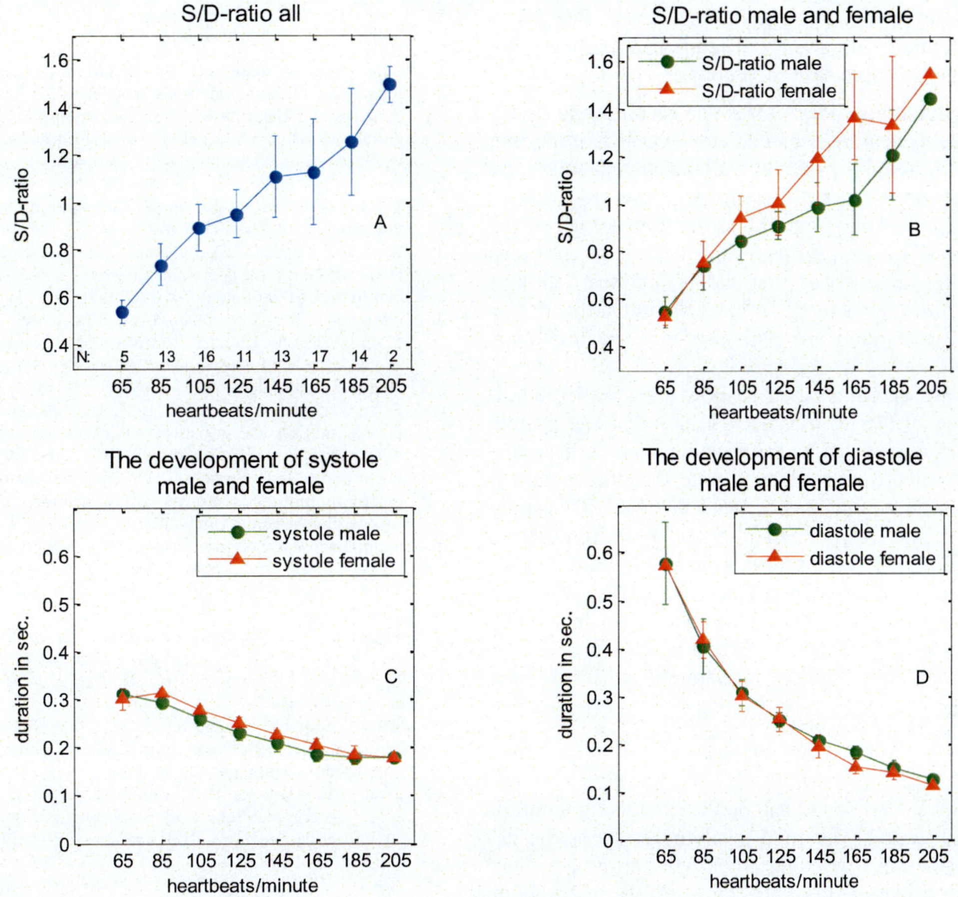

Figure 3 A: The development of S/D-ratio for all participants as a function of heart rate. B: The development of S/D-ratio for male and female as a function of heart rate. C: The development of systole duration as a function of heart rate. D: The development of diastole duration as a function of heart rate.

a difference in quality of the recordings between male and female subjects. Often recordings from female subjects where easier to interpret than those of male subjects – even at similar heart rate.

The recordings were used to calculate S/D-ratios for all subjects. Because of the gender differences noticed in the interpretation of recordings, S/D-ratios were also calculated for male and female subjects separately. Figure 3A shows the development of S/D-ratios as a function of the heart rate. It is evident that the S/D-ratios changes as the heart rate increases. The S/D-ratio as a function of heart rate develops similarly for men and women. Only at pulse interval 155-165 the development differs markedly, as shown in figure 3B. When looked at separately, the development of systole and diastole duration is very similar between healthy male and female subjects, see figure 3C and D.

IV. Discussion

The objective of this study was to examine whether a microphone could be used to detect first and second heart sound during an exercise test, and if these recordings could be used to calculate S/D-ratios in healthy subjects. The study showed that it is possible to obtain usable recordings of first and second heart sound by phonocardiography. However there are difficulties connected to using a microphone during an exercise test. It becomes more difficult to interpret the recording with higher heart rates in spite of the noise-reducing filter developed to reduce excess noise in the recordings. Respiratory sounds and movements in the cords disturb the recording. Furthermore the subjects perspire during the exercise test, which loosens the tape that holds the microphone in place. It is therefore necessary to modify the method of attachment of the microphone prior to further studies in order to make it more resistant to perspiration. These factors must be sought minimized in order to obtain clear heart sound recordings at higher load levels. Possibly a further development of signal filtering can increase the interpretability of heart sound recordings.

It was possible to use the heart sound recording to calculate S/D-ratios in healthy subjects. However it is important to note that the reliability of the calculated S/D-ratios are heavily dependent of the precision of the detected onset of the first and second heart sound. In order to make use of the calculation of S/D-ratios in diagnosing IHD, it is necessary to automate the detection of first and second heart sounds, since it is to time demanding to have a clinician to define them manually. Automation will also increase the inter-tester reliability.

The obtained recordings clearly show that systole and diastole duration differs at increased heart rate. The systole duration is not as effected as the diastole. This is in accordance with Bombardini et al. who note that the systole is relatively consistent at increasing heart rate [16]. This study showed that at increasing heart rate, the diastolic duration decreased significantly. This results in an increase in S/D-ratio for healthy subjects. In subjects with IHD, the S/D-ratio is expected to increase more rapidly than in healthy subjects because of a relatively longer systole duration as earlier mentioned and therefore a relatively shorter diastole duration.

V. Conclusions

First and second heart sound can be detected using a microphone, but noise at higher load levels necessitates the development of a noise-reducing filter. Furthermore a more perspirant resistant method of attachment of the microphone is needed. For this detection method to be used clinically, an automatically detection method for onset of first and second heart sound must be developed and validated. This study is limited by the small number of subjects, but the results indicate that recording heart sounds can be used in a clinical setting if the method gets refined. Furthermore this study found that duration of systole and diastole duration altered during stress. The systole duration only decreased minimally whilst the diastole duration decreased markedly as a function of higher heart rate. There is need of further studies in order to determine the normal development of systole and diastole duration in healthy subjects. When the normal development is determined, it is possible to test if subjects with IHD differ significantly from these normal values, and assess whether or not an integration of S/D-ratio calculations will enhance the accuracy of an exercise test.

Acknowledgment

Thanks to John Hansen, M.Sc.EE, Ph.D. and assistant professor at Department of Health and Science Technologies at Aalborg University for technical assistance and guidance.

References

1. Lloyd-Jones D, Adams RJ, Brown TM et al. (2010) Executive summary: heart disease and stroke statistics--2010 update: a report from the American Heart Association. Circulation 121(12):e259.
2. Gibbons L, Blair SN, Kohl HW et al. (1989) The safety of maximal exercise testing. Cirkulation 80 (4): 846-852
3. Gianrossi R, Detrano R, Mulvihill D et al. (1989) Exercise-induced ST depression in the diagnosis of coronary artery disease: A meta-analysis. Cirkulation 80 (1): 87-98.
4. Sansoy V, Watson DD, Beller BA (1997) Significance of slow up-sloping ST-segment depression on exercise stress testing. The American Journal of Cardiology 79 (6): 709-712.
5. Høilund-Carlsen PF, Johansen A, Christensen HW et al. (2005) Usefulness of the exercise electrocardiogram in diagnosing ischaemic or coronary heart disease in patients with chest pain. The American Journal of Cardiology 95 (1): 96-99.
6. Ferro G, Giunta A, Maione S (1984) Diastolic time during exercise in normal subjects and in patients with coronary artery disease. A plethysmographic study. Cardiology 71 (5): 266-272.
7. Ferro G, Duilio C, Spinelli L et al. (1995) Relation between diastolic perfusion time and coronary artery stenosis during stress-induced myocardial ischemia. Circulation 95 (3): 342-347.
8. Gemignani V, Bianchini E, Faita F et al. (2008) Assessment of Cardiologic Systole and Diastole Duration in Exercise Stress Test with a Transcutaneus Accelerometer Sensor. Computers in Cardiology 35: 153-156.
9. Phan DH, Bonnet S, Guillemaud R et al. (2008) Estimation of respiratory waveform and heart rate using an accelerometer. Engineering in Medicine and Biology Society. 30th Annual International Conference of the IEEE: 4916-4949.
10. Bito Y, Soeta Y, Nakagawa S et al. (2005) The optimal method for recording prosthetic heart valve sounds in clinical situations. Osaka City Medical Journal 51 (2): 73-81.
11. Akay YM, Akay M, Welkowitz W et al. (1993) Noninvasive acoustical detection of coronary artery disease: a comparative study of signal processing methods. IEEE Transactions on Biomedical Engineering 40 (6): 571-578.
12. Schmidt SE, Holst-Hansen C, Graff C et al. (2007) Detection of Coronary Artery Disease with an Electronic Stethoscope. Computers in Cardiology 34: 757-760.
13. Tilkian AG, Conover MB (2001) Understanding Heart Sounds and Murmurs – With an Introduction to Lung Sounds. 4th ed. W. B. Saunders Company, Philadelphia Pennsylvania.
14. Vermarien H, van Vollenhoven E (1984) The recording of heart vibrations: a problem of vibration measurements on soft tissue. Medical & Biomechanical Engineering & Computing 22: 168-178.
15. Saunamäki K, Egstrup K, Krusell L et al.(2001) Vejledende retningslinjer for kliniak arbejdstest i relation til iskæmisk hjertesygdom. Dansk Cardiologisk Selskab.
16. Bombardini T, Gemignani V, Bianchini E et al. (2008) Diastolic time - frequency relation in the stress echo lab: filling timing and flow at different heart rates. Cardiovascular ultrasound 6 (15).

Corresponding author

Author: Sidsel Maria Monrad Rønved
Institute: Department of Health and Science Technologies, Aalborg University
Street: Frederik Bajers Vej 7
City: 9220 Aalborg
Country: Denmark
Email: sidselronved@gmail.com

An Approach to a Multiple Channel Oximetry System

A.G. Mohammedani[1], K. Mankodiya[1,5], A. Opp[2], H. Gehring[3], M. Klinger[4], and U.G. Hofmann[1]

[1] Institute of Signal Processing, University of Luebeck, Luebeck, Germany
[2] Institute of Medical Engineering, University of Luebeck, Luebeck, Germany
[3] Department of Anesthesiology, University Hospital of Schleswig-Holstein, Campus Luebeck, Luebeck, Germany
[4] Institute For Anatomy, University of Luebeck, Luebeck, Germany
[5] Dept. of Rehabilitation Science & Technology, University of Pittsburgh, Pittsburgh, USA

Abstract— **Monitoring the oxygen saturation is a standard procedure during surgical intervention and on intensive care units. A common way is to measure the arterial saturation at various locations such as the finger or the forehead. Cardio vascular interventions such as coronary bypass surgery require additional monitoring of the cerebral oxygenation. Commercial devices are bulky and allow for processing one measurement parameter only. The multiple channel oximetry system introduced here has the ability to non-invasively measure and process both the arterial and the cerebral oxygen saturation. It is an optical method that utilizes absorption of light at specific wavelength emitted by LEDs. The present work introduces such a device as well as two different sensor positions for the arterial saturation. Here the signal quality in terms of SNR is compared at the different locations. Heart of this device is an embedded smart phone processor which allows for both driving the LED circuit and processing data.**

Keywords— **NIRS, cerebral, pulse, oximetry, brain.**

I. Introduction

Brain blood supply depends on a complex network of arteries and veins to deliver enough oxygen supply for a healthy brain metabolism [1]. Although adults' brain is only about 2% of the total body weight it consumes about 15% of the cardiac output and about 18-20% of total body oxygen consumption [1-3]. One safe and cost effective method used in monitoring the brain is Near Infrared Spectroscopy (NIRS). This method was introduced for the first time in 1977 by Jöbsis [4], and since that time it has seen great developments. It has been used in different procedures and surgeries to reflect the cerebro-vascular status during anaesthesia [5].

In the NIR spectral region of 700-1000nm the biological tissue including the skull is translucent, photons are scattered and absorbed inside the scalp but few survive and exit the skin back to the detectors [6].

Some commercial models are already available on the market for monitoring cerebral oximetry, however in these models the data is usually collected from the forehead while in this study we are aiming to investigate the frontal region and the temple at different penetration depths and different wavelengths. The Signal to Noise Ratio (SNR) is calculated for these regions to investigate the best locations for conducting optical measurements on the adult's head. Monitoring the brain oxygenation in adults is very challenging since the incident light can undergo multiple scattering events which in the end lead to a low detected signal on the photodiodes [5,7].

Unlike infants who show a high SNR [8], the SNR in the adults' head is substantially low, due to the thick bone and different layers surrounding adults' brain [7] therefore a satisfying signal to noise ratio is hard to obtain [9].

Based on the initial measurements we observed in this study a multiple channel oximetry system could be investigated in the future to give a good estimate to both pulse oximetry and cerebral oximetry at the hemisphere monitored using multiple wavelengths and detectors. In pulse oximetry the arterial saturation is monitored. Monitoring the arterial saturation at the head region is particularly useful especially in patients with poor peripheral circulation [10] while cerebral oximetry monitoring provides global assessment to the tissue oxygenation and brain hemodynamic metabolism [7,11,12].

By integrating both systems it is possible to monitor the arterial supply on the head region and the brain hemodynamic metabolism by utilizing only one system. This will help in a better estimation of cerebral blood flow than using two individual systems [13]. Additionally and by utilizing the smartphone processor features we will be able to have a low cost miniaturized system where the signal can be send wirelessly to any physician's PC and by using such a miniaturized way of monitoring the system could be integrated in the future into daily-life used devices such as a cap or a band especially while monitoring elderly with previously diagnosed cardiovascular diseases.

K. Dremstrup, S. Rees, M.Ø. Jensen (Eds.): 15th NBC on Biomedical Engineering & Medical Physics, IFMBE Proceedings 34, pp. 89–92, 2011.
www.springerlink.com

Figure 1 shows a conceptual diagram of the sensors locations on the adult's head for one channel of the multi-channel oximetry system.

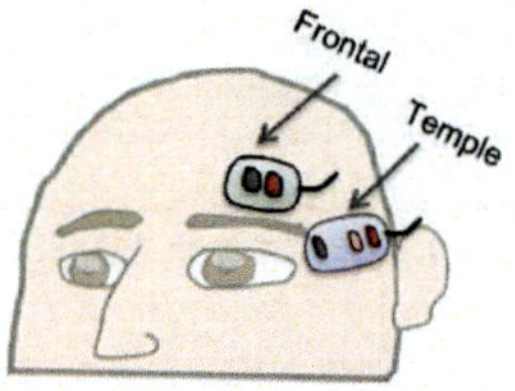

Fig. 1 Conceptual diagram of sensors locations

II. MATERIAL AND METHODS

In order to validate the use of NIR light for cerebral oximetry measurement at different skull regions an SNR test of the detected signal quality was conducted on three African male volunteers in their early thirties. We confined ourselves to the regions mentioned earlier to find the optimum locations on the adult's head for optical measurements. To carry out the measurements two different adhesive patch sensors were used. For cerebral monitoring we used the commercial sensor (4100-SAF, Somanetics, USA), the sensor includes two LEDs (730,810nm) and two photodiodes differentially spaced at 3 and 4cm from the light sources. For pulse oximetry another commercial sensor (MaxFast, Nellcor, USA) was used. The sensor includes two LEDs (660,940nm) and one photodiode at 1cm distance from the light sources. Both sensors – shown in figure 2 – were connected to a custom build analog hardware system to process the raw data.

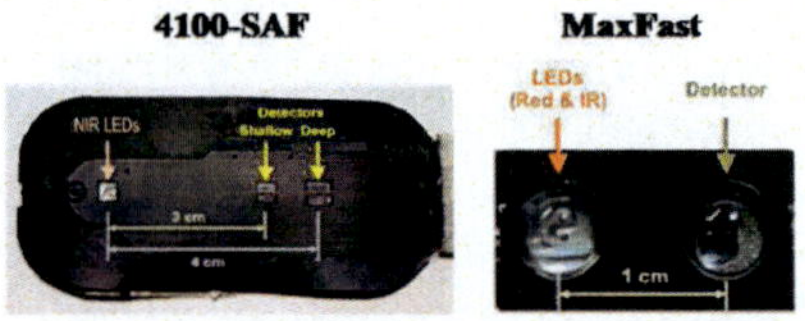

Fig. 2 A picture of sensors used in the measurements

In order to illuminate the targeted region a controlled current source circuit was designed to control the current flowing to the LEDs. A typical configuration is shown in Figure 3. This driver configuration is usually used on oximetry circuits since it is highly reliable and has a very small error compared to other driver circuits like bipolar transistor driven source [14]. The MOSFET passes the load current while the opamp compares the sampled voltage across R_s on the inverting input, to the non-inverting input, then the opamp will respond by increasing or decreasing its output depending on the drain current amount.

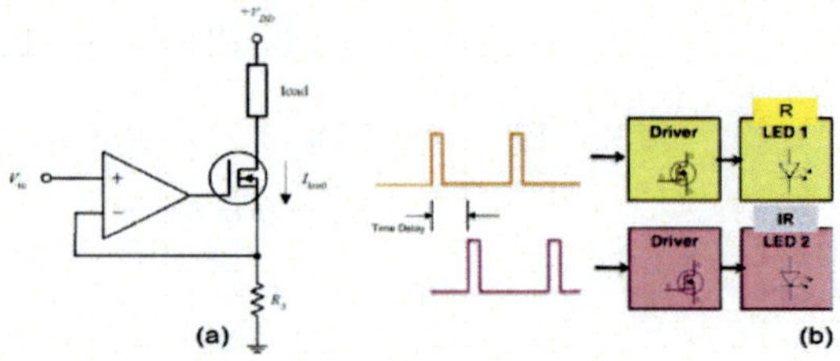

Fig. 3 Typical configuration for a constant current source (a) Circuit configuration and (b) Timing control.

The raw signal detected by the photodiodes is processed by a transimpedance amplifier to amplify the signals resulting from the photons impinging on the detectors to a useful level, since photodiodes produce a very little current from their exposure to light [15]. Consecutively a 5th order Butterworth low pass filter with an upper cut-off frequency of 20Hz was used to smooth the signal and eliminate the noise. Figure 4 shows a simplified block diagram of the hardware used to study the optical properties of the adult's head. Based on the same design principle the plethysmography signal was investigated using the wavelengths 660 and 940nm. However the low pass filter

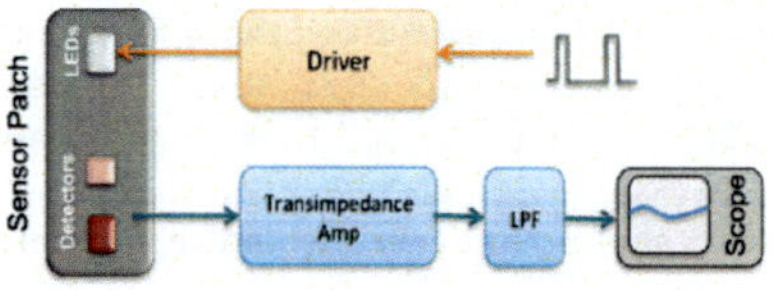

Fig. 4 Simplified block of the hardware used in measuring optical signal from the head

was replaced with a band pass filter with a cut-off frequency of 0.3 - 10Hz.

The timing control for the LEDs pulsation was managed by the embedded processor's developers board www.beagleboard.org/ using its General Purpose Input Output (GPIO). The developers board is called the BeagleBoard (BB) and it is shown in figure 5.

It is a 3"x3" smartphone processor board, featuring Texas instruments' (TI) OMAP3530 application processor.

Fig. 5 Image of the BeagleBoard and its components

This gives the BB a laptop like performance in a single board [16]. Like any computer the BB accepts only digital data, therefore an ADC is needed to digitize the analog data for further processing and display.

III. Results and Discussion

A. Pulse Oximetry

Figure 6a shows an oscilloscope screenshot of the pleth-wave while locating the MaxFast sensor at the temple region. While figure 6b shows the measurement at the frontal region directly above the eyebrow. These measurements were performed while the volunteers were in a well seated position.

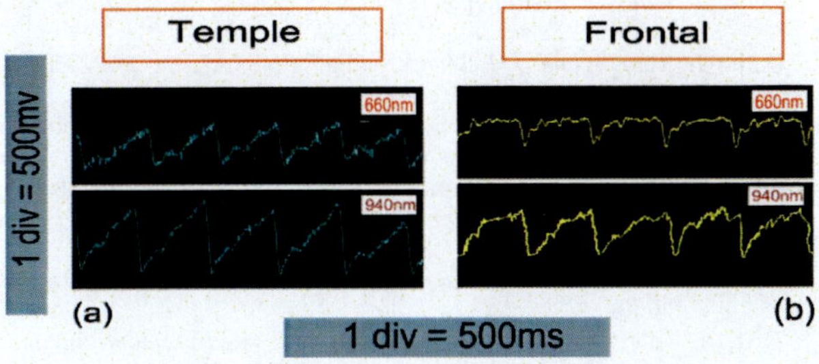

Fig. 6 Screenshot of the pleth-wave on (a) frontal and (b) temple region.

B. Cerebral Oximetry

The signal quality was investigated at the previously mentioned locations using 4100-SAF commercial sensor. The sensor utilizes two NIRS wavelengths 730 and 810nm. These wavelengths monitor the absorption spectra of deoxygenated hemoglobin and total hemoglobin respectively [17,18]. Additionally and by using two photodiodes distanced at 3 and 4cm from the light sources the extracerebral and intracerebral signals can be distinguished [19] and by subtracting the shallow detector signal from the deep one the extracerebral signals are reduced and we have a probability of 85% that the resulting signal is coming from the brain tissue [5]. Table 1 shows the SNR after locating the 4100-SAF sensor at the frontal region. The measurements show the SNR calculated at the 3cm and 4cm distanced photodiodes and the mean value is shown in figure 7.

Table 1 SNR measurements at the forehead (cerebral data)

Forehead	**Volunteer A**					
	10%		20%		30%	
Wavelength	3cm	4cm	3cm	4cm	3cm	4cm
730 nm	8dB	3dB	5dB	2dB	5dB	2dB
810 nm	12dB	8dB	12dB	3dB	10dB	3dB
	Volunteer B					
	10%		20%		30%	
Wavelength	3cm	4cm	3cm	4cm	3cm	4cm
730 nm	5dB	2dB	4dB	1dB	3dB	1dB
810 nm	10dB	3dB	11dB	3dB	8dB	3dB
	Volunteer C					
	10%		20%		30%	
Wavelength	3cm	4cm	3cm	4cm	3cm	4cm
730 nm	7dB	1dB	5dB	1dB	5dB	1dB
810 nm	14dB	4dB	10dB	3dB	10dB	2dB

In the same manner the SNR was calculated for the temple region as observed in table 3 and it showed an increase in SNR value as observed in Figure 7. This is probably a result of its lower thickness compared to other locations [20]. As seen in figure 7 the SNR strongly decreases while increasing the duty cycle which is probably due to the increased temperature in the p-n junction of the LEDs [8,21].

Table 2 SNR measurements at the temple (cerebral data)

Temple	**Volunteer A**					
	10%		20%		30%	
Wavelength	3cm	4cm	3cm	4cm	3cm	4cm
730 nm	14dB	7dB	10dB	7dB	10dB	5dB
810 nm	20dB	15dB	18dB	12dB	18dB	10dB
	Volunteer B					
	10%		20%		30%	
Wavelength	3cm	4cm	3cm	4cm	3cm	4cm
730 nm	8dB	5dB	5dB	3dB	6dB	3dB
810 nm	12dB	7dB	10dB	7dB	10dB	5dB
	Volunteer C					
	10%		20%		30%	
Wavelength	3cm	4cm	3cm	4cm	3cm	4cm
730 nm	9dB	5dB	4dB	3dB	4dB	3dB
810 nm	15dB	7dB	8dB	6dB	6dB	5dB

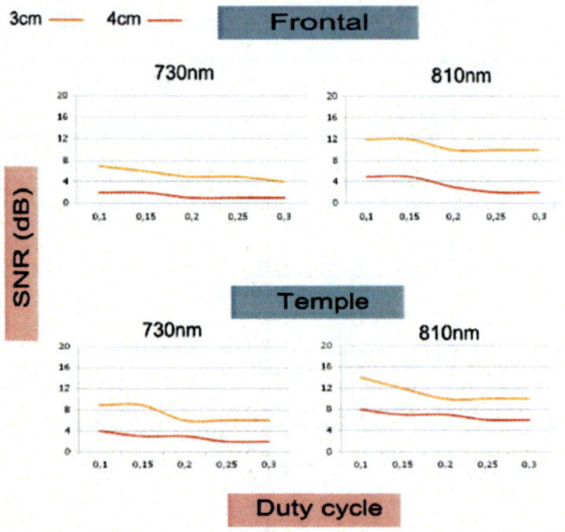

Fig. 6 Calculated mean SNR at the forehead and the temple.

The best SNR was observed at 10 to 15% duty cycle. Using the same amplification factor for both photodiodes resulted in a lower signal quality at the 4cm distanced photodiode. However, The SNR was significantly enhanced by applying pressure on the sensor. We used for that a black rubber band tied around the head.

IV. Conclusion

In this study we presented an initial measurements for cerebral and pulse oximetry with the aim of integrating both systems in the future. The hypothesis of having a better signal quality at the temple region has been proven, however further research is needed to employ more filtering techniques to enhance the signal quality.

Acknowledgment

We wish to express our gratitude to Hafiz Hamid, Ahmed Bernawi , Yousuf Alkannan and Mutasim Khalil Hassan for their support and contribution in this study.

References

[1] G.J. Tortora and B. Derrickson, eds., *PRINCIPLES OF ANATOMY AND PHYSIOLOGY*, John Wiley and Sons, 2009.

[2] A. Zauner, W. Daugherty, M. Bullock, and D. Warner, "Brain Oxygenation and Energy Metabolism: Part I-Biological Function and Pathophysiology," *Neurosurgery*, vol. 51, 2002, pp. 289-302.

[3] N.P. Mitagvaria and H.I. Bicher, eds., *Cerebral Blood flow regulation*, Nova Biomedical, 2009.

[4] F. Jöbsis, "Noninvasive infrared monitoring of cerebral and myocardial sufficiency and circulatory parameters," *Science*, vol. 198, 1977, pp. 1264-1267.

[5] M.A.A.M. Schepens and F.G.J. Waanders, "Monitoring the brain: near-infrared spectroscopy," *Aortic Arch Surgery*, J.S. Coselli and S.A. LeMaire, eds., Wiley-Blackwell, 2008.

[6] A. Villringer, J. Planck, C. Hock, L. Schleinkofer, and U. Dirnagl, "Near infrared spectroscopy (NIRS): a new tool to study hemodynamic changes during activation of brain function in human adults.," *Neuroscience letters*, vol. 154, May. 1993, pp. 101-104.

[7] H. Owen-Reece, M. Smith, C.E. Elwell, and J.C. Goldstone, "Near infrared spectroscopy.," *British journal of anaesthesia*, vol. 82, Mar. 1999.

[8] A. Bozkurt, A. Rosen, H. Rosen, and B. Onaral, "A portable near infrared spectroscopy system for bedside monitoring of newborn brain," *BioMedical Engineering OnLine*, vol. 11, 2005, pp. 1-11.

[9] Q. Zhang, H. Ma, S. Nioka, and B. Chance, "Study of near infrared technology for intracranial hematoma detection," *Journal of Biomedical Optics*, vol. 5, 2000, pp. 206-213.

[10] M. Fernandez, K. Burns, B. Calhoun, S. George, B. Martin, and C. Weaver, "Evaluation of a new pulse oximeter sensor.," *American journal of critical care : an official publication, American Association of Critical-Care Nurses*, vol. 16, Mar. 2007, pp. 146-152.

[11] T. Kato, "Principle and technique of NIRS-Imaging for human brain FORCE : fast-oxygen response in capillary event," *International Congress Series*, vol. 1270, 2004, pp. 85 - 90.

[12] P.L. Madsen and N.H. Secher, "NEAR-INFRARED OXIMETRY OF THE BRAIN," *Progress in Neurobiology*, vol. 58, 1999.

[13] C.E. Elwell, M. Cope, A.D. Edwards, J. Wyatt, E.O.R. Reynolds, and D.T. Delpy, "Measurements of cerebral blood flow in adult humans using nearinfrared spectroscopy - Methodology and possible errors," *Advances in experimental medicine and biology*, vol. 317, 1993.

[14] P. Scherz, ed., *Practical Electronics for Inventors*, McGraw-Hill, 2000.

[15] M. Johnson, ed., *Photodetection and measurement maximizing performance in optical systems*, McGraw-Hill, .

[16] Texas Instruments, "BeagleBoard System Reference Manual, Revision c2," 2009.

[17] D.B. Andropoulos, S.A. Stayer, and E.D. Mckenzie, "Regional low-flow perfusion provides comparable blood flow and oxygenation to both cerebral hemispheres during neonatal aortic arch reconstruction," *Thoracic and cardiovascular Surgery*, 2003.

[18] A. Deschamps, J.M. Murkin, and A. Denault, "A Proposed Algorithm for the Intraoperative Use of Cerebral Near-Infrared Spectroscopy," *Seminars in Cardiothoracic and Vascular Anesthesia*, vol. 11, 2007, pp. 274-281.

[19] M.J.T.V.D. Ven, W.N.J.M. Colier, M.C.V.D. Sluijs, D. Walraven, B. Oeseburg, and H. Folgering, "Can Cerebral Blood Volume Be Measured Reproducibly With an Improved Near Infrared Spectroscopy System ?," *Journal of Cerebral Blood Flow and Metabolism*, 2001, pp. 110-113.

[20] S. Standring, ed., *Gray's anatomy: the anatomical basis of clinical practice*, 2008.

[21] J.G. Webster, ed., *Design of pulse oximetrs*, 1997.

The address of the corresponding author:

Author: Ulrich G. Hofmann.
Institute: Institute of Signal Processing, University of Luebeck.
Street: Ratzeburger Allee 160, Building 64.
City: 23562 Luebeck.
Country: Germany
Email: hofmann@isip.uni-luebeck.de

Muscle Strength as a Predictor of the Magnitude of Multidirectional Force Fluctuations during Steady Contractions

S.E. Salomoni and T. Graven-Nielsen

Center for Sensory-Motor Interaction (SMI), Department of Health Science and Technology, Aalborg University, Denmark

Abstract— **It has been shown that, during static contractions, the standard deviation of force increases linearly as the target force increases due to the orderly recruitment of motor units. In addition, previous studies suggested that larger muscle groups present lower normalized force fluctuations. The aim of the present study was to test the generalization of these two principles by the assessment of three-dimensional force fluctuations of different muscle groups during steady contractions of dorsiflexors, elbow flexors, trunk extensors, knee extensors, and plantarflexors. According to the results, although significant linear correlations exist between MVC forces and the different parameters of variability assessed, no clear relationship was found when the mean values were compared across muscle groups. Moreover, the present data demonstrate that high fluctuations of the task-related force component do not necessarily imply in high fluctuations of tangential components. Hence, factors such as co-contraction and the number of degrees of freedom in the joints might also account for the variability of multidirectional forces.**

Keywords— **Force steadiness, three-dimensional forces, signal-dependent-noise, isometric contractions.**

I. Introduction

Variability is an inherent characteristic of the force output [1] which affects accuracy of movements, fine manipulative capability, and may be associated with a history of falls [2]. Although fluctuations in force output during voluntary contractions are influenced by the descending drive onto the motor neuron pool, afferent feedback from the periphery, and the intrinsic properties of the motor neurons, they are ultimately a consequence of the behavior of the motor units that innervate the muscles [3].

Previous studies demonstrated that, due to the orderly recruitment of motor units, the standard deviation (SD) of the force produced by a subject increases linearly as the mean force output increases [4], a relation often termed signal-dependent-noise, or SDN. However, because the relative contributions of individual motor units to the global force output decrease in larger motor unit populations [5], it has been suggested that the normalized force variability, assessed by the coefficient of variation of force (CV=SD/Mean), is inversely related to the number of motor units in the muscle, i.e. larger muscles produce lower force fluctuations. This relation between muscle size and force variability has been experimentally observed across different arm and hand muscles [6] and between dorsi- and plantarflexors [7], but not consistently between elbow flexors and knee extensors [3], indicating the need of a methodological assessment across muscle groups of different sizes, functions, and neuromuscular properties.

In fact, force fluctuations are profoundly influenced by the architecture of the observed muscle group, depending on the distribution of activity within and between muscles [7]. Each muscle acting alone induces fluctuations that are directionally aligned with the muscle's direction of action, and cooperation among muscles leads to load sharing variations, eliciting fluctuations that reflect in multiple (task-related and tangential) directions of action [8], although very few studies have investigated three-dimensional force fluctuations.

The current study employed three-dimensional force analysis to assess the relationship between muscle strength and force fluctuations during isometric contractions of five different muscle groups. Force fluctuations were assessed in the normal, task-related direction (Fz) and also in tangential directions (Fx and Fy). Based on previous studies, stronger muscle groups are expected to produce higher SD of force compared with weaker muscle groups. However, the opposite is expected when normalized parameters are used.

II. Methods

A. *Experimental Protocol*

Fifteen subjects (12 males, age 23.8 ± 4.7 yr; mean ± SD) with no known musculoskeletal disorder participated in this study. It was conducted in accordance with the Declaration of Helsinki, approved by the local Ethics Committee (N-20090036), and written informed consent was obtained from all participants prior to inclusion.

Subjects participated in a total of five experiments, corresponding to contractions of different muscle groups, assigned in random order: Dorsiflexors (DF), elbow flexors

K. Dremstrup, S. Rees, M.Ø. Jensen (Eds.): 15th NBC on Biomedical Engineering & Medical Physics, IFMBE Proceedings 34, pp. 93–96, 2011.
www.springerlink.com

(EF), trunk extensors (TE), knee extensors (KE), and plantarflexors (PF).

For each muscle group, subjects performed 3 trials (5s) of isometric maximal voluntary contraction (MVC) force followed by 4 submaximal isometric contractions (12s) at 30% of MVC force. Feedback of the task-related force component (Fz, normal to the surface of the sensor) was provided on an oscilloscope.

B. Force Recordings

A six-axis sensor, yielding three force components and three moment components (MC3A, AMTI, USA), was mounted in custom-made setups for each experiment. The analogue output of the sensor was amplified (MSA-6, AMTI, USA), sampled at 1 kHz, and stored after 12 bits A/D conversion. Before each experiment, subjects were properly seated and familiarized with the protocol. Straps were used to avoid movements not related to the required tasks.

C. Data Analysis

The force and moment signals were low-pass filtered using a cut-off frequency of 20Hz. The absolute peak force in the task-related direction was extracted from the MVC trials and used as a reference for the submaximal trials. From the submaximal contractions, the SD of the three force components and the CV of force were calculated using a time window from 5 to 10 seconds, avoiding excessive fluctuations at the beginning and end of contractions. However, tangential force magnitudes very close to zero resulted in inconsistently high values of the CV of tangential forces during some trials. Hence, variability of tangential forces was indirectly assessed by total excursions of the center of pressure (CoP). In this analysis, the CoP represents the point of application of the resultant force on the surface of the sensor, normalized by the magnitude of the normal (task-related) force [9]. Thus, the total excursions of the CoP provide an indirect and normalized measurement of lateral displacements of quasi-static forces.

For each parameter extracted from the submaximal trials (SD, CV, total excursions of CoP) a linear regression analysis was performed between the parameter and the MVC force using the least mean squares method.

D. Statistical Analysis

All parameters were assessed using one-way repeated measures analysis of variance (RM-ANOVA) with muscle group as within-subjects factor. In addition, for each sub-

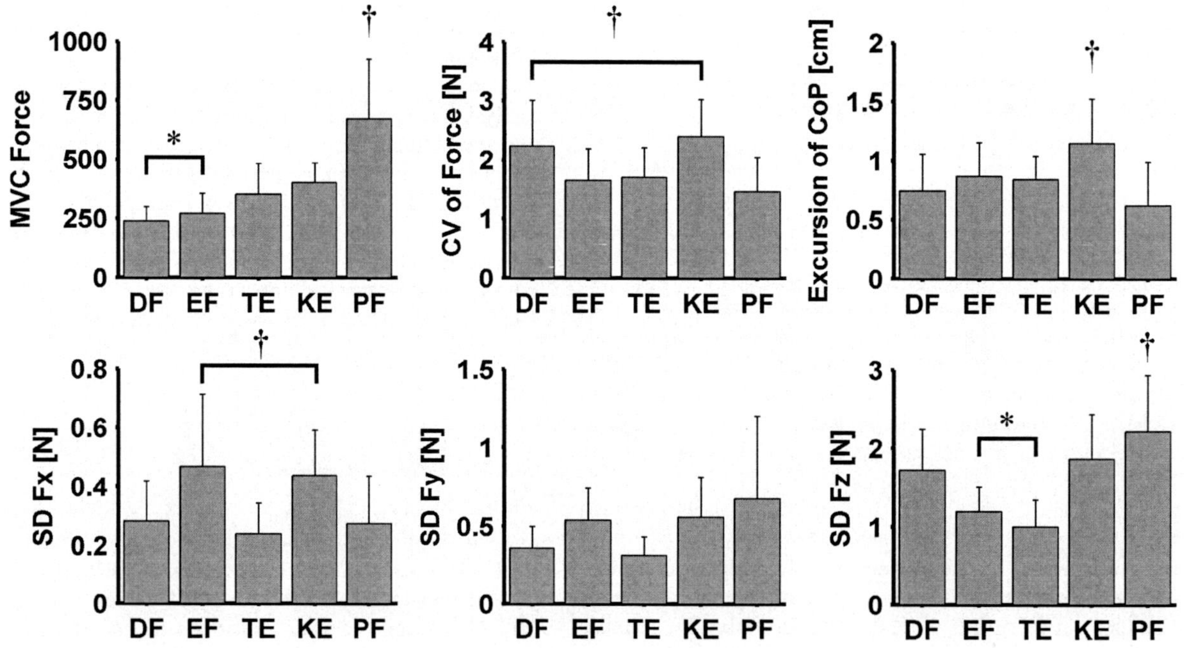

Fig. 1 Mean (± SEM) of the assessed force parameters: Peak force from maximal voluntary contraction (MVC) force trials and, from the submaximal trials, coefficient of variation (CV) of force, total excursions of the center of pressure (CoP), and standard deviation (SD) of the three force components. Recordings were performed during isometric contractions of the dorsiflexors (DF), elbow flexors (EF), trunk extensors (TE), knee extensors (KE), and plantarflexors (PF). †: Higher than all other muscle groups (P<0.05). *: Lower than all other muscle groups (P<0.05)

maximal force parameter, the coefficients (slope and intercept) of the regression line and the Pearson correlation coefficient were calculated and the regression was tested for significance. Statistical significance was considered for P-values lower than 0.05 and Newman-Keuls (NK) post-hoc test was applied when appropriate. All values are reported as mean ± standard error of the mean (SEM).

III. Results

A. Force Parameters

A significant main effect of muscle group was identified in the analysis of all force parameters for all muscle groups (Fig. 1, RM-ANOVA: $F_{4,56}>6.21$, $P<0.001$ for all tests). During the MVC trials, the highest peak force between muscle groups was observed for PF (NK: $P<0.001$), and the lowest for DF and EF (NK: $P<0.02$). DF and KE exhibited higher CV of force than all other muscle groups (NK: $P<0.02$), and KE also showed the highest total excursions of the CoP between muscle groups (NK: $P<0.02$). In the Fx (medial-lateral) direction, higher SDs of force were observed for EF and KE compared with the other muscle groups (NK: $P<0.01$). Finally, the highest SD of force in the Fz (task-related) direction was observed for PF (NK: $P<0.04$), and the lowest for EF and TE (NK: $P<0.03$).

B. Linear Regression Analysis

Table 1 summarizes the results of the linear regression analysis. Weak, albeit statistically significant ($P<0.003$) linear correlations were observed between MVC force and all submaximal force parameters, except for SD of force in the Fx direction. Moreover, the analysis revealed positive slopes and positive correlations between MVC force and the SD of the three force components, but negative correlations with CV and total excursions of the CoP.

Table 1 Summary of linear regression analysis between MVC forces and each submaximal force parameters: Slope and intercept of the regression lines, Pearson correlation coefficients and P-values. Positive slopes and positive correlations were observed for the SD of force, while negative correlations were obtained for the normalized parameters. Slopes are shown multiplied by 1,000 for visualization purposes

Parameter	Slope (×1,000)	Intercept	Pearson	P-value
SD Fx	0.03	0.33	0.016	0.777
SD Fy	0.60	0.26	0.171	< 0.003
SD Fz	2.03	0.81	0.262	< 0.001
CV of force	-0.99	2.27	-0.226	< 0.001
Excursions of CoP	-0.51	1.04	-0.237	< 0.001

IV. Discussion

A. Signal-Dependent Noise (SDN)

The presence of SDN in the task-related force component has been demonstrated over a wide range of target forces within the same muscle group [4;6]. The present study attempted to generalize this principle by investigating if stronger muscle groups produce higher SD of force. The results presented here only partially support the concept of SDN in the force output. Although a significant and positive correlation was observed between MVC forces and SD of the task-related force component, it is clear from the results shown in Fig. 1 that comparison of muscle strength alone is not sufficient to infer about the variability of the force produced by different muscle groups (e.g. KE showed higher MVC force than DF, with no significant difference in their SD of Fz force). Furthermore, weak correlations were identified between MVC forces and the SD of Fx and Fy components. So, other factors, such as co-contraction and the number of degrees of freedom in the joints, might also account for the variability of tangential forces [9].

It has been shown that the SD of task-related force is proportional to the mean exerted force when subjects are required to match different force levels using the same muscle group [4]. According to the size principle, as muscle activity increases, larger motor units are recruited [11], which have lower firing rates and produce unfused twitches, resulting in increased fluctuations [4]. Moreover, individual motor units produce forces at different orientations, depending on muscle fiber angle and attachments [12]. Thus, it is plausible to speculate that, for the same muscle group, contractions at higher force levels would lead to the activation of larger muscle areas and increased force fluctuations in multiple directions compared with lower force levels.

In other words, the present data show that the SDN cannot be generalized for comparisons between muscle groups, but they do not rule out the possibility that the model holds for multidirectional force assessments of the same muscle group at different target forces.

B. Normalized Assessment of Force Fluctuations

In line with previous investigations, a negative correlation was observed between MVC forces and normalized force fluctuation parameters (CV of force and total excursions of the CoP). However, the results in Fig. 1 show no clear relationship when the mean values are compared across muscle groups. For example, KE produced the highest normalized fluctuations between all muscle groups, although it was not the weakest muscle group. Thus, the collective results of the present study do not support the theory that stronger muscles, with a greater number of motor units, present lower normalized force fluctuations [3;6].

Interestingly, the assessment of a limited subset of the muscle groups in the current experiment could have lead to different conclusions. For example, lower MVC forces and higher CV of force were observed for DF compared with PF, supposedly supporting the negative correlation between muscle strength and normalized force fluctuations. Corroborating, a published study in the literature reported similar results comparing force fluctuations of only DF and PF muscle groups [7]. Based on this limited comparison, the author suggested, as a general rule, that smaller muscles exhibit greater normalized fluctuations compared to larger muscles when the gross level of activation is similar.

In the present study, although the CV of force was similar between DF and KE, the same was not true for the total excursions of the CoP, demonstrating that high fluctuations of the task-related force component do not necessarily imply in high fluctuations of tangential components. It is possible that the number of joints involved in a given contraction is more relevant to lateral force displacements than muscle strength.

V. Conclusions

Although significant correlations were observed between MVC forces and the different variability parameters assessed, the present data revealed that muscle strength alone cannot be used to predict the magnitude of force fluctuations between different muscle groups. In addition, it was found that high fluctuations of the task-related force component during static contractions of a given muscle group do not necessarily imply in high fluctuations of tangential components, which may be influenced by co-contraction of different muscles and the number of joints involved. Hence, comparison of the variability of different muscle groups should be interpreted with care, as they may lead to an oversimplification of the underlying physiological mechanisms.

Acknowledgment

The present study was financed by Svend Andersen Fonden (Aalborg, Denmark).

References

1. Stein RB, Gossen ER, Jones KE (2005) Neuronal variability: noise or part of the signal? Nature Rev Neurosci 6:389-397
2. Carville SF, Perry MC, Rutherford OM et al. (2007) Steadiness of quadriceps contractions in young and older adults with and without a history of falling. Eur J Appl Physiol 100:527-533
3. Tracy BL, Mehoudar PD, Ortega JD (2007) The amplitude of force variability is correlated in the knee extensor and elbow flexor muscles. Exp Brain Res 176:448-464
4. Jones KE, Hamilton AFC, Wolpert DM (2002) Sources of signal-dependent noise during isometric force production. J Neurophysiol 88:1533-1544
5. Fuglevand AJ, Winter DA, Patla AE (1993) Models of recruitment and rate coding organization in motor-unit pools. J Neurophysiol 70:2470-2488
6. Hamilton AFC, Jones EJ, Wolpert DM (2004) The scaling of motor noise with muscle strength and motor unit number in humans. Exp Brain Res 157:417-430
7. Tracy BL (2007) Force control is impaired in the ankle plantarflexors of elderly adults. Eur J Appl Physiol 101:629-636
8. Graves AE, Kornatz KW, Enoka RM (2000) Older adults use a unique strategy to lift inertial loads with the elbow flexor muscles. J Neurophysiol 83:2030-2039
9. Kutch JJ, Kuo AD, Bloch AM et al. (2008) Endpoint force fluctuations reveal flexible rather than synergistic patterns of muscle cooperation. J Neurophysiol 100:2455-2471
10. Seigle B, Ramdani S, Bernard PL (2009) Dynamical structure of center of pressure fluctuations in elderly people. Gait Posture 30:223-226
11. Henneman E, Somjen G, Carpenter DO (1965) Functional significance of cell size in spinal motoneurons. J Neurophysiol 28:560-580
12. Herrmann U, Flanders M (1998) Directional tuning of single motor units. J Neurosci 18:8402-8416

Corresponding author:

Author: Thomas Graven-Nielsen
Institute: Center for Sensory-Motor Interaction, Aalborg University
Street: Fredrik Bajers Vej 7D-3
City: Aalborg
Country: Denmark
Email: tgn@hst.aau.dk

Postural Variability during Pursuit Tracking in Low-Back Pain Patients

J.H. Svendsen[1,2], H. Svarrer[3], M. Vollenbroek-Hutten[2], and P. Madeleine[1]

[1] Laboratory for Ergonomics and Work-related Disorders, Center for Sensory-Motor Interaction (SMI), Dept. of Health Science and Technology, Aalborg University, Aalborg, Denmark

[2] Roessingh Research and Development, Enschede, The Netherlands

[3] Department of Rheumatology, Aalborg University Hospital, Aalborg, Denmark

Abstract— **In low-back pain patients the somato-sensory regulation of quiet stance is altered compared to healthy subjects which in turn results in larger variation in the postural control. In this study, 8 low back pain patients performed postural sways on a force plate to follow a squared shaped track on visual feedback corresponding to 20% of maximum sway in the 4 directions of forward, backward, left and right. The tracking was repeated 4 times and variability- and complexity measures of the center of pressure were calculated and the motor control learning effect was analyzed. The effect of epoch size on the statistical result was also investigated. As the main result through the four tracking trials, variability (calculated as standard deviation of the force output) decreased in both the anterior-posterior direction and the medial lateral direction ($p<0.05$) and complexity measures (calculated as both approximate entropy and sample entropy) increased for anterior-posterior direction and medial-lateral direction ($p<0.05$). Secondary, no effect was found for the analysis of different epoch sizes. Results show that a learning effect is present even for 4 repetitions of a postural tracking task. The results presented can be of relevance in training sessions for patients with impaired postural control.**

Keywords— **Balance postural control, tracking, learning effect, sample entropy.**

I. INTRODUCTION

In quiet stance a complex regulation of the somatosensory, visual and vestibular afferent inputs is used to stabilize body posture [1]. In low-back pain patients, several studies have reported that this regulation is altered as shown by delayed onset of the stabilizing muscles and a larger fluctuation in the postural control [2,3]. During the last decade the variation of the postural control and how the regulation of the erected posture results in fluctuations have been investigated. The fluctuations of the postural control have been considered as noise [4]. However, a number of recent studies have mentioned the functional relevance of these fluctuations [3,5]. The variability of quiet stance can be computed in different ways. The linear measure (usually calculated as the standard deviation or coefficient of variation) gives a good insight into how well subjects maintain their balance. But by looking into the nonlinear measures, complexity of the variation can be assessed and structural behaviors in the movement or quiet stance can be outlined [6].

To assess the complexity in the variation of physiological signals, a measure like approximate entropy was introduced two decades ago and has shown promising findings of structural changes. The exact interpretation of the measure is still a matter of debate, but the relevance of assessing nonlinear structures in physiological recordings has been demonstrated thoroughly.

Sample entropy is an extension of approximate entropy, as approximate entropy can be inaccurate in small sample sizes [7]. In recordings with lower frequencies, like force recordings, this can be an issue and therefore sample entropy can be a tool to assess entropy measures for smaller sample sizes and epochs.

The analysis of erected balance has enabled the detection of subtle changes in the postural control system found in low-back pain patients or whiplash patients i.e. decrease in signal complexity. Such analysis conducted on low-back pain patients can actually contribute to a better quantitative characterization of their deficit in postural control. Low-back pain patients are suspected to make use of a different postural motor strategy during quiet stance [8] and different parameters have been suggested to be useful in the investigation of postural control [9,10].

One of the main effects of learning in motor control is the discovery and strengthening of motor synergies to stabilize salient performance variables [11,12]. For subjects with impaired postural control (like low-back pain patients), this effect would be expected to be less potent when performing repetitive tasks.

With training sessions, individuals can improve their performance but when looking at a simple parameter as variability in the balance, how fast can an improvement be observed? In this study we investigated the ability to control balance in 8 low-back pain patients. Referring to the effect of learning and strengthening of motor synergies [11,12], it was hypothesized that the subjects would improve their performance, i.e. decrease SD and increase in entropy during the session of 4 repeated trials.

II. Methods and Materials

Eight low-back pain patients participated in this experiment (mean age ± standard deviation: 40.7 ± 9.9 years, BMI: 24.1 ± 2.5). The subjects were standing on a force plate where they were given visual feedback on a visual display unit. Prior to the tracking sessions, the limit of stability was recorded as the maximum sway the subjects could do in their forward/backward direction and left/right direction. The tracking then consists of quiet stance with sways in the 4 direction.

The feedback was presented as a cursor on the monitor corresponding to the subjects' center of pressure (COP) on the force plate. On the monitor a square-shaped track was displayed with a moving target following the square (speed 0.95 cm/s). Subjects were instructed to sway their body and thereby displace the COP to put the cursor (representing COP) on the moving target. The track was set to correspond to 20% of maximum limit of stability sway in the four directions, anterior, posterior, medial and lateral directions sway. The tracking session was repeated 4 times and each session lasted for 90 seconds. For every time subjects managed to put the cursor on top of the moving target, they were given a score.

Force plate signals were low-pass filtered and amplified 2000 times. Signals were recorded through a custom made program in LabVIEW 8.0, which also was used for feedback. Force plate signals were sampled at 100 Hz. The change in performance over the 4 repetitions was evaluated by measuring the score as the number times the cursor was put on top of the moving target (hits). Force signals in the anterior-posterior (AP) direction and medial-lateral (ML) direction were analyzed by both linear and nonlinear measures to quantify variability in the force recorded. Standard deviation (SD) of the force was used to represent the size of variability and both approximate entropy (ApEn) and sample entropy (SaEn) were computed as two almost identical measures of the nonlinear dynamics within the force signals. Variability measures were computed for 3 different epoch lengths (5, 10 and 15 s) throughout the recorded signals in the AP- and ML-directions. Measures of all 4 trackings were compared in a repeated measure analysis of variance. Independent variables were SD, ApEn and SaEn for the 4 trackings and epoch lengths was used as fixed factor. Bonferroni was used for post-hoc analysis. The significance level was set to $p<0.05$.

III. Results

All results computed for SD, ApEn and SaEn in the tracking sessions are presented in figure 1.

A. Tracking sessions effects

The number of hits did not change significantly over the 4 tracking sessions ($p=0.470$).

In the comparison of the 4 repeated trackings, size of variability (estimated by SD) and structure of variability (computed as ApEn and SaEn) resulted the changes reported below.

In the AP direction:

SD decreased significantly for 5, 10 and 15 s epoch lengths (respectively, $F_{3,21}=6.22$, $p=0.034$; $F_{3,21}=6.04$, $p=0.034$ and $F_{3,21}=5.52$, $p=0.042$). SaEn increased only for the 5 s epochs ($F_{3,21}=3.83$, $p=0.025$), but ApEn increased for both the 5 s ($F_{3,21}=5.04$, $p=0.022$) and 10 s epochs ($F_{3,21}=3.89$, $p=0.026$).

In the ML direction:

SD also decreased for 5, 10 and 15 s epoch lengths (respectively, $F_{3,21}=8.27$, $p=0.009$; $F_{3,21}=9.32$, $p=0.005$ and $F_{3,21}=10.65$, $p<0.001$). SaEn only increased for 15 s epochs ($F_{3,21}=3.35$, $p=0.038$), while ApEn increased for both 5 s ($F_{3,21}=3.64$, $p=0.034$) and 15 s epochs ($F_{3,21}=4.20$, $p=0.038$).

B. Epoch length effects

For the comparison of epoch length, no differences were found between 5, 10 and 15 s ($F = 0.09$, $p = 0.918$). No significance was found for the post-hoc test.

IV. Discussion

During the session of 4 repeated trials, it was demonstrated in line with the hypothesis that the variability measures changed and the subjects altered their postural control, even though the performance in hits did not change significantly.

Postural mechanisms in the AP and ML directions are generally used to describe postural control [13,14]. In the AP, the ankle strategy, i.e. torque production around the ankle joints, is used for the control of posture. While in the ML direction, postural control is achieved by limb unloading-loading mechanism performed by hip abductors and adductors [14].

Previous studies have reported postural control deficit in whiplash and low-back pain patients and highlighted a misleading in the dynamic weighting of the sensory inputs [6,8,9]. Thus, as expected, standard deviation decreases during the four trials showing that subjects alter their variability even though it is not demonstrated in the performance represented by the number of hits. The use of standard deviation to quantify the variability in the postural control during the tracking is a common measure and it demonstrates

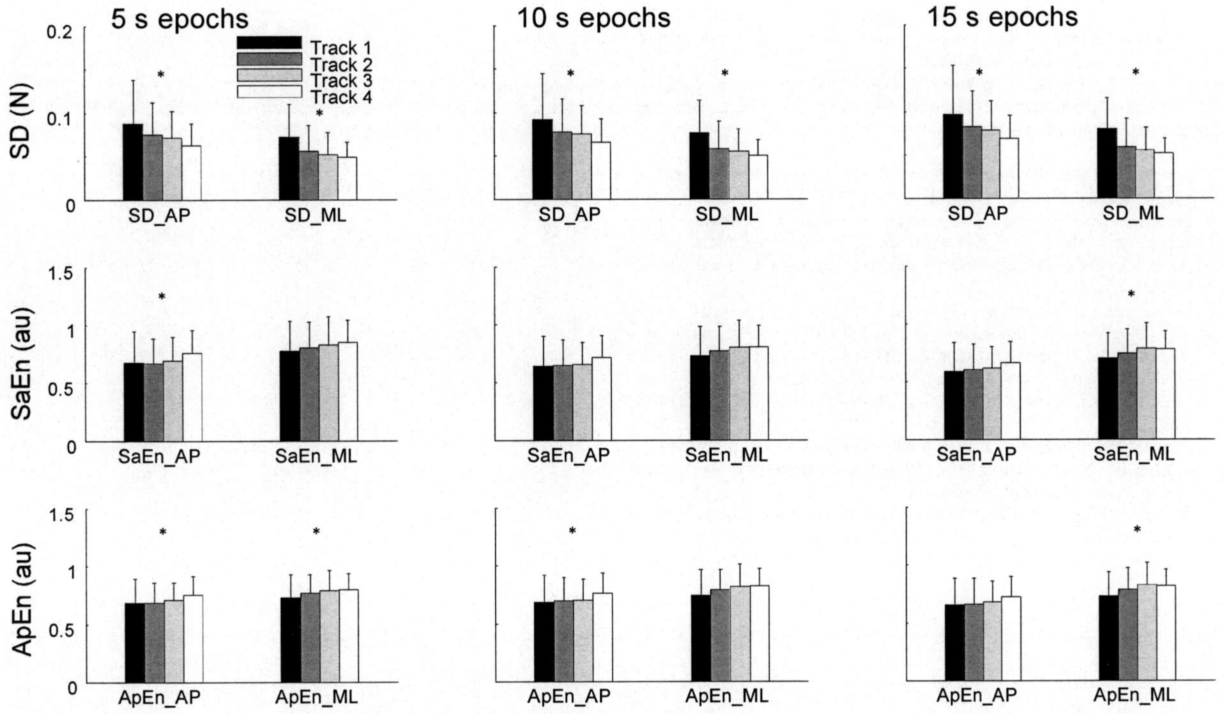

Fig 1. Results in anterior-posterior and medial-lateral direction for the tracking task. Tracking was repeated 4 times and variability and complexity measures were calculated for 3 different epoch sizes, 5, 10 and 15 s. Standard deviation (SD) was computed as linear variability, and both approximate entropy (ApEn) and sample entropy (SaEn) were computed as measures of nonlinear structures in the force plate recordings. Significant findings for the repeated ANOVA are marked by *. No significance was found between the different epoch lengths.

well how subjects fluctuate in the postural control. In the analysis of structural behavior, the computations of approximate and sample entropy are increasing as speculated, but as stated previously [7] the epoch length used for the computation of the entropy measures can be crucial for the results.

From this study it is difficult to conclude for how small sample sizes the use of SaEn will be superior to ApEn. In our results both ApEn and SaEn are increasing and only minor differences are shown. Presumably, if epoch lengths were shortened even further down to 1 or 2 s, a difference between the 2 entropy computations could be seen.

With the results gained in this study, it is demonstrated that even for a session of only 4 repetitions of the same trial, a learning effect is present. But from these results it is unclear whether a possible long-term effect can be achieved with an extended tracking task. The decrease in size of variability and increase in structure of variability reflect that a short term learning effect is achieved. Such effects can be of relevance in training sessions aiming at regaining motor confidence in low-back patient pain patients.

References

1. Massion J. Movement, posture and equilibrium: interaction and coordination. Prog Neurobiol 1992;38(1):35-56.
2. Hodges PW, Richardson CA. Inefficient muscular stabilization of the lumbar spine associated with low back pain. A motor control evaluation of transversus abdominis. Spine (Phila Pa 1976) 1996 Nov 15;21(22):2640-2650.
3. Brumagne S, Janssens L, Knapen S, Claeys K, Suuden-Johanson E. Persons with recurrent low back pain exhibit a rigid postural control strategy. Eur Spine J 2008 Sep;17(9):1177-1184.
4. Schmidt RA, Zelaznik H, Hawkins B, Frank JS, Quinn JT,Jr. Motor-output variability: a theory for the accuracy of rapid motor acts. Psychol Rev 1979 Sep;47(5):415-451.
5. Morrison S, Hong SL, Newell KM. Inverse relations in the patterns of muscle and center of pressure dynamics during standing still and movement postures. Exp Brain Res 2007 Aug;181(2):347-358.
6. Madeleine P, Nielsen M, Arendt-Nielsen L. Characterization of postural control deficit in whiplash patients by means of linear and non-

linear analyses - A pilot study. J Electromyogr Kinesiol 2011 Jun 15;21:297.
7. Richman JS, Moorman JR. Physiological time-series analysis using approximate entropy and sample entropy. Am J Physiol Heart Circ Physiol 2000 Jun;278(6):H2039-49.
8. Popa T, Bonifazi M, Della Volpe R, Rossi A, Mazzocchio R. Adaptive changes in postural strategy selection in chronic low back pain. Exp Brain Res 2007 Mar;177(3):411-418.
9. Baratto L, Morasso PG, Re C, Spada G. A new look at posturographic analysis in the clinical context: sway-density versus other parameterization techniques. Motor Control 2002 Jul;6(3):246-270.
10. Harringe ML, Halvorsen K, Renstrom P, Werner S. Postural control measured as the center of pressure excursion in young female gymnasts with low back pain or lower extremity injury. Gait Posture 2008 Jul;28(1):38-45.
11. Latash ML. Stages in learning motor synergies: a view based on the equilibrium-point hypothesis. Hum Mov Sci 2010 Oct;29(5):642-654.
12. Roerdink M, De Haart M, Daffertshofer A, Donker SF, Geurts AC, Beek PJ. Dynamical structure of center-of-pressure trajectories in patients recovering from stroke. Exp Brain Res 2006 Sep;174(2):256-269.
13. Horak FB, Macpherson JM. Postural orientation and equilibrium. In: Rowell LB, Shepherd J, editors. Handbook of Physiology. 1st ed. New York: Oxford University Press; 1996. p. 255-292.
14. Winter D. Human balance and posture control during standing and walking. Gait Posture 1995 12;3(4):193-214.

Non-linear Imaging Using an Experimental Synthetic Aperture Real Time Ultrasound Scanner

Joachim Rasmussen[1], Yigang Du[1,2], and Jørgen Arendt Jensen[1]

[1] Center for Fast Ultrasound Imaging, Department of Electrical Engineering, Technical University of Denmark, Kgs. Lyngby, Denmark

[2] BK Medical Aps, Mileparken 34, Herlev, Denmark

***Abstract*— This paper presents the first non-linear B-mode image of a wire phantom using pulse inversion attained via an experimental synthetic aperture real-time ultrasound scanner (SARUS). The purpose of this study is to implement and validate non-linear imaging on SARUS for the further development of new non-linear techniques. This study presents non-linear and linear B-mode images attained via SARUS and an existing ultrasound system as well as a Field II simulation. The non-linear image shows an improved spatial resolution and lower full width half max and -20 dB resolution values compared to linear B-mode imaging on the other systems. For the second scatterer at 47 mm depth the -20 dB resolution value for the non-linear SARUS image is 0.9907 mm and 1.1970 mm for the linear image from SARUS.**

***Keywords*— non-linear imaging, pulse inversion, synthetic aperture real time ultrasound scanner.**

I. Introduction

One way to improve the spatial resolution of a B-mode ultrasound image is to perform non-linear imaging. The pulse inversion (PI) technique [1] has for many years been an easy method to perform non-linear imaging. This technique acquires data in the same direction twice, where the second emitted pulse is phase shifted 180° compared to the first pulse. Adding the two received signals will cancel the 1st harmonic component of the received summed pulse due to the 180° phase shift. The 2nd harmonic component is phase shifted 2 · 180° and will therefore add constructively and be amplified. The technique can thus isolate the 2nd harmonic component even for broad band signals.

While non-linear imaging benefits from a good spatial resolution and low side lobes, PI suffers from lower penetration depth and a loss in frame rate. At the Center for Fast Ultrasound Imaging (CFU) a new fast non-linear imaging technique aimed at solving these issues is being developed using the experimental synthetic aperture real-time ultrasound scanner (SARUS) [2]. The purpose of this paper is to document the first non-linear imaging attempts using PI on SARUS and compare the results to existing ultrasound imaging systems and simulations.

II. Pulse inversion

In PI two consecutive waveforms, x_1 and x_2, that are identical except for a 180° phase shift are emitted [1],[3]. That is, $x_1=-x_2$ (see Fig. 1). The received signals, y_1 and y_2, contain higher order harmonics due to the non-linear propagation of sound waves in tissue. That is,

$$y_1 = a_1x_1+a_2x_1^2+\ldots \quad (1)$$

$$y_2= a_1x_2+a_2x_2^2+\ldots= a_1(-x_1)+a_2(-x_1)^2+\ldots, \quad (2)$$

where a_i are non-linear constants.

When the received waveforms, y_1 and y_2, are summed the out of phase odd number harmonics (1st, 3rd, ...) cancel out while the even in phase harmonics (2nd, 4th, ...) add. The amplitude of the even harmonics in the summed signal is twice that of the amplitude seen in either of the two received signals (see bottom of Fig. 2).

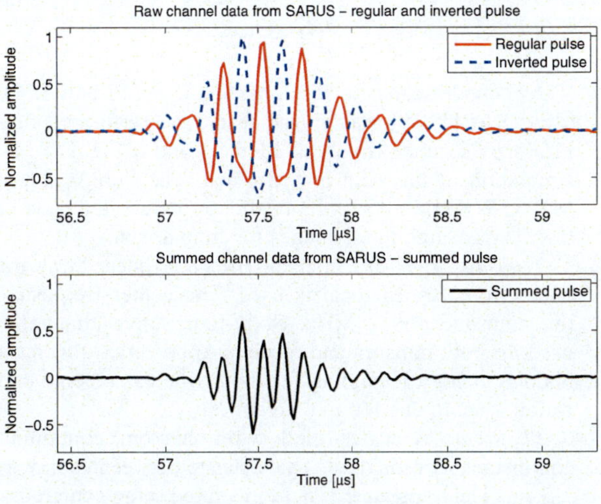

Fig. 1 Raw channel data from SARUS showing the normalized amplitude of the received regular pulse, the received inverted pulse, and the summed pulse used for PI imaging.

K. Dremstrup, S. Rees, M.Ø. Jensen (Eds.): 15th NBC on Biomedical Engineering & Medical Physics, IFMBE Proceedings 34, pp. 101–104, 2011.
www.springerlink.com

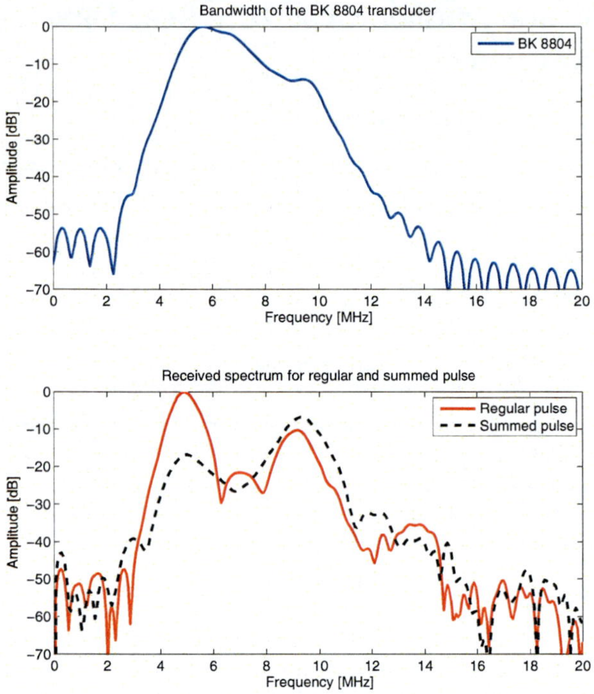

Fig. 2 Bandwidth of the BK 8804 transducer (top) and the spectrum of the received regular pulse and summed pulse (bottom). The fundamental frequency of 5 MHz is easily detected for the regular pulse as well as the 2nd harmonic component at 10 MHz. Both are well within the bandwidth of the transducer. For the summed pulse the fundamental frequency is suppressed by 17 dB while the 2nd harmonic component is enhanced by 4 dB compared to the 2nd harmonic component of the regular pulse.

III. Setup

A B-mode scan of a wire phantom using PI is performed using SARUS. Similar linear scans are performed on the ProFocus scanner from BK Medical and a Field II [4],[5] simulation of the scan is performed. The Full Width Half Max (FWHM) and -20 dB resolution values for each scatterer is measured in all images for comparison.

Transducer: A 192 element BK 8804 linear array transducer from BK Medical is used. The center frequency of this transducer is 7 MHz. Sixty-four active channels are used for both transmit and receive. Apodization for transmit is done using a Hamming function, whereas receive apodization is set to one for all 64 elements.

Phantom: A water-filled wire phantom containing 6 equidistant wires is used. The distance between each wire is 2.5 cm. The transducer is held in a fixed setup centered over the wires in the phantom with the surface of the transducer slightly submerged in water.

SARUS setup: For the SARUS scan 129 dual emissions are obtained to derive the non-linear image. The total collection of received data can be used for linear B-mode images (one for the regular pulse; one also for the inverted pulse) and one PI non-linear B-mode image (from the summed pulse).

Since the transducer of the system has a limited band width, the transmitted center frequency must be chosen such that it allows for the detection of the 2nd harmonic component in the received signal. From the bandwidth plot of the transducer in Fig. 2, a 5 MHz center frequency for the excitation pulse with a 10 MHz 2nd harmonic component is chosen. Both frequencies are well within the bandwidth of the BK 8804 transducer.

A fixed focal depth of 40 mm is used and the 64 channel received data are beam formed using the BFT3 toolbox [6].

ProFocus scanner setup: Twenty consecutive B-mode scans using 129 emissions with a 7 MHz center frequency are obtained using the ProFocus system from BK Medical. A fixed focal depth of 40 mm is set for all scans.

Field II setup: A simulation of a B-mode scan of the phantom is obtained using Field II with a 70 MHz sampling frequency, 5 MHz center frequency, 129 emissions, and a 40 mm fixed focal depth.

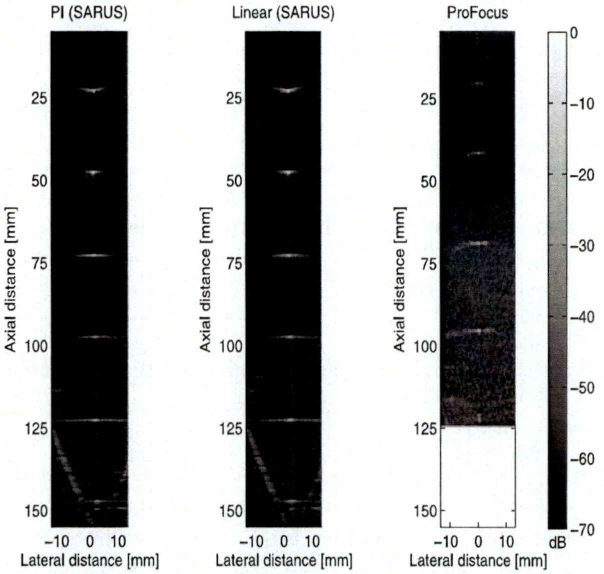

Fig. 3 B-mode images obtained from SARUS. Left shows the non-linear B-mode image of the wire phantom created via PI. Middle shows the same image obtained via linear B-mode imaging. Right shows the B-mode image obtained using the ProFocus scanner. Note that the ProFocus image depth is only 125 mm.

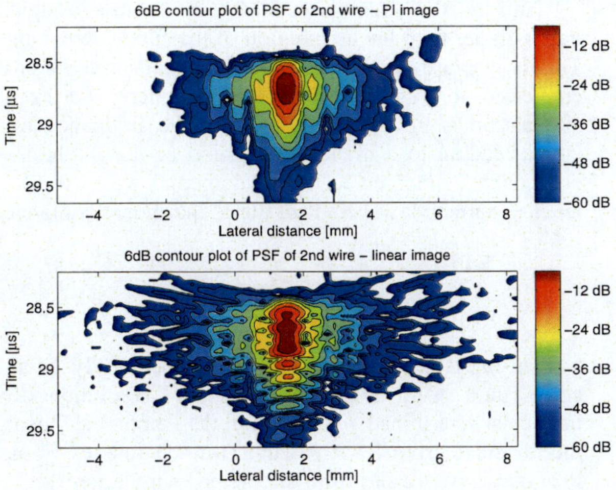

Fig. 4 Six dB contour plot of the PSF around the 2nd wire. Top shows the PSF for the non-linear PI B-mode image, bottom shows the PSF for the linear B-mode image.

IV. Results

B-mode images are obtained from all scans and from the Field II simulation. Fig. 3 shows the non-linear B-mode image from the PI scan, the linear B-mode image from a linear scan on the SARUS system, and a linear B-mode image from the ProFocus system. All 6 wires in the phantom are detectable in both the SARUS images as well as the structure of the bottom of the phantom at 150 mm depth. The ProFocus image depth is only 125 mm due to the settings on the system. Consequently, only the first 5 wires are seen and the bottom of the phantom can only just be perceived.

The point spread functions (PSF) for both the PI signal and the regular linear signal from the SARUS images around the 2nd wire in the phantom are shown in Fig. 4. In this figure it is clearly seen that the spatial resolution in the PI B-mode image is improved compared to the linear B-mode image. The PSF for the 2nd harmonic pulse has more narrow side lobes than the linear fundamental pulse.

Another quantitative measure for the spatial resolution of the B-mode image is the FWHM and -20 dB resolution values for each of the wires as shown in Fig. 5. From the top view of Fig. 5 it is seen that both SARUS imaging modalities have almost same FWHM values for all depths. The ProFocus system, however, has a lower FWHM resolution value than any of the other imaging system for the 4th wire indicating a better spatial resolution at this point.

In the bottom view of Fig. 5 it is seen that both imaging modalities on the SARUS system and the Field II simulation have generally lower -20 dB resolution values than the ProFocus system indicating an overall better spatial resolution. Especially at the 1st and 5th wire the ProFocus system is outperformed by all of the other imaging modalities. Furthermore, it is seen that SARUS PI and SARUS linear imaging have almost same resolution values except at the 4th and 6th wire where PI has lower -20 dB resolutions. At the 3rd wire the -20dB resolution of both SARUS modalities is lower than both Field II and ProFocus and at the 6th wire SARUS PI outperforms all other modalities.

On close inspection of the second wire at 47 mm depth, close to the focal point, the FWHM value of the non-linear SARUS image is found to be 0.7017 mm and the -20 dB resolution value to be 0.9907 mm. In comparison the FWHM of the linear SARUS image is 0.6604 mm and 1.1970 mm. These values indicate that although linear SARUS has lower FWHM value than non-linear SARUS, the shape of the PSF of the non-linear SARUS scatterer has more narrow side lobes due to the lower -20 dB resolution value. This is further verified by the PSF plot in Fig. 4 which also shows an improved spatial resolution of the non-linear image.

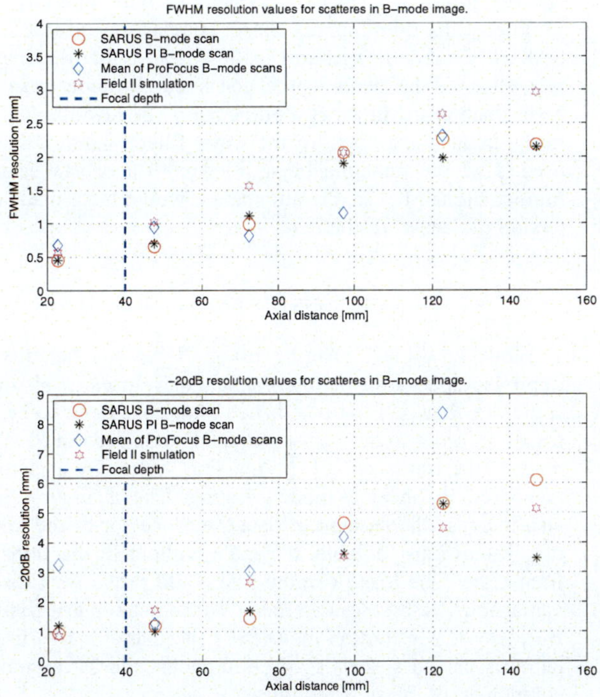

Fig. 5 FWHM (top) and -20 dB (bottom) resolution values for each wire in the phantom obtained using different imaging systems. Values for linear SARUS B-mode, PI SARUS B-mode, linear ProFocus, and Field II imaging are shown. Notice that only the first 5 wires of the phantom were imaged using the ProFocus system.

V. Discussion

From the results in Fig. 5 it is seen that the imaging modalities on SARUS generally has lower -20 dB resolution values than the ProFocus system. The very high -20 dB resolution values for the 1st and 5th wire in the ProFocus image could indicate a poor spatial resolution of the wire. While this is the case for the 5th wire the spatial resolution of the 1st wire in the image is in fact not as bad as indicated. This is due to the shape of the PSF around the 1st wire. Here the PSF takes a very pointy appearance with a very high maximum value and steep slopes, but with very wide low-level side lobes. This leads to a high -20 dB resolution value while the spatial resolution remains good. Accordingly, had the PSF taken the appearance of a hump with low maximum value, but with narrow side lobes, the -20 dB resolution value would be lower but the spatial resolution poor. When determining the spatial resolution of an imaging modality the FWHM and -20dB resolution values cannot be used alone, but must be compared to the actual image of the scanning before conclusions can be made.

Both SARUS imaging modalities have low FWHM and -20 dB resolution values compared to both Field II and ProFocus. In addition, in close comparison of the two SARUS B-mode images in Fig. 3, it is seen that the spatial resolution of the wires in the non-linear B-mode image is better than in the linear B-mode image. The low attenuation in the images is a result of the water filled phantom that is used. Had the phantom been filled with a substance that mimics human tissue, the attenuation would have made detection of deep wires harder.

VI. Pros and cons for non-linar imaging

While non-linear imaging using PI has the benefits of improved spatial resolution and low side lobes, it also has some drawbacks. First of all, two emissions need to be received in order to derive the summed pulse used in PI. This reduces the frame rate of the imaging system by a factor 2 compared to linear B-mode imaging. The dual emissions also increase the amount of data the processor of the imaging system must be able to handle without further loosing frame rate. The loss of frame rate could prove very disadvantageous, if the scan is made on non-stationary tissues. Here any tissue motion may lead to a phase change in the paired received signals severely reducing the 2nd harmonic component of the summed pulse.

Secondly, the transducer must function optimally over a broad spectrum to be able to transmit and receive a maximum energy at the fundamental and 2nd harmonic center frequency. If there is an energy loss at either frequencies the signal to noise ratio (SNR) in the image will decrease.

Thirdly, the attenuation of the 2nd harmonic signal is much higher than the attenuation of the fundamental signal. This is because only a fraction of the transmitted signal is converted to the 2nd harmonic component. Further, the attenuation is proportional in dB to the frequency of the signal leading to a higher attenuation of the 2nd harmonic component compared to the fundamental wave. In all this leads to a much lower SNR of the 2nd harmonic component.

VII. Conclusion

Non-linear B-mode imaging has successfully been accomplished using SARUS. The spatial resolution of the image is determined to be better than both the linear B-mode image from SARUS, the B-mode images from the ProFocus system, and from the Field II simulation.

Acknowledgement

This work was supported by grant 024-2008-3 from the Danish Advanced Technology Foundation and BK Medical Aps, Denmark.

References

1. Chapman C S and Lazenby J C (1997) Ultrasound imaging system employing phase inversion subtraction to enhance the image. Official Gazette of the United States Patent and Trademark Office Patents, Vol. 1198, Issue 4, pp. 2249.
2. Jensen J A, Tomov B G, Nikolov S I, Hansen M and Holten-Lund H (2007) System architecture of an experimental synthetic aperture real-time ultrasound system. Proceedings IEEE Ultrasonics Symposium, 2007, pp. 636-640.
3. Jiang P, Mao Z and Lazenby J C (1998) A new tissue harmonic imaging scheme with better fundamental frequency cancellation and higher signal-to-noise ratio. Proceedings IEEE Ultrasonics Symposium, 1998, Vol. 2, pp. 1589-1594.
4. Jensen J A (1996) Field: A program for simulating ultrasound systems. 10th Nordic-Baltic Conference on Biomedical Imaging vol. 4 supplement 1, part 1:351-353.
5. Jensen J A and Svendsen N B (1992) Calculation of pressure fields from arbitrarily shaped, apodized, and excited ultrasound transducers. IEEE Trans. Ultrason., Ferroelec., Freq. Contr., 39:262-267.
6. Hansen J M, Hemmsen M C and Jensen J A (2011) An object-oriented multi-threaded software beam formation toolbox. SPIE, Medical Imaging, Ultrasonic Imaging and Signal Processing, 2011.

Corresponding author: Joachim Rasmussen
Institute: Technical University of Denmark
Street: Oersteds Plads 349
City: Kgs. Lyngby
Country: Denmark
Email: jr@elektro.dtu.dk

Stable Hydrophilic Polydimethylsiloxane Surfaces Produced by Plasma Treatment for Enhanced Cell Adhesion

C. Jensen[1], L. Gurevich[2], A. Patriciu[3], J. Struijk[1], V. Zachar[1], and C.P. Pennisi[1]

[1] Department of Health Science and Technology, Aalborg University, Aalborg, Denmark
[2] Department of Physics and Nanotechnology, Aalborg University, Aalborg, Denmark
[3] Neurodan A/S, Aalborg, Denmark

Abstract— **Polydimethylsiloxane (PDMS) is a widely used polymer for medical implants due to its excellent physical properties, low cost and ease of fabrication. However, in some applications the hydrophobic nature of the material remains an issue. To increase PDMS hydrophillicity, a variety of surface treatments based on plasma discharge have been proposed. In this study, we investigated the effect of water-vapor based plasma on PDMS surfaces. Surface topography was analyzed by means of atomic force microscopy (AFM) while surface chemistry was obtained by Fourier transform infrared spectroscopy (FTIR). To analyze the stability of the treatment, surface wettability was assessed over a period of seven months by contact angle measurement. Furthermore, using primary human fibroblasts, *in vitro* cell growth and morphology was investigated. It was found that plasma treament produced long-term stable hydrophillic surfaces (contact angle between 70° to 80°). This property was correlated with hydroxylation of the surface and was accompanied by a slight increase in RMS roughness. Concomitantly, there was a significant increase in the number of cells growing on the plasma-treated surfaces, which was linked with a more spread cellular morphology. The results presented here suggest that water-vapor plasma treatment may be useful to enhance cell adhesion on PDMS implants.**

Keywords— **Biomedical polymers, surface modification, surface chemistry, wettability, cell compatibility.**

I. INTRODUCTION

Polydimethylsiloxane (PDMS), also known as silicone, is a polymer consisting of repeating inorganic siloxane groups (O-Si) as backbone with two organic methyl groups (CH_3) attached to the silicon. These methyl groups give the surface a low interfacial free energy, which makes the polymer hydrophobic (water contact angle around 109°) and chemically inert [1]. Due to its ease of fabrication, low cost, and biocompatibility, PDMS is widely used as biomaterial in applications such as reconstructive implants, catheters and insulation for implantable electrodes [1]. However, due to its surface hydrophobicity, significant protein adsorption occurs when the material is in contact with biological fluids. In some cases, this could lead to problems such as accumulation of inflammatory cells or bacterial film adhesion, which ultimately would cause failure of the implant [2].

Several surface treatments have been applied to reduce hydrophobicity of PDMS [3]. A useful technique to increase PDMS wettability consists in exposing the material to a plasma discharge, which is simple and easy compared to other methods, allowing simultaneous adjustment of surface topography and chemistry. A common approach is the application of an oxygen-based plasma, which forms silanol groups (Si-OH) on the PDMS surface [3]. However, this surface modification is unstable and hydrophobicity is recovered sooner or later after the treatment, depending upon the storage conditions [4].

As an alternative to oxygen plasma, water-vapor plasma is an attractive technology to increase surface wettability of diverse materials. In this modality, high concentrations of reactive, excited water molecules and hydroxyl radicals are present in the atmosphere during the treatment. This approach has shown to produce densely hydroxylated surfaces, in which hydrophobic recovery is slower [5]. This increased wettability has shown to enhance cell compatibility of various polymers of biomedical interest [6]. The aim of this study was to apply water-vapor plasma based treatment to PDMS, and analyze changes in wettability in relation to surface chemistry and topography. In addition, we aimed to investigate the effect of these surface modifications on *in vitro* cell growth and morphology.

II. MATERIALS AND METHODS

A. *Fabrication of Samples and Surface Treatment*

The experimental samples were fabricated with a medical grade silicone polymer (MED-4211, NuSil) mixed in a 1:10 ratio (w/w). A volume of the mixture was poured on a glass mold to achieve a thickness of 1.5±0.5 mm and degassed in

K. Dremstrup, S. Rees, M.Ø. Jensen (Eds.): 15th NBC on Biomedical Engineering & Medical Physics, IFMBE Proceedings 34, pp. 105–108, 2011.
www.springerlink.com

vacuum for an hour to remove air bubbles. After curing it overnight in an oven at 100 °C, the material was detached from the molds and annealed at 100 °C for a week to reduce the number of low molecular weight chains [7]. The substrate was cut into 1×1 cm samples and placed in a reactive ion etching system (STS RIE 320, Surface Technology Systems) for the surface treatment. The treatment consisted of a cleaning step in oxygen plasma at a pressure of 50 mTorr, RF power of 400 W, gas flow rate of 20 sccm (standard cubic centimeters per minute) for three minutes, followed by an etch in water vapor at a pressure of 50 mTorr, 400W, for eight minutes. After the plasma treatment, the samples were stored in ultrapure water.

B. *Surface Chemistry*

Attenuated total reflectance Fourier transform infrared (ATR-FTIR) spectroscopy (FT-IR 1750, Perkin Elmer) was used for analysis of the surface chemistry. Briefly, the samples were mounted into the spectrometer and the IR absorption spectra was obtained by the Perkin-Elmer's IRDM software (v3.3). Spectra were obtained by averaging twenty scans.

C. *Surface Wettability*

The wettability of different surfaces was monitored on a daily basis over a period of 18 days after etching. To assess long-term stability, wettability was measured at 147 and 233 days after treatment. A 3 µL droplet of deionized water was deposited onto a test sample and a picture was taken within 30 s after deposition. This procedure was repeated in at least three different locations for each surface. Images were captured using a digital camera (Camedia C3030, Olympus) attached to an upright microscope (BX41, Olympus), which was tilted to obtain the images of the droplet. The average contact angle was determined using the Drop Shape Analysis plugin for the image processing program Image J version 1.42q (available at http://rsb.info.nih.gov/ij).

D. *Surface Topography*

Surface topography of the samples was obtained by atomic force microscopy (AFM) in tapping mode (SMENA Scanning Probe Microscopy, NT-MDT). Silicon cantilevers (AC-160TS, Olympus) with a tip radius of 10 nm, a resonance frequency around 300 kHz and a spring constant between 50.3 to 69.3 N/m were employed. Sample scanning was controlled by software (Nova 1.0.26, NT-MDT). Scans were performed on a 10×10 µm area at a resolution of 512×512 pixels. Image processing and calculation of root-mean-square (RMS) roughness were done by using the WSxM 5.0 software (Nanotec Electronica).

E. *Cell Culture and Assessment of Cell Growth*

Each specimen consisted of an untreated and a plasma treated sample placed in the same culture dish. Samples were seeded with primary human fibroblasts (ATCC CRL2429) at a density of 4,000 cells/cm^2. Growth medium consisted of Iscove's Modified Dulbecco's Medium (IMDM) supplemented with 10% fetal calf serum, 10% penicillin/streptomycin and 0.5% gentamycin. To assess cell growth, two dishes were selected per day on the following three days. Cells were washed in phosphate buffered saline (PBS), fixed in 4% formaldehyde and stained with Hoechst 33342 dye. For cytoskeletal and morphological analysis, samples from day 2 were permeabilized with 0.1% Triton X-100 in PBS, incubated with PBS containing 1% of bovine serum albumin and Bodipy 558/568 Phalloidin (Invitrogen). Image sampling and counting was performed by means of fluorescence microscopy (Observer.Z1, Carl Zeiss) using the software AxioVision v2.7. For cell counting, a total of 51 images were collected on each surface.

III. Results

A. *Plasma Treatment Induced Surface Hydroxylation*

The infrared spectra of untreated and plasma treated samples are shown in figure 1. The spectrum for the untreated sample contains five main peaks, which are correlated to Si-$(CH_3)_2$ (788 cm^{-1}), Si-O-Si asymmetric deformation (1010 cm^{-1}), Si-CH_3 symmetric deformation (1259 cm^{-1}), Si-CH_3 asymmetric deformation (1451 cm^{-1}), and methyl CH bonds (2965 cm^{-1}). On the plasma treated sample there is a broad absorption band between 3000 and 3700 cm^{-1}, which corresponds to the OH bond of water molecules.

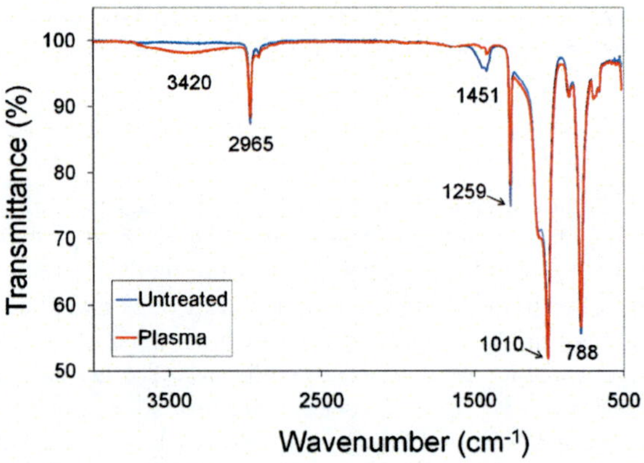

Fig 1 Infrared absorption spectra showing the effect of plasma treatment on the surface chemistry of PDMS samples

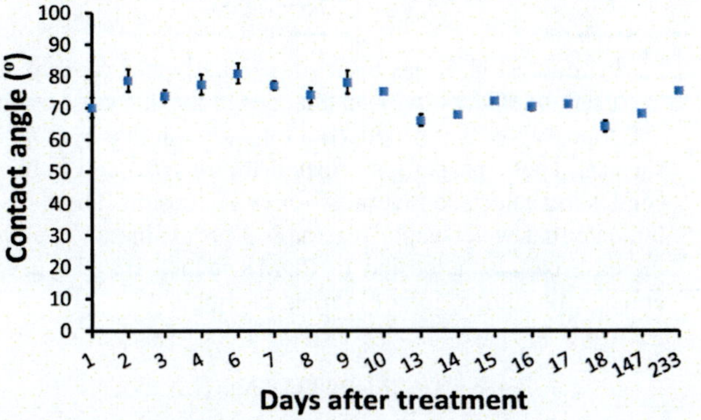

Fig 2 Contact angle measurements vs. time for the plasma treated sample. Data is presented as mean value ± standard error of the mean

B. *Long-Term Stable Contact Angles on the Plasma Treated Samples*

The contact angle of untreated PDMS was constant over the entire observation period (103±3°). Plasma treatment resulted in an increased hydrophillicity of the surfaces, which was relatively stable over the observation period (Figure 2). When measured on a daily basis, the mean contact angle displayed day-to-day variations. However, these variations were stable within a narrow range and wettability did not show an evident trend to increase or decrease with time. Long-term stability was confirmed by measurements performed at 147 and 233 days after treatment.

C. *Plasma Treatment Increased Surface Topography*

Representative AFM images depicting surface topography of the samples is shown in Figure 3. It can be seen that untreated samples present a relatively smooth surface, with height fluctuations in the range of 90 nm. On the other hand, plasma treated samples display an increased roughness, with surface features in the range of 400 to 500 nm. AFM analysis revealed significant differences in the surface root mean square (RMS) roughness values, which were 15.4±6.9 nm and 37.5±12.8 for untreated and plasma treated samples ($p<0.05$).

D. *Increased Cell Numbers and Spread Cellular Morphology on Plasma Treated Samples*

Figure 4 shows the morphology and distribution of cells on the different samples. DIC micrographs reveal evident differences in fibroblast morphology on the different surfaces, with cells on the plasma treated surfaces that were clearly more spread.

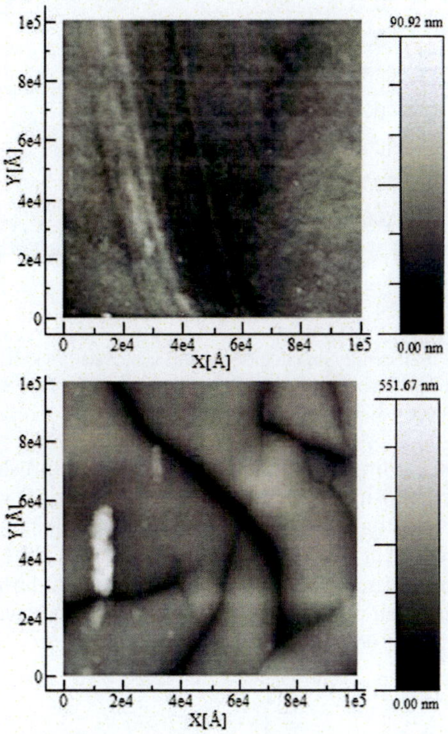

Fig 3 Representative AFM images of the PDMS samples. Top: Untreated. Bottom: Plasma treated

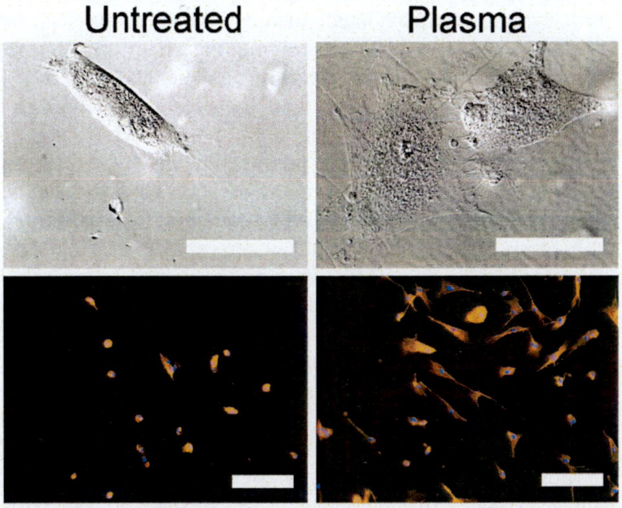

Fig 4 Photomicrographs of cells growing on the different surfaces. Top: DIC images showing cellular morphology. Scale bar represents 50 μm. Bottom: Representative fluorescence micrographs of fibroblasts at day 2 showing the general morphology of the cells on the surfaces. Cells were stained with Hoechst 33342 nuclear stain (blue) and Bodipy 558/568 phalloidin (orange). Scale bar represents 100 μm

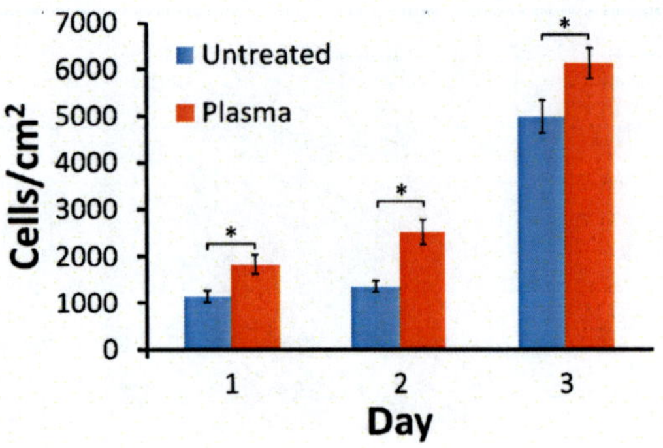

Fig 5 Cell density on the different surfaces. Data is shown as mean value ± standard error of the mean. Asterisks indicate a statistically significant difference ($p<0.05$)

Regarding cell numbers, figure 4 (bottom) shows a higher cell density on the plasma treated surface at day 2, which was quantitatively confirmed after cell counting (figure 5). Moreover, it was found that the number of cells on the plasma treated samples was significantly higher during the entire observation period ($p<0.05$). The growth rate was, however, similar for both untreated and treated samples.

IV. Dicussion

While the FTIR spectrum of untreated samples agrees excellently with published spectra of PDMS [8,9], the spectrum of plasma treated samples indicates the development of a highly hydroxylated surface. Hydroxylation is consistent with previous reports showing the effect of water-vapor plasma in other biomedical polymers such as PE and PP [6]. In addition, it was shown that the presence of OH groups caused an increase in surface wettability, which was relatively stable over a long period. This is in contrast with samples that are treated using oxygen-based plasmas, in which hydrophobic recovery can be retarded but not avoided [5]. It seems that water-vapor plasma does not create oxidized surface oligomers suspected to be involved in the process of hydrophobic recovery [5]. Regarding surface topography, although a significant increase in the RMS roughness values was found, estimates were within the same order of magnitude, suggesting a slight etching of the polymer surface. We consider this change is milder than other types of plasma treatments, which produce RMS roughness values in the order of 200 nm that turn the surfaces superhydrophobic [10]. Regarding cell behavior, it can be seen that although the cell growth rate was similar on both surface types, increased cell numbers and spread morphologies suggested a better cell adhesion to plasma treated PDMS. This is also supported by the fact that cell attachment is optimal in surfaces with a water contact angle about 70° [6].

V. Conclusion

In conclusion, it has been shown that water-vapor plasma treatment of PDMS induced changes in the surface chemistry and topography that produced a long-term stable hydrophillic surface. These properties affected the *in vitro* cell behavior, suggesting a surface favorable for cell adhesion. These results indicate that water-vapor plasma is a promising alternative to increase cellular adhesion to PDMS, which may be used to enhance biointegration of biomedical implants.

Acknowledgment

Authors would like to thank Jesper de Claville Christiansen for assistance with ATR-FTIR measurements, Neurodan A/S for providing the silicone polymer, and Lisa Crosato, Enrico Camatti, Mark Lillelund Rousing and Kristian Kjær Petersen for help with obtaining data. This work was supported by the Danish National Advanced Technology Fund, project Implantable Neural Prostheses.

References

1. Ratner BD (2004) Biomaterials Science: An Introduction to Materials in Medicine. Elsevier Academic Press, Amsterdam
2. Grill WM, Mortimer JT (1994) Electrical properties of implant encapsulation tissue. Ann Biomed Eng, 22:23-33
3. Zhou JW, Ellis AV, Voelcker NH (2010) Recent developments in PDMS surface modification for microfluidic devices. Electrophoresis 31:2-16
4. Kim B, Peterson ETK, Papautsky I (2004) Long-term stability of plasma oxidized PDMS surfaces. Conf Proc IEEE Eng Med Biol Soc, San Francisco, CA, 2004, pp. 5013-5016
5. Weikart CM, Yasuda HK (2000) Modification, Degradation, and Stability of Polymeric Surfaces Treated with Reactive Plasmas. J Polym Sci A, 38:3028-3042
6. Lee JH, Park JW, Lee HB (1991) Cell adhesion and growth on polymer surfaces with hydroxyl groups prepared by water vapour plasma treatment. Biomaterials, 12:443-448
7. Eddington DT, Puccinelli JP, Beebe DJ (2006) Thermal aging and reduced hydrophobic recovery of polydimethylsiloxane. Sens Actuators B, 114: 170-172
8. Bodas D, Khan-Malek C (2006) Formation of more stable hydrophilic surfaces of PDMS by plasma and chemical treatments. Microelectron Eng, 83:1277-1279
9. He Q, Liu Z, Xiao P et al. (2003) Preparation of Hydrophilic Poly(dimethylsiloxane) Stamps by Plasma-Induced Grafting. Langmuir, 19: 6982–6986
10. Pennisi CP, Zachar V, Gurevich L et al (2010) The influence of surface properties of plasma-etched polydimethylsiloxane (PDMS) on cell growth and morphology. Conf Proc IEEE Eng Med Biol Soc, Buenos Aires, Argentina, 2010, pp. 3804-3807

Author: Carina Jensen
Institute: Department of Health Science and Technology, Aalborg University
Street: Frederik Bajers vej 7
City: Aalborg, DK-9220
Country: Denmark
E-mail: cjen07@student.aau.dk

EMG Analysis of Level and Incline Walking in Reebok EasyTone ET Calibrator

E.F. Elkjær, A. Kromann, B. Larsen, E.L. Andresen, M.K. Jensen, P.J. Veng, and M. de Zee

Department of Health Science and Technology, Aalborg University, Aalborg, Denmark

Abstract— **This study examines (1) the effect of the Reebok Easytone walking shoe on muscle activity of three large extrinsic muscles compared to walking in a neutral running shoe, and (2) a comparison of muscle activity when walking on an inclined slope and walking in the EasyTone shoe on level ground. Methods. Ten male subjects participated in a cross-over study. A comparative analysis of treadmill walking in a Reebok EasyTone ET Calibrator and a Nike Lunarglide +2 at zero and 12% gradient was conducted. Electromyographic (EMG) signals from m. gastrocnemius lateral head (GL), m. biceps femoris (BF), and m. gluteus maximus (GM) of the left leg were processed. Results. This study found no significant (P < 0.05) increase in muscle activity between the two types of shoes. An expected significant increase (P < 0.05) from level to incline walking in GL, BF and GM was observed. Conclusion. Our study did not observe an effect of the Reebok EasyTone shoe on larger extrinsic muscles. Further testing should focus on the shoe's effect on smaller extrinsic muscles.**

Keywords— **Reebok EasyTone, EMG, Level walking, Incline walking.**

I. Introduction

Traditionally, shoes have been designed to stabilize and support the foot during walking. This image, however, has been challenged or nuanced by the concept of, for example, barefoot running, and thus alternative types of shoes are found in stores. According to Landry et al. [1], the idea of shoes being potentially 'over-protective' and causing reduced utilization of relatively small extrinsic muscles is a prevalent one among physical trainers and in the biomechanical milieu. Running and training barefoot is therefore seen as a means to strengthen the muscles in the lower leg. Other studies [2, 3, 4] have shown the efficiency of balance training devices such as the wobble board and the BOSU-ball in preventing sports-related injuries and improving ankle proprioception. The fitness shoes seek to combine and simulate these effects and others. By alternating the traditional design of soles, the shoes aim to destabilize the support phase during gait. Skechers and Masai Barefoot Technology (MBT) shoes are examples of shoes with a rounded sole, providing instability primarily in the posterior-anterior direction. Another example is the Reebok EasyTone shoe with its BOSU-ball inspired soles, with a potential instability manifesting in both posterior-anterior and medial-lateral directions (figure 1).

According to our knowledge and conviction, the MBT has been the most extensively studied shoe among the mentioned fitness products. However, its effects are not consistent across all studies. Nigg et al. [5] examined the electromyographic (EMG) activity of m. tibialis anterior, m. gastrocnemius, m. vastus medialis, m. biceps femoris, and m. gluteus medius in the MBT. They found no significant difference in muscle activity wearing the MBT or a control shoe during gait. Another study by Romkes and colleagues [6] found significant increases in m. tibialis anterior and the gastrocnemius muscle during contact and swing phases. In a systematic review of literature concerning the MBT and other instability devices, Murley et al. concludes by the following quote on [5, 6]: "we found no evidence to suggest this type of footwear has a systematic effect." [7: p. 179]. These contradictory results lay grounds to the primary hypothesis of this study.

Landry et al. [1] found significant values when examining the EMG-activity of smaller extrinsic muscles, namely: flexor digitorum longus, peroneus brevis and longus, tibialis anterior, extensor hallucis longus, and extensor digitorum longus.

Even though the effects of using this shoe are not consistent, the studies described above, especially Landry et al. [1], point out interesting tendencies. Further, the fact that the shoe is utilized by physiotherapists [6] indicates a belief in its effects among professionals.

In contrast to the MBT, the Reebok EasyTone shoe and its effects on muscle activity are more or less undescribed in literature. "Take the gym with you," a slogan used in the marketing of the product, appeals to practical individuals seeking an easy and effective exercise session with toning of muscles as a main aspiration. 'Toning' may refer to different processes within or the shape and firmness of the muscle but no definition of the word is provided by Reebok. In this study, we stipulate toning as the muscle tone – the residual muscle tension, which depends on the recruitment of motor units.

What is the immediate training response of the EasyTone shoe? In a study funded by Reebok, increases in muscle activity of 11%, 11% and 28% in calves, hamstrings, and m. gluteus maximus (GM), respectively, were observed. However, the inaccessibility and partiality of this study taken into account, we consider it an invalid scientific reference. Our study examined the activity of the select 'representatives' of the muscles monitored in the Reebook study, namely m. gastrocnemius lateral head (GL), m. biceps femoris (BF) and GM through EMG signal

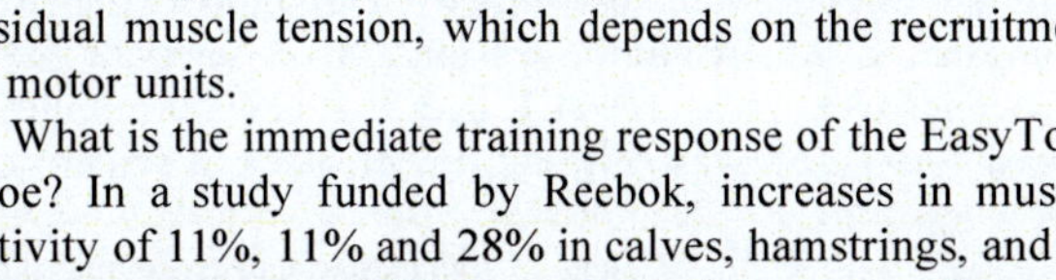

Fig. 1. Reebook EasyTone ET Calibrator

K. Dremstrup, S. Rees, M.Ø. Jensen (Eds.): 15th NBC on Biomedical Engineering & Medical Physics, IFMBE Proceedings 34, pp. 109–112, 2011.
www.springerlink.com

processing. The neutral Nike Lunarglide +2 was chosen as control shoe.

Based on the studies mentioned above [5, 6, 7], the following hypothesis is presented: there will be no significant change in the EMG activity between these two types of footwear.

In studies of human gait [11, 12], increases in muscle activity in the transition from level to incline walking were observed. On the basis hereof we hypothesize that walking on a 12% gradient leads to an equivalent or greater muscle activity than the values postulated by Reebok when walking on level ground in an EasyTone shoe.

II. Methods

A. Subjects

Ten healthy males participated in the study (age = 24.5 ± 3.8 years, BMI = 24.03 ± 1.09, and self-assessed training condition = 6.9 ± 0.88 on a scale from 1-10). Inclusion and exclusion criteria for participation were shoe size 43-45 (US size 10-11), no previous usage of the Reebok EasyTone shoe, regular level of physical activity, and a BMI within the norm as defined by the WHO [13]. For practical purposes, males rather than females were chosen. Subjects gave their informed consent and were orally and in writing informed of the possibility of withdrawing from participation at any given time.

B. Data collection

EMG measurements were carried out on the left leg with a reference electrode positioned on tibia. Surface electrodes (Ambu Neuroline 720) were attached to GL, BF, and GM according to the recommendations of electrode placement set by Konrad [14]. The EMG signal was hardware filtered at high and low pass settings of 5 – 1000 Hz before the collection of data was conducted at a sample rate of 2000 Hz. For every muscle, three maximal voluntary contraction tests (MVC) were performed. Throughout the testing, consisting of treadmill sessions, five intervals of five seconds were measured at 1, 3, 3.5, 4, and 4.5 minutes. The raw EMG data was collected in Mr. Kick, a scientific data acquisition software [14].

C. Protocol

In this cross-over design study, every subject was tested in the Reebok EasyTone ET Calibrator and the Nike Lunarglide +2 control shoe at a 0% and 12% gradient (level and incline walking, respectively). The sequence of gradient and shoe type was randomized.

Initially, the participant was orally introduced to the procedure of testing, and had electrodes placed on the left leg. Following, the subject warmed up for three minutes on the treadmill at a test speed of 4.86 km/h, followed by MVC tests of the three muscles. The MVC setups are similar to those utilized in Konrad [15]. However, the GL setup deviated from [15] and will be discussed later. Finally, the subject walked for 4 x 5 minutes on the treadmill at 4.86 km/h. Between every five-minute walking test, two minutes were added for adjustments of gradient and/or change of shoes.

D. Data analysis

Analysis of data was done in MatLab. All negative signals from walking and MVC tests were rectified and filtered at Butterworth 4th order. Within all five-second intervals, three peaks were defined. The mean of these 15 peaks was used to express the muscle activity (referred to as 'peak calculation'). For the MVC test, a mean of three steady-state windows of a 1000 samples was calculated for every muscle. Further, MVC normalization was calculated for the mean peaks from treadmill walking test. To obtain the muscle activity in relation to time, the integrated muscle activity was calculated via the trapezoidal rule (referred to as 'integration calculation'). A window of 100 samples for every burst defined the area in question. A burst was taken into account when 50 samples were present on both sides of its peak value. MVC normalization was carried out here as well for later comparison of results. Before peak calculation of the mean muscle activity for GL, BF, and GM, four outliers were identified and discarded.

Table 1 Mean muscle activity, computed by either the peak (A) or integration (B) calculation, in percent of MVC (±) standard deviation.

A	Level Lunarglide	Level EasyTone	Incline Lunarglide	Incline EasyTone
GL	48,68 ± 13,26	51,68 ± 8,45	123,35 ± 26,73	124,46 ± 22,53
BF	35.12 ± 17.65	33.68 ± 16.28	55.03 ± 23.00	56.67 ± 26.14
GM	22.57 ± 9.01	22.54 ± 10.30	41.02 ± 19.29	42.08 ± 19.56

B	Level Lunarglide	Level EasyTone	Incline Lunarglide	Incline EasyTone
GL	44,26 ± 10,95	46,95 ± 8,93	110,29 ± 29,06	110,62 ± 24,71
BF	33,78 ± 23,7	32,86 ± 22,29	51,43 ± 29,24	53,19 ± 29,4
GM	34,41 ± 16,61	34,05 ÷ 19,67	61,2 ± 26,14	70,55 ÷ 39,58

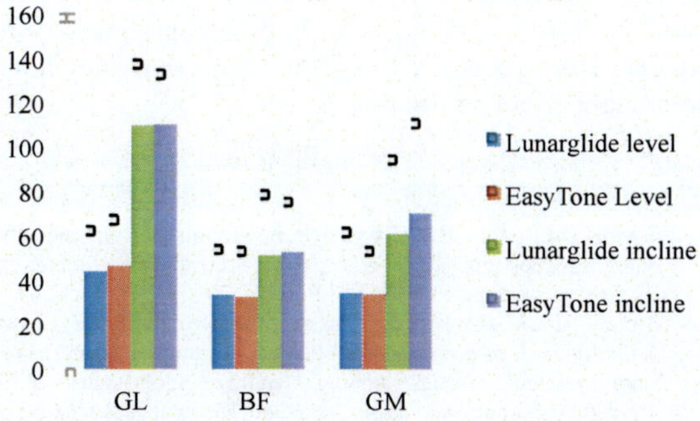

Fig. 2 Peak computed means in percent of MVC (y-axis) during level and incline walking in Lunarglide and EasyTone. Signficant differences between level and incline walking were observed for all three muscles.

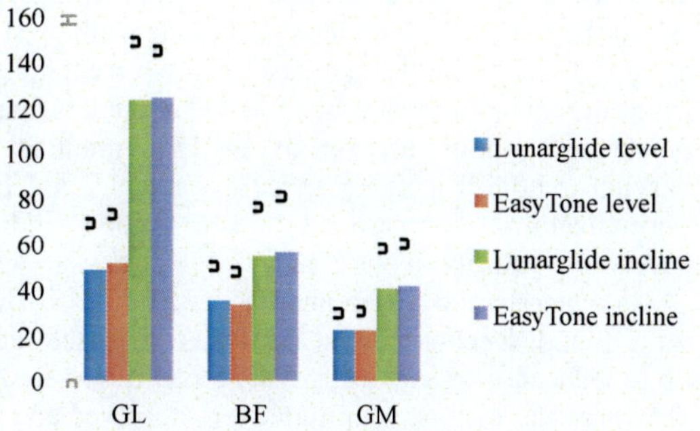

Fig. 3 Integration computed means in percent of MVC.

E. Statistical analysis

The normal distribution for all samples was tested using the Shapiro-Wilk test at a significance level of 0.05 in SPSS 18. In 12 out of 12 cases, the measurements were normally distributed. Two-way repeated measures ANOVA was employed to test for a significant difference between shoe type and gradients and performed in MS Excel 2007. Level of significance was set at $P < 0.05$ for all tests.

III. Results

MVC-normalized results calculated from both mean peaks and the integrated muscle activity show no significant($P < 0.05$) difference between Lunarglide and EasyTone at level and incline walking, respectively. However, the results show a significant increase in muscle activity from level to incline walking (figures 2 and 3) for GL ($P < 0.001$), BF ($P = 0.003$), and GM ($P < 0.001$). The P values apply to both calculating methods. MVC-normalized results calculated from both mean peaks and the integrated muscle activity show no significant ($P < 0.05$) difference between Lunarglide and EasyTone at level and incline walking, respectively. However, the results show a significant increase in muscle activity from level to incline walking (figures 2 and 3) for GL ($P < 0.001$), BF ($P = 0.003$), and GM ($P < 0.001$). The P values apply to both calculating methods.

A. Level and incline walking

The following means are based on the values listed in Table 1. For the peak calculation of GL in Lunarglide and EasyTone during level walking, we found muscle activity levels to be 50-60% of MVC. A corresponding 50% of MVC was found using the integration calculation. Relative similar values for BF when taking both calculation methods into account were found at 30-35%. For GM the results varied between the two methods, thus we found 20-25% muscle activity when using peak values and 30-40% for the integration calculation. No significant difference in muscle activity for GL, BF and GM was found between EasyTone and Lunarglide.

At a 12% gradient, differences in the means of both calculation methods were observed. For GL, these differences were 120-125% and 110-115% of MVC, respectively. All results for BF were between 50 and 60%. The peak calculated value of GM was 41% and 60-70% of MVC for the integration method. Furthermore, both calculations showed significant differences between level and incline walking in GL ($P < 0.001$), BF ($P = 0.003$) and GM ($P < 0.001$) with values above 75%, 55%, and 140% of MVC, respectively.

IV. Discussion

The Reebok EasyTone shoe is a newly launched product sold in various sports stores. Through advertisements on TV and internet, the alleged effects of the shoes on lower leg muscles are presented. However, no scientific study supports these claims. Thus, it is important to highlight the actual benefits, if any, for a given consumer buying and wearing this kind of walking shoe. This study compared the EasyTone to a neutral running shoe and found no significant difference ($P < 0.05$) between the two.

EMG activity of the three muscles at level walking corresponds to findings of other studies. Ferraro [16] found a muscle activity of 60% of MVC at a 0% gradient in m. gastrocnemius. Chuansi et al. [11] observed a muscle activity of 24% in BF. This study produced muscle activities of 55 and 35%, respectively. For GM, Mangold et al. [17] measured an activity of 27% of MVC against 23% found in this study.

When walking on an inclined surface (0 to 14% gradients) Wall-Scheffler et al. [12] observed increases of 15 and 15-20 percentage points for GM and BF, respectively. The corresponding values of these muscles from our study are 18 and 20 percentage points. We observed a significant increase of 70 percentage points in GL in the transition from level to incline walking. Leroux et al. [18] found a similar increase of 80 percentage points in m. gastrocnemius medial head at a 9% gradient, and a 120 percentage points increase at an 18% gradient. Despite varying treadmill walking speeds, gradients, and factors of normalization, comparisons made above indicate an acceptable validity in our results.

As expected, no differences between the two shoe types were found. Reebok claims increases in muscle activity of up to 28% and 11% by using EasyTone. In our study, only a few increases of this magnitude were observed and in just as many cases as negative values, i.e. decreasing muscle activity. We consider these values arbitrary and not indicative of an actual level of muscle activity of 11% and 28%. As a consequence, we use means to express the influence of the shoe on muscle activity during walking.

According to the findings of this study, greater effects are gained from incline walking than level walking in the EasyTone shoe. This effect potentially has a hypertrophic impact on the three muscles as it surpasses 40% of MVC [15] in all cases (Table 1). Hypertrophy may tone muscles [9] but, as indicated by our results, only at incline walking and not at level walking in EasyTone.

Before discarding the four outliers mentioned above, the mean muscle activity level of GL during level walking in Lunarglide was found to be 54.55%. We consider the actual mean value to be lower, as a single subject's mean reading deviated sharply from the rest. In this case, we observed an activity of 107% of MVC (not included in Results). This deviation may indicate improper execution of the MVC test due to the subject (1) not having sufficient time to get familiar with the MVC procedure or (2) manual fixation may have led to mobility of the MVC setup. The deviation appears only for GL, which may indicate an external influence (2).

Our results show no increase in muscle activity in the relatively large extrinsic muscles when walking in the Reebok EasyTone shoe. In addition to the findings of other studies of instability shoes [1, 5], these results indicate that the potential effects are not to be found in the larger lower extremity muscles, but a change in the smaller extrinsic muscles cannot be ruled out. For example, Landry et al. [1] found increased muscle activity in m. flexor digitorum longus, peroneal, and muscles in the anterior compartment, but no increase in the larger m. soleus for standing in an unstable shoe. Further investigation of the effects of EasyTone on smaller extrinsic muscles is necessary to determine whether this assumption holds true.

V. Conclusion

No significant increase in muscle activity in GL, BF, or GM was observed, confirming the primary hypothesis of this paper. In terms of muscle activity, incline walking proved a more effective alternative to the EasyTone shoe. Further testing on more parameters, and of EMG on smaller extrinsic muscles, is required, but these preliminary results indicate that Reebok's EasyTone shoe does not have any immediate effect on training.

References

1. Landry SC, Nigg BM, Tecante KE. Standing in an unstable shoe increases postural sway and muscle activity of selected smaller extrinsic foot muscles. Gait and Posture, 2010, 32, p. 215-219.
2. Emery CA, Cassidy JD, Klassen TP, Rosychuk RJ, Rowe BH. Effectiveness of a home-based balance-training program in reducing sports-related injuries among healthy adolescents: a cluster randomized controlled trial. Canadian Medical Association Journal 2005, 172 (6), p. 749-754.
3. Waddington GS, Adams RD. The effect of a 5-week wobble-board exercise intervention on ability to discriminate different degrees of ankle inversion, barefoot and wearing shoes: a study in healthy elderly. Journal of the American Geriatrics Society, 2004, 52, p. 573-576.
4. Yaggie JA, Campbell BM. Effects of balance training on selected skills. Journal of Strength and Conditioning Research, 2006, 20(2), p. 422–428
5. Nigg BM, Hintzen S, Ferber R. Effect of an unstable shoe construction on lower extremity gait characteristics. Clinical Biomechanics, 2006, 21, p. 82-88.
6. Romkes J, Rudmann C, Brunner R. Changes in gait and EMG when walking with the Masai Barefoot Technique. Clinical Biomechanics, 2005, 21, p. 75–81.
7. Murley GS, Landorf KB, Menz HB, Bird AR. Effect of foot posture, foot orthoses and footwear on lower limb muscle activity during walking and running: A systematic review. Gait & Posture, 2009, 29, p. 172-187.
8. http://www.reebok.com/DK/womens/easytone, 25/9-2010.
9. Martini FH, Nath JL. Fundamentals of Anatomy and Physiology. 2008. Forlag: Pearson Education (Us), 8th edition, p. 316.
10. Reebok og University of Delaware. No further information available.
11. Chuansi G, Oksa J, Rintamäki H, Holmér I. Gait Muscle Activity during Walking on an Inclined Icy Surface. Industrial Health, 2008, 46, p. 15-22.
12. Wall-Scheffler CM, Chumanov E, Steudel-Numbers K, Heiderscheit B. Electromyography Activity Across Gait and Incline: The Impact of Muscular Activity on Human Morphology. American Journal of Physical Anthropology, 2010, 143, p. 601-611.
13. http://apps.who.int/bmi/index.jsp?introPage=intro_3.html, 4/10-2010.
14. http://www.smi.hst.aau.dk/~knl/mk/introspec.htm, 2/10-2010.
15. Peter Konrad. The ABC of EMG - a practical introduction to kinesiological electromyography. Noraxon INC. USA, 2005.
16. Ferraro RA. The Effect of an incline walking surface and the contribution of balance on spatiotemporal gait parameters of older adults. Ph.D. thesis, 2010, p. 23.
17. Mangold S, Keller T, Popovic MR. Muscle activity during normal walking and its relevance for the functional electrical stimulation applications. No further information available.
18. Leroux A, Fung J, Barbeau H. Adaptation of the walking pattern to uphill walking in normal and spinal-cord injured subjects. Experimental Brain Research, 1999, 126, p. 359-368.

Author: Mark de Zee
Institute: Aalborg University, Dept. of Health Science and Technology
Street: Fredrik Bajers Vej 7-D3
City: Aalborg
Country: Denmark
Email: mdz@hst.aau.dk

In vivo Impedance Characterization of a Monopolar Extra-Neural Electrode

S. Meijs[1], M. Fjorback[1], and N.J.M. Rijkhoff[2]

[1] Neurodan A/S, Aalborg, Denmark
[2] Center for Sensory Motor Interaction (SMI), Aalborg University, Aalborg, Denmark

Abstract— **The impedance of 4 titanium nitride (TiN) coated monopolar extra-neural electrodes has been measured *in vivo* for a period of 3 weeks. The objectives of the study were to quantify both the electrode-electrolyte interface, as well as the tissue resistance *in vivo* as a function of time after implantation. Different currents (0.05, 0.1, 0.5 and 5.0 mA) have been used at frequencies ranging from 0.1 Hz-100 kHz for extensive measurements once a week, while the animals were under anesthesia. The tissue resistance, R_{tissue}, the faradic resistance, R_f, double layer capacitance, C_{dl}, and charge transfer ratio between capacitive and faradic processes, Q_c/Q_f, of the electrode were estimated. For 3 of 4 electrodes, R_{tissue} could be reliably estimated. The tissue impedance was low in the first week, after which it increased and stabilized. Using an amplitude of 0.1 mA, charge was transferred predominantly via a capacitive pathway. With increasing current density, the faradic pathways became more dominant and the frequency at which the faradic pathways became more dominant decreased with increasing current density. R_f was higher at higher currents, whereas C_{dl} was lower at higher currents. The results indicate that when the phase angle approaches 0, R_{tissue} provides reliable information regarding the healing process. The Q_c/Q_f curves confirm that the charge transfer mechanism of the TiN electrode interface is mainly capacitive. The mechanism of charge transfer changes towards faradic charge transfer for increasing current density. Both the increasing trend in R_f as well as the decreasing trend in C_{dl} reaches a plateau at the 0.5 mA. This might indicate that equilibrium is reached between the surface area used for faradic and capacitive charge transfer. The performance of the electrode is comparable to a Platinum/Iridium (Pt/Ir) electrode.**

Keywords— **Neural prostheses, electrochemical impedance spectroscopy, electrical stimulation.**

I. Introduction

Neural prostheses are currently used for treatment of e.g. chronic pain, incontinence, neurological and movement disorders [1, 2]. Neural electrodes can be used for stimulation of and recording from nervous tissue, both in the central and peripheral nervous system. The functionality of the electrode (stimulating, recording or both), its placement and its mode of operation impose different requirements on the electrode design [3].

In this study, a monopolar extra-neural electrode is considered, which is to be implanted in adipose tissue. The electrode will therefore use a relatively high current in order to achieve nerve fiber excitation. The materials used for the electrode are chosen based on their charge-injection capacity, biocompatibility and chemical stability [3].

The performance of the electrode is investigated *in vivo* using electrochemical impedance spectroscopy (EIS). The impedance spectrum is used both to investigate the tissue impedance, as well as the electrical properties and performance of the electrode over time. In order to make such analysis, the electrode interface is described by a circuit model. The values of the circuit components describing the physical characteristics of the interface are derived from the impedance spectrum [3, 4].

II. Materials and Methods

The impedance of 4 monopolar extra-neural electrodes was measured *in vivo* for a period of 3 weeks following implantation in the adipose tissue of the pubo-genital region of 2 Göttingen minipigs. The spherical electrode contact was made of Platinum/Iridium (Pt/Ir) semi-sphere (diameter: 1.00 mm), coated with titanium nitride (TiN) to enhance charge injection capacity [3]. A TiN coated titanium plate (diameter: 50.0 mm) was used as counter electrode.

A. Impedance measurements

Impedance measurements were performed using Solartron, Model 1294 in conjunction with 1260 Impedance/gain-phase Analyzer (Solartron Analytical, UK). Accompanying SMaRT software was used to run the system, specifying the root mean square (rms) current amplitude, the frequencies at which measurements are made and the integration level.

Measurements were performed both during anesthesia and while the animals were awake. During measurements in awake animals, the animals were restrained using a custom built sling. The impedance gain-phase analyzer was connected to the implanted electrodes via percutaneous wires.

Impedance was measured at frequencies ranging from 0.1Hz-100kHz. At low frequencies, fewer measurements were made and the number of cycles used for integration was low (see Table 1) to restrict experiment duration. Measurements were performed using sinusoidal currents with a rms amplitude of 0.1 mA.

K. Dremstrup, S. Rees, M.Ø. Jensen (Eds.): 15th NBC on Biomedical Engineering & Medical Physics, IFMBE Proceedings 34, pp. 113–116, 2011.
www.springerlink.com

During anesthesia, the animals were already immobilized, which allowed for more extensive impedance measurements. Measurements were performed once a week using different currents: 0.05, 0.1, 0.5 and 5.0 mA.

Table 1 Impedance measurement settings for different frequencies

Frequency	Points per decade	No. of integration cycles
< 1 Hz	2 - 0.1 & 0.5 Hz	3
1 – 100 Hz	5	5
> 100 Hz	5	10

B. Impedance characterization

In order to analyze the impedance data, a circuit model of the electrode interface was specified. Due to the TiN coating, the electrode was expected to operate mainly in a capacitive way [3], which was described by the Helmholz or double layer capacitance (C_{dl}) [5]. To allow the passage of direct current, a Faradic resistance (R_f) was assumed in parallel with C_{dl} [5]. The Warburg impedance [6] was neglected, because of the high current densities applied. In series with the parallel combination of C_{dl} and R_f, a resistance (R_{tissue}) was modeled to account for the electrolyte or tissue resistance, as shown in Figure 1. The surface of the counter electrode was approximately 300 times larger than the stimulation electrode's surface. It was therefore assumed that it would not have a significant influence on the measured impedance.

The values of the different circuit elements were derived from the impedance spectrum using the following mathematical description:

$$Z(\omega) = R_{tissue} + \frac{R_f}{1 + j\omega \cdot R_f \cdot C_{dl}} \quad (1)$$

Where ω is the angular frequency and Z(ω) the measured impedance. C_{dl} was also measured by the impedance device.

The charge transfer through the electrode-electrolyte interface is shunted by C_{dl} at high frequencies, R_{tissue} is therefore determined as the measured impedance at the highest frequency, $Z(\omega \rightarrow \infty)$. R_f can then be determined as follows:

$$R_f(\omega) = \frac{R_{tissue} - Z(\omega)}{-j\omega \cdot C_{dl}(\omega) \cdot (R_{tissue} - Z(\omega)) + 1} \quad (2)$$

Lastly, the ratio of charge delivery via capacitive and faradic pathways is computed as follows:

$$\frac{Q_c}{Q_f} = \frac{I_c}{I_f} = j\omega \cdot R_f \cdot C_{dl} \quad (3)$$

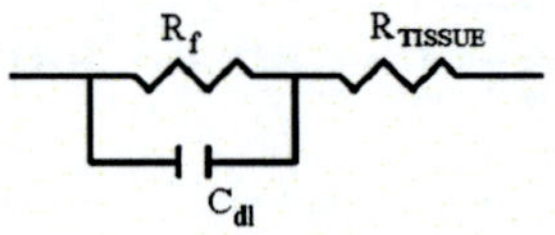

Fig. 1 Circuit model of the electrode interface and the tissue impedance.

III. Results

The impedance of a TiN coated monopolar extra-neural electrode implanted in adipose tissue was measured. C_{dl} and R_f were quantified as a function of frequency, current density and time after implantation. R_{tissue} was quantified only as a function of time after implantation.

A. Impedance characteristics

The impedance magnitude and phase spectra of all electrodes showed the same trend. The impedance magnitude at low frequencies is high, decreasing to the asymptotic high frequency impedance. The impedance phase angle starts at a high negative value and has two local minima. For most electrodes the phase angle approaches 0 at high frequencies. When the phase angle did not approach 0, no reliable value was established for R_{tissue}. This was the case for the right electrode of pig 2 and the last week of the right electrode of pig 1. R_{tissue} is shown as a function of time after implantation for all electrodes for which a reliable estimate was made in Figure 2. In pig 1, both electrodes showed relatively low tissue impedances in the first week after implantation, where after the impedance increased and stabilized.

The asymptotic high frequency impedances using different currents were the same. For higher currents, the low frequency impedance is lower and the impedance drops to approach R_{tissue} at lower frequencies (see Figure 3a). Furthermore, the phase angles are closer to 0 at high frequencies and increase towards 0 at lower frequencies (see Figure 3b). The second local minimum disappeared at 0.5 mA.

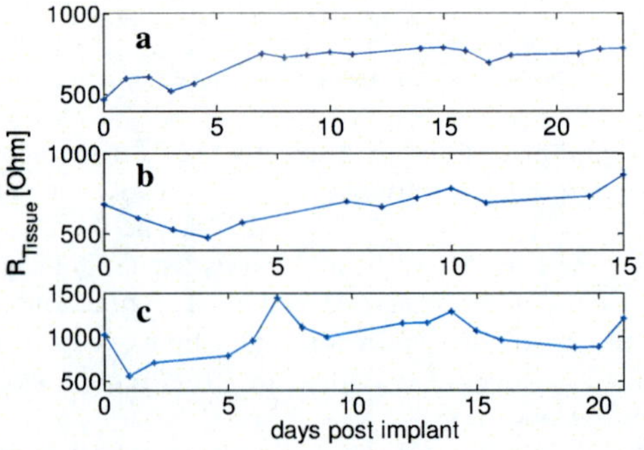

Fig. 2 The tissue impedances The left (a) and right (b) electrode of pig 1 and the left electrode of pig 2 (c) as a function of time post implant

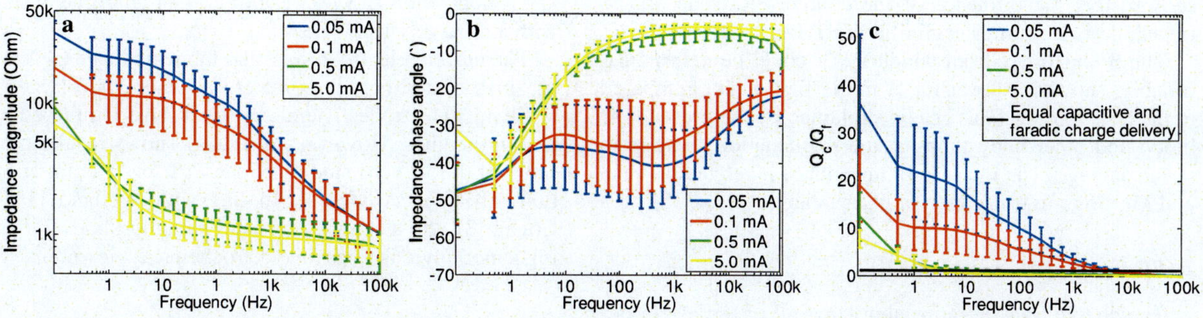

Fig. 3 The average impedance magnitude (a) and phase angle (b) of all electrodes measured using different currents during anesthesia. The average charge transfer ratios (c) are shown for the same currents. The ratio of capacitive charge transfer decreases with increasing current and frequency.

B. Electrode characteristics

The ratio of charge delivery via capacitive and faradic pathways is shown in Figure 3c. At low currents (<0.5mA) and frequencies (<10kHz), charge is transferred mainly via capacitive pathways. With increasing current, faradic processes became dominant at lower frequencies. The median frequency at which the faradic charge transfer becomes dominant was 10 kHz. This remained approximately constant during the course of this study.

Both C_{dl} and R_f were dependent on frequency and current density. The average values of all electrodes of R_f and C_{dl} measured during anesthesia are shown as a function of current density in figure 4. R_f decreases with increasing current density and reaches a plateau at 0.5 mA. C_{dl} increases with increasing current density and also reaches a plateau at 0.5 mA. With increasing frequency, these trends are smaller. The current density linearity limit is established as the current density at which the values of R and C deviate by 10% from their value at low current density [7]. This limit could not be established, as the values of C_{dl} and R_f increased from the lowest current density.

IV. Discussion

The aim of this study was to use EIS to quantify the impedance of a TiN coated monopolar extra-neural electrode in terms of C_{dl}, R_f as a function of frequency, current density and time after implantation. R_{tissue} was evaluated as a function of time after implantation.

A. Impedance characteristics

The impedance and phase spectra of all electrodes show the same trend. However, the variation in the impedances measured using the different electrodes is high, which is partly due to the differences in R_{tissue}. This also influences C_{dl} and $R_{f.}$, as a larger series resistance with the same applied current, will lead to higher potentials at the electrode-electrolyte interface.

All phase spectra show two minima, around 1 Hz and around 1 kHz. In vitro measurements using 2 counter electrodes also showed a minimum at 1 Hz. Additional in vitro experiments using the stimulation and counter electrode

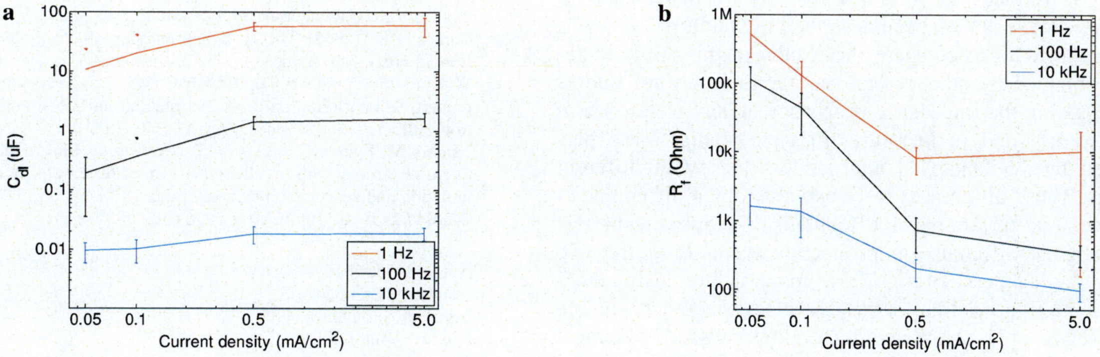

Fig. 4 The average R_f (a) and C_{dl} (b) of all electrodes as a function of current density for different frequencies.

showed that the influence of the counter electrode on the impedance spectrum is negligible (*f*>1Hz).

The 3 electrodes from which R_{tissue} could be determined reliably, showed a lower R_{tissue} in the first week, compared to the last 2 weeks. This can be explained by the presence of blood and other body fluids at the implantation site due to tissue damage. The increase in impedance is related by Grill and Mortimer to the growth of encapsulation tissue. [8]

B. Electrode characteristics

The charge transfer ratios show that all of the electrodes operate mainly in a capacitive way. This was expected, due to the TiN coated interface [3]. Impedance differences were noticed between all of the electrodes, especially between the two animals. Slight differences between the electrodes can be explained by the individual manufacturing of the electrodes, resulting in slightly different surface areas and structures. But there is a major difference between the two pigs as well. This can be explained by the difference in R_{tissue} [9]. Dissection showed an infection affecting both electrodes in pig 2. The development of R_{tissue} over time (Figure 2) shows an increase followed by a decrease in impedance and thus no stabilization. This causes continuous presence of fluids instead of tissue around the electrode, which makes transport of chemical species and direct current passage across the electrode interface easier [9].

The current density linearity limit could not be established, meaning that the electrodes did not operate in a linear fashion during the measurements [7]. The decrease in R_f is caused by a different type of reaction taking place at the electrode surface [4]. The increase in capacitance as a function of current density was explained by Conway as an increase in the fraction of the electrode surface that is covered by adsorbed charged chemicals partaking in charge transport [10]. The plateaus in Figure 4, occurring at the same (high) current density, may indicate that equilibrium between the two charge transfer mechanisms covering the electrode surface has been reached. Such a plateau has to our knowledge not yet been described in literature.

It is difficult to compare the results of this study to results obtained by other researchers, as there are no studies investigating the impedance properties of TiN electrodes *in vivo* were found in literature. *In vitro* studies were performed by Weiland [11] and Janders [12] with different results. These differences were explained by a difference in surface area (65 μm^2 vs. 4000 μm^2). [11] However, there is also a great difference in production methods of the two electrodes. Janders' electrode was manufactured using reactive sputtering, yielding a greater electrochemical surface area (ESA). [3, 12] Weiland used low-pressure chemical vapor deposition (CVD), resulting in an amorphous surface with a relatively small ESA [3].

The current electrode was also fabricated using CVD, but its surface area is approximately a factor 1000 larger than Weiland's electrode. Therefore it is still difficult to compare the two studies. However, this study shows results comparable to the in vivo studies of the Pt/Ir deep brain stimulation (DBS) electrode [4]. In line with Weiland [11] and Cogan [3], the advantage of using an amorphous TiN coating is not obvious from an electrochemical viewpoint.

V. CONCLUSION

The performance of the current electrode is similar to a Pt/Ir DBS electrode. [4] The amplitudes used in this study were too high, as the electrode does not operate in a linear fashion. In a subsequent study, the charge injection limits will be measured and high currents will not be used until the end of the study. In order to enhance the performance of the electrode, a fractal coating will be used. [3]

REFERENCES

1. Zhou D, Greenbaum E (2009) Implantable neural prostheses 1: devices and applications. Springer, New York
2. Bhadra N, Chae J (2009) Implantable neuroprosthetic technology. NeuroRehabilitation 25: 69-83
3. Cogan SF (2008) Neural stimulation and recording electrodes. Annu. Rev. Biomed. Eng. 10:275-309
4. Wei XF, Grill WM (2009) Impedance characteristics of deep brain stimulation electrodes in vitro and in vivo. J. Neural Eng. 6:9pp
5. Geddes LA (1997) Historical evolution of circuit models for the electrode-electrolyte interface. Annals of Biomed. Eng. 25:1-14
6. Randles JEB (1947) Kinetics of rapid electrode reactions. Discuss. Faraday Soc 1:11-19
7. Ragheb T, Geddes LA (1990) Electrical properties of metallic electrodes. Med. & Biol. Eng & Comput. 28:182-186
8. Grill WM, Mortimer T (1994) Electrical properties of implant encapsulation tissue. Ann. Of Biomed. Eng. 22:23-33
9. Duan YY, Clark GM, Cowan RSC (2004) A study of intra-cochlear electrodes and tissue impedance by electrochemical impedance methods in vivo. Biomaterials 25:3813-3828
10. Conway BE (1965) Theory and principles of electrode processes. Ronald press, New York.
11. Weiland JD, Anderson DJ, Humayun MS (2002) In vitro electrical properties for iridium oxide versus titanium nitride stimulating electrodes. IEEE Trans. Biomed. Eng. 49:1574-1579
12. Janders M, Egert M, et al. (1996) Novel thing film titanium nitride micro-electrodes with excellent charge transfer capability for cell stimulation and sensing applications. Proc. 18th Int. Conf. IEEE/EMBS: 245-247 DOI 10.1109/IEMBS.1996.656936

Author: Morten Fjorback
Institute: Neurodan A/S
Street: Sofiendalsvej 85
City: Aalborg
Country: Denmark
Email: mf@neurodan.dk

Telemedicine for Rural and Underserved Communities of Nepal

Ramesh R. Subedi, Carrie B. Peterson, and Sofoklis Kyriazakos

[1] Center for TeleInFrastruktur (CTIF), Aalborg University, Aalborg, Denmark

Abstract— **Health workers in rural health care serve most of the population in Nepal, but are isolated from specialist support and access to current medical information. Fortunately, the advent of Information and Communication Technologies (ICT) has unleashed new opportunities for the delivery of health services. In Nepal, there are very remote and less developed communities with limited access to roads and poor infrastructure to access direct health services; here, telemedicine can be taken as the best alternative form to physically travelling and treating people. The strengths of telemedicine (TM) for remote populations include making specialty care more accessible, eliminating lengthy travel and costly transportation, and reducing the cost of some medical services in rural settings. This paper will focus on implications, barriers, proposed solutions, and future extensions of telemedicine in rural and remote places as well as a review on the kinds of services which are most appropriate in the context of Nepal. The main purpose of this paper is to explore practicability of telemedicine in Nepal and its scope of implementation and use.**

Keywords— **Telemedicine, Nepal, Remote and Underserved Populations, ICT, Health care.**

I. Introduction

Nepal is a hilly country consisting of around 4000 village development committees (VDCs), around 60 municipalities, and 75 districts. It is a poor country having low GNPs, low per capita income, and low literacy rates but higher population density. There is also an acute shortage of doctors with the person to doctor ratio at approximately 4800:1 [1].These factors have contributed to the prevalence of communicable, respiratory, and nutrition deficiency diseases, which are among the most common disorders seen in hospital outpatient departments. Telemedicine is therefore an attractive potential means of improving health services in Nepal.

Telemedicine as a service is the process of providing medical expertise and health services to remote, rural, and underserved communities in primary care, secondary care, and in emergency conditions with the help of telecommunications. It is particularly helpful to deliver health care to remote and rural areas, and is therefore very useful in Nepal where there is an acute storage of medical specialists separated from most of the population in remote places [2]. Since people in all parts of the country need proper health care, telemedicine can be used as an alternative form of treating people in the absence of medical facilities in the area [3]. At the district level, many diseases can be diagnosed and treated via specialists' advice through telemedicine. Using telemedicine for surgeries and complex examinations may prove to be more difficult, but common diseases and ailments can be diagnosed and treated in a timely manner; furthermore, follow-up treatment and routine health checks are made easier by this technology [3].

Telemedicine as a system allows the integration of technology and sharing of information, enhances accessibility of health service to people and health providers, increases efficiency of treatment, lower health related costs, and improves patient care [4]. TM is beginning to have an important impact on many aspects of health care in developing countries. When implemented well, telemedicine may allow developing countries to leapfrog over their developed neighbors in successful health care delivery [5]. Places such as Nepal, Bangladesh, and Pakistan may find that local practitioners can provide the best advice to their patients without having to send them from small communities to large urban centres.

One of the major problems that rural and remote communities of Nepal have been facing is information poverty. Since the IT revolution is limited to a smaller percentage of population, there is huge digital divide within the country. More than 80 percent of the computers and Internet connections are located in Kathmandu city only. This kind of disparity should be amended for the overall development of the country. For this, TM could also be a very good potential mode. Since telemedicine is possible only when ICT infrastructure is well developed, it will make ICT accessible to remote places and reduce the digital divide. Telemedicine may in fact have a more profound impact in developing countries than in developed countries [6].

The development of ICT infrastructure and access capabilities is vital for the overall development of the country. Nepal has availed the opportunity to develop various sectors such as health, education, agriculture, tourism, and trade, among others, using information technology [7]. It should be developed and expanded as a foundation for the enhancement of public awareness, development of knowledge based societies, temporal and monetary savings, proper dissemination of information, as well as for merging into the mainstream of globalization. The establishment of vibrant information technology will mitigate some of the disadvantages Nepal faces on account of its geographical condition [7].

K. Dremstrup, S. Rees, M.Ø. Jensen (Eds.): 15th NBC on Biomedical Engineering & Medical Physics, IFMBE Proceedings 34, pp. 117–120, 2011.
www.springerlink.com

Currently, some TM projects are running at different places in Nepal. One of them is from Kathmandu Model Hospital (KMH) to Dolakha Hospital, and this project seems to be the most suitable project to extend further due to its location and current status of telemedicine. There is a uniformity of health problems and diseases among most of the rural communities where similar kinds of methodologies can be used to deliver health services. In the first phase, it is better to cover most of the mountainous districts and some hilly districts. Once TM is utilized and people start seeing and accepting the benefits, the service can be expanded gradually to more districts. In fact, it is possible to extend TM services to all the 75 districts to provide health services to people across the country. This paper will discuss some implications, barriers, proposed solutions, and future extensions of TM services throughout the country as well as a review on the kinds of services which are most appropriate in the context of Nepal.

II. Barriers

Today, a lack of electric power is the biggest problem in Nepal. This problem has slowed down the overall development of the country. Specifically, the lack of energy supplies in rural and remote places is remaining as a chronic problem. Other major hindrances are due to technological development, remote and inaccessible geographic terrain of the country, non-uniformity in the construction of infrastructure over all the regions, lack of skilled human resources to adopt newer technology, lack of timely supply of required man powers to particularly underserved areas, lack of motivation to the available human resources (lack of incentives, trainings etc.), and disturbances in developmental works due to the political conflicts.

In Nepal, most of the diseases in underserved societies are related to communicable and infectious diseases. About 70% of all health problems and deaths in Nepal are attributed to infectious diseases [8]. People still die from simple diseases like diarrhoea, malaria, encephalitis, dengue fever, hepatitis A, etc. due to lack of knowledge or inaccessibility to medicinal services. Many children die from easily preventable and treatable diseases such as malnutrition, dysentery, acute respiratory infections, etc. [9]. Skin problems are also common in rural and remote areas of the country as is diabetes, asthma, high blood pressure, and some chronic diseases are increasing both in rural as well as urban areas of Nepal. These types of diseases can be diagnosed, treated, and managed with the help of TM services like teleradiology, telepathology, and teledermatology. If facilities could deliver these services successfully to the remote and underserved people, the government will be able to solve most of the health problems and prevent many deaths in Nepal.

Additionally, many doctors and medical specialists are neither fully convinced nor familiar with telemedicine; the very thought of diagnosing and treating a patient when physically absent, solely on basis of data provided through the TM system, is not one that sits well with medical professionals. Furthermore, studies suggest that patients from remote places feel uncertain about using technology-based remote health services and are reluctant to use telemedicine, decreasing the opportunity for providers to utilize the technology [10]. In particular, people who believe in a hands-on approach to healing are more comfortable with traditional medicine which comes from the laying of hands. These kinds of hesitancy on the side of both patient and physician toward the use of technology are remaining as a barrier to successfully adopt TM services.

III. Appropriate Solutions

In any country, economic health has a direct impact on the implementation and sustainability of new technology, which, in turn, are affected by the geographical conditions, transportation and communication systems, socio-cultural factors, and the political situation of the nation. In addition, the technological development depends upon pre-existing or old infrastructures (installed bases) into which new infrastructures are designed and developed.

Nepal is a hilly and mountainous country, so the establishment of a good telecommunication network is the biggest physical challenge to deliver TM services. For the connectivity of remote villages, there are several requirements like low installation and maintenance costs, low power, robustness, scalability, and ease of use. It is not feasible to connect mountains and hilly areas by the use of wired communication because the installation cost is too high and maintenance and reliability are also not cost-effective.

Now-a-days, optical fiber networks have reached most of the cities and some of the district headquarters (towns). Figure 1 illustrates the national optical fiber backbone of the country. This study substantiates that it is economically viable to extend this optical fiber network using microwave links to connect remote and rural villages. The city or town connected to the optical fiber network should have a good line-of-sight (LOS) and minimal distance from remote villages. If there is a clear LOS, a microwave link is a more appropriate and cost-effective technology than satellite communications would be. From this figure, it can be observed that it is better to connect mountainous and hilly districts from Kathmandu and other cities using microwave links. The optimal way of connecting several villages is to use point to point (P2P) connectivity from optical fiber (city) to distribution point and point to multi-point (P2MP) from distribution point to multiple villages. In this network, optical fiber has reached Pokhara city and is used to connect distribution point of some villages using P2P connectivity. This technology is suitable for speedy and movable setup, last-mile communications, and mobility. It further has the advantages of relatively high bandwidth, low cost (for both provider and user), ease of setup, use, and maintenance.

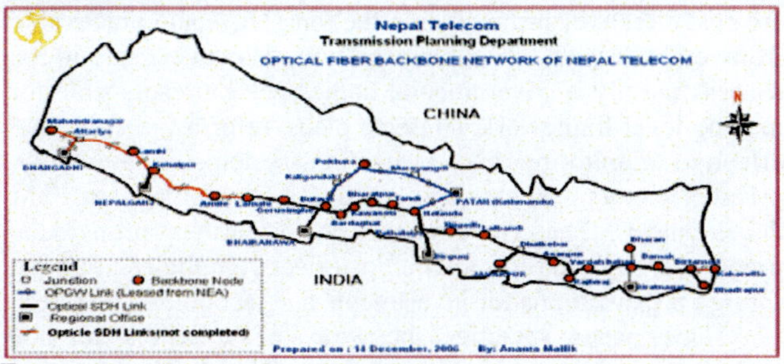

Figure 1: National optical fiber backbone [11]

For the transfer of information, the Internet may be the best option. It allows all kinds of data communication (e.g. audio, video, text, image, etc.) and is inexpensive and easy to use for health workers. Wireless extensions using microwave links from national optical fibre backbone is an appropriate technology to connect remote VDCs of Nepal. Moreover, there are some installed bases of wireless connectivity from Kathmandu and other cities to many districts for the purpose of communication, banking, insurance, and more. The most viable economic solution to establishing a network is to use pre-installed bases as much as possible.

Telemedicine is possible only when there is required amount of electric power is accessible. Fortunately, alternative energy resources like micro hydro, solar panels have the potential of greatly improving the living conditions of rural and underserved communities. They are perhaps the most appropriate, modern, small-scale, decentralised energy supply technologies.

The most cost-effective way of capturing images suitable for diagnosis and management of cases is to use high pixel digital cameras and inexpensive equipment already available on the market. Digital cameras have reasonable prices, very high resolution, and vibration-reduction ability, all of which are suitable to use in, at least, telepathology, teleradiology, and teledermatology. Today, an advantage to implementing TM services is the affordable price of electronic devices (e.g. camera, scanner, etc.) which is decreasing even though their specifications and functionalities are increasing.

It has been mentioned that the beginning of suitable TM services in Nepal is based upon the store and forward method, which is cheaper, easy to sustain, and suits most of the health problems of remote and underserved societies. The modality of telemedicine services proposed in this paper is very simple. If we have reliable connectivity we can send all the patients' data online which will then be studied by the specialists at the centre and provide advice according to which health care workers at the remote places take action.

An appropriate implementation of telemedicine in the context of Nepal is the balanced approach, a combination of horizontal and vertical. It is better to establish a limited number of reasonably capable sites in the beginning for the effective management of training and technical support. This type of implementation maximizes likelihood of success in initial sites and allows participants to have a hands-on understanding of the vision and scope of what TM can achieve with further deployment.

In order for patients to feel comfortable using telemedicine, health care practitioners must first learn how to communicate effectively using the technology. If health care workers and medical specialists make effective use of technology, the resulting better communication will encourage people to adopt the technology as well. In order to address uncertainty and build confidence, it is critical that both medical specialists and patients have access to quality training in how to use technologies to deliver health services. If doctors and patients are aware of what TM is and how it is used to deliver health services, this will reduce uncertainty in many ways. Moreover, it will help understand how the use of TM services impacts relationships with doctors, which in turn, can influence a patient's health. In addition, it is exceedingly important to provide effective training and better incentives to the health workers utilizing TM services. As they have to work in rural and remote places, they need more incentives to work happily, and developing training and retaining incentives is not costly for the majority of organizations, especially when benefits often outweigh costs. It is important that local health care workers are trained and encouraged to take a lead in developing and operating telemedicine projects as far as possible. This will assist in further economic stability as more health professionals feel adequately trained in their field of work by reducing typically high turnover rates in the health care sector. What's more is that the sites will save time and money by interviewing, hiring, and training new employees less frequently and with a more structured organizational process.

IV. FUNDING AND SUSTAINABILITY

One of the major problems of any development project in Nepal is its sustainability. Usually, after execution of donor support, the program is eventually terminated due to the lack of financial resources. Nepal is a poor country and, in many instances, cannot afford to pay for implementing newer technologies, even if installation costs can be managed. It appears to be better to take financial help from donor agencies to setup and run TM projects as they are willing to support rural health programs; but for sustainability, programs should not depend fully on donor agencies for financial support throughout the lifetime of the system. Additionally, large supporters of TM initiatives, like local or national governments will be more likely to invest in and continue to support telemedicine programs that have a set plan for long-term sustainability and can produce results affirming the cost-benefit trade-off. Since there will be a larger number of patients hoped to use TM services, a service fee will also be a good income to support the long-term sustainability of the project.

In the context of Nepal, one way to achieve sustainability is to focus on TM services having less running and maintenance costs, such as services that are based on asynchronous and/or remote monitoring techniques. These kinds of services have low running costs in comparison to that of real-time, or synchronous, services where there must be the presence of both parties at the same time while using high bandwidth connectivity. Although store-and-forward is an older form of TM, it is suitable to utilize in the first phase of implementation. After utilizing and understanding asynchronous transfer modes, which is often the most common transfer of information in TM systems, clinical practitioners will have an easier time adjusting to TM methods and it will be easier to exploit synchronous transfer modes (real-time).

V. Conclusion

Telemedicine is the process of providing medical expertise to remote places with the help of telecommunication which indicates that it has a great potential to deliver health services for the people of underserved communities. Telemedicine is a concept which allows very remote and less developed places that have limited access to roads and poor infrastructure to receive direct health services. Access to specialist services, especially for remote areas, is of great importance, especially where there are few or no health workers available to provide quality health care. Fortunately, Nepal has availed from the opportunity to deliver health services to the remote places using Information and Communication Technologies.

This paper covers the background study of telemedicine in Nepal, the review of currently running TM projects, and the analysis of their implications for future works. We have discussed the highest reported diseases in rural and underserved communities that contribute most of the health problems and deaths in Nepal and how they can be diagnosed, treated, and managed with the help of TM. Similarly, the methodologies to connect rural and remote places, the kinds of telemedicine services, as well as the implementation approach appropriate in the context of Nepal are well covered. Moreover, the analysis of funding and sustainability of TM projects as well as the proposed extension of TM services and the technology is performed.

This paper specifies delivering TM services for rural and underserved communities of Nepal and their importance. The implementation barriers are analyzed in depth and the solutions are proposed for all of them. The proposed solutions are cost effective, sustainable in the long-term, and practicable from every aspect of implementation. However, it will be successful only if governmental bodies will be responsible for making legal framework, strategic plans, reforms, visions, and effective monitoring and evaluation systems; additionally, private sectors should also actively participate in TM development. The Government could further propel the process through the formulation of effective policies to attract foreign organizations for investments and technology transfer.

This paper specifies the array of activities for the successful revolution of telemedicine in Nepal. The purposed model of telemedicine introduced in this paper is found to be the most appropriate model to deliver health services and to make familiar with the importance and potential of telemedicine technology. When well implemented, the vast population living in rural and remote places will benefit from quality health services and ICT will be accessible to the general public, which will greatly reduce information poverty and the digital divide, at least within national borders. It is concluded that this will be a strong foundation for the growth and future extension of telemedicine in Nepal.

References

1. Frank Jacobs. (2007, Oct) big think. [Online]. http://bigthink.com/ideas/21237
2. MR Pradhan, "Telemedicine in Nepal: a Pilot Project, Project Proposal (ict r&D grants, 2004 April), HealthNet Nepal, Nepal.," 2004.
3. Dr Mingmar G Sherpa, "Telemedicine: Healthcare in times of high technology," March 2010.
4. WHO, "Information Technology: In Support of Health Care," 2009.
5. Rajiv Ulpe. (2010, March) PHI Wiki Project. [Online]. http://phiwiki.wetpaint.com/page/Telemedicine+in+Health+Care
6. Steven M Edworthy. (2001, September) BMJ. [Online]. http://www.bmj.com/content/323/7312/524.extract
7. HLCIT, "Information Technology Policy, 2000," 2000.
8. National Planning Commission, GON, "The Ninth Plan," 1998.
9. Kazuko Hirai, Ayako Abe & Yoshimi Ohno Shiba K Rai, "Infectious Diseases and Malnutrition Status in Nepal: an Overview," 2002.
10. Kelly E. Tenzek Ali Gattoni. (2010, Aug.) COMMUNICATION CURRENTS. [Online]. http://www.natcom.org/CommCurrentsArticle.aspx?id=1348
11. Prashant Manandhar, "Current NREN Network and activities," 2008.

Author: Ramesh R Subedi
Institute: Center for TeleInFrastruktur (CTIF), Aalborg University
Street: Fredrik Bajers Vej 7, 9220, Aalborg Ø
City: Aalborg
Country: Denmark
Email: ramesh533@yahoo.com

Investigation of the Linear Relationship between Grasping Force and Features of Intramuscular EMG

M.F. Bøg[1], E. Erkocevic[1], M.J. Niemeier[1], J.R. Mathiesen[1], A. Smidstrup[1], and E.N. Kamavuako[2]

[1] Department of Health Science and Technology, Aalborg University, Aalborg, Denmark
[2] Center for Sensory-Motor Interaction, Department of Health Science and Technology, Aalborg University, Aalborg, Denmark

Abstract— **Surface electromyography (sEMG) can be used to control prosthetic devices. However, intramuscular EMG (iEMG) has been proposed as an alternative control signal, since it provides advantages such as electrode implantation and more selective recordings. iEMG can potentially be used to develop more intuitive prosthetic devices but since only limited research is available within this area, further investigation is needed on the relationship between iEMG and force.**

An earlier study quantified the linear relationship between iEMG and grasping force, however, this was solely based on one feature and force ranging from 0-50 N. Therefore the aim of the present study was to quantify the linear relationship between grasping force and 14 different EMG features using the entire force range from 0 to 100 % Maximum Voluntary Contraction (MVC). Single-channel iEMG and sEMG were recorded concurrently from the muscle Flexor Digitorum Profundus (FDP) from 11 subjects who exerted four force profiles during power grasping. The Wilson Amplitude (WAMP) feature showed the best results for both sEMG and iEMG ($R^2 > 0.9$), where sEMG had a significantly higher mean R^2-value than iEMG (P = 0.044). However, the potential of using iEMG should be investigated further based on the predictive capabilities of the features.

Keywords— **Surface EMG, Intramuscular EMG, Power grip, Flexor digitorum profundus, Linear relationship.**

I. Introduction

It is well known that surface EMG (sEMG) is used for control of myoelectric prosthetic devices, where the applied force is predicted proportionally to muscle activity. Several studies have shown that there is a monotonic relationship between features extracted from sEMG and force [1,2].

sEMG is noninvasive, but it is sensitive to crosstalk and limited to a few Degrees-of-Freedom (DoF) since it can only be measured from superficial muscles. Although iEMG is invasive, which causes a small risk of infection, iEMG may provide more stable and selective recordings than sEMG, and may be used for achieving effective control of multiple DoFs. Furthermore, it might be possible to implant iEMG electrodes chronically. Therefore, the use of intramuscular EMG (iEMG) for prosthetic devices has been proposed [1].

However, only few studies on iEMG recordings for linear relationship have been published. Thus, only two features: Global Discharge Rate (GDR), and Integrated EMG have been investigated for iEMG [1,3]. Kamavuako et al. [1] has shown that there is a high correlation (linear correlation coefficient of ~0.9) between the GDR feature of iEMG recordings and force. However, force was limited to 50 N and an indirect muscle was used to measure sEMG. Onishi et al. [3] showed a coefficient of determination (R^2) above 0.85 between the integrated EMG feature of iEMG and force, though, this was done for knee extension. No studies have shown whether the used features proposed for sEMG can be applied for iEMG in the entire range of force from 0 to 100% Maximum Voluntary Contraction (MVC).

Therefore, the aim of this study was to quantify the linear relation between grasping force and a library of 14 EMG features using the entire force range from 0 to 100 % MVC. Furthermore, the aim was to show if the used features for sEMG can be applied for iEMG.

II. Methods

A. Experiment

Subjects: The experiment included 11 healthy subjects (4 w / 7 m) in the age of 22 to 26 years, with a mean of 23.8 years. The experiment was approved by the Danish local ethical committee (approval no.: N-20080045). All subjects received both written and oral information about the experiment and gave written consent prior to the experiment.

Procedure: The subjects exerted force while seated in a chair with their right arm placed in an armrest (Fig. 1). First, the subjects exerted MVC force three times with a 3 minutes rest between the trials. Afterwards the subjects were asked to follow four different force profiles:

1. A *step* profile of 9 s with force increasing in 6 steps.
2. A *double ramp* profile of 9 s.
3. A *bell* profile of 9 s.
4. A *free varying* profile of 9 s with the only constraint to reach the MVC force within this time.

The order of the profiles was randomized. The *step*, *double ramp* and *bell* profiles were recorded two times and

K. Dremstrup, S. Rees, M.Ø. Jensen (Eds.): 15th NBC on Biomedical Engineering & Medical Physics, IFMBE Proceedings 34, pp. 121–124, 2011.
www.springerlink.com

the level of force spanned from 0 to 100 % MVC. The *free varying profile* was only recorded once. The force was shown on an oscilloscope in order to provide the subject with visual feedback during each profile. Each trial was followed by a 3 minutes rest, and all subjects were provided with adequate time to practice matching the profile before the actual recording.

Data recording: In the experiment, a Jamar compatible handgrip dynamometer (Noraxon) with an adjustable grip size was used in order to measure the grasping force. The grip size was set according to the maximum force of each subject. The iEMG electrodes (custom-made by use of hypodermic needles and Teflon coated wires (A-M Systems, Carlsberg, WA; diameter 50 μm)) were placed in a bipolar configuration, in the muscle Flexor Digitorum Profundus (FDP), a direct finger flexor. The needle was placed in the middle one-third of the forearm ventral to the ulnar shaft. The iEMG signals were amplified with a factor of 1000 and filtered with a band pass of 20-5000 Hz. Simultaneously, sEMG was recorded in a bipolar configuration (Ambu Neuroline 720) from the same muscle. The sEMG signals were amplified with a factor of 2000 and filtered with a band pass of 20-500 Hz. A wristband was used as a common reference electrode. Force, iEMG and sEMG signals were sampled by use of a 16 bit AD converter (NI-DAQ USB-6259) with a sampling frequency of 20 kHz.

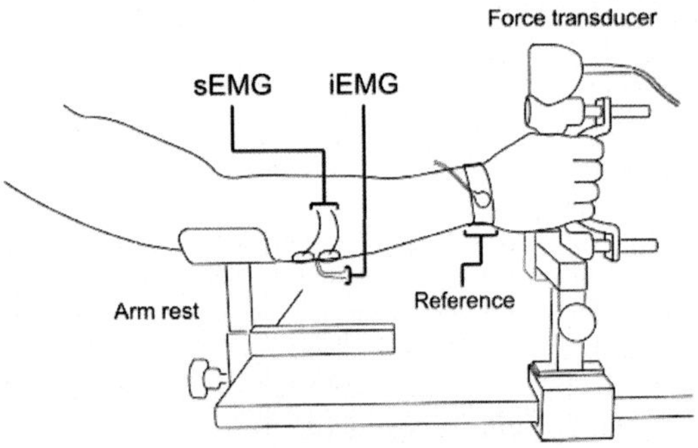

Fig. 1 Sketch of the experimental setup.

B. Signal processing

Digital filters: A 4th order Butterworth filter was applied for each signal. The force was low pass filtered with a cut-off frequency of 20 Hz. The iEMG and sEMG signals were band pass filtered with frequencies of 100-2500 Hz and 20-500 Hz, respectively. Furthermore, a 2nd order Butterworth filter with a cut-off frequency of 1 Hz was applied in order to smooth the features.

Extracted features: In total 14 features were chosen to represent the iEMG and sEMG signals. These 14 features have shown good potential and/or were commonly used in literature. A moving window of 200 ms was applied to the EMG signals with a step size of 50 ms. The features were calculated for each of these windows.

The same moving window was applied on the force signal where the mean was calculated for each window. Thresholds were found by visually inspecting the performance of the features. The used threshold levels were the same for all subjects and profiles. The following 14 features were extracted from the EMG signals:

- Waveform Length (WL), Slope Sign Changes (SSC), Mean Absolute Value (MAV) and Variance (VAR) were implemented as in Phinyomark et al. [4], where SSC was applied with a threshold of 0.2 nV for sEMG and 0.01 mV for iEMG.
- Modified Mean Absolute Value (MMAV) and Mean Absolute Value Slope (MAVSLP) were implemented as modified versions of the definitions in Phinyomark et al. [4]. For MMAV a Hanning window was applied and for MAVSLP the absolute value of MAV_{i+1} - MAV_i was taken.
- Zero Crossing (ZC) and Wilson Amplitude (WAMP) were implemented as in Huang and Chen [5]. ZC was modified as each zero crossing results in an increment. A threshold of 0.01 mV and 0.05 mV were implemented for iEMG and sEMG, respectively. WAMP was implemented with a threshold of 0.005 mV for sEMG and 0.02 mV for iEMG.
- EMG Envelope Energy (EMG_env_energy) and EMG Envelope (EMG_Env) were implemented as in Du et al. [6]. EMG_env_energy was implemented with a threshold of 0.5 μV for sEMG and 20 μV for iEMG. EMG_env was modified by taking the absolute value. A threshold of 0.025 mV for sEMG and 0.01 mV for iEMG was implemented.
- Constraint Sample Entropy (CSE) was defined as in Kamavuako [7], with the tolerance *r* being 0.2 times the standard deviation of the EMG signals during the MVC profiles.
- Autoregressive model (AR-model) of order 4 was used by representing the signal within a given window by the RMS of the AR-coefficients. These were implemented using the Yule-Walker approach.
- Histogram EMG (HEMG) was implemented as a normalized histogram where the distribution of a given window was represented by the amplitude level for the median of the distribution.
- Root mean square (RMS)

C. Data analysis

The linear relationship was found between grasping force and the extracted features by use of seven different profile combinations: Single profiles (*step, bell and double ramp*), combination of two profiles (*step+bell, step+ramp, bell+double ramp*) and combination of all profiles (*step+bell+double ramp*).

The performance measure for each relationship was the R^2-value.

D. Statistical analysis

For each signal type (iEMG or sEMG) two-way ANOVA (with factors features and profiles) was performed in order to compare the features and associated profile combinations. The feature with the highest mean R^2-value was selected for each signal and one-way ANOVA (with factor profiles) was performed for each signal in order to compare the profiles for the specific feature. Additionally, two-way ANOVA (with factors signals and profiles) was performed in order to compare the features with the highest R^2-value between sEMG and iEMG. P-values less than 0.05 were considered significant. The Bonferroni–Dunn test was used for pair-wise comparisons if the ANOVA was significant.

III. Results

In Table 1 the results are summarized for the feature with the highest R^2-value for each signal along with the profile with the highest R^2-value for this feature. In Fig. 2 the R^2-values for the different features are depicted. WAMP showed the highest R^2-value for both iEMG and sEMG.

The two-way ANOVA (with factors signals and profiles) showed that WAMP for sEMG had a significantly higher R^2-value than WAMP for iEMG ($P = 0.044$). For iEMG, WAMP was significantly better than MAVLSP, VAR, HEMG and AR-model ($P \leq 0.001$). For sEMG, WAMP was significantly better than SSC, VAR and EMG_Env ($P < 0.04$), and MAVLSP, ZC, HEMG and AR-model ($P < 0.001$).

When comparing the profiles for WAMP with the one-way ANOVA (with factor profiles) for each signal, *bell* showed the highest R^2-value for both sEMG and iEMG. However, *bell* was not significantly different from the other profiles for iEMG and only significantly different from the combination of *bell+step* profile for sEMG ($P = 0.039$).

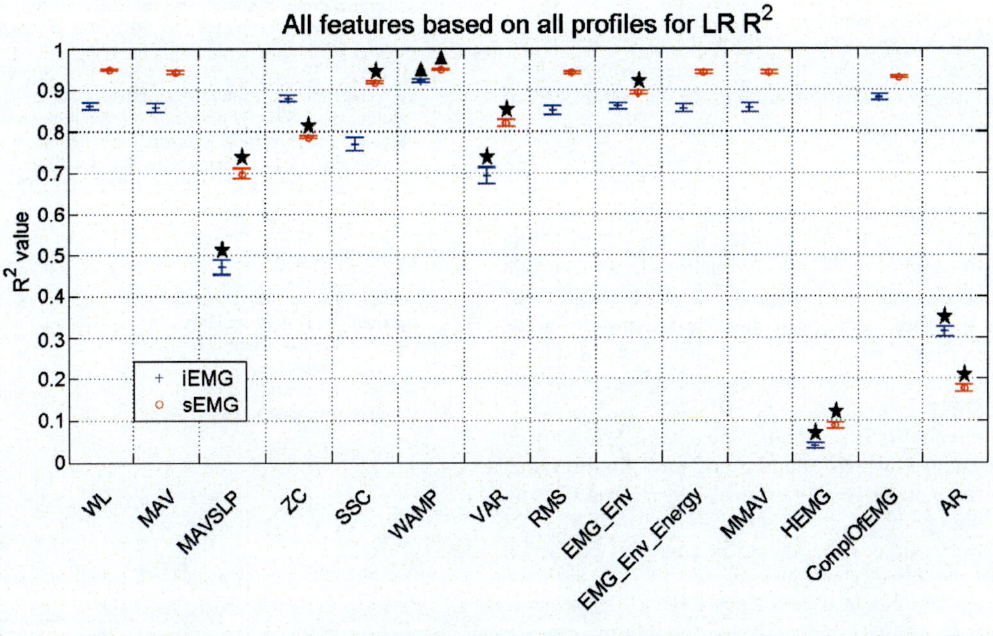

Fig. 2 All resulting features from the linear relationship for all profiles for iEMG and sEMG. The x-axis represents the 14 features. The y-axis represents the R^2-values with the standard error (SE). The circles and pluses represent sEMG and iEMG respectively. The stars represent R^2-values that were significantly different from the feature with the highest R^2-value which is represented with a triangle.

Table 1 Results from the linear relationship (LR) for each EMG-signal with standard error (SE) and confidence interval (CI). The P-value (P) is given for the difference between the signals regarding the chosen features. Only the feature with the highest $\boldsymbol{R^2}$-value and the corresponding profile with the highest $\boldsymbol{R^2}$-values are listed.

LR	Feature/Profile with highest R^2	R^2	SE	CI	P
Feature for:					
iEMG	WAMP	0.921	0.010	[0.898 , 0.944]	
sEMG	WAMP	0.949	0.004	[0.940 , 0.959]	0.044
Profile for:					
iEMG	Bell	0.946	0.009	[0.927 , 0.965]	
sEMG	Bell	0.966	0.004	[0.957 , 0.975]	

IV. Discussion

The results showed that the linear relationship was dependent on the type of feature. The WAMP feature showed to have the highest R^2-value for both sEMG and iEMG with sEMG significantly higher than iEMG. However, Kamavuako et al. [1] showed no significant difference between iEMG and sEMG for a linear relationship, which might be caused by the difference in the used force range and in the choice of feature. Nevertheless, the obtained degree of relationship was similar although our investigation was based on the entire range of force and for 14 different features. However, these results may not apply if another relationship than the linear is used, where features with bad performance might prove good results.

When using WAMP, *bell* showed the highest absolute value for both sEMG and iEMG. We believe that this might be due to the fact that the *bell* profile includes continuous changes in the force slope that may be lacking from the remaining profiles. Based on this, *bell* should be considered for training the association between grasping force and features of EMG.

Even though sEMG had the best results, iEMG showed good performance, which support previous findings that iEMG may be suitable in proportional myoelectric control. Therefore, the potential of using features of iEMG for control of prosthetic devices should be investigated further based on its predictive capabilities.

Acknowledgement

This study was supported by a grant from and the Danish Agency for Science, Technology and Innovation (Council for Independent Research | Technology and Production Sciences, Grant number 10-080813).

References

1. Kamavuako, E.N., Farina, D., Yoshida, K., Jensen, W.. Relationship between grasping force and features of single-channel intramuscular EMG signals. Journal of Neuroscience Methods 2009;185:143–150.
2. Hoozemans, M.J., van Dieën, J.H.. Prediction of handgrip forces using surface EMG of forearm muscles. Journal of electromyography and kinesiology 2005;15:358–366.
3. Onishi, H., Yagi, R., Akasaka, K., Momose, K., Ihashi, K., Handa, Y.. Relationship between EMG signals and force in human vastus lateralis muscle using multiple bipolar wire electrodes. Journal of Electromyography and Kinesiology 2000;10:59–67.
4. Phinyomark, A., Hirunviriya, S., Limsakul, C., Phukpattaranont, P.. Evaluation of EMG feature extraction for hand movement recognition based on euclidean distance and standard deviation. Electrical Engineering /Electronics Computer Telecommunications and Information Technology (ECTI-CON), International Conference 2010;1:856 – 860.
5. Huang, H.P., Chen, C.Y.. Development of a myoelectric discrimination system for a multi-degree prosthetic hand. International Conference on Robotics & Automation, Detroit 1999;1:2392–2397.
6. Du, Y.C., Shyu, L.Y., Hu, W.. The effect of combining stationary wavelet transform and independent component analysis in the multichannel SEMGs hand motion identification system. Journal of Medical and Biological Engineering 2006;26(1):9–14.
7. Kamavuako, E.N. Intramuscular and intrafascicular recordings for proportional control of prostheses. SMI - PhD Thesis 2010;

Corresponding author

Author: Ernest Nlandu Kamavuako
Institute: HST, Aalborg University
Street: Fredrik Bajers Vej 7D3
City: Aalborg
Country: Denmark
Email: enk@hst.aau.dk

Use of Sample Entropy Extracted from Intramuscular EMG Signals for the Estimation of Force

E.N. Kamavuako[1], D. Farina[1,2], and W. Jensen[1]

[1]Center for Sensory-Motor Interaction, Department of Health Science and Technology, Aalborg University, Aalborg
[2]Department of Neurorehabilitation Engineering, Bernstein Center for Computational Neuroscience, University Medical Center Göttingen, Georg-August University of Göttingen, Göttingen, Germany,

Abstract— **This study investigates the use of sample entropy as a feature extracted from intramuscular electromyography (EMG) for the estimation of force. Grasping force and intramuscular EMG) signals were measured in 10 able-bodied subjects. Constraint sample entropy (CSE) was extracted from the EMG signal (window size of 200 ms). The association between the CSE and force was modeled using an artificial neural network. The accuracy of estimation was on average R^2 = 0.89 ± 0.05 and root mean square difference (*RMSD*) = 6.67 ± 2.17 N). It was concluded that sample entropy does capture the dynamics in the intramuscular EMG, and that a single channel of intramuscular EMG can be used for muscle force estimation. The information of muscle force is necessary in proportional myoelectric control.**

Keywords— **grasping force, intramuscular EMG, sample entropy, proportional control.**

I. Introduction

Myoelectric upper limb prostheses are typically controlled by the use of surface electromyography (sEMG), where features of the surface EMG signal are widely used to determine the speed and strength (force) to open or close hand prostheses. An alternative control signal can be derived from EMG signals recorded from electrodes placed within the muscle. Intramuscular EMG may offer advantages in comparison to surface EMG for the control of active prostheses such as chronic implantation, more selective recordings and access to deep muscles [1].

It is well known that proportional control is based on the extraction of features of the signal that carry information on the kinematics or dynamics of the task, e.g. force [2]. Therefore, this study investigates if there is a relation between grasping force and entropy of the intramuscular EMG.

Features of the intramuscular EMG previously investigated in relation to muscle force include the signal energy and the number of motor unit action potentials per second. These approaches are based on the rationale that force control is achieved by the central nervous system with recruitment of motor neurons and modulation of their discharge rates [3, 4]. Based on this rationale, in this study we hypothesize that the complexity of intramuscular EMG signals increases with increasing force due to the increased number of action potentials. Therefore, measures of complexity, such as sample entropy of the EMG signal may be associated with muscle force. In this study we propose the use of sample entropy of the intramuscular EMG (a non-linear complexity feature) for predicting grasping force.

II. Methods

A. Experiment

Ten right handed able-bodied human subjects (mean age 26.9 yrs) were included in the study, and all experimental procedures were approved by the Danish local ethical committee (approval no.: N-20080045). The subjects had no history of upper extremity or other musculoskeletal disorders.

The grip force was measured using a hand grip dynamometer (Vernier Software & Technology, accuracy ±0.6N, operational range 0-600N, grip size 50 x 25 mm). Intramuscular EMG was recorded using bipolar wire electrodes from the m. extensor carpi radialis. This muscle does not act directly on grasping since it is a wrist extensor. However, Hoozemans and Van Dieën [7] have shown that extensor muscle activity is highly associated with grip force for counteracting the wrist flexion torque caused by the finger flexor tendons. A pair of sterilized wire electrodes made of Teflon-coated stainless steel (A-M Systems, Carlsborg, WA; diameter 50 µm) was inserted into the muscle with a sterile 25-gauge hypodermic needle. The insulated wires were cut to expose only the cross section at the tip. The needle was inserted to a depth of a few millimeters below the muscle fascia and then removed to leave the wire electrodes inside the muscle. The intramuscular EMG signals were amplified and provided a bipolar recording (Counterpoint EMG, Dantec Medical, Skovlunde, Denmark) that was band-pass filtered (500 Hz - 5 kHz). A reference electrode was placed around the wrist. Intramuscular EMG and force were A/D converted on 12 bits, and sampled at 10 kHz. The force signal was displayed on an oscilloscope for online feedback.

K. Dremstrup, S. Rees, M.Ø. Jensen (Eds.): 15th NBC on Biomedical Engineering & Medical Physics, IFMBE Proceedings 34, pp. 125–128, 2011.
www.springerlink.com

Subjects exerted handgrip forces with their right hand (dominat) while seated comfortably with both arms placed on a table in front of them. The subject's elbows were flexed at approximately 90° and the forearms were supported. The subjects were asked to produce two maximum grip force (MGF) contractions in 3 s and to maintain the MGF for 3 s. The two contractions were separated by 3 min of rest. The subjects were then asked to follow five grip force profiles in random order. Each profile was repeated twice and was followed by a rest of 1 min. The force profiles were defined as follows [4]:

1. Five step increases of 10 s duration in static grip force by 10 N (*step*, Fig. 1a)
2. A Gaussian-shaped grip force ranging from 0- 50 N in 10 s (*bell*, Fig. 1b).
3. Two linear ramps of 10-s duration (*saw10*, Fig. 1c)
4. Two linear ramps of 5-s duration (*saw5*, Fig. 1d).
5. Two linear ramps of 4-s duration (*saw4*, Fig. 1e).

After these calibration recordings, the subjects performed twice freely varying grip force for 15 s with the only constraint to keep the force within 50 N (this task will be indicated as "*vol*").

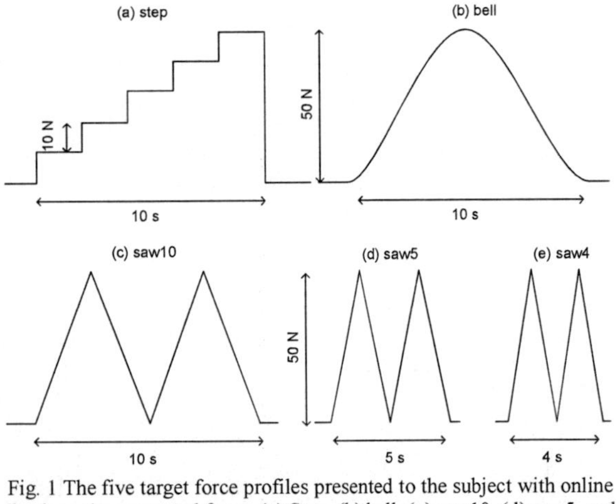

Fig. 1 The five target force profiles presented to the subject with online feedback on the generated force. (a) Step, (b) bell, (c) saw10, (d) saw5, and (e) saw4.Inspired from [4].

B. Signal processing

The force signals were low-pass filtered (4th order Butterworth filter, f^{3dB}= 20Hz). The Constraint Sample Entropy (CSE) was extracted from intervals of 200-ms duration of intramuscular EMG as a measure of signal complexity.

Sample entropy (SampEn) quantifies the complexity and regularity of a system [5]. It is defined as the logarithmic likelihood that patterns in a data set that are similar to each other will remain similar for the next comparison within a longer pattern [6]. For a time series of N points, m-dimensional vector sequences $y_m(i)$ (for $i = 1 \dots N\text{-}m+1$) were generated. A vector $y_m(j)$ is defined to match another vector $y_m(i)$ if the distance between these two vectors lies within a predefined tolerance r, i.e. $|y_m(j)- y_m(i)| < r$. Let's define B_i as the number of vectors $y_m(j)$ within r of $y_m(i)$ and A_i as the number of vectors $y_{m+1}(j)$ within r of $y_{m+1}(i)$ for $m+1$ dimension. Then $C_i^m(r) = \frac{B_i}{N-m+1}$ is the probability that any vector $y_m(j)$ is within r of $y_m(i)$, where m specifies the pattern length and r defines the criterion of similarity. The sampEn is computed as:

$$SampEn(N, m, r) = ln\left[\frac{\sum_{i=1}^{N-(m-1)} C_i^m(r)}{\sum_{i=1}^{N-m} C_i^{m+1}(r)}\right] \quad (1)$$

We will refer to standard sample entropy (SSE) as the sample entropy with tolerance r computed as 0.2 times the standard deviation (SD) of the signal window [5]. We propose the Constraint Sample Entropy (CSE) as the sample entropy with tolerance r equal to 0.2 times the standard deviation of the signal measured during MGF. Thus, contrary to SSE, in the CSE computation, the tolerance value was the same for all signals in each subject.

C. Data analysis

The data analysis consisted of two steps. In the first step, a suitable dimension m for the computation of sampEn was found using calibration data. For this purpose, we computed *CSE* using m values ranging from 1 to 9, which is a suitable range for the duration of the decision window (200 ms). The optimal m value was the one leading to the greatest value of the Pearson correlation coefficient between the *CSE* and the measured force. The second step was the force estimation, where calibration measurements (*step, bell, saw10, saw5, saw4*) (training data) were used to train the model to estimate freely varying profiles (*vol*) (test data).

An artificial neural network (*ANN*) model was fitted to each subject, using a two-layer feed-forward network consisting of one hidden layer with a tangent sigmoid function and one output layer with a linear function. The network used batch training based on the Levenberg-Marquardt algorithm and had one output neuron that provided the estimated force. For each subject, the network was optimized according to the number of neurons in the hidden layer (between 1 and 10) based on the least mean square error criterion. The performance of the prediction was in all cases assessed using the coefficient of determination (R^2) and the root mean square difference (*RMSD*) between the estimated and the measured force.

III. RESULTS

Fig. 2 depicts an example of a raw intramuscular EMG recording (Fig. 2a) from one subject during the bell force profile (Fig. 2b). CSE feature for this signal was computed for different windows and the results are presented in Fig 2c.

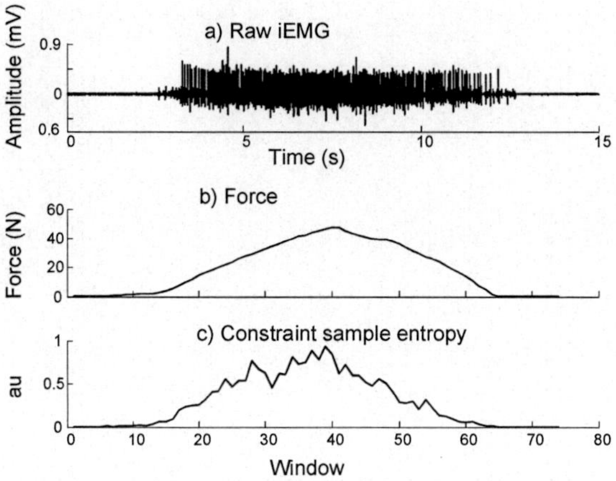

Fig. 2 (a) Intramuscular EMG signals recorded during a bell contraction with (b) the corresponding force signal. (c) Constraint sample entropy (CSE) without applying any threshold. '*au*' stands for arbitrary unit.

Fig. 3 shows the effect of the *m* value [Eq. (1)] on the correlation of *CSE* with force for the calibration data. The results did not statistically depend on the *m*-value (≤ 1 % of difference) therefore *m* was fixed to 2 for all the subsequent analyses, as also used for other biological signals [5, 6]. This result indicates that the dimension of the embedded space did not substantially influence the performance.

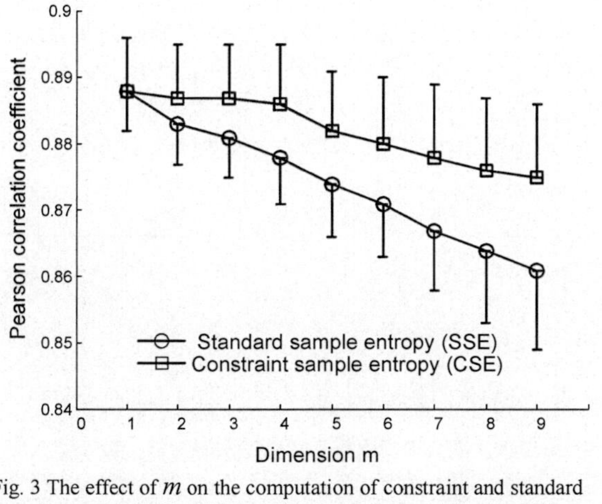

Fig. 3 The effect of *m* on the computation of constraint and standard sample entropy (CSE, SSE) based on Pearson correlation coefficient between with force. Data are given as mean ± standard error (SE).

The training of the *ANN* was performed with the data from the five calibration profiles (*step, bell, saw10, saw5, saw4*). The optimal number of neurons in the hidden layer had a median value of 7. Results of the estimation of freely varying profiles (*vol*) were on average $R^2 = 0.892 \pm 0.05$ and $RMSD = 6.67 \pm 2.17$ N as shown in Table 1 for all subjects.

Table 1 Performance of force estimation based on constraint sample entropy for all subjects. Results are provided for the coefficient (R^2) of determination and root mean square difference (RMDS).

Subjects	R^2	RMDS (N)
Subject 1	0.906	5.03
Subject 2	0.893	5.41
Subject 3	0.936	9.91
Subject 4	0.779	7.92
Subject 5	0.849	6.08
Subject 6	0.924	7.47
Subject 7	0.888	8.96
Subject 8	0.872	8.29
Subject 9	0.933	4.48
Subject 10	0.941	3.18
Average (mean ± SD)	**0.892 ± 0.05**	**6.67 ± 2.17**

IV. DISCUSSIONS

Our aim was to investigate a measure of signal complexity for predicting grasping force from intramuscular EMG signals. The results of this study showed that grasping force may be predicted with high accuracy based on features extracted from a single channel of intramuscular EMG. The results confirmed the hypothesis that the degree of complexity of the EMG signal can be used for grasping force prediction with relatively high accuracy ($R^2 > 0.88$).

Constraint Sample Entropy was proposed as a new non-linear feature for the investigation of intramuscular EMG for force estimation. In general, this feature showed to be suitable for force estimation up to 50 N. Furthermore this feature may be suitable for physiological investigation of multi unit's signals (from nerve or muscle), where decomposition to single unit action potential is not possible. Thus a selective EMG recording is representative of the applied grasping force and can potentially be suitable for proportional control of prosthetic devices.

ACKNOWLEDGMENT

This study was supported by the Danish National Advanced Technology Foundation and the Danish Agency for Science, Technology and Innovation (Council for Independent Research | Technology and Production Sciences, Grant number 10-080813).

REFERENCES

1. Herberts P, Kadefors R, Petersen I (1968) Implantation of microcircuits for myoelectric control of prosthetises, J. *Bone Joint Surg.*, vol. 50B, pp. 780-791.
2. Jiang N, Englehart KB, Parker PA (2009) Extracting simultaneous and proportional neural control information for multiple degree of freedom prostheses from the surface electromyographic signal, *IEEE Trans. Biomed. Eng.*, vol. 56, no. 4, pp. 1070-1080.
3. De Luca C, LeFever R, McCue M et al. (1982) Xenakis, Behaviour of human motor units in different muscles during linearly varying contractions *J. Physiol.*, vol. 329, pp. 113-28.
4. Kamavuako EN, Farina D, Yoshida K et al (2009) Relationship between grasping force and features of single-channel intramuscular EMG signals, *J. Neurosci. Methods,* vol. 85, no. 1, pp. 143-150.
5. Richman JS, Moorman JR (2000) Physiological time-series analysis using approximate entropy and sample entropy, *Am. J. Physiol.*, vol 278, no. 6, pp. H2039-H2049.
6. Khandoker AH, Jelinek HF, Palaniswami M (2009) Identifying diabetic patients with cardiac autonomic neuropathy by heart rate complexity analysis, *BioMed. Eng. Online,* vol. 8, no. 3.
7. Hoozemans MJ and van Dieën JH (2005) Prediction of handgrip forces using surface EMG of forearm muscles*J. Electromyogr. Kinesiol.*15(4), pp. 358-366.

Corresponding author

Author: Ernest Nlandu Kamavuako
Institute: HST, Aalborg University
Street: Fredrik bajers vej 7D3
City: Aalborg
Country: Denmark
Email: enk@hst.aau.dk

Leased Line via Mobile Infrastructure for Telemedicine in India

Ujjwal Bania, Carrie Beth Peterson, and Sofoklis Kyriazokos

Center for TeleInFrastruktur (CTIF), Aalborg University, Aalborg, Denmark

Abstract— **Telemedicine is the use of information and communication technologies (ICT) to exchange medical information for the purpose of health care and health education. In the context of developing countries, good health care facilities are concentrated in the urban cities, while they are still lacking in rural communities with lower economies. Telemedicine provides a best solution to solve this disparity of health sectors between urban and rural areas. In rural areas of developing countries, a reliable communication link for telemedicine is one of the key challenges. In the recent years, there is an increasing growth of mobile communication in developing countries that has saturated in urban cities and now growing towards the rural areas. This article focuses in India as a developing nation and discusses the cost effective use of widespread mobile communications infrastructure for communication link for telemedicine in rural areas.**

Keywords — **Telemedicine, lease line, India, mobile communication.**

I. Introduction

Health care is a basic necessity for human beings and can be divided into primary care, secondary care, and tertiary care. Primary health care refers to the point of first consultation by the patient. Secondary health care refers to specialized medical care, such as cardiologists for heart related diseases. Tertiary health care deals with highly specialized care systems that require sophisticated equipment and multiple specialists. In the context of developing and under developed countries, secondary and tertiary services are concentrated in the urban cities and the rural areas that cover most part of the country are still served by minimal primary care health services. Medical professionals and doctors prefer to stay in the urban cities and hesitate to serve in the rural areas, due to the lack of medical resources for them to practice rural areas and the lack of incentive for them. In urban areas with more hospitals and care facilities, doctors can achieve the full financial and reputational gain as well as practice more professionally with the breadth of their medical expertise. This has resulted in the rural areas lacking basic health care and suffering due to simple diseases.

Telemedicine provides a good solution for this kind of scenario. Telemedicine connects the doctors and medical facilities to the patient at a distance through communication media. The mobile telecommunications services and internet bandwidths are becoming more available and affordable; in developing countries, these communication services have almost completely saturated the urban cities. Prompted by lower costs and increasing demand, telecommunications operators are expanding their networks towards rural villages to obtain more customers. In India, there is pervasive deployment of telecommunications and the tele-density (telephone per person) has already crossed 50% with the majority (93%) of the telecom market share covered by wireless mobiles [1].

India has a very good reputation in the medical sector in the subcontinent. In India, a new concept known as medical tourism has developed [2]. It attracts people from different parts of the world, both developing and developed countries, for pursuing medical treatment in India. People who can afford quality health care in the neighboring developing countries like Nepal often go to India for treatment. While from the developed countries, people go to India for quality medical treatment at cheaper rates.

On one side, India is attracting people from outside the country for quality healthcare, while on the other side, India is not able to provide primary healthcare to its own citizens in some of the poorer villages. This paper briefly explores different types of telemedicine that have been in used in India and gives a cost effective solution to use mobile communication network for telemedicine in the rural parts of India. The rest of the paper is structured as follows: Section II describes the brief background of India, Section III describes standardization, Section IV proposes a solution for communication link in telemedicine, Section V lists the benefits of the proposed solution, and Section VI contains the discussion.

II. Background of India

A. Demographic figures of India

India is a developing country covering a total area of 3,287,268 square kilometers with around 1.1 billion living in it. India has a rich, ancient history of medical and holistic health. Before the modern medical facility, *ayurbedic* medicine was practiced in India. *Ayurbedic* medicine is popular for its minimal side effects and, over the years, western countries have also taken interest in traditional health practices.

In India, nearly 70% of the population live in rural areas that lack proper health care services; 27.5% of the population lives under the poverty line (earning less than 1$

K. Dremstrup, S. Rees, M.Ø. Jensen (Eds.): 15th NBC on Biomedical Engineering & Medical Physics, IFMBE Proceedings 34, pp. 129–132, 2011.
www.springerlink.com

per day per person)[3]; and 90% of secondary and tertiary healthcare facilities are in cities away from rural areas. Furthermore, the doctor to population ratio in India is an unacceptable 1:1722 as per Medical Council of India [4].

B. Overview of telemedicine in India

Telemedicine is not new for India. There are many telemedicine systems running in India, several of which are described in this section. In India, telemedicine programs are supported by both governmental and private parties. Government bodies include Department of Information Technology, Indian Space Research Organization (ISRO), NEC Telemedicine program for North-Eastern states, State governments etc. And private parties include Apollo Hospitals Group, Sanjay Gandhi Postgraduate Institute of Medical Sciences (SGPGIMS), Asia Heart Foundation etc.

Apollo Hospital Group started in 1987 and is now one of largest private healthcare groups in Asia. It delivers turnkey healthcare facilities, like building a small primary care hospital to super specialty hospitals. Its network is wide spread with 50 hospitals in and outside India. Apollo Telemedicine Networking Foundation (ATNF) is the telemedicine branch of the Apollo Hospitals Group. It is credited with being the first to setup a Rural Telemedicine centre in 1999 in Aragonda (a remote village in mid India). The telemedicine services provided by ATNF are Tele-Radiology, Tele-Dermatology, Tele-Pathology, Tele-Cardiology, Remote ICU Monitoring, Ambulance Monitoring, Mobile Telemedicine Unit, Electronic Health Record, etc [5]. ATNF has collaborated with CISCO to expand the telemedicine services in India [6] in May, 2010. Collaboration basically uses HealthPresence™, a product of CISCO, for the telemedicine services that will be provided by ANTF. In this system, doctors do not have to go to the telemedicine centers; rather, the doctors can use their laptop through the internet to check up their patient at remote telemedicine center assisted by a nurse from anywhere.

Indian Space Research Organization (ISRO) is a government organization dealing with space technologies in India. ISRO started a telemedicine project in 2001 to introduce the telemedicine facility to the rural areas. ISRO mainly uses INSAT Satellites as a means of communication for telemedicine. Satellites provide two main advantages: (1) it is reliable and (2) easy to reach in remote places. Though satellite is costly solution, government support has made it possible to connect to the rural areas. Using satellite, ISRO's Telemedicine Network has connected 306 rural hospitals and 16 mobile telemedicine units to 60 super specialty hospitals located in metropolitan cities [7].

C. Telecom Sector in India

The Department of Telecommunications under the Ministry of Telecommunication and IT targeted the deployment of 500 million mobile telephones by 2010 and this was achieved in September, 2009. This prior achievement of tele-density is due to the involvement of the private sector in the telecom market.

Wired line services like POTS and ISDN do not have much penetration in the rural areas of India. It consists of only 7% of the total tele-density over the entire country. Landline telephone lines are decreasing in India as is the trend in the rest of the world, largely due to the high cost of copper cables and their fixed nature, and furthermore, due to the low cost deployment in wireless telephone services.

On the other hand, there is an exponential penetration of wireless systems like GSM and CDMA networks in India and mobile tele-density is expected to reach 100% by the year 2015. The mobile tele-density in the urban areas has already saturated to 110 % and in the rural area it is 21% as of December, 2009. There are 10 different telecom operators all over the country [8], and they should now target the rural areas for the new customers.

III. Standardization

Telemedicine faces technological issues in facilitating healthcare solutions that are easily accessible and available to cover most of the country. Telemedicine facilities have been developed by different vendors using various types of software and hardware, many of which were created specifically for that facility or project. Like any other technology, telemedicine needs to be standardized by a governing body for interoperability among different vendors for the correct representation and utilization of medical services.

In India, the Ministry of Health and Family Welfare has formed a National Task Force that includes the Department of Information Technology, Union Ministries of Health, ICT, the Indian Space Research Organization (ISRO), the Medical Council of India, and various hospitals that practice telemedicine to address the standardization issues of telemedicine in India[9]. Standardization will benefit all the stakeholders in telemedicine field by facilitating regulations for interoperability among different vendors, which in turn allows the telemedicine users to choose the best suited vendor for their purposes. However, standardization is a much larger issue than the scope of this paper as there has to be standards in both in the medical ontology used and the protocols used in communication.

IV. Overview of Mobile Infrastructure

A. Mobile Network

Mobile networks consist of two parts, namely Base station subsystem (BSS) and Network Switching Subsystem (NSS). NSS also known as the core part which performs the call control function and service control function for the entire mobile network. NSS is normally located in urban areas and consists of many network elements like HLR, MSC, VLR, SMSC, SGSN, and GGSN, etc.

BSS is also known as the radio part and consists of two types of network elements: Base Station Controller (BSC) and Base Transceiver Station (BTS). BSC connects to the NSS and controls the BTS that are scattered over a region. BTS communicates with the user's mobile handset at frequencies 900, 1800, or 1900 MHz. Each BTS site consists of antenna mounted on mast. BTS are scattered over different regions for mobile service coverage. BTS are separated at 8 to 10 km to provide good wireless services in sparsely populated rural areas, while in densely populated urban areas, they may be as close as 300 meters for higher numbers of users. Expansion of mobile service coverage requires expansion in the number of BTS sites over the region.

B. Transmission network

Besides the various mobile network elements discussed briefly above, a transmission network is required that can provide a very reliable communication link between different sites in the network. The transmission network usually consists of equipment like SDH and PDH that use microwave and optical fiber; sometimes satellite is also used as the mode of transmission. All the BTS sites are connected to BSC and to the core network through communication links provided by this transmission network. Expansion of BTS sites for mobile service coverage in an area requires the side by side expansion of transmission network. This transmission network can be shared for other purposes like leased line for telemedicine.

C. Leased line service

Leased line is a communication link between any two places. Normally for telemedicine, lease line service is required between the hospitals for exchanging medical information. Hospitals may be located in rural areas with primary healthcare and urban areas with secondary and tertiary healthcare.

The expansion of mobile services requires the expansion of BTS sites with a well-established transmission network. Telecom service providers should enhance transmission networks in each BTS site with the provision of leased services. These leased services can be used to connect the hospitals or healthcare units of the remote areas. Since the BTS sites in remote villages are scattered 8 to 10 km, it will be easy to connect the hospital or healthcare unit and the BTS site via copper, optical, or microwave. Once the remote hospital is connected to the telecom service provider's network, the leased line connectivity should be made to the specialty hospitals in metropolitan cities through the transmission network.

The authors would recommend leased bandwidth connectivity from telecom service providers for the purpose of telemedicine as the most cost-effective solution for India. Telemedicine equipment often requires a longer, stable connection to a fixed number of systems, for example, a rural hospital connected to a super specialty hospital; leased line connections suit this type of network requirement for both real-time and store and forward type of telemedicine applications.

V. Financial Benefits

A. Satellite

The proposed solution for using the mobile network will significantly reduce the cost of communication links as it will be an option to choose from the satellite link in the rural areas. A satellite communication link costs a higher monthly fee for satellite bandwidth and a higher amount of initial investment. While the cost of leased line via mobile network for the same bandwidth will be significantly less monthly cost and even lesser initial investment if the mobile BTS has reached the premises.

B. Business model for telecom operators

The proposed solution will be a good business model for telecom operators. Besides getting the mobile subscribers, the mobile operators will also be able to serve the corporate customers like hospitals and clinics for the leased service. Technological changes in the wireless mobile network from 2G, 3G, 4G to IMS will add more benefit to the operators as it will be easier to provide new services like Centrex systems and VPNs for the purpose of telemedicine.

C. *Electricity*

Electricity is highly inconsistent in rural areas of India [10] with voltage fluctuation and daily load shedding (black outs for several hours). This is due to insufficient electricity production demanded by the consumers of the country. Operation of any electronic equipment requires constant power supply. For this huge investment on a power backup system such as Uninterrupted Power Supplies (UPS) with batteries, standby generators and solar power systems will be required at the equipment sites. These power backup systems are mandatory for mobile telecommunication operators in rural areas to keep their BTSs up and running 24 hours a day. Hence, for the leased line communication, the hospitals do not need to worry about their communication system going out of service due to power outages. The rural hospitals will still need to prepare their own power backup system for their telemedicine equipment; however, many do not require constant a power supply. For example, computers, cameras, microscopes, etc may be switched off during hours when no one is using the telemedicine service. Thus, lower investment on power backup system will suffice in the hospitals. The heavy cost burden on power backup system will be faced by the mobile operators instead of the hospital.

VI. DISCUSSION

Communication networks play a vital role in all the different types of telemedicine systems. It is not possible for medical entities to build their own communication networks for the purpose of telemedicine. Network infrastructure sharing should be done with telecom service providers.

Leased bandwidth connectivity using expanding mobile infrastructure for telemedicine in rural areas is the most cost-effective solution for the present scenario in India. Telemedicine equipment often requires a longer, stable connection to a fixed number of systems, for example, a rural hospital connected to a super specialty hospital; leased line connections suit this type of network requirement for both real-time and store and forward types of telemedicine applications. Since mobile networks are growing in the rural areas in India, further utilization of telemedicine services will be easier if leased connections can be provided through the BTS sites in the rural areas.

For this type of leased line service for a telemedicine network, a major role will be played by the telecom operator. Telecom operators should be ready for providing leased bandwidth service through their transmission network from their BTS sites. There will be issues of reliability in the leased network that will additionally require maintenance by the telecom operator.

India has 10 different telecom operators all over the country, but not all the operators will reach all the rural villages. Hence, different hospitals at different rural places will get connected with different telecom operators. Interoperability between the operators for leased connections will also be a key issue. A governing body should play a role in standardization and ensuring interoperability regulations between the operators for leased connections. Telecom Regulatory Authority of India (TRAI) should play a role in the provision of the leased connections for health services at fair prices by the telecom operators at rural places.

VII. CONCLUSION

Telemedicine has a wide range of applications in developing countries like India where the medical resources and professionals are insufficient. As medical facilities are centralized in highly populated cities, telemedicine provides easily accessed, quality medical services to rural areas. Quite possibly, the biggest advantage for growth in telemedicine is the boom in IT sectors worldwide. In India, the IT industry is booming as mobile networks and high bandwidth optical links will reach most of the remote villages in very near future. Using mobile infrastructure for telemedicine in rural areas will be the most beneficial solution. With the help of telemedicine, better health facility can be served to the poor communities of the rural villages and enhance their living standard.

REFERENCES

1. Department of Telecommunications, Ministry of Communication & IT, Government of India (2010) Annual Report 2010, New Delhi.
2. Health Line, Okhla. Medical tourism India. at http://www.medical-tourism-india.com
3. PricewaterhouseCoopers (2007) Emerging Market Report: Health in India 2007
4. The financial express (July 2005) at http://www.financialexpress.com/news/doctorpopulation-ratio-stands-at-11-722/139534
5. Apollo Telemedicine Network Foundation. Services at http://www.telemedicineindia.com/Services.htm
6. CISCO newsroom (May 2010) at http://newsroom.cisco.com/dlls/2010/prod_050710b.html
7. ISRO (2010) at http://www.isro.org/scripts/telemedicine.aspx
8. Cellular Operators Association of India. (November 2010) Annual Report on Cellular Operators Association of India 2009-10 at http://www.coai.in
9. S.K. Mishra, D. Gupta, and J. Kaur (June 2007) Telemedicine in India: Initiatives and vision, e-Health Networking, Application and Services, pp 81-83, DOI. 10.1109/HEALTH.2007.381608
10. S. Surana, R. Patra, S. Nedevschi, and E. Brewer, (2008) Deploying a Rural Wireless Telemedicine System: Experiences in Sustainability, vol. 41, Computer, no. 6, pp 48-65, DOI: 10.1109/MC.2008.184

Author: Ujjwal Bania
Institute: Center for TeleInFrastruktur (CTIF), Aalborg University
Street: Fredrik Bajers Vej 7, 9220, Aalborg
City: Aalborg
Country: Denmark
Email: ujjwal.bania@gmail.com

PbS Nanodots For Ultraviolet Radiation Dosimetry

Yu. Dekhtyar[1], M. Romanova[1], A. Anischenko[1], A. Sudnikovich[1], N. Polyaka[1], R. Reisfeld[2], T. Saraidarov[2], and B. Polyakov[3]

[1] Institute of Biological Engineering and Nanotechnology, Riga Technical University, Riga, Latvia
[2] Hebrew University of Jerusalem, Jerusalem, Israel
[3] Institute of Solid State Physics, University of Latvia, Riga, Latvia

Abstract— Lead sulfide **(PbS) nanodots in Zirconia (ZrO_2) thin film matrix (ZrO_2:PbS films) were investigated for UV radiation dosimetry purposes. Samples were fabricated using sol-gel technique. ZrO_2:PbS films were irradiated with UV light with wavelengths 250 – 400 nm during 50 minutes. Photoelectron emission spectra of ZrO_2:PbS films were recorded and band structure for nonradiated and UV irradiated samples was calculated. It was found that quantity of localized states decreased after UV irradiation while density of localized states was dependent on concentration of PbS nanodots. The observed changes in band structure of ZrO_2:PbS films after UV irradiation suggest that the films may be considered as an effective material for UV radiation dosimetry, PbS nanodots being the UV sensitive substance of such a dosimeter.**

Keywords— **PbS nanodots, ultraviolet radiation, dosimetry, photoelectron emission.**

I. Introduction

Direct or indirect damage of biological structures caused by ultraviolet (UV) radiation depends on interaction of UV photons with either DNA or other biomolecular structures. These structures are scaled to nanodimensions, therefore, it is necessary to have an UV sensor of corresponding nano volume. This experimental work offers to use a thin film dosimeter, which consists of nanodots embedded in a solid thin film matrix. The nanodots are supposed to be a radiation-sensitive substance. The signal can be detected by measuring emission of low energy photo excited electrons (~ 1 eV) which have a mean free path of the order of several nanometers.

The target of the research was to examine changes in photoemission properties of PbS nanodots embedded in ZrO_2 thin film matrix under the influence of UV radiation. PbS nanodots were chosen for their possible application as an UV radiation dosimeter because it has been reported that they have emission and absorption lines in a large spectral region [1]. To eliminate the influence of ZrO_2 matrix and prove that the nanodots not the matrix are sensitive to radiation, samples with polyvinyl alcohol (PVA) matrix were studied as well.

II. Samples

Both ZrO_2:PbS and and PVA:PbS films were made using sol-gel technology [1]. PbS nanocrystals were embedded in ZrO_2 and PVA matrixes. Samples with 10%, 20% and 50% concentration of PbS in ZrO_2 matrix and 20% PbS in PVA matrix were studied. All films were deposited on a glass substrate. Thickness of the films was in a range of 0.1-1 µm. Typical size of the PbS nanodots was 2-4 nm in ZrO_2 matrix and 2-3 nm in PVA matrix. To verify size of nanodots, the atomic force microscope Solver P-47 PRO was employed.

III. Methods

The samples were UV irradiated with HAMAMATSU PHOTONICS xenon-mercury lamp L8222 (250 – 400 nm) during 0-50 min. The photoemission (PE) current of irradiated and nonirradiated samples was recorded, a handmade spectrometer was used. PE was excited by photons in energy range 4-6 eV from the deuterium lamp source (LOT-Oriel Europe). Emitted electrons were detected using the secondary electron multiplier (VEU-6, Russia) in vacuum condition 10^5 torr.

IV. Results and discussion

Derivatives of PE current of nonirradiated ZrO_2:PbS and PVA:PbS films are shown in Fig. 1.

K. Dremstrup, S. Rees, M.Ø. Jensen (Eds.): 15th NBC on Biomedical Engineering & Medical Physics, IFMBE Proceedings 34, pp. 133 – 136, 2011.
www.springerlink.com

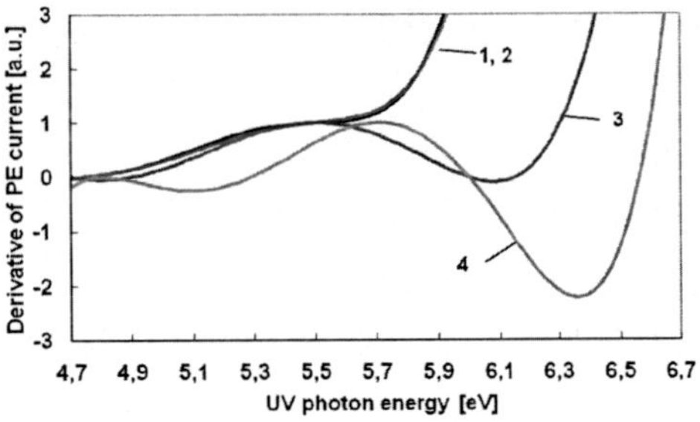

Fig. 1 Derivatives of PE current of nonirradiated films: (1) ZrO_2:10%PbS; (2) ZrO_2:20%PbS; (3) ZrO_2:50%PbS; (4) PVA:20%PbS

ZrO_2:50%PbS film has an emission interval 4.8 – 6 eV with a maximum at 5.5 eV. Interval width is 1.2 eV, which is several times more the uncertainty 0.02 eV of the photon energy hν. Therefore, the interval can be associated with emission from localized states inside the energy gap. The derivative curve goes up after 6 eV that might belong to the edge region of the valence band.

Derivatives of ZrO_2 films with smaller PbS concentration (10% and 20%) have an inflection point at the same photon energy as the maximum detected for ZrO_2:50%PbS films (5.5 eV). The derivative curve starts to go up after 5.7 eV that is similar to the behavior of ZrO_2:50%PbS derivative after 6 eV. Considered features of the spectra allow to suppose that localized states and the valence band are overlapped for ZrO_2:10%PbS and ZrO_2:20%PbS films, and concentration of localized states is not sufficient to provide a clear emission maximum.

The shape of PVA:20%PbS derivative is similar to ZrO_2:50%PbS derivative but the maximum is shifted to 5.7 eV and the curve starts to go up after 6.35 eV. However, the shape of PVA:20%PbS derivative differs from that of ZrO_2:20%PbS derivative. Taking into account the similarities in the shape of derivatives, it is possible to suppose that localized states are created by PbS nanodots but PVA matrix has strong influence on localized states.

Fig. 2 shows derivatives of PE current of 50 minutes UV irradiated films.

To calculate the band structure of the films before and after UV irradiation (Fig. 4) the scheme shown in Figure 3 was employed. All band structure parameters were calculated using the spectra shown in Figures 1 and 2.

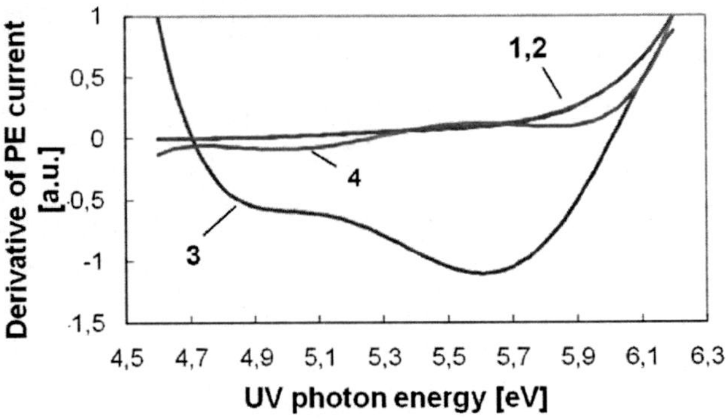

Fig. 2 Derivatives of PE current of 50 min UV irradiated films: (1) ZrO_2:10%PbS; (2) ZrO_2:20%PbS; (3) ZrO_2:50%PbS (4) PVA:20%PbS

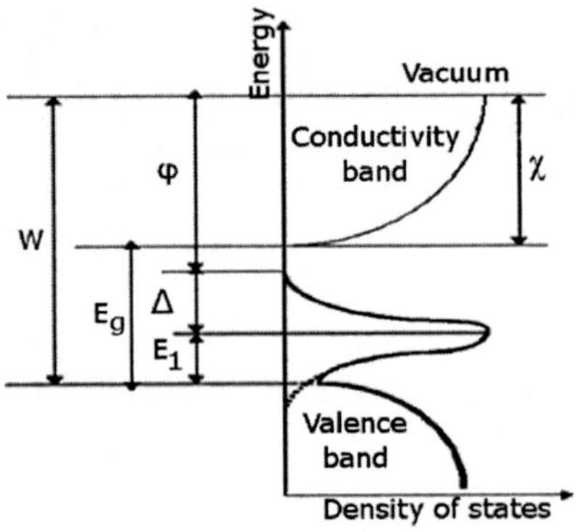

Fig. 3 Scheme for band structure calculation of ZrO2:PbS and PVA:PbS films: φ – electron work function; Eg – energy gap; W –energy needed to release an electron from the valence band; Δ – half width of localized states; E1 – distance between the valence band and the midpoint of localized states; χ – electron affinity. W and Δ were calculated using the recorded photoemission current. To calculate W, the tail of a spectrum which is related to the valence band was approximated using MS Excel and further extrapolated to I=0. χ=W-Eg; E1=W-(φ+Δ)

Figure 4 shows:

1. Density of localized states decreases under the influence of UV radiation for both ZrO_2:PbS and PVA:PbS films. Tails appear at the edge of the valence band for nonirradiated ZrO_2:10%PbS and ZrO_2:20%PbS films. Localized states of ZrO_2:50%PbS film disappear completely after 50 minutes of UV irradiation. This can indicate that interaction between localized states increases with increase of PbS nanodot concentration.
2. The position E_1 of localized states has a trend to be closer to the edge of the valence band when concentration of PbS increases.

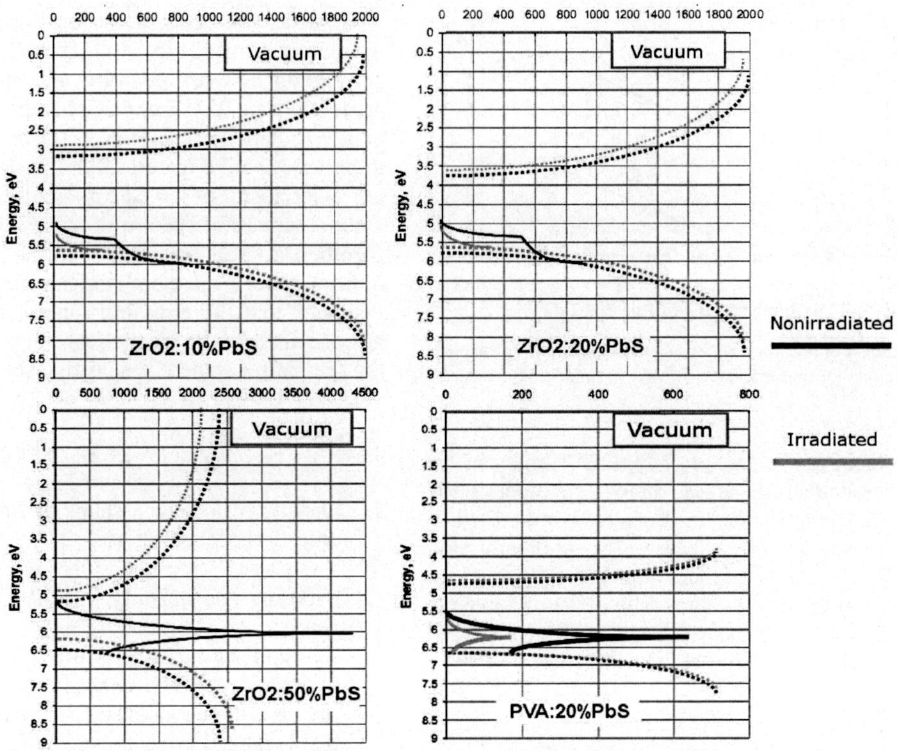

Fig. 4 Band structure of the ZrO_2:PbS and PVA:PbS films. The horizontal axis shows density of states, the vertical axis shows energy [eV]

3. The electron affinity χ increases with increase in PbS concentration in ZrO_2 matrix (Fig. 5). According to [2], increase in PbS concentration results in formation of larger PbS nanodots. Therefore, it is possible that larger nanodots are characterized with higher χ values.

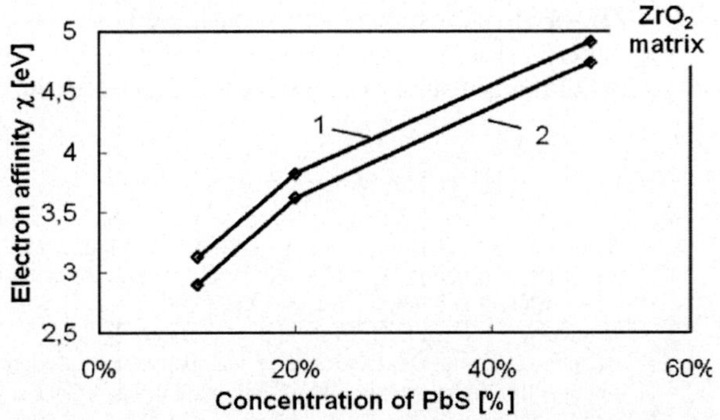

Fig. 5 Electron affinity of the ZrO_2:PbS films as a function of PbS concentration: (1) nonirradiated films; (2) irradiated films (50 minutes)

The electron affinity χ of the irradiated ZrO_2:PbS films is smaller than of nonirradiated films (Fig. 5). However, the difference $\Delta\chi$ between irradiated and nonirradiated films depends on PbS concentration (Fig. 6). $\Delta\chi$ value is smaller for films with higher PbS concentration. At the same time $\Delta\chi$ of PVA:20%PbS films (0.75 eV) is significantly higher than of ZrO_2:20%PbS films (0.2 eV), not shown in the Figure. It means that the matrix has strong influence on χ.

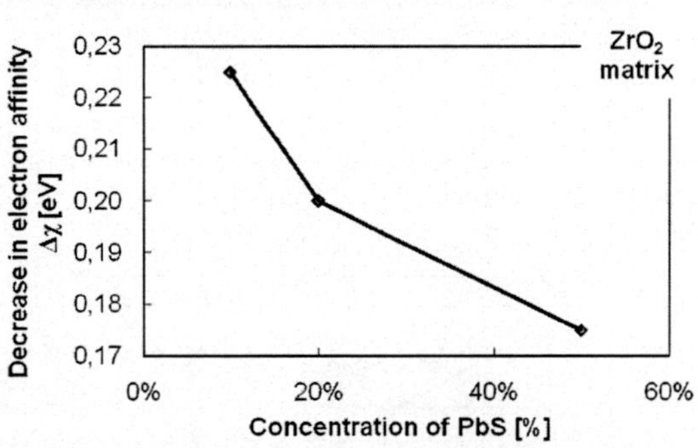

Fig. 6 Decrease in electron affinity after irradiation for the ZrO_2: PbS films

4. The half width Δ of localized states increases with increase in PbS concentration in ZrO_2:PbS films (Fig. 7). It might mean that density of localized states increases with increase in PbS concentration. It also might evidence that the detected emission was provided by localized states created by PbS nanodots.

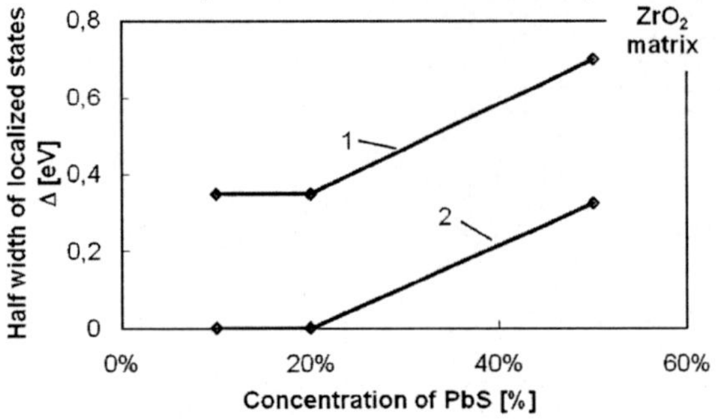

Fig. 7 The half width Δ of localized states of ZrO_2:PbS films as a function of PbS concentration: (1) nonirradiated films; (2) irradiated films (50 minutes)

Density of localized states of ZrO_2:PbS films can be evaluated by calculating areas below photoemission maximums at 5.5 eV shown in Figures 1 and 2. UV irradiation decreases the area below the photoemission maximums (Fig. 8). That evidences that UV light releases electrons from localized states. Films with 50%PbS have the largest value of the area. It means that density of localized states increases with increase in PbS concentration.

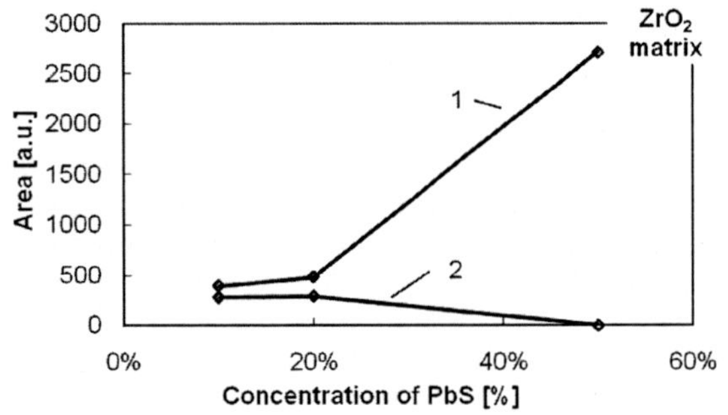

Fig. 8 The area below the photoemission maximum (5.5 eV) of ZrO_2:PbS films as a function of PbS concentration: (1) nonirradiated films; (2) irradiated films (50 minutes)

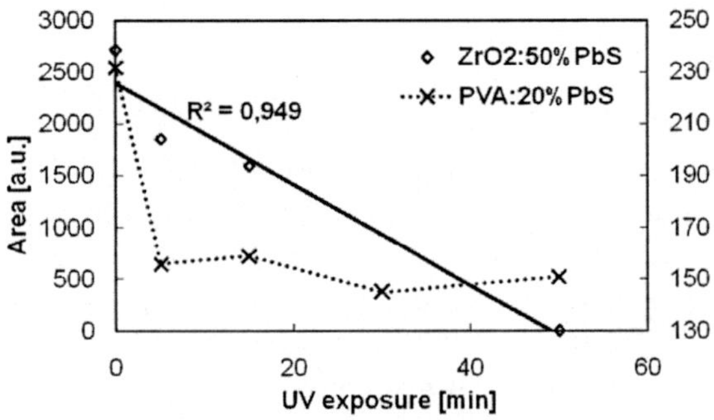

Fig. 9 Dependence of the area below the photoemission maximum of ZrO_2:50%PbS and PVA:20%PbS (secondary axis) films on UV exposure

Quantity of occupied localized states depends on irradiation time (Fig. 9). ZrO_2:PbS films have linear correlation between quantity of states (represented by the area) and time of UV exposure (representing different doses of UV radiation). Therefore, ZrO_2:PbS films are considered to be an effective material for UV dosimetry purposes. The correlation isn't observed for PVA:PbS films.

Photoemission spectra of ZrO_2 and PVA matrixes were recorded as well (not shown in the Figures). UV radiation did not change these spectra significantly. This gives more evidence that the emission maximums shown in Figure 1 are provided by PbS nanodots, meaning that PbS nanodots not the matrix is the UV sensitive substance of the film.

V. Conclusions

1. Density of localized states of ZrO_2:PbS films increases with PbS concentration.
2. There is a linear correlation between UV radiation exposure and quantity of localized states for ZrO_2:PbS films.
3. The half width Δ of localized states of ZrO_2:PbS films increases with PbS concentration that might depend on increase in nanodots size. Increase of Δ might evidence that localized states are created by PbS nanodots.
4. The electron affinity χ of ZrO_2:PbS films decreases after UV irradiation and increases with increase in PbS concentration. However, higher concentrations of PbS nanodots lead to smaller difference $\Delta\chi$ between nonirradiated and irradiated ZrO_2:PbS films.
5. The matrix of films influences χ value and density of localized states. Density of localized states is higher in ZrO_2 matrix than in PVA matrix.
6. ZrO_2:PbS films may be considered as a suitable material for UV radiation dosimetry, PbS nanodots being the UV sensitive substance of such a dosimeter.

References

1. Sashchiuk A, Lifshitz E, Reisfeld R et al. (2002) Optical and conductivity properties of PbS nanocrystals in amorphous zirconia sol-gel films. J Sol-Gel Sci Technol 24:31–38.
2. Saraidarov T, Reisfeld R, Sashchiuk A, Lifshitz E. (2003) Synthesis and characterization of lead sulfide nanoparticles in zirconia-silica-urethane thin films prepared by sol-gel method. J Sol-Gel Sci Technol 26:533–540.

Author: Marina Romanova
Institute: Riga Technical University
Street: Ezermalas 6b, 231
City: Riga
Country: Latvia
Email: marina.romanova@inbox.lv

Developments towards a Psychophysical Testing Platform – A Computerized Tool to Control, Deliver and Evaluate Electrical Stimulation to Relieve Phantom Limb Pain

B. Geng[1], K.R. Harreby[1], A. Kundu[1], K. Yoshida[1,2], T. Boretius[3], T. Stieglitz[3], R. Passama[4], D. Guiraud[4], J.L. Divoux[5], A. Benvenuto[6], G. Di Pino[6], E. Guglielmelli[6], P.M. Rossini[6,7], and W. Jensen[1]

[1]Department of Health Science and Technology, Aalborg University, DK [2]Biomedical Engineering Department, Indiana University-Purdue University Indianapolis, USA [3]Department of Microsystems Engineering, University of Freiburg, Germany [4]Laboratoire d'Informatique, de Robotique et de Microelectronique de Montpellier, France [5]MXM Neuromedics, France [6]Università Campus Bio-Medico di Roma, Italy [7]IRCCS S. Raffaele-Pisana, Italy

Abstract— **Phantom limb pain (PLP) frequently follows amputation. Artificially inducing phantom hand sensations by electrical stimulation may reduce PLP. The use of implantable, multi-channel microelectrodes provides the opportunity to selectively activate sensory fibres. However, combinations of variables from a multichannel stimulation system can produce a huge number of possible stimulation paradigms. It makes the use of psychophysical evaluation of the evoked sensations an impractical and time-consuming task in the clinical setting. Our aim is to develop a computerized, automatic, psychophysical testing platform to support control, delivery and evaluation of the electrical stimulation for PLP relief.**

Keywords— **Phantom limb pain, psychophysical test, multichannel electrical stimulation.**

I. INTRODUCTION

Amputation of a limb involves the complete truncation of all afferent and efferent nerves, and it is usually followed by the sensation that the lost body part is still present and kinaesthetically perceived. These phenomena are called phantom awareness and phantom sensation [1]. In 50-80% amputees, phantom limb pain (PLP) develops in the lost limb [2]. Today, it is not completely understood why the pain develops, and there are no fully effective treatments.

The role of cortical neuroplasticity in phantom sensations and phantom pain has been examined by several groups. For example, Flor and colleagues reported a strong association between changes in the S1 area of the brain cortex with PLP [3;4]. Also Jenkins et al. found that the cortical representation of the amputated limb invaded the mouth region, i.e., a cortical shift of the limb representation [5].

Several studies have demonstrated the favorable effect of enhancing the sensory feedback to the missing limb on PLP relief. For instance, patients with PLP, who intensively used myoelectric prosthesis [4] or received daily discrimination training of surface electrical stimuli applied to the stump experienced significant reduction of PLP [6]. Also, Dhillon et al. demonstrated the use of intrafascicular, electrical stimulation (through an implanted neural interface) proved to be capable of eliciting tactile or proprioceptive sensations [7;8]. Finally, Rossini et al. demonstrated that training for control of a robotic hand with limited amount of sensory feedback significantly reduced PLP in a human amputee implanted with four LIFE electrodes. The reduction in PLP lasted several weeks after the electrodes were removed and changes in sensorimotor cortex topography were shown [9].

These previous studies provide evidence that it may be possible to relieve PLP by providing appropriate sensory feedback to the amputee subject. The aim of the EU consortium 'TIME' (www.project-time.eu) is to develop a complete, implantable neural prosthesis system with sufficient stimulation selectivity to manipulate phantom sensations and explore the possibility of using the method as a treatment for PLP. The key technological developments include a multi-channel implanted stimulator, a peripheral nerve interface, and a psychophysical testing platform (Fig. 1).

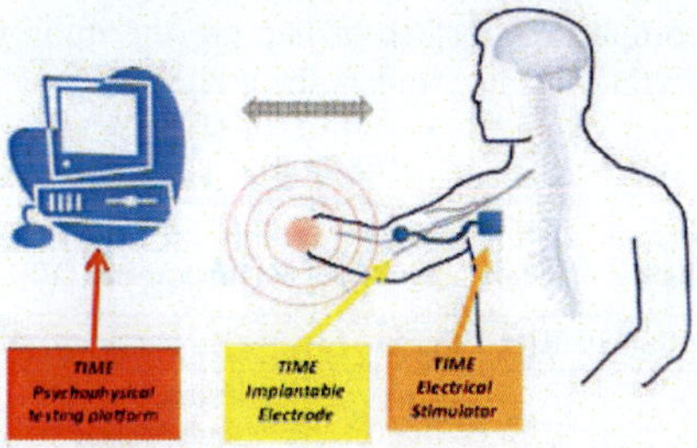

Fig. 1 The 'TIME' system for relieving phantom limb pain.

In a multichannel stimulation system (i.e., the TIME electrodes include up to 12 active sites [10]) various combinations of stimulation parameters will generate a large amount of possible stimulation paradigms. For example, the stimulation can be presented from any combination of electrode sites, and pulse duration, amplitude and frequency must be defined. Also, experimental paradigms that involve the quantification of subjective sensations involve the use of psychophysical measures. The goals of the psychophysical testing experiments are: a) to locate evoked sensations; b) evaluate the type of evoked sensations; c) quantify the strength of evoked sensations. As such, an automated, com-

K. Dremstrup, S. Rees, M.Ø. Jensen (Eds.): 15th NBC on Biomedical Engineering & Medical Physics, IFMBE Proceedings 34, pp. 137–140, 2011.
www.springerlink.com

puterized approach is necessary to minimize the time needed to collect and evaluate the data from a subject. The objective of the present work is to design and develop a platform for carrying out psychophysical testing in human volunteer subjects. In the present paper we report on our developments and tests so far.

II. METHODS

A. Overall design strategy

To evaluate the effect of intrafascicular stimulation on relieving PLP in future clinical trials, we identified two main experiments to be carried out: 1) sensation threshold measurement and 2) mapping of sensation location, type and strength. To perform these experiments on the platform, specific functionalities were defined:

- *Functionality I. Configuration of stimulation variables.* A stimulation sequence is defined by a number of parameters: waveform, amplitude, pulse duration, pulse rate and number of pulses (see Table 1 the range and the step size of the stimulation parameters). Since the TIME electrode has 8-12 active sites [10], a choice of which site to stimulate must be made, and possibly timing between sites, if pairs of active sites are used.
- *Functionality II. Measurement of sensation location, type and strength.* To evaluate evoked sensations a psychophysical questionnaire is presented on the computer interface to capture the quality of perceived phantom sensations in the subject.

Table 1 Summary of the various stimulation parameters that the TIME psychophysical testing platform incorporates. The stimulation parameters were chosen based on previous animal experiments [12].

Stimulation parameter	Range	Step Size
Waveform	Square wave, monophasic Square wave, biphasic	N/A
Amplitude (μA)	1-500	5
Pulse Duration (μs)	10-500	5
Time between two subsequent stimuli	Determined by the subject	N/A
Number of pulses in a pulse train	1-250	5-10
Pulse Rate (Hz)	1-500	5-10
Number of sites to be activated simultaneously	1 or 2	N/A
Time between active sites (ms)	min 2	5-10

These two functionalities are implemented through specific hardware (HW) and software (SW) components. The platform is implemented through the following two subsystems (Fig. 2):

- *Stimulator and Experiment Control (SEC) subsystem.* The main HW component in the SEC subsystem consists of a computer (*Computer #1*) to be operated by the experimenter. Through the software distributed in this computer, the experimenter will be able to control, define and deliver stimulation sequences, and will also be able to monitor the experimental progress. Also, the custom-designed bench-top stimulator is defined to be part of the SEC subsystem [11]. The miniaturized version of the stimulator prefigures the final implantable system for which further optimizations in size and power consumption have to be carried out.
- *Interactive Subject Interface (ISI) subsystem.* The main HW component in the ISI subsystem consists of a computer (*Computer #2*) to be operated by the amputee subject. The software distributed in this computer allows to collect a series of psychophysical measures, i.e., the subject interactively responds to the stimulation by answering a questionnaire. The TIME electrodes are defined to be part of the ISI subsystem.

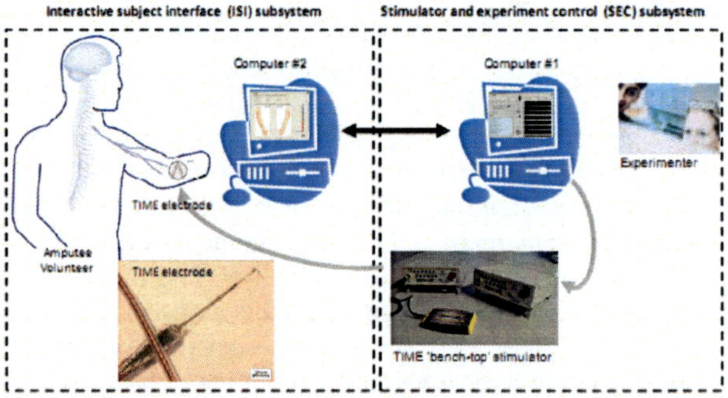

Fig. 2 Schematic drawing of the hardware and software components involved in the TIME psychophysical testing platform.

B. Design and implementation of the SEC subsystem

The SEC subsystem is used to control, define and deliver stimulation sequences, and to monitor the experimental progress. The software is developed under the LabVIEW programming environment, and it controls the TIME stimulator through a low level driver. The low level driver consists of a set of application programming interface (API) functions. The API functions are built in dynamic linked libraries, serving as interfaces for the software to access the hardware. Since the experimenter needs to set stimulation parameters and control a number of functions, these are grouped in different physical locations in the graphical user interface (GUI) according to their function (Fig. 3):

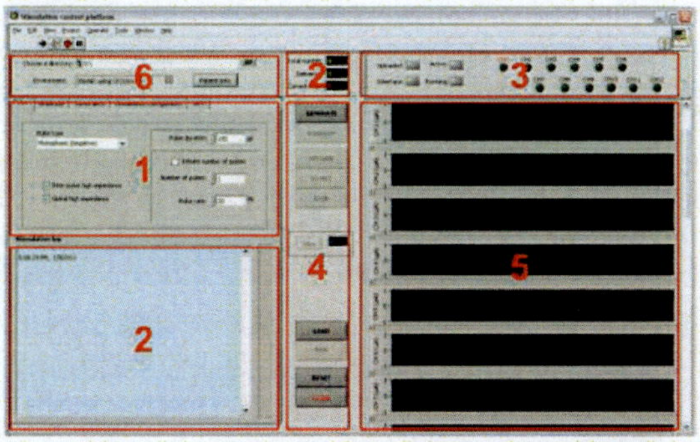

Fig. 3 The main GUI for the SEC subsystem. The stimulation parameters and functions to be controlled are divided into six groups.

- *Group 1: Stimulation parameter and constraint configuration.* The stimulation sequence to be applied is configured here. To ensure the safety of the subject during stimulation, a number of safety constraints can be defined, including: maximum accumulated charge, maximum global discharge, duration of limited passive discharge and duration of global passive discharge. If the stimulation sequences violate the constraints, the stimulation will automatically stop.
- *Group 2: Progress monitoring.* The experiment progress is monitored by a 'progress bar', as well as numeric numbers indicating the total number of stimuli to be delivered, the number of stimuli delivered and the current amplitude of the pulse being delivered. In addition, the communication with the stimulator is displayed and automatically saved in a log file to an assigned directory.
- *Group 3: Status monitoring.* The status of the stimulation is indicated by a series of LEDs: 1) stimulation sequence uploaded to the stimulator, 2) communication interface ready, 3) stimulator active, and 4) stimulator running. Another twelve LEDs indicate active cathode channels (i.e., they light up, when they are active).
- *Group 4: Experimental control.* This panel allows to control the way that the defined stimulation sequences are delivered to the human volunteer subject, including generate/randomize stimuli, start/stop stimulation, and display a particular stimulation sequence.
- *Group 5: Graph display.* A graphical display of the stimulation sequences being delivered to each of the 12 channels of the stimulator.
- *Group 6: Information management.* This function is designed to track experiment and patient information.

C. Design and implementation of the ISI subsystem

The ISI subsystem is used to evaluate the evoked phantom sensation in the amputee subject. The GUI comprising three psychophysical questions were implemented (Fig.4):

- *Question 1*: *What type of sensation did you feel*? The subject can choose between the following words: touch/pressure, vibration, tugging, spider crawling, finger flexion/extension, wrist flextion/extension, cold, warm, pinch and pain.
- *Question 2*: *Where did you feel the sensation*? To locate the sensation an illustration of an arm/hand is presented to the subject on the interface.
- *Question 3*: *How strong was the sensation*? A visual analogue scale (VAS) is presented on the GUI to quantize the strength of the sensation.

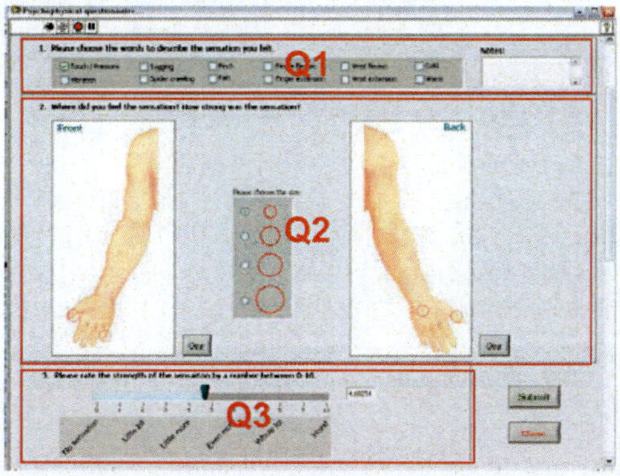

Fig. 4 The GUI for the ISI subsystem. The subject will need to answer the computerized questionnaire immediately after each stimulation.

D. Communication between SEC and ISI subsystems

The ISI and SEC subsystems need to communicate during the automated psychophysical testing procedure. The purpose of the communication is to send specific control commands and acknowledgement messages between two subsystems distributed in the two computers. *Computer #1* (experimenter side) sends an acknowledgement to *Computer #2* (subject side) when a stimulus is successfully delivered, that allows *Computer #2* to proceed. Following this, *Computer #2* sends an acknowledgement to *Computer #1* when the questionnaire is answered, and the subject is ready for next stimulus. The communication has been implemented by an Ethernet crossover cable on the hardware level. The cable directly connects the two computers allowing data transfer between the two computers across the network. On the software level, the LabVIEW DataSocket has been used. The technique allows data communication between two applications residing in different computers.

III. Results

A. Evaluation of the SEC subsystem in acute animal experiments

The SEC subsystem was successfully tested in two acute animal experiments. The stimulation protocol consisted of a series of monopolar, negative, single-pulse stimulation of a TIME electrode implanted into the median nerve in the pig forelimb (see [12] for a description of the animal preparation and TIME electrode implant procedures). The specific stimulation parameters tested are listed in Table 2.

Table 2 Stimulation parameters used in two acute animal experiments to test the SEC subsystem.

Stimulation parameter	First animal	Second animal
Waveform	Square wave, monophasic	Square wave, monophasic
Amplitude (μA)	10-800	20-800 80-3200
Step size (μA)	10	20/80
Pulse Duration (μs)	100	100
Number of pulses	1	40
Pulse Rate (Hz)	N/A	2
Number of sites to be activated simultaneously	1	2
Randomized?	No	No
Automated delivering?	Yes	Yes

B. Evaluation of the ISI subsystem and communication protocol with able-bodied, human volunteers

A preliminary version of the ISI subsystem and the communication protocol was tested in able-bodied subjects using electrodes placed on the surface of the skin. The preliminary version of the ISI included a simplified but comparable questionaire, and the communication protocol between two subsystems was working successfully (see [13] for a description of the experiments performed).

IV. Discussions and conclusions

Within the TIME project, the ultimate aim of the TIME project this work is to test the psychophysical testing platform and the integrated system in an amputee volunteer. Currently, work on designing, optimizing and testing the TIME electrode, the TIME stimulator together with theoretical stimulations and animal experimental work is being carried out. Although previous experience with eliciting sensations by applying electrical stimulation through an implanted, neural interface does exist, it is expected that specific stimulation patterns must be identified for the individual subject to elicit specific sensations. As such, to choose a set of 'optimal' subject-specific stimulation parameters', it may be necessary to add a new functionality to automatically quantify and statistically select an appropriate subset of stimulation parameters. It is expected that the subset of stimulation parameters that generate distinct and/or unique sensation types will be chosen. It is also expected that several, repeated stimulation sessions with 'optimal' parameters must be carried out further.

References

1. Navarro X, Vivó M, Valero-Cabré A (2007) Neural plasticity after peripheral nerve injury and regeneration. Prog Neurobiol 82:163-201
2. Ephraim PL, Wegener ST, MacKenzie EJ et al (2005) Phantom pain, residual limb pain, and back pain in amputees: results of a national survey. Arch Phys Med Rehabil 86:1910-1919
3. Flor H, Nikolajsen L, Jensen TS (2006) Phantom limb pain: a case of maladaptive CNS plasticity. Neuroscience 7:873-881
4. Flor H, Dencke C, Schaefer M et al (2001) Effect of sensory discrimination training on cortical reorganization and phantom limb pain. Lancet 357: 1763-1764
5. Jenkins WM, Merzenich MM, Ochs MT et al (1990) Functional reorganization of primary somatosensory cortex in adult owl monkeys after behaviorally controlled tactile stimulation. J Neurophysiol 63:82-104
6. Lotze M, Grodd W, Birbaumer N et al (1999) Does use of myoelectric prosthesis reduce cortical reorganization and phantom limb pain? Nat Neurosci 2:501-502
7. Dhillon GS, Krüger TB, Sandhu JS et al (2005) Effects of short-term training on sensory and motor function in severed nerves of long-term human amputees. J Neurophysiol 93: 2625-2633
8. Dhillon GS, Horch K (2005) Direct neural sensory feedback and control of prosthetic arm. IEEE Trans Neural Syst Rehabil Eng 13:468-472
9. Rossini PM, Micera S, Benvenuto A et al (2010) Double nerve intraneural interface implant on a human amputee for robotic hand control. Clin Neurophysiol 121:777-783
10. Boretius T, Pascal-Font A, Schuettler M, et al (2010) A Transverse intrafascicular multichannel electrode (TIME) to interface with the peripheral nerve. Biosens Bioelectron 26:62-69
11. Andreu D, Guiraud D, Souquet G (2009) A distributed architecture for activating the peripheral nervous system. Journal of Neural Engineering 6, 001-018
12. Kundu A, Jensen W, Kurstjens M, et al (2010) Dependence of implantation angle of the transverse, intrafascicular electrode (TIME) on selective activation of pig forelimb muscles. Artif Organs. 34, 8, s. A43, No. 92
13. Geng B, Yoshida K, Jensen W (2010) Effects of the number of pulses on evoked sensations in pairwise electrocutaneous stimulation. Artif Organs. 34, 8, s. A39, No. 67

Address of the corresponding author:

Author: Bo Geng
Institute: Dept. Health Science and Technology, Aalborg University
Street: Fredrik Bajersvej 7D
City: Aalborg
Country: Denmark
Email: bogeng@hst.aau.dk

Comparing MRCP of Healthy Subjects with That of ALS Patients

Ying Gu and Kim Dremstrup

Center for Sensory-Motor Interaction (SMI), Department of Health Science and Technology, Aalborg University, Aalborg, Denmark

Abstract— **The study compared the movement related cortical potential (MRCP) of healthy subjects with that of amyotrophic lateral sclerosis (ALS) patients. We applied the same experimental and analytical methods to 7 healthy subjects and 4 ALS patients. They were asked to imagine right wrist extension at two speeds (fast and slow). The peak negativity and rebound rate were extracted from MRCP. Significance test showed that the healthy presented higher peak negativity during fast movement imagination than ALS. In addition, the healthy showed stronger rebound rate than ALS during both fast and slow movement imagination. Weak rebound rate might reflect the impairment of motor output pathway.**

Keywords— **Electroencephalography (EEG), Movement related cortical potential (MRCP), motor imagery, Brain computer interface (BCI) and amyotrophic lateral sclerosis (ALS).**

I. Introduction

MRCPs are the electroencephalographic (EEG) evidence of motor cortical involvement during movement preparation and execution (1). They have been studied for decades mostly in motor control physiology and psychophysiology (2-5). It is known that MRCPs occur in association with both executed and imaginary movements and that their magnitude and latency are modulated by the participants' psychological status and the characteristics of the movement performed, such as speed, precision and movement repetition (6-10). Recently, efforts have been devoted to identify MRCP in single trail basis for their application in BCI (11-14). BCI aims to provide no-muscular communication and control channel for severely disabled patients. ALS patients are among those groups. Specifics in ALS patient, the disease progresses from the first symptoms of muscular or respiratory weakness to the locked in state. In these patients, sensory, emotional and cognitive processing often remains largely intact despite extensive degeneration of the motor system (15). Modern life support technology can allow the most individuals to live long lives. Paralysis greatly limits the independence and communication. Comparing the characteristics of MRCP from healthy subjects with those from ALS is expected to contribute to a better understanding of brain motor functions. It could help to transfer BCI developed based on healthy subjects to BCI for ALS patients.

We have conducted two separate motor imagery studies with similar experimental tasks setting and signal analysis. The aim of the study was to find significant differences between healthy subjects and ALS patients based on analysis of MRCP.

II. Methods

A. Subjects

In the studies, there were 7 healthy volunteers and 4 ALS patients. None of healthy volunteers had known sensory-motor diseases or any history of psychological disorders. In ALS patient 1, 2 and 3, limbs movements were severely impaired. ALS patient 3 was in locked-in state and artificially fed and ventilated. ALS patient 4 was in very early stage of disease and he had intact limbs movement except slight weakness of right index finger.

B. Experiment Procedure

Participants was seated in a comfortable chair and asked to imagine right wrist extension at two speeds (fast and slow). The fast speed corresponded to a movement executed as fast as possible whereas slow speed was associated to a movement performed in approximately 3s. The tasks were randomly presented to the participants, controlled by a computer program. The EEG signals were amplified with a digital DC EEG amplifier (Neuro Scan Labs, NuAmps), low-pass filtered with cut-off frequency 200Hz and sampled at 500Hz using a 22-bit A/D converter. The participants were asked to avoid eye blinking, slow eye movement and facial movement during motor imagery. We conducted two separate studies as follows:

a) EEG recording from 7 healthy volunteers: The subjects executed the tasks for approximately 3 minutes to acquire experience on the movements to be imagined and on the experimental procedure. Then they were instructed to recall the experience of wrist movements and to perceive the movements while physically relaxing during motor imagery.
b) EEG recording from 4 ALS patients: To instruct patients on the tasks, the experimenter described the

K. Dremstrup, S. Rees, M.Ø. Jensen (Eds.): 15th NBC on Biomedical Engineering & Medical Physics, IFMBE Proceedings 34, pp. 141 – 143, 2011.
www.springerlink.com

movement while performing them in front of the patients. In addition, the experimenter passively executed the movements with patients' right wrist. The subjects were asked to feel themselves performing the movements instead of visualizing movements during recording.

C. MRCP Analysis and Statistical Analysis

Epochs starting 2s before the imagination onset and 2s after were extracted by the software EEGlab (16). Trials contaminated by EOG signal exceeding 75uV and facial EMG were discarded visually from further analysis. The baseline was corrected on each EEG channel by subtracting the mean amplitude value in the interval -2s to -1.8s with reference to the imagination onset (time 0).

The peak negativity (PN) and the rebound rate (RR) were identified from single trial. First, the EEG signals were smoothed by moving average over 400 time samples. Then, the PN was calculated as the lowest value between -1 and 2s. The RR was calculated as difference between amplitude of potential at 0.4s after the peak negativity and the peak negativity, divided by 0.4s. Finally, the averaged PN and RR were calculated for each subject for statistical analysis. Unbalanced two-sample T-test was performed to test significant difference of PN and RR between healthy subjects and ALS patients. Outcomes were considered significant at $p<0.05$.

III. Results

Fig. 1 and 2 show averaged MRCP from representative healthy subject and ALS patient respectively. These two figures show typical time course of MRCP. The negativity started increasing around -2s. After movement onset (time 0), the potential rounded. In Fig.1, rebound rates between fast and slow speed were quite different for the healthy, while it was not the case for ALS in Fig.2.

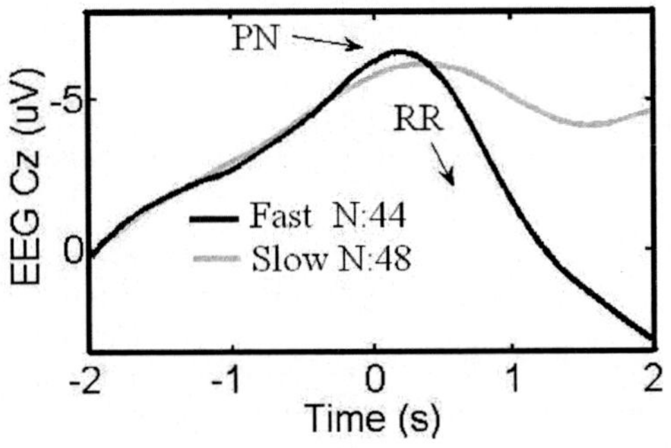

Fig. 1 Average MRCP from one representative healthy subject at channel Cz during the fast and slow speed tasks. Imaginary movement onset is represented by time 0s; N: number of averaged trials; PN: peak negativity; RR: rebound rate

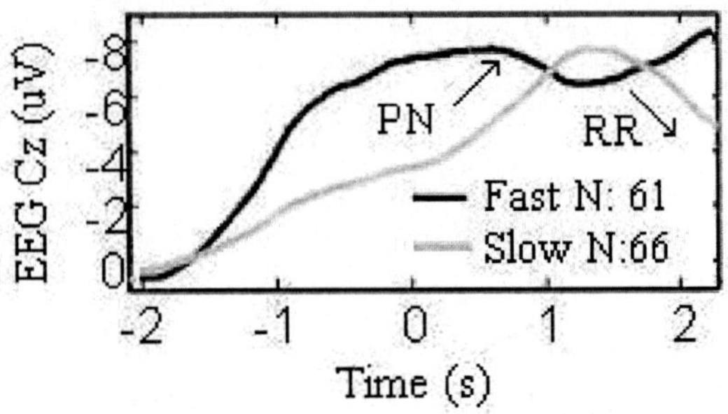

Fig. 2 Average MRCP from one representative ALS patient at channel Cz during the fast and slow speed tasks

Table 1 shows the average values for peak negativity and rebound rate associated to two speeds for each healthy subject and ALS patient. The mean ±Std were also shown. Table 2 reports the corresponding statistical analysis. There were significant differences on FastPN, FastRR and SlowRR between healthy subjects and ALS patients. However, there was no significant difference on SlowPN between the healthy and ALS. Statistically, healthy subjects showed higher peak negativity than ALS during fast speed task and they also showed stronger rebound rate than ALS during both fast and slow speed task. Interestedly, patients 4 showed quite similar rebound rate as the healthy.

Table 1 Average PN, RR at Cz for each healthy subject and ALS patient

	Fast PN (μV)	Slow PN (μV)	Fast RR (μV/s)	Slow RR (μV/s)
Healthy subjects				
1	-11.36	-9.86	10.57	6.73
2	-15.05	-15.96	11.05	8.32
3	-15.54	-20.90	12.05	11.12
4	-18.87	-31.15	13.85	10.42
5	-14.03	-13.41	12.52	10.11
6	-11.82	-10.28	10.17	11.66
7	-10.40	-13.96	8.93	6.07
Mean ±Std	-13.87±2.93	-16.50±7.45	11.31±1.64	9.2±2.19
ALS patients				
1	-8.43	-9.42	3.60	3.63
2	-5.94	-8.48	3.19	1.47
3	-13.57	-10.53	4.60	2.34
4	-7.80	-8.26	10.77	6.74
Mean ±Std	-8.94±3.27	-9.17±1.04	5.54±3.54	3.54±2.31

Table 2 Statistical analysis on FastPN, SlowPN, FastRR and Slow RR between healthy subject and ALS. ns: non-significant. For significant difference, p-value is shown

	Fast PN	Slow PN	Fast RR	Slow RR
Healthy ALS	p=0.03	ns	p=0.004	p=0.003

IV. Conclusions

MRCPs have been compared between actual movement and motor imagery (7, 9) and analyzed among different psychological status and movement parameters (5-15). This study first compared the features of MRCP of healthy subjects with those of ALS. Statistical analysis showed that the healthy presented higher peak negativity than ALS during fast movement, while it is not the case for slow movement. The rebound rate was stronger for the healthy than ALS during both fast and slow movement. The rebound rate reflects the characteristic of post-movement, which is highly dependent on cortical afferent response. The weaker rebound rate on ALS patients might reflect impairment of motor output pathway. In ALS patients in table 1, patient 1-3 had weak rebound rate, while patient 4 had similar rebound rate as the healthy. Patient 4 was in very early stage of ALS and he had normal limbs' movement except slight weakness of right index finger, while other 3 patients were severely disabled. Therefore, the rebound rate might also reflect the degree of motor degeneration. The analysis of MRCPs should be examined on more patients to provide strong evidence.

The comparison between the healthy and the patients might provide clues explaining the movement disorders. For motor rehabilitation, relevant features of MRCP could be examined to see their improvements in parallel with motor control performance.

References

1. Kornhuber HH, Deecke L. Hirnpotentiala¨nderungen bei Willkurbewegungen und passiven Bewegungen des Menschen: Bereitschaftspotential und reafferente Potentiale. Pflugers Archiv 1965; 284: 1-17.
2. Libet B, Gleason CA, Wright EW, Pearl DK. Time of conscious intention to act in relation to onset of cerebral activity (readiness-potential). The unconscious initiation of a freely voluntary act. Brain. 1983a;106 (3):623-42.
3. Libet B, Wright EW Jr, Gleason CA. Preparation- or intention-to-act, in relation to pre-event potentials recorded at the vertex. Electroencephalogr Clin Neurophysiol. 1983b; 56(4):367-72.
4. Shibasaki H, Barrett G, Halliday E, Halliday AM. Cortical potentials associated with voluntary foot movement in man. Electroencephalogr. Clin. Neurophysiol. 1981; 52:507–516.
5. Shibasaki H, Hallett M. What is the Bereitschaftspotential? Clin. Neurophysiol. 2006; 117: 2341-2356.
6. Slobounov SM, Ray WJ, Simon RF. Movement-related potentials accompanying unilateral finger movement with special reference to rate of force development. Psychophysiology 1998; 35: 537-548.
7. Romero DH, Lacourse MG, Lawrence KE, Schandler S, Cohen MJ. Event-related potentials as a function of movement parameter variations during motor imagery and isometric action. Behav. Brain Res. 2000; 117:83–96.
8. do Nascimento OF, Nielsen KD, Voigt M. Relationship between plantar-flexor torque generation and the magnitude of the movement-related potentials. Exp. Brain Res. 2005; 160: 154-165.
9. do Nascimento OF, Nielsen KD, Voigt M. Movement-related parameters modulate cortical activity during imaginary isometric plantar-flexions. Exp. Brain Res. 2006a; 171: 78-90.
10. Nielsen KD, Cabrera AF, do Nascimento OF. EEG based BCI-towards a better control. Brain-computer interface research at Aalborg University. IEEE Trans. Rehabil. Eng. 2006; 14: 202-204.
11. Farina D, do Nascimento OF, Lucas MF, Doncarli C. Optimization of wavelets for classification of movement-related cortical potentials generated by variation of force-related parameters. J. Neurosci. Methods. 2007; 162: 357-363.
12. do Nascimento OF, Farina D. Movement-Related Cortical Potentials Allow Discrimination of Rate of Torque Development in Imaginary Isometric Plantar Flexion. IEEE Trans. Biomedical Eng. 2008a; 55: 2675 - 2678.
13. Gu Y, do Nascimento OF, Lucas M-F, Farina D. Identification of task parameters from movement-related cortical potentials. Medical & Biological Engineering & Computing 2009a; 47(12): 1257-1264.
14. Gu Y, Dremstrup K, Farina D. Single-trial discrimination of type and speed of wrist movements from EEG recordings. Clinical Neurophysiology 2009b; 120:1596-1600.
15. Kübler A, Nijboer F, Mellinger J, Vaughan TM, Pawelzik H, Schalk G, McFarland DJ, Birbaumer N, Wolpaw JR. Patients with ALS can use sensorimotor rhythms to operate a brain–computer interface. Neurology 2005; 64: 1775-1777.
16. Delorme A, Makeig S. EEGLAB: an open source toolbox for analysis of single-trial EEG dynamics. *J Neurosci Methods* 2004;134 : 9-21.

Meat Cutting Tasks Analysis Using 3D Instrumented Knife and Motion Capture

C. Pontonnier[1,2], M. de Zee[1], A. Samani[1], G. Dumont[2,3], and P. Madeleine[1]

[1] Center for Sensory-Motor Interaction (SMI), Department of Health Science and Technology, DK-9220 Aalborg, Denmark
[2] VR4I project team, IRISA, Campus de Beaulieu, F-35042 Rennes Cédex, France
[3] ENS Cachan Antenne de Bretagne, Campus de Ker Lann, F-35170 Bruz, France

Abstract— **This article presents a complete experimental setup and pipeline designed to analyze kinematics and dynamics of meat cutting tasks in terms of musculoskeletal disorder appearance risk. An instrumented knife records the cutting force in 3D. A motion capture system assesses trunk and arm motion in 3D. Finally, the AnyBody software is used to run an inverse dynamics analysis on the recorded motions to obtain muscle forces. At last, EMG records are set up on most of relevant superficial muscles of the shoulder-neck area in order to validate the results. Sample results are proposed for standard workspace parameters, showing the relevance of the information for assessing the risk of developing work-related musculo-skeletal disorders.**

Keywords— **Movement analysis, Inverse Dynamics, External Forces, Muscle Forces, Ergonomics.**

I. INTRODUCTION

Meat cutting tasks as performed in slaughter industry often provoke musculoskeletal disorders (MSD) among butchers [1]. This can be explained by the presence of many of the well-acknowledged physical risk factors like repetitive arm movements, strenuous, relative short work cycle duration, insufficient rest, static posture and cold (e.g. [2,3]). The most common location for these disorders among butchers is the neck-shoulder region. The etiology of MSD is poorly understood and studies assessing the biomechanics of cutting tasks in greater detail have potential to understand the complex inter-relations between internal and external risk factors [4, 5]. The main part of the literature analyzing the cutting force is based on relative simple force assessment as only the resulting force was measured [6,7]. Indeed assessing the cutting forces in 3D exerted by the worker is necessary. McGorry [8] has designed a more advanced knife instrumented with strain gauges to record reactive forces and grip moments. This knife led to a more precise estimation of the intensity level of the task. However, no knife measuring force in 3D has to our knowledge been used in combination with 3D movement analysis. Such an experimental design has the potential to estimate the muscle forces generated during a cutting task.

In the present study we want to obtain a complete image of the cutting force, to be used as input in a detailed musculoskeletal model of the arm designed in the AnyBody Modeling System [9]. The force has to be precise in intensity and direction in order to be used as input in the 3D model. This is why we propose the design of an instrumented knife that can measure forces exerted on the handle during the cutting task. Once, the 3D model established, it can be used to investigate working situations. In this article we will investigate two different cutting tasks in terms of muscle load and posture.

II. MATERIALS AND METHODS

A. 3D instrumented knife

The knife is based on the instrumentation of a 3D force sensor (FS6, AMTI, Watertown, MA, USA). The applied forces to the force transducer were recorded as sketched in figure 1. The directions of force exertion were denoted as the "x-direction" for Fx, the "y-direction" for Fy and the "z-direction" for Fz (Figure 1). Force signals were low-pass filtered (10.5 Hz) and amplified 2000 times. The signals were A/D converted and sampled at 60 Hz (12 bits A/D converter, Nidaq 6024, National Instruments, Austin, TX, USA) and recorded through a custom made program in LabView 8.2 (National Instruments, Austin, TX,USA), which also provided feedback to the experimenter..

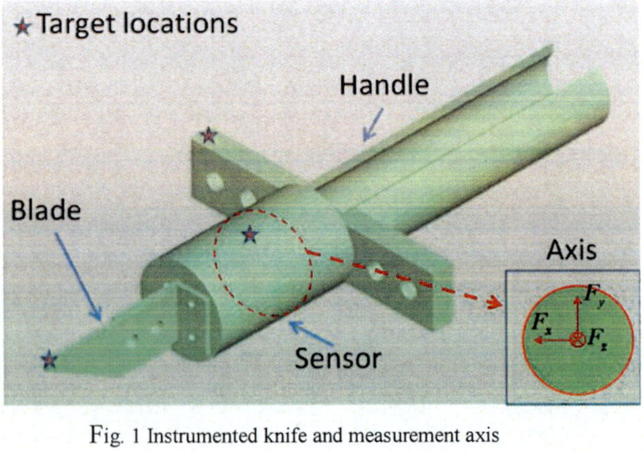

Fig. 1 Instrumented knife and measurement axis

K. Dremstrup, S. Rees, M.Ø. Jensen (Eds.): 15th NBC on Biomedical Engineering & Medical Physics, IFMBE Proceedings 34, pp. 144–147, 2011.
www.springerlink.com

The knife has been designed, as shown in figure 1, with respect to the following constraints:

- The blade should have the shape of a real knife (we hypothesize that only the tip of the blade is used during cutting);
- The knife has to be handled as a real knife (the handle diameter is the same as the one of real slaughter knife and a tennis grip is added to facilitate grasping);
- The knife should have the same weight as a real knife; however this was not possible due to the weight of the 3D force sensor. The instrumented knife will be about 0.3 to 0.7 kg heavier than a real knife. The mass difference will be taken into account in the simulation;
- The knife has to be rigid enough in order to avoid bending issues during the cutting task. Most parts of the blade are realized in steel ;
- In the same way, all parts are designed to be easily replaced. The blade can be replaced in a very quick time for example;
- At last the motion of the knife has to be captured, so we have defined three target locations on it, allowing the definition of the global location and orientation of the knife.

B. Experimental setup

The basic idea is to analyze natural variations of the motion during cutting tasks to determine the most damageable kinematical patterns that would lead to an increased risk of developing MSD in the shoulder/neck area. The data obtained are used to estimate muscle forces in relation with the motion. These muscle forces are then analyzed in order to detect the most strenuous patterns. The experimental setup used to collect the data is fully described in this part.

Figure 2 shows an overview of the experimental setup. The subject (Age: 39 yrs; Mass: 72 kg; Height: 1.86 m) was standing in front of a work bench that represents the workspace. The table height was set at the subject's elbow height for a neutral position of the arm (recommended height for light work [10]). Two main directions of cutting were designed on the work plane, i.e. arm flexion and a combination of arm internal rotation and abduction. The subject was asked to perform cutting tasks in these two directions for 20 seconds corresponding to 10 cycles in line with [6]. The data collected during the motion consists in motion capture data (using the Visualeyez II™ system set up with two VZ4000 trackers, Phoenix Technologies Inc., BC, Canada) sampled synchronously with the forces at 60 Hz. Twelve markers were used to collect trunk, right arm and knife motion, as shown. Finally, in order to validate the results in future studies; we collected surface electromyographic (EMG) data. A 64-channel matrix was used to collect the muscle activity of the upper-middle trapezius.

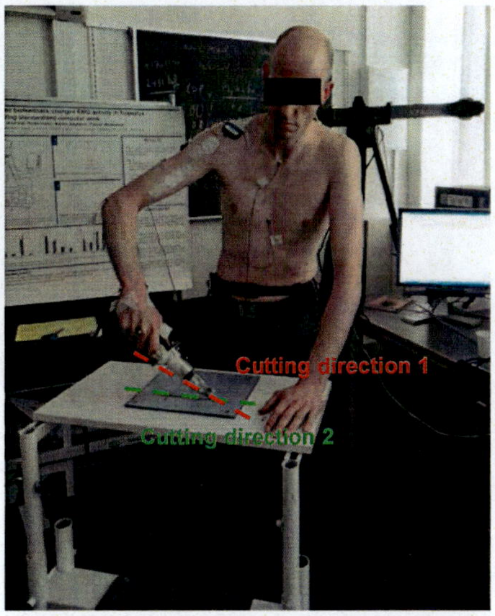

Fig. 2 Experimental situation with the instrumented knife and the active markers for movement assessments

The part of the skin covered by the grid was slightly abraded with abrasive paste (Medic-Every, Parma, Italy). The grid was then placed on the muscle with the 4th row aligned on the cervical vertebra C7-acromion line, parallel to the muscle fibre direction [11]. The lateral edge of the grid was 10 mm medial to the identified innervation zone. 30µl of conductive gel was inserted with a gel dispenser (model Eppendorf, Multiette plus, Hamburg, Germany) into the cavities of the adhesive electrode grid to ensure proper electrode-skin contact. A reference electrode was placed at C7. Moreover, 4 bipolar channels were used to collect EMG from the Deltoideus Medialis, Deltoideus Anterior, Biceps Brachii and Triceps Long head (figure 3) with bipolar surface electrodes (Neuroline 720, Ambu, Denmark). Bipolar surface electrodes were aligned (inter-electrodes distance: 2 cm) on abraded ethanol-cleaned skin along the direction of the muscle fibers. Bipolar electrodes were placed with respect to anatomical landmarks. The EMG signals were amplified 2000 times (64-channel surface EMG amplifier, SEA64, LISiN-OT Bioelectronica, Torino, Italy) and sampled at 2048 Hz (National Instrument, 12 bits acquisition board, Austin, USA).

The recording time (20 seconds) allows the appearance of natural variations during repetitive motions (marker trajectories are moving slightly around a mean). In order to obtain realistic cutting forces, the subject has to pass the knife in a slot that leads to a force level close from a real meat cutting task, i.e. 30-40 N [7,12].

All data acquisition were synchronized using an end-of-frame signal generated by the motion capture system.

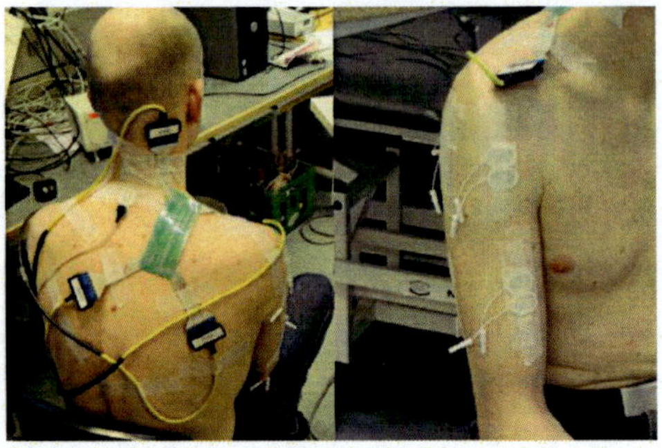

Fig. 3 EMG measurement

C. Analysis pipeline

The collected data were processed in order to drive a detailed musculoskeletal model of the trunk and the right arm, close from the one developed in [13], as shown in figure 4. The musculoskeletal model was scaled to fit the morphology of the subject and marker positions are optimized to fit real landmarks, using an optimization routine described in [14]. The 3D cutting force is then used to define the external forces applied to the model.

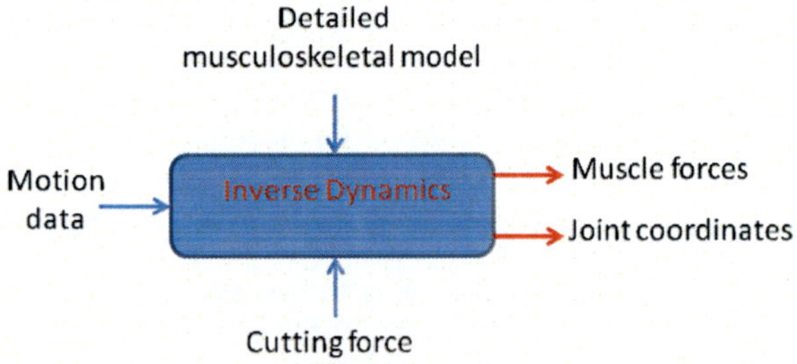

Fig. 4 AnyBody pipeline

A classic inverse dynamics method is used to generate muscle forces from computed joint torques. The objective function used for the muscle recruitment step is a standard quadratic criterion:

$$\min f = \sum_{k=1}^{n} \left(\frac{F_k(i)}{F_k^{\max}(i)} \right)^2 \tag{1}$$

III. Results and Discussion

Results focus on 4 main muscles: Deltoideus ('Del'), Biceps Brachii ('Bic'), Triceps Long head ('Tri'), and Trapezius ('Tra'). Each muscle in the model is defined as an ensemble of sub-muscles, having different action lines and different capabilities (maximum isometric forces). Figure 5 shows the mean of the tensions for each muscle group during the motion, regarding the cutting force intensity, designated by 'Cut' in the figure. This intensity is the norm of the cutting force vector recorded by the knife. As can be seen, mean forces developed by these muscles are deeply related with the cutting force intensity. Indeed, for both cutting directions biceps and triceps forces are similar. Mechanically, the triceps is the only muscle activated during the cutting event, because its action counterbalances the resulting cutting force. Triceps force is also quite important in regard of the cutting force. This is due to the moment arm that allows the triceps to contribute to the elbow joint torque. In reality, EMG shows that biceps and triceps are activated simultaneously to improve the elbow mechanical impedance [15, 16] during a position-controlled task. At the beginning of each cutting event in the cutting direction 2 the subject is performing a slight flexion of the elbow (highlighted by increased activity of biceps) to get an accurate position in the slot. This explains why the cutting force increases without any triceps contraction.

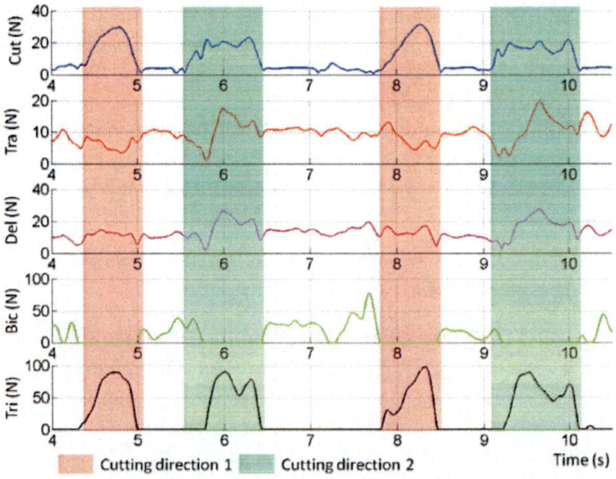

Fig. 5 Evolution of mean muscle forces during the recorded motions

Deltoidus and Trapezius forces highlight that the second cutting direction led to a higher shoulder tension for a lower cutting force intensity. The motion seems to be less natural for this height of table. Figure 6 shows the evolution of the gleno-humeral rotations during the motion. We can see that the second cutting task leads to a higher rotation in the "x- and z- directions" that corresponds to a greater extent of flexion of the upper arm and elevation of the shoulder. This can partially explain the increase of the global tension in the shoulder during this task.

It is important to mention that the 3D cutting force allows the model to counterbalance both cutting and lateral bending effects issued from the cutting task.

For further investigation, we will use specific action lines corresponding to the EMG recordings realized during the experimentation; in order to validate the muscle recruitment estimated by the AnyBody software.

If the mean muscle forces give a global idea of the behavior of the shoulder and the neck during the task, it is not sufficient to determinate risks in terms of MSD. The definition of more relevant biomechanical criteria to investigate the motions recorded is the next step of the work.

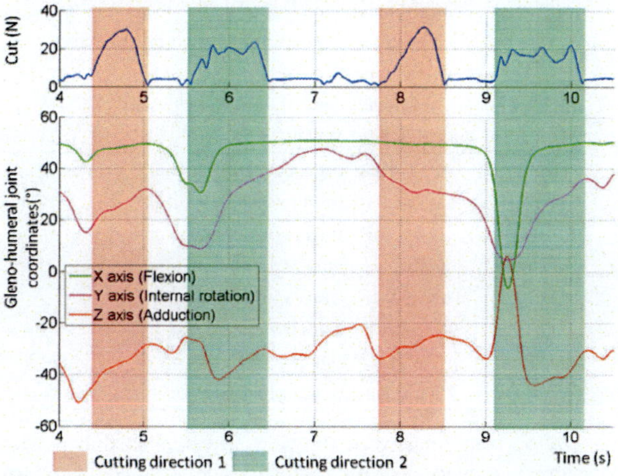

Fig. 6 Evolution of gleno-humeral joint coordinates during the recorded motions

IV. Conclusion

This study shows the potential for estimating muscle forces for cutting tasks using an instrumented knife, and motion capture. The next step will be to validate the models using the EMG recordings. The EMG matrix can also be used to see the evolution of the spatial activation (fatigue effects) of the trapezius during the cutting task. It is anticipated that this model can be used to quantify the musculoskeletal loading during cutting tasks and evaluate the impact of workspace parameters (such as table height) on the working conditions for this kind of task.

References

1. OSHA - Europa (2007) Lighten the load. Magazine of the European Agency for Safety and Health at Work
2. Sommerich C M, Mc Glothin J, Marras W S (1993) Occupational risk factors associated with soft tissue disorders of the shoulder: a review of recent investigations in the literature. Ergonomics 36:697-717
3. Bernard BP (1997) Musculoskeletal disorders and workplace factors : a critical review of epidemiologic evidence for work-related musculoskeletal disorders of the neck, upper extremity and low back. Cincinnati, OH: US Department of Health and Human Services
4. Madeleine P, Farina D (2008) Time to task failure in shoulder elevation is associated to increase in amplitude and to spatial heterogeneity of upper trapezius mechanomiographic signals. European Journal of Applied Physiology 102:325-333
5. Madeleine P, Mathiassen S E, Arendt-Nielsen L (2008) Change in the degree of motor variability associated with experimental and chronic neck-shoulder pain during a standardized repetitive arm movement. Experimental Brain Research 185:689-698
6. Madeleine P, Lundager B, Voigt M and Arendt-Nielsen L (1999) Shoulder muscle coordination under chronic and experimental neck-shoulder pain: An occupational pain study. European Journal of Applied Physiology 79:127-140
7. Jull-Kristensen B, Fallentin N, Hansson G A, Madeleine P, Andersen J H, Ekdahl C (2002) Physical Workload during manual and mechanical deboning of poultry. International Journal of Industrial Ergonomics 29:107-115
8. McGorry R W (2001) A system for the measurement of grip forces and applied moments during hand tool use. Applied Ergonomics 32:271-279
9. Damsgaard M, Rasmussen J, Christensen S T, Surma E, de Zee M (2006) Analysis of musculoskeletal systems in the AnyBody modeling system. Simulation Modeling Practice and Theory 14:1100:1111
10. McCormik E J, Sanders M S (1987) Human Factors in engineering and design.
11. Jensen C, Westgaard R H (1997) Functional subdivision of the upper trapezius during low-level activation. European Journal of Applied Physiology 76:335-339
12. McGorry R W, Dowd P C, Dempsey P G (2003) Cutting moments and grip forces in meat cutting operations and the effect of knife sharpness. Applied Ergonomics 34:375-382
13. Rasmussen J and de Zee M (2010) Computational investigation of two interventions for neck and upper extremity pain in office workers. IFMBE Proceedings 31:64-66
14. Andersen M S, Damsgaard M, MacWilliams B and Rasmussen J (2010) A computationally efficient optimisation-based method for parameter identification of kinematically determinate and over-determinate biomechanical systems. Computer Methods in Biomechanics and Biomedical Engineering 13:171-183
15. Pontonnier C and Dumont G (2009) Inverse dynamics method using optimization techniques for estimation of muscle forces involved in the elbow motion. IJIDeM3:227-236
16. Pontonnier C and Dumont G (2010) From motion capture to muscle forces in the human elbow aimed at improving ergonomics of workstations. Virtual and Physical prototyping 5:113-122

Author: Charles Pontonnier
Institute: Center of Sensory-Motor Interaction (SMI), Department of Health Science and Technology
Street: Frederiks Bajers Vej 7 DK-9000
City: Aalborg
Country: Denmark
Email: cpontonn@hst.aau.dk

A Highly Integrated Wearable Multi-parameter Monitoring System for Athletes

O. Chételat[1], J. Oster[1], O. Grossenbacher[1], A. Hutter[1], J. Krauss[1], and A. Giannakis[2]

[1] CSEM, Neuchâtel, Switzerland
[2] SenseCore, Zurich, Switzerland

Abstract— **A new system for the monitoring of a large set of physiological, kinetics, and environmental parameters is presented. In particular, this paper focuses on a first small-scale verification of the ECG and chest impedance signals. Comparative results with gold-reference devices on three subjects following a protocol including resting, walking, cycling, jogging and fast running are presented. The initial results showed that the developed system provides high quality signals comparable with gold-reference devices but in a highly integrated and wearable way. Moreover, qualitative assessment confirmed the high comfort of the system that features dry electrode sensors easily detachable from a layer-1 shirt for maintenance (e.g., shirt washing, recharge of sensors).**

Keywords— **ECG, respiration, dry electrodes, sports, multi-parameter, monitoring.**

I. INTRODUCTION

The Swiss Center for Electronics and Microtechnology (CSEM, www.csem.ch) developed in cooperation with Sense (www.sense-core.com) a new system for monitoring sportsmen. The system is highly integrated and consists of simply two electrode sensors clipped in a layer-1 shirt (see Fig. 1). Nevertheless, a large set of physiological, kinetics and environmental signals (see Table 1) are measured.

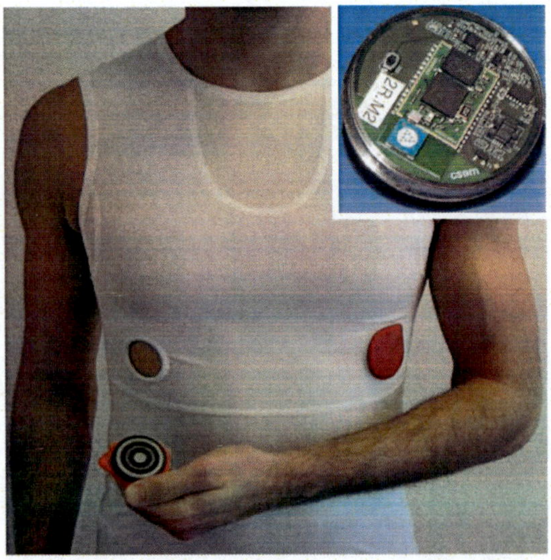

Fig. 1 The system featuring two electrode sensors clipped in a layer 1-shirt to monitor a large set of physiological parameters

The Sense system is able to process in real-time and on-board several parameters, including heart rate (HR) and breath rate (BR). All measured and processed signals are stored in the local memory for later download and analysis. They can also be transmitted wirelessly in real-time via Bluetooth to a 3rd party device, e.g., a smart phone, for instantaneous display or other functions.

Table 1 Measured and processed signals
(in bold those that are highlighted in this paper)

	signal name	remark
measured signals	**ECG**	compliant with IEC 60601-2-47
	impedance	compliant with IEC 60601-1
	body temperature	skin temperature at electrode sensor
	ambient temperature	at the sensor back
	altitude	derived from pressure sensor
	acceleration	3D
processed signal	**IHR**	instantaneous heart rate
	HR	heart rate = filtered IHR
	BR	filtered breath rate
	activity class	lying, standing, walking, running, others
	activity level	filtered intensity
	cadence	steps per minute
	speed	running speed
	distance	running distance

One particularity of the Sense system is that it was designed so as to have the highest signal quality for the lowest obtrusiveness. In particular, a special technology [1] that combines ECG (electrocardiogram) and impedance measurements from two dry electrode sensors connected by only one unshielded wire was implemented. Moreover, this technology allows measuring one lead ECG with only two electrode sensors while keeping the same signal quality as the classical approach which uses an additional guard electrode (also sometimes called right-leg electrode). Finally, this technology is extensible in the sense that an arbitrary number of electrodes can be used (all connected to the same unshielded connection embedded in the shirt). This allows measuring as many ECG leads as desired.

In the sequel we present the results of an initial small-scale verification regarding ECG, IHR (instantaneous heart rate defined as the inverse of the time between two consecutive R-waves), HR (IHR filtered, in particular without outliers), and BR (breath rate). This verification was performed in a clinical unit equipped for effort tests at the Landeyeux hospital.

K. Dremstrup, S. Rees, M.Ø. Jensen (Eds.): 15th NBC on Biomedical Engineering & Medical Physics, IFMBE Proceedings 34, pp. 148–151, 2011.
www.springerlink.com

II. MATERIAL AND METHOD

A. Verification setup

Fig. 2 describes the setup used for the verification. The ECG, IHR and HR are simultaneously measured by the Sense system and by a reference Holter (*LifeCard CF* from SpaceLabs) using three disposable adhesive gel electrodes (two for the measured lead and one as guard). Additionally, the HR is also measured by an HR chest belt (*RS800* from Polar) for comparison.

The chosen reference for the BR is a spirometer mask (*Metamax 3B* from Cortex-medical). The results for the other reference parameters, such as the skin temperature, are not detailed in this paper for brevity reasons.

Fig. 2 also shows the contrast between the high integration of the Sense system and the reference instruments which require a lot of cables. Note also the hairy chest of this subject who was selected to check for possible disturbance due to hairs.

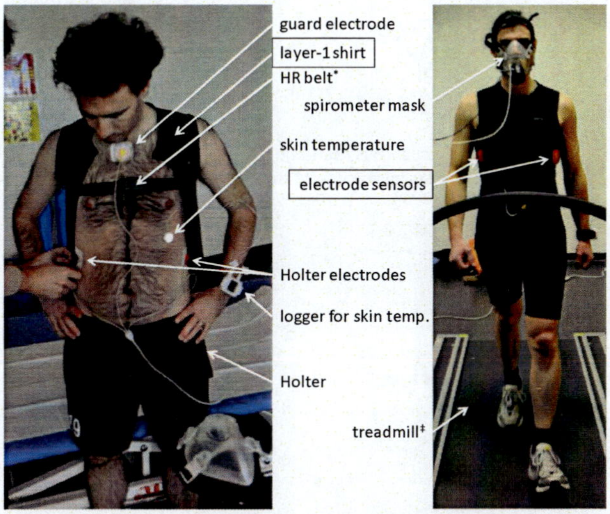

Fig. 2 Verification setup (framed: Sense's system, *HR belt for comparison, ‡treadmill for controlled activities, all others: reference systems)

B. Verification protocol

Three subjects (young, healthy, male) participated to the measurement campaign. Each of them followed a strict protocol (see Table 2) including various activities, including resting at different postures (lying, sitting, standing), cycling, walking, and running at four different speeds precisely controlled by the use of a treadmill. The activities are separated by short breaks. The last activity is identical to the first one, i.e., lying, but the conditions are not identical since the electrode sensors were applied dry on the skin and sweat appeared when the efforts started.

The running speed was set by the subject depending on his running level and expressed in percent of his MAS (maximum aerobic speed).

Table 2 Verification protocol

#	activity	duration (min.)
1	Lying	5
2	Sitting	2
3	Standing	2
4	Cycling	5
5	Lying	2
6	walking at 5 km/h	5
7	jogging at 60% MAS (8.5…10.5 km/h)	3
8	Standing	1
9	running at 70% MAS (10.5…12.5 km/h)	3
10	Standing	1
11	running at 80% MAS (12…14 km/h)	2
12	Standing	1
13	fast running at 90% MAS (14…16 km/h)	1
14	Lying	3

III. RESULTS & DISCUSSION

A. ECG and impedance measured signals

Fig. 3 shows the comparison between the Sense system and the reference of some typical ECG samples of the three subjects at rest and when running. One can observe that, at rest, the quality of the Sense system (which uses two dry electrode sensors) is excellent and as good as the reference (which uses three adhesive gel electrodes). In particular, advanced ECG analysis can be performed on such signals since the different waves are clearly visible.

When running, the Sense signals are qualitatively slightly worse than the one of the reference, but both are similarly affected by EMG (electromyogram) and motion artifacts. Nevertheless, the R-waves (ECG peaks) are clearly visible. Information on the other waves, if desired, could be classically obtained by ensemble averaging.

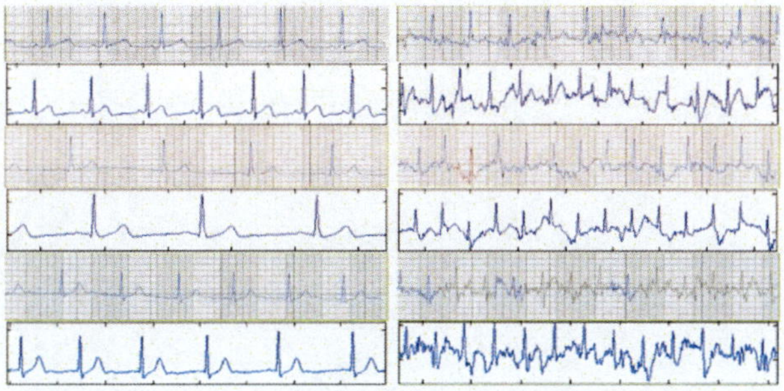

Fig. 3 ECG (left: resting, right: running; first row: subject 1, reference on top, Sense system below; subsequent rows: same for subject 2 and 3). Note that the signals are manually and approximately synchronized.

Fig. 4 shows the impedance signal for subject 1 (the signal of the other subjects is similar) in three different activities (sitting, standing, and fast running). One can see that the respiration waves are clearly visible. Some motion artifact spikes occur during posture change and sometimes during exercise.

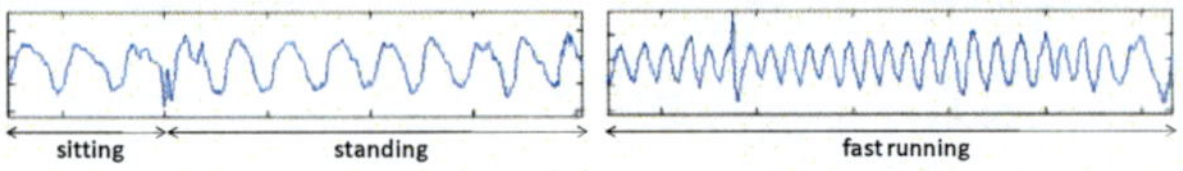

Fig. 4 Typical impedance signal

B. IHR and HR processed signals

The results for IHR (instantaneous heat rate, i.e., inverse of time interval between R-waves) and the HR (filtered heart rate, i.e., automatic removal of outliers) are shown in Fig. 5 (subject 1) to Fig. 7 (subject 3), where the performances of the Sense system are compared to the reference. Three classes of error are considered: 'good' for ±3 bpm (beat per minute), 'not too bad' up to ±10 bpm, and 'bad' beyond ±10 bpm. The Sense values are displayed differently depending on which class they belong to, and the scoring is performed according to the percentage of time the Sense values belong to a given class. For example, 95% of the IHR Sense values of subject 1 (Fig. 5, top panel) are identical at ±3 bpm to the reference, while 4% are further but still within ±10 bpm, and only 1% exceeds ±10 bpm.

The 4% of the 'not too bad' class occur principally during the end of the recording, while the subject's IHR is relatively high (over 100 bpm). The limitation of the reference Holter can explain the presence of these errors, since the time resolution of the R-R intervals (= inverse of IHR) estimated by the Holter is low (16 ms), which corresponds to a quantization of, e.g., 3 bpm at 105 bpm and of 6 bpm at 150 bpm. Therefore, the true performances of the Sense system are most likely better.

The number of outliers is low (1%). As they are isolated, they are easily rejected by the filter used for the HR estimation (Fig. 5, bottom panel). The origin of these outliers is due to motion artifact.

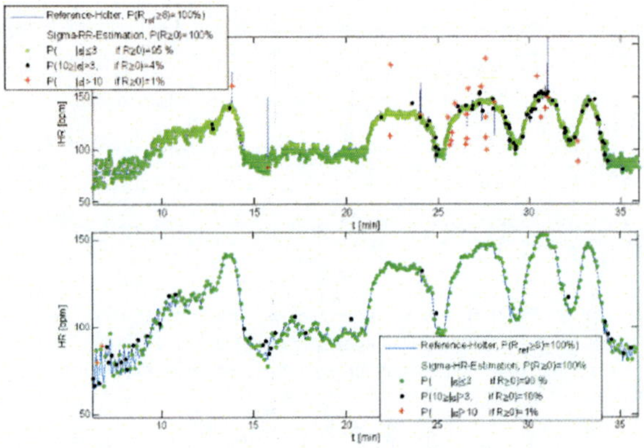

Fig. 5 IHR (instantaneous heart rate) and HR (filtered heart rate) signals processed from ECG (subject 1)

The HR estimation (Fig. 5, bottom panel) is in good agreement with the same signal obtained from the reference Holter. The percentage of large differences (more than 10 bpm) is very low (< 1%). About 10% of the values are classed as 'not too bad' instead of 'good' mostly due to a high heart rate variability, mainly during low activity.

The results for the two other subjects are shown in Fig. 6 and 7. They show similarly good performances.

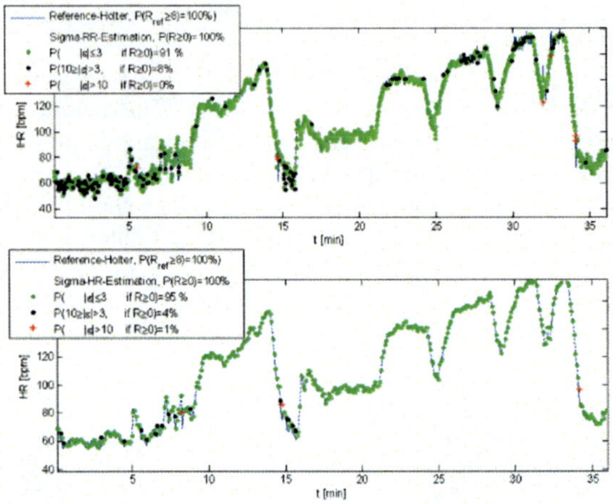

Fig. 6 IHR (instantaneous heart rate = 1/RR) and HR (filtered heart rate) signals processed from ECG (subject 2)

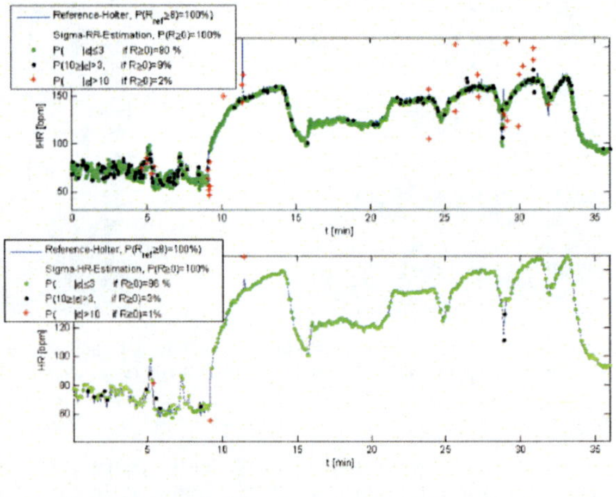

Fig. 7 IHR (instantaneous heart rate = 1/RR) and HR (filtered heart rate) signals processed from ECG (subject 3)

Table 3 gives a summary of the scoring obtained for the three subjects for the IHR, and Table 4 shows the same for the HR. The average μ and standard deviation σ, as well as the boundary of the spanned interval (minimum and maximum values) are also given. However, the number of subjects in this first low-scale verification is so low that the pertinence of these statistical parameters is limited and the whole span of inter-subject variability may be not covered.

Table 3 Summary of IHR (instantaneous heart rate) performances of Sense's system with respect to reference Holter (LifeCard CF from SpaceLabs)

IHR		subject ID				
		1	2	3	$\mu \pm \sigma$	[min, max]
$P_{\|\varepsilon\| \leq 3}$	(%)	95	91	90	**92** ± 2.6	[90, 95]
$P_{3 < \|\varepsilon\| \leq 10}$	(%)	4	8	9	7 ± 2.6	[4, 9]
$P_{\|\varepsilon\| > 10}$	(%)	1	0	2	1 ± 1	[0, 2]

Table 4 Summary of HR (filtered heart rate) performances of Sense's system with respect to reference Holter (LifeCard CF from SpaceLabs)

HR		subject ID				
		1	2	3	$\mu \pm \sigma$	[min, max]
$P_{\|\varepsilon\| \leq 3}$	(%)	90	95	96	**93.7** ± 3.2	[90, 96]
$P_{3 < \|\varepsilon\| \leq 10}$	(%)	10	4	3	5.7 ± 3.8	[3, 10]
$P_{\|\varepsilon\| > 10}$	(%)	1	1	1	1 ± 0	[1, 1]

For comparison, the same was performed with a commercial HR chest belt in place of the Sense system (i.e., when compared to the same reference Holter). The results are shown in Table 5. For subject 3, the number of outliers is abnormal. The belt placement or a potential low tightness could be one explanation of the HR estimation failure for this subject.

Table 5 For comparison, the same HR verification performed with HR belt (RS800 from Polar with respect to reference Holter)

HR/Polar		subject ID				
		1	2	3	$\mu \pm \sigma$	[min, max]
$P_{\|\varepsilon\| \leq 3}$	(%)	78	90	66	78 ± 12	[66, 90]
$P_{3 < \|\varepsilon\| \leq 10}$	(%)	19	9	13	13.7 ± 5	[9, 19]
$P_{\|\varepsilon\| > 10}$	(%)	3	0	20	7.7 ± 10.8	[0, 20]

C. BR processed signal

In addition to a better ECG and HR values compared to usual HR chest belts, the Sense system offers many other signals (see Table 1), one of them being the trans-thoracic impedance used for BR (breath rate) values. Fig. 8 shows the results obtained for the three subjects as compared to the reference (which is a spirometer mask). For the first subject (top panel), the Sense system estimation is 68% of the time within ±3 bpm. Having a closer look, it is obvious that the main contribution to larger errors comes from the first 10 s and from the periods with rapid changes. This subject has revealed two weak points of the algorithm used for this first small-scale verification. The first is a problem during initialization for very low breath rates around or below 4 bpm, and the second is the update rate of the algorithm which is not sufficient to follow rapid changes. The results for the other two subjects (middle and bottom panels of Fig. 8) are much better.

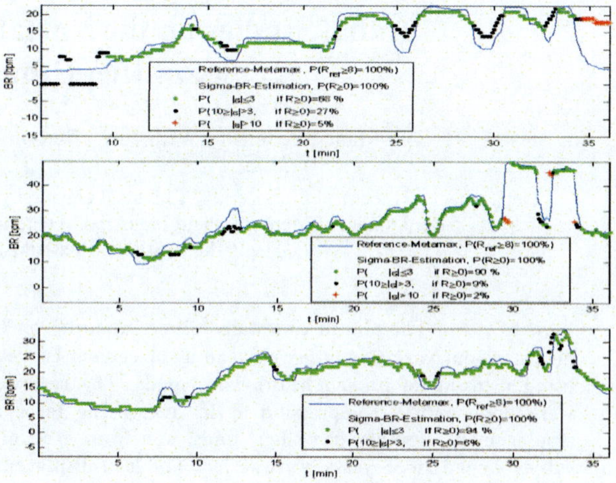

Fig. 8 BR signal processed from impedance (subject 1, 2, and 3)

Table 6 shows the summary for the BR, including some statistical figures. The results are already good. They are expected to be better with the upgraded algorithm that takes into account the observe limitations described above.

Table 6 Summary of BR (filtered breath rate) performances of Sense's system with respect to reference spirometer (Metamax 3B)

BR		subject ID				
		1	2	3	$\mu \pm \sigma$	[min, max]
$P_{\|\varepsilon\| \leq 3}$	(%)	68	90	94	**84** ± 14	[68, 94]
$P_{3 < \|\varepsilon\| \leq 10}$	(%)	27	9	6	14 ± 11.4	[6, 27]
$P_{\|\varepsilon\| > 10}$	(%)	5	1	0	2 ± 2.6	[0, 5]

IV. Conclusion

A first small-scale verification showed that the Sense system exhibits promising performances and signal quality for ECG, IHR, HR, and BR while being very comfortable with minimum obtrusiveness.

References

1. Chételat O, Gentsch R, Krauss J et al. (2008) Getting rid of the wires and connectors in physiological monitoring. EMBC Int. conf. of the IEEE Eng. in Med. and Biology Society, Vancouver.

The corresponding author's address is:

Olivier Chételat
CSEM
Jaquet-Droz 1
2002 Neuchâtel
Switzerland
och@csem.ch

Initial Studies on the Variations of Load-Displacement Curves of *in vivo* Human Healthy Heel Pads

Sara Matteoli[1], Jens E. Wilhjelm[1], Antonio Virga[2], Andrea Corvi[2], and Søren T. Torp-Perdersen[3]

[1] Department of Electrical Engineering, Technical University of Denmark, Ørsteds Plads, Building 349, DK-2800 Kgs. Lyngby, Denmark
[2] Department of Mechanics and Industrial Technologies, University of Florence, via S. Marta 3, 50139 Florence, Italy
[3] The Parker Institute, Frederiksberg Hospital, DK-2000, Frederiksberg, Denmark

Abstract— **The aim of this study was to quantify on the measurement variation of *in vivo* load-displacement curves by using a group of human healthy heel pads. The recordings were done with a compression device measuring force and displacement. Twenty three heel pads, one from each of 23 subjects aged 20-35 years, were tested. The load-displacement curves showed the hysteresis, typical for a visco-elastic tissue. Seven load-displacement curves were measured for each subject. Each hysteresis was approximated by a 3rd degree polynomial, which in turn was described by two parameters: the slope and the average curvature. No statistically significant tendency (increasing or decreasing) were found for the seven polynomials (χ^2 test, P-values of 0.81 and 0.17 for the two parameters, respectively). The study revealed no systematic error in the recorded load-displacement curves. The mean slope and the average curvature for the 23 subjects were found to be 6.02±1.54 N/mm and 0.02±0.01, respectively. The new apparatus shows its reliability for further clinical investigations.**

Keywords— **Compressibility, hysteresis, reproducibility.**

I. Introduction

Quantitative measurements of the biomechanical properties of heel pad tissue are an important component in development of methods for diagnosis of heel pad diseases.

The human heel pad tissue is located between the calcaneus and the skin on the posterior part of the foot. It is a highly specialized visco-elastic structure that provides shock absorption during gait. Due to the nature of the heel pad tissue and its capability to deform under a load, when a loading/unloading cycle is applied, a characteristic load-displacement curve (hysteresis) is obtained. An obvious way to perform a loading/unloading cycle [1]-[3] is to use a compression device. Use of these hysteresis curves requires, however, knowledge of their variation from measurement to measurement. The literature shows a few studies dealing with the behavior of the load-deformation of *in vivo* heel pads [4]-[8]. Unfortunately, these studies are not comparable as the methodologies used are different (equipment, applied load, impact velocity, etc.) and none of the papers describes the variation in the data recorded.

This study proposes a new compression device which is not meant to reproduce the physiological condition of walking, but to be a possible clinical device capable to characterize the biomechanics of injured heel pads. The study attempts to investigate the measurement variation in load-displacement curves of *in vivo* human healthy heel pads. Each curve was described by two parameters: the slope and the average curvature. These were investigated for systematic errors and variation in order to assess the reliability of the apparatus and the consistency of the results.

II. Material and methods

B. Subjects

Twenty-three healthy subjects (11 males, 12 females) were enrolled. Table 1 shows the anthropometric characteristics. Only one foot was tested, the foot normally used to hit the ball during football. All subjects declared to be in healthy conditions, and have ever had injuries/trauma to any of the feet. The enrolled subjects had different lifestyles, including some being sporty and some following a more sedentary routine. Subjects engaged in professional sport were not included in this study. All participants were volunteers and were informed about the conditions of the test that involved no harmful procedures or physical pain. The weight and height of the subject were measured. Before starting the compression test the subject was asked to give information about age, nature of physical activity and hours per week, as well as size of shoe.

Table 1 Anthropometric characteristics of subjects under investigation grouped according to the gender given as mean plus/minus one standard deviation

	Females	Males
Subjects	12	11
Age (years)	24.7±2.7	24.9±3.9
Body Mass (kg)	61.9±6.1	70.5±9.0
Height (cm)	167.1±5.7	173.5±7.3

K. Dremstrup, S. Rees, M.Ø. Jensen (Eds.): 15th NBC on Biomedical Engineering & Medical Physics, IFMBE Proceedings 34, pp. 152–155, 2011.
www.springerlink.com

A. Equipment

The measurement part of the device consisted of a load cell (model 31, RDP Electronics Ltd, UK) and a linear transducer (LVDT, RDP Electronics Ltd, UK), both connected to an amplifier (E725, RDP Electronics Ltd, UK). The load cell and the linear transducer were both assembled in a cylindrical aluminium body. One end was fixed to a vertical aluminium plate, as shown in Fig. 1. The other end of the cylinder consisted of a threaded shaft which was connected to a stepper motor (PK245-03A, Oriental motor, Japan) by a shaft and a flexible joint, also visible in Fig.1. The more the threaded shaft was tightened by the stepper motor, the more compression was applied to the heel pad by a flat cylinder (diameter of 40 mm) guided by the threaded shaft. The sole of the foot under investigation was in contact with the aluminium vertical plate covered by an electrically insulating layer of rubber, while the heel pad touched the cylinder during the compression and decompression. A hole in the vertical plate allowed the cylinder to be in contact with the heel pad. The stepper motor, the load cell and the linear transducer were connected to PC through a digital acquisition board (NI USB-6009, National Instruments). The sampling frequency used was 10 Hz.

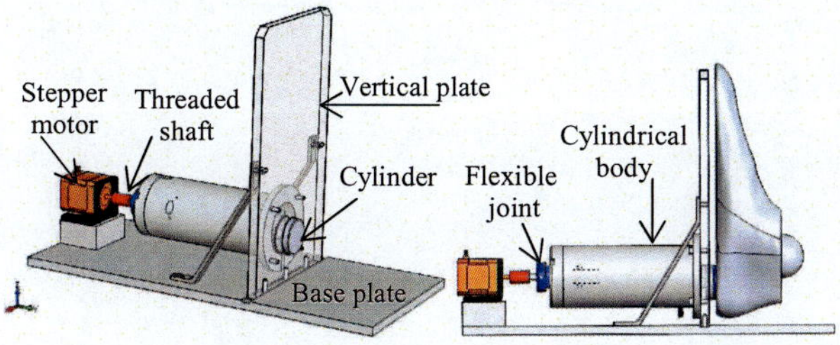

Fig. 1 The compression device. The stepper motor is connected to the threaded shaft with a shaft and flexible joint.

C. Procedure for compression test

The same procedure was applied to each volunteer. The subject removed the shoe and the sock from the foot to be investigated, and then laid down on an adjustable hospital bed with both legs completely straight and relaxed. The compression device was fixed on an appropriate table in front of the hospital bed.

The selected foot was positioned in such a way that the anterior part touched the vertical aluminium plate, with the heel pad in front of the cylinder. Specifically, the heel pad was placed with the center almost coincident with the center of the cylinder, as shown in Fig. 2, on the right.

Once the foot was well positioned, it is blocked with four Velcro fasteners (two to strap down the anterior part of the foot, one to keep the heel in front of the cylinder, and one to stabilize the ankle), as shown in Fig 2, on the left. When necessary, a cushion was positioned under the calf in order to better position the heel pad in front of the cylinder.

The subject was asked to remain as relaxed as possible and to maintain the foot in the same position for the duration of the test.

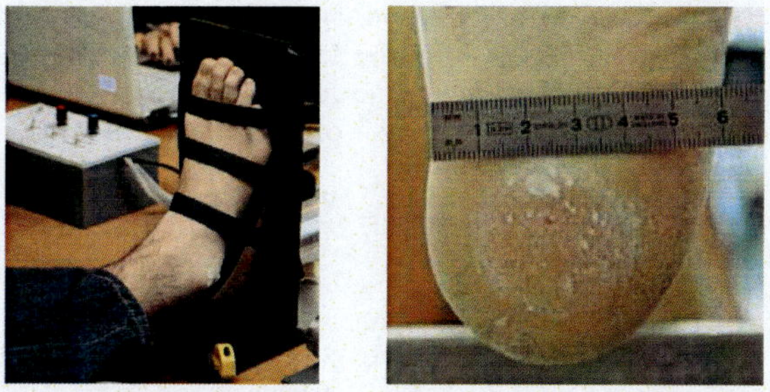

Fig. 2 Position of the foot on the vertical plate, on the left. Position of the cylinder on the heel pad, on the right. The print of the cylinder is made by placing some talcum powder on its surface

A program made with LabView (version 2009, National Instruments) was used to control the entire measurement (on/off of system, start/stop of stepper motor, direction of rotation of the stepper motor). The velocity used for applying the compression was 1.7 mm/s. Before starting the compression test on the heel pad, an idling test was done in order to verify the system functionality.

As soon as the subject was completely relaxed and ready to be tested, the examiner ran the LabView program controlling each step of the measurement procedure. For each subject, the compression test was repeated M=7 times with a one-minute break between each trial, to allow the heel pad tissue to return to its initial shape.

The value of the displacement determining the point of inversion of rotation of the stepper motor was fixed at 9 mm (to avoid arriving at the end of the thread of the shaft that guides the cylinder), while the superior limit of the force was fixed at 40 N.

The entire approach was designed to minimize any discomfort and any sensation of being strapped in order to be applicable to victims of torture.

D. Parameterization of hysteresis curves

In order to analyze the variation in the data of each subject, the best fit to a 3rd degree polynomial curve, y, was calculated for each load-displacement curve, $F(x)$, of each subject

$$y = a \cdot x^3 + b \cdot x^2 + c \cdot x + d \tag{1}$$

where x is the displacement, y is the force, a, b, c are coefficients, and d is a constant. The polynomial is exemplified in Fig. 3. In this study, the curves start at very near (0 mm, 0 N). Any distance interval where the load was not increasing, was subtracted from all measurement points, so that $F(x)$ starts at x=0 ($d = 0$ in (1)).

For each 3rd degree polynomial curve two parameters were chosen to describe the load-displacement curves. The slope, $\alpha_{n,m}$, representing the inclination of the straight line connecting the two extremes of the polynomial curve

$$\alpha_{n,m} = \frac{y(x_{\max})}{x_{\max}} \tag{2}$$

where n is the subject number, m is the measurements number for subject n and x_{max} is the value of x corresponding to the maximum of the polynomial curve.

The other parameter is the average curvature, $\bar{k}_{m,n}$, obtained by averaging the local curvatures ($k_{m,n,q}$). The latter was calculated by dividing the polynomial curve into Q equidistant intervals. At each interval, an expression for the curvature was found by

$$k_{m,n,q} = \frac{1}{R_q} = \frac{\left|\frac{d^2y}{dx^2}\right|}{\left[1+\left(\frac{dy}{dx}\right)^2\right]^{\frac{3}{2}}} \tag{3}$$

where R_q is the radius of the local curvature. Q ≈ 80 - 90.

III. Results

Considering the seven polynomial curves for a typical subject, Fig. 4 shows the trend of $\alpha_{n,m}$ and $\bar{k}_{m,n}$. For each parameter the tendency line is drawn as well.

The tendency lines for the 23 subjects were next analyzed to see if there was a prevailing slope. The slope of the tendency line of $\alpha_{n,m}$ is denoted β_α while the one of $\bar{k}_{m,n}$ is denoted β_k. These are shown in Fig. 5. By assuming that the variability can be approximated with a Gaussian distribution characterized by a mean of zero and a given standard deviation ($SD = 0.213$ for $\alpha_{n,m}$, and $SD = 0.00043$ for $\bar{k}_{m,n}$), the χ^2 test was applied. A P-value < 0.1 was chosen to indicate a statistically significant tendency [9]. As seen from Table 2, no statistically significant tendency was identified.

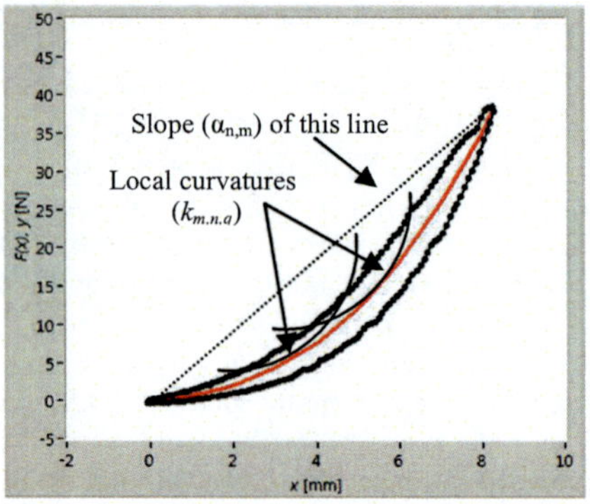

Fig. 3 Typical load-displacement curve with inside its best fitting 3rd degree polynomial curve.

Table 3 shows the typical variation of the hysteresis curves over the seven measurements. Calculating the final hysteresis curve for a subject by averaging the parameters for the seven measurements, Table 4 shows the variation over the 23 subjects.

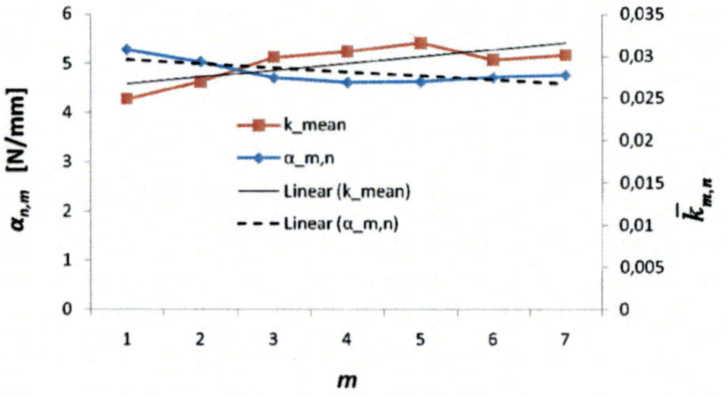

Fig. 4 Trend of $\alpha_{n,m}$ and $\bar{k}_{m,n}$ of all 7 polynomial curves for a typical subject. The correlation lines are drawn.

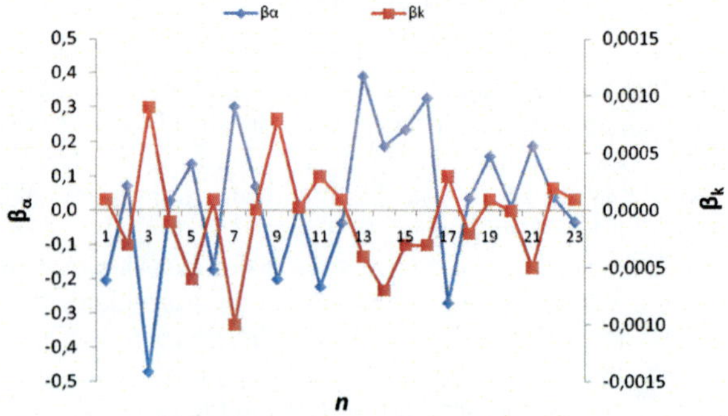

Fig. 5 Trend of β_α and β_k for all subjects

Table 2: Results obtained from the χ^2 test on $\alpha_{n,n}$ and $\bar{k}_{m,n}$

Parameter	χ^2	P-value
$\alpha_{n,n}$	0.81	0.85
$\bar{k}_{m,n}$	5.07	0.17

Table 3 Typical variation of hysteresis over the seven measurements of a subject. Specifically, for each subject the *SD* of a parameter was calculated over the seven measurements. Then the mean value of all these *SD* was calculated.

$\underset{n}{mean}\{\underset{m}{SD}\{\alpha_{n,m}\}\}$	0.78 [N/mm]
$\underset{n}{mean}\{\underset{m}{SD}\{\bar{k}_{n,m}\}\}$	0.0015

Table 4 Typical variation of hysteresis over the 23 subjects. Specifically, for each subject the mean value of a parameter was calculated over the seven measurements, and then the mean value ± *SD* was calculated over the 23 subject.

$\underset{n}{mean}\{\underset{m}{mean}\{\alpha_{n,m}\}\}$	6.02±1.54 [N/mm]
$\underset{n}{mean}\{\underset{m}{mean}\{\bar{k}_{n,m}\}\}$	0.02±0.01

IV. Discussion

The load-displacement curves exhibited the hysteresis behavior, typical for a visco-elastic tissue. The hysteresis curves obtained for each subject were not completely overlapping, which is reflected by the deviations in Fig. 4. Such spread in data might be due to:

- Rotation and translation of the structure consisting of the heel pad tissue and its support apparatus (the calcaneus and the soft tissue surrounding the heel pad tissue). The support apparatus cannot be completely fixed relative to the measurement device, thus both rotation and translation of the heel pad tissue can occur.
- Muscles in the leg might not be completely relaxed during all measurements. An unconscious pressure against *e.g.* the approaching piston cannot be completely ruled out.

The two points above could theoretically be further reduced by *e.g.* a vacuum cushion, but the effect on the tissue by a more tight support is unknown. On a more speculative basis, it is not known if the individual components of the heel pad tissue returns to exactly the same position after every trial and neither are the influence of possible variations in liquid content (*e.g.* blood perfusion) known.

Nevertheless, the result of Fig. 5 and the P-values in Table 2 indicates that no systematic error is taking place during the seven individual recordings. Likewise, the variation among the seven hysteresis curves found in Table 3 is smaller than among the twenty-three subjects.

V. Conclusions

The study reveals no systematic error in the recorded load-displacement curves of a group of twenty-three healthy humans. The mean slope and the average curvature for a polynomial fit to these curves were found to be 6.02±1.54 N/mm and 0.02±0.01, respectively. The new apparatus shows its reliability for further clinical investigations.

Acknowledgments

We would like to thank the anonymous reviewers. We really appreciated the helpful comments.

References

[1] Hsu T.C., Wang C.L., Tsai W.C., et al. (1998). Comparison of the mechanical properties of the heel pad between young and elderly adults. Arch Phys Med Rehab 79(9):1101-1104.

[2] Jørgensen, U., Larsen, E., & Varmarken, J. E. (1989). The HPC-device: a method to quantify the heel pad shock absorbency, Foot Ankle,10(2):93-98.

[3] Aerts P, Ker RF, De Clercq D, Ilsley DW, Alexander RM. (1995). The mechanical properties of the human heel pad: a paradox resolved. J Biomech. 28(11):1299-308.

[4] Challis J.H., Murdoch C., Winter S.L. (2008). Mechanical properties of the human heel pad: a comparison between populations. J Appl Biomech 24(4):377-381.

[5] Tsai W.C., Wang C.L., Hsu T.C., et al. (1999). The mechanical properties of the heel pad in unilateral plantar heel pain syndrome. Foot Ankle Int 20(10):663-668.

[6] Erdemir A., Viveiros M.L., Ulbrecht J.S., Cavanagh P.R. (2006). An inverse finite-element model of heel-pad indentation. J Biomech, 39(7): 1279-1286.

[7] Rome K., Webb P., Unsworth A., et al. (2001). Heel pad stiffness in runners with plantar heel pain. Clin Biomech, 16(10):901-905.

[8] Tong J., Lim C.S., Goh O.L. (2003). Technique to study the biomechanical properties of the human calcaneal heel pad. The Foot, 13:83-91.

[9] Navidi W., Statistics for Engineers and Scientists, McGraw-Hill 2006

Prediction of Alzheimer's Disease in Subjects with Mild Cognitive Impairment Using Structural Patterns of Cortical Thinning*

S.F. Eskildsen[1,2], V. Fonov[2], P. Coupé[2], L.R. Østergaard[1],
D.L. Collins[2], and the Alzheimer's Disease Neuroimaging Initiative

[1] Department of Health Science and Technology, Aalborg University, Aalborg, Denmark
[2] McConnell Brain Imaging Centre, Montreal Neurological Institute, McGill University, Montreal, Canada

Abstract— **Predicting Alzheimer's disease (AD) in patients exhibiting early symptoms of cognitive decline may have great influence on treatment and drug discovery. Structural magnetic resonance imaging (MRI) has the potential of revealing early signs of neuro-degeneration in the human brain and may thus aid in predicting and diagnosing AD. Surface-based cortical thickness measurements from T1w MRI have demonstrated high sensitivity to cortical gray matter changes. In this study we investigated the possibility for using patterns of cortical thickness measurements for predicting AD in patients with mild cognitive impairment (MCI). We used a novel technique for identifying cortical regions potentially discriminative for separating subjects with MCI, which progress to AD, from subjects with MCI, which do not progress to AD. Cortical thickness measurements from these selected regions were used in a classifier for testing the ability to predict AD. The classification showed an overall accuracy of 72% for predicting AD conversion in MCI patients 12 months in advance, which is better than recently published results on similar data.**

Keywords— **AD, MCI, MRI, cortical thickness, prediction.**

I. Introduction

The ability to diagnose Alzheimer's disease (AD) at an early or even pre-clinical stage of the disease has great impact on the possibility for improving treatment of the disease. This type of early diagnostic may also reduce costs associated with recruiting subjects for pharmaceutical trials when performing large scale tests on specific drugs targeting AD, since false positives can be excluded in the initial stage. The high tissue contrast offered by magnetic resonance imaging (MRI) enables accurate structural neuroimaging which could be used as a possible surrogate biomarker for diagnosing and predicting AD [1]. However, image processing techniques have so far not been able to accurately predict AD in patients with prodromal AD (also known as mild cognitive impairment or MCI) [2]. Surface-based measurements of cortical thickness based on MRI are highly sensitive to small structural changes in the cortex and have been widely used to investigate cortical structural changes and differences in various diseases [3-5]. However, a recent study indicated that cortical thickness measurements do not perform better than other techniques when trying to predict AD in subjects with MCI, a condition related to the stages before early AD [2]. Cortical thickness is usually measured at a very high resolution (tens of thousands of points on the cerebral cortex). Using such high numbers of measurements in prediction may lead to overfitting the model. The dimensionality can be reduced by defining regions of interests (ROI) in which measurements are averaged. Usually, such ROIs are defined from a structural or functional perspective. However, the pattern of neurodegeneration may not follow anatomical or functional regions, thus such ROIs may lead to loss of discriminative information. Therefore, a way to select the most sensitive features of cortical thickness measurements may lead to better prediction results.

In this study, we investigated the possibility for predicting which patients with prodromal Alzheimer's disease (MCI) convert to clinically definite Alzheimer's disease by selecting cortical measurements based on cortical thinning patterns characteristic for prodromal AD.

II. Methods

A. Subjects and Acquisition

Data used in the preparation of this article were obtained from the Alzheimer's Disease Neuroimaging Initiative (ADNI) database (www.loni.ucla.edu/ADNI). The ADNI

* Data used in the preparation of this article were obtained from the Alzheimer's Disease Neuroimaging Initiative (ADNI) database (www.loni.ucla.edu/ADNI). As such, the investigators within the ADNI contributed to the design and implementation of ADNI and/or provided data but did not participate in analysis or writing of this report. A complete listing of ADNI investigators can be found at: http://loni.ucla.edu//ADNI//Collaboration//ADNI_Authorship_list.pdf

K. Dremstrup, S. Rees, M.Ø. Jensen (Eds.): 15th NBC on Biomedical Engineering & Medical Physics, IFMBE Proceedings 34, pp. 156 – 159, 2011.
www.springerlink.com

was launched in 2003 by the National Institute on Aging, the National Institute of Biomedical Imaging and Bioengineering, the Food and Drug Administration, private pharmaceutical companies and non-profit organizations, as a $60 million, 5-year public-private partnership. The primary goal of ADNI has been to test whether serial MRI, positron emission tomography, other biological markers, and clinical and neuropsychological assessment can be combined to measure the progression of MCI and early AD. Determination of sensitive and specific markers of very early AD progression is intended to aid researchers and clinicians to develop new treatments and monitor their effectiveness, as well as lessen the time and cost of clinical trials.

The ADNI database contains 1.5T and 3.0T T1w MRI scans for AD, MCI, and cognitively normal controls (CN) at several time points. MCI patients were scanned at baseline, 6 months, 12 months, 18 months, 24 months, 36 months and 48 months. At each time point a clinical diagnosis was made to identify MCI subjects who converted to clinically definite AD.

In this study, scans of all MCI subjects were selected 12 months prior to the time an AD diagnosis was made. This collection of MCI converters consisted of scans at baseline (n=46), 6 months (n=32), 12 months (n=32), 24 months (n=21) and 36 months (n=2), which formed the 12m MCI converter (MCIc) group (n=133). To identify characteristic traits in the MCIc group, we selected scans at baseline for all MCI subjects, which did not progress to AD within 48 months. This MCI non converter (MCInc) group consisted of 221 scans.

In addition, baseline data for AD patients (n=200) and CN subject (n=233) were selected to compare prediction rates to classifications of clinically definite AD over CN. Only 1.5T T1w scans were used in this study.

B. Image Processing

T1w images were bias field corrected [6], intensity normalized, registered to ICBM space [7] and skull stripped [8]. Cortical thickness was calculated using FACE (fast accurate cortex extraction) [9] and mapped to an average cortical surface of 100 AD patients (template surface) [10]. Cortical segmentations were manually checked for errors and subjects were excluded if errors were found in one of the image processing steps mentioned above. After quality control the group sizes were by exclusion reduced to MCIc=102, MCInc=184, AD=150, and CN=189.

C. Feature Generation

Statistical parametric maps of differences in cortical thickness between MCIc and MCInc were constructed by one-sided t-tests per vertex of the template surface (from a total of 146,122 vertices). Maps were generated in a leave one out cross validation (LOOCV) fashion and merged to obtain a stable p-value map of significant differences between groups. The merged map was thresholded at p=0.001 and local minima were detected. Cortical features were determined as the mean cortical thickness in a circular neighborhood around the local minima in the *p*-value map (see Fig. 1).

D. Classification

The feature pattern generated from the statistical differences between MCIc and MCInc was applied to extract cortical thickness from all four groups (MCIc, MCInc, AD and NC).

Linear discriminant analysis (LDA) was used for the classification. All classifications were done in a LOOCV fashion. The correct classification rate, the sensitivity, and specificity of the classifier were calculated from the resulting classifications on the test sets. Furthermore, McNemar's chi-square test was used to assess whether the classifier performed better than a random classifier.

III. Results

The feature selection method generated nine regions for measuring cortical thickness (see Fig. 1). Seven regions were identified in the right hemisphere while only two in the left hemisphere. Sensitive regions were located bilaterally along the middle temporal sulcus and the posterior cingulate gyrus. Additional regions in the right hemisphere were found in the medial temporal lobe along the parahippocampal gyrus.

Table 1 Classification results for MCI converters (MCIc), MCI non converters (MCInc), patients with Alzheimer's disease (AD), and cognitive normal subjects (CN)

Classification	Correct Rate	Sensitivity	Specificity	McNemar's test
MCIc vs MCInc	72%	75%	70%	P<0.001
AD vs CN	85%	81%	88%	P<0.001
AD vs MCInc	76%	77%	75%	P<0.001
AD vs MCIc	55%	54%	56%	P=0.326
MCInc vs CN	62%	58%	65%	P=0.002
MCIc vs CN	81%	78%	83%	P<0.001

Table 1 shows the correct classification rate, the sensitivity, and the specificity for the classifiers along with results of McNemar's chi-square test (α=0.05). MCIc and MCInc were classified with an accuracy of 72%, a result significantly

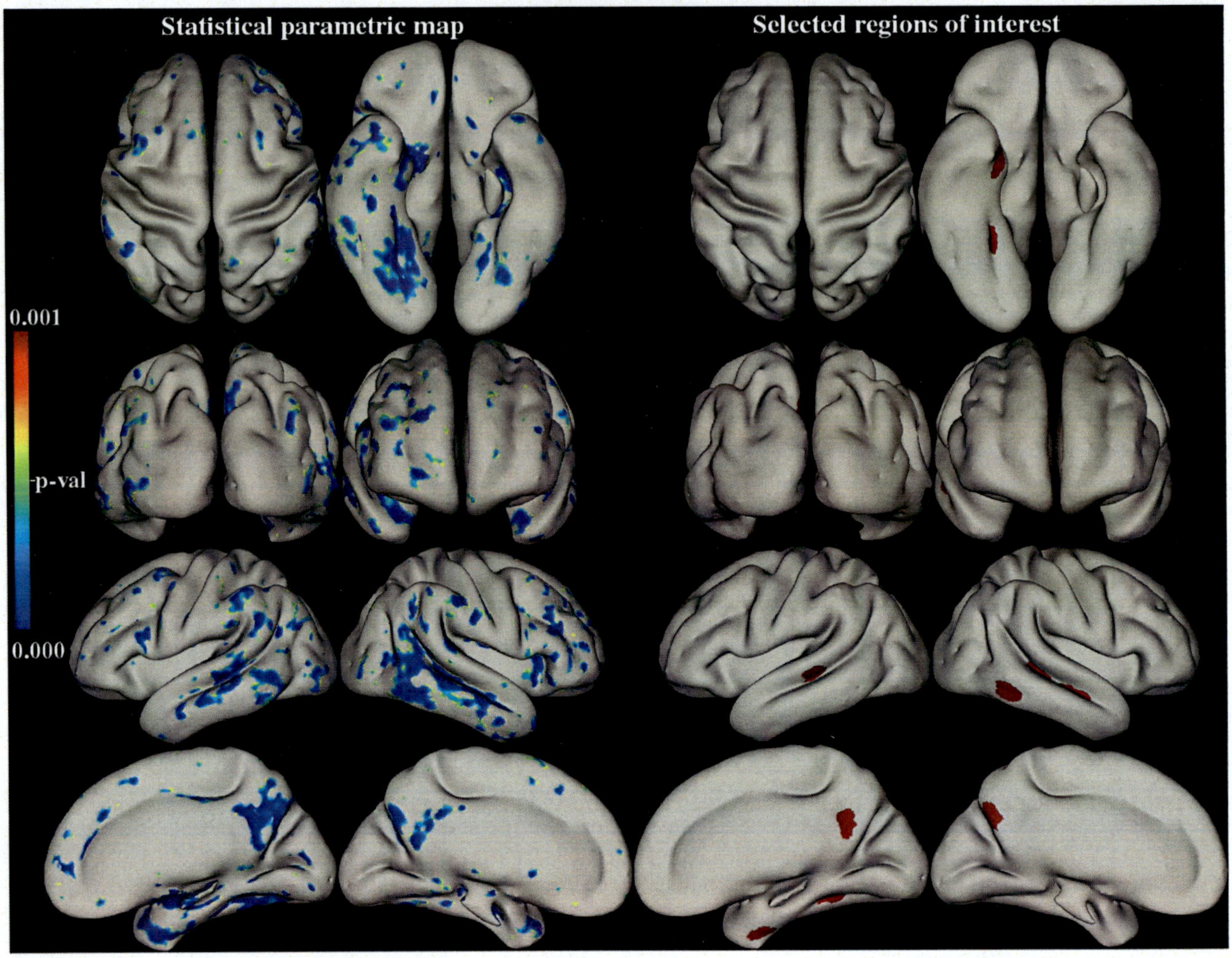

Fig. 1 Left: Statistical parametric map of significant differences between MCI converters and non-converters. Right: Resulting pattern of regions of interest after performing automatic selection

better than a random classifier. Higher accuracy was obtained when classifying AD vs CN, AD vs MCInc, and MCIc vs CN using the same method. Lower accuracy was obtained when classifying AD and MCIc and MCInc and CN. Classification of AD and MCIc was the only classifier that did not perform significantly better than a random classifier.

IV. CONCLUSIONS

Using patterns of characteristic cortical thinning in MCI converters compared to MCI non-converters demonstrated promising results for the prediction of patients with prodromal AD progressing to clinically definite AD. These classification results are better than other comparable methods recently published [2]. Nevertheless, the sensitivity of the method needs to be increased to be clinically applicable. The specificity was slightly lower than the sensitivity. From an economic perspective a high specificity is very important in clinical trials when recruiting subjects. Reducing the number of false positives in trials may save time and reduce costs.

The results showed that the pattern discriminating between MCIc and MCInc also could be used to classify AD vs CN with high accuracy (85%) and separate MCIc from CN with almost as high accuracy (81%). This suggests that the pattern found is not specific for the population that generated the pattern, but can be generalized to be sensitive to early signs of AD. Generating cortical patterns for each specific classification may improve the accuracy for the individual classifier, but may not be as discriminative for other groups as the MCI pattern found in this study.

Although the classification based on patterns of cortical thickness demonstrated promising results, the method is limited by the high number of excluded subjects due to errors in the processing pipeline. Twenty percent of all subjects were excluded based on the manual quality control. The majority of these failed due to incorrect brain extraction as the brain extraction has a tendency to cut off gyri on atrophic brains This can explain why more AD and MCIc tend to be excluded than NC and MCInc. Therefore, an improved brain extraction algorithm will not only greatly reduce the number of subjects failing the processing pipeline, but may also improve the classification accuracy of those that survive the processing.

Further work should investigate new methods to obtain the full discriminative information of the sensitive cortical thickness measurements. Improved selection of the cortical thickness pattern and inclusion of other structural measurements, such as hippocampal and ventricular volumes, and clinical data, such as neuro-psychological test scores, may improve the classification. Choice and optimization of feature set and/or classification algorithm may also influence the results. Further investigations should be made to explore these possibilities.

ACKNOWLEDGMENT

Data collection and sharing for this project was funded by the Alzheimer's Disease Neuroimaging Initiative (ADNI) (National Institutes of Health Grant U01 AG024904). ADNI is funded by the National Institute on Aging, the National Institute of Biomedical Imaging and Bioengineering, and through generous contributions from the following: Abbott, AstraZeneca AB, Bayer Schering Pharma AG, Bristol-Myers Squibb, Eisai Global Clinical Development, Elan Corporation, Genentech, GE Healthcare, GlaxoSmithKline, Innogenetics, Johnson and Johnson, Eli Lilly and Co., Medpace, Inc., Merck and Co., Inc., Novartis AG, Pfizer Inc, F. Hoffman-La Roche, Schering-Plough, Synarc, Inc., as well as non-profit partners the Alzheimer's Association and Alzheimer's Drug Discovery Foundation, with participation from the U.S. Food and Drug Administration. Private sector contributions to ADNI are facilitated by the Foundation for the National Institutes of Health (www.fnih.org). The grantee organization is the Northern California Institute for Research and Education, and the study is coordinated by the Alzheimer's Disease Cooperative Study at the University of California, San Diego. ADNI data are disseminated by the Laboratory for Neuro Imaging at the University of California, Los Angeles. This research was also supported by NIH grants P30AG010129, K01 AG030514, and the Dana Foundation.

REFERENCES

1. Ritchie K, Lovestone S (2002) The dementias. The Lancet 360:1759-1766.
2. Cuingnet R, Gerardin E, Tessieras J et al. (in press) Automatic classification of patients with Alzheimer's disease from structural MRI: A comparison of ten methods using the ADNI database. NeuroImage.
3. Narr KL, Bilder RM, Toga AW et al. (2005) Mapping cortical thickness and gray matter concentration in first episode schizophrenia. Cerebral Cortex 15(6):708-719.
4. Shin Y, So YY, Jun KL et al. (2007) Cortical thinning in obsessive compulsive disorder. Human Brain Mapping 28(11):1128-1135.
5. Lerch JP, Pruessner J, Zijdenbos AP et al. (2008) Automated cortical thickness measurements from MRI can accurately separate Alzheimer's patients from normal elderly controls. Neurobiology of Aging 29(1):23-30.
6. Sled JG, Zijdenbos AP, Evans AC (1998) A nonparametric method for automatic correction of intensity nonuniformity in mri data. IEEE Transactions on Medical Imaging 17(1):87-97.
7. Collins DL, Neelin P, Peters TM, Evans AC (1994) Automatic 3D intersubject registration of MR volumetric data in standardized talairach space. Journal of Computer Assisted Tomography 18(2):192-205.
8. Smith SM (2002) Fast robust automated brain extraction. Human Brain Mapping 17(3):143-155.
9. Eskildsen SF, Østergaard LR, Rodell AB et al. (2009) Cortical volumes and atrophy rates in FTD-3 CHMP2B mutation carriers and related non-carriers, NeuroImage 45(3):713-721.
10. Fonov V, Evans AC, Botteron K et al. (2011) Unbiased average age-appropriate atlases for pediatric studies, NeuroImage 54(1):313-327.

Address of the corresponding author:

Author: Simon Fristed Eskildsen
Institute: Montreal Neurological Institute
Street: 3801 University Street
City: Montreal
Country: Canada
Email: simon.eskildsen@mcgill.ca

Performance Evaluation of a Synthetic Aperture Real-Time Ultrasound System

M.B. Stuart, B.G. Tomov, and J.A. Jensen

Center for Fast Ultrasound Imaging, Department of Electrical Engineering, Technical University of Denmark

Abstract— **This paper evaluates the signal-to-noise ratio, the time stability, and the phase difference of the sampling in the experimental ultrasound scanner SARUS: A synthetic aperture, real-time ultrasound system. SARUS has 1024 independent transmit and receive channels and is capable of handling 2D probes for 3D ultrasound imaging. It samples at 12 bits per sample and has a sampling rate of 70 MHz with the possibility of decimating the sampling frequency at the input. SARUS is capable of advanced real-time computations such as synthetic aperture imaging. The system is built using field-programmable gate arrays (FPGAs) making it very flexible and allowing implementation of other real-time ultrasound processing methods in the future.**

For conventional B-mode imaging, a penetration depth around 7 cm for a 7 MHz transducer is obtained (signal-to-noise ratio of 0 dB), which is comparable to commercial ultrasound scanners. Furthermore, the jitter between successive acquisitions for flow estimation is around 1.41 ps with a standard deviation of 48.3 ps. This has a negligible impact (0.03%) on the flow measurement. Additionally, for the phase of the sampling, it is shown that the small differences between different channels (on average 111 ps for a 70 MHz sampling clock) are deterministic and can therefore be compensated for.

Keywords— **Ultrasound, hardware, performance, beamformer, scanner.**

I. INTRODUCTION

In modern medical ultrasound scanners, multi-element transducers and advanced beamforming are used to obtain high quality images. Typical commercial scanners acquire and process RF data on 64 to 128 channels at sampling rates of 40 to 70 MHz, giving data rates of approximately 5 to 18 GB/s. For 3D imaging, the required number of channels increases, driving up the data rate proportionally. Dedicated integrated circuits must be used to satisfy the real-time processing requirements of medical ultrasound with reasonable power consumption. However, this makes it difficult - if not impossible - to use these systems to obtain in-vivo data for research purposes. Such data is needed to develop and evaluate advanced data processing methods such as synthetic aperture imaging, adaptive beamforming or vector flow imaging. There is, thus, a need for ultrasound systems that are capable of acquiring and storing this data.

A number of systems with this capability have previously been built. The RASMUS system [1] has 128 independent transmit channels and 64 independent receive channels. The receive channels are fitted with 2-to-1 multiplexers to allow the use of larger apertures. More than 3 seconds of data can be stored, and real-time conventional beamforming can be performed, but other real-time processing capabilities are limited. Storage of echo data in RASMUS has been instrumental in developing new ultrasound imaging and processing methods and pre-clinical trials of vector flow imaging [2,3], synthetic aperture imaging [4,5], and imaging using coded excitations [6,7]. A system by Lu et al. [8] can handle 128 channels and can perform real time imaging, but has limited capability for storing long data sequences. An affordable sampling system is created by Tortoli et al. [9], that handles 64 channels and samples a depth range of 3 cm, saving the data in 1 GB of shared on-board RAM. A number of other systems have been described in a special UFFC issue on ultrasound systems [10].

None of the above systems can handle 3D imaging and real-time processing of synthetic aperture data or advanced vector flow imaging. The synthetic aperture, real-time ultrasound scanner (SARUS) has been presented previously [11]. It has 1024 independent transmit and receive channels, a sampling rate of 70 MHz, and storage for 3 to 10 seconds of all 1024 receive channels. In this paper, more details on the scanner performance are presented with focus on time stability and signal-to-noise ratio.

The rest of the paper is structured as follows: In Section II the architecture of the system is presented, and Section III gives the performance measures that are used. Section IV presents the methods used for evaluating the scanner performance, while Section V and Section VI present the results. Finally, Section VII contains the conclusion.

II. ARCHITECTURE

The scanner consists of an analog front-end with 1024 independent transmit and receive channels and a digital back-end. The back-end consists of 64 digital acquisition and ultrasound processing (DAUP) boards. Each DAUP board has 16 independent transmit and receive channels and hosts 5 field-programmable gate arrays (FPGAs) that implement the various functions of a scanner capable of synthetic aperture imaging: Transmitting, filtering, beamforming, summing, and control. The 64 DAUP boards are organized in 4 racks with 16 DAUP boards in each as

K. Dremstrup, S. Rees, M.Ø. Jensen (Eds.): 15th NBC on Biomedical Engineering & Medical Physics, IFMBE Proceedings 34, pp. 160–163, 2011.
www.springerlink.com

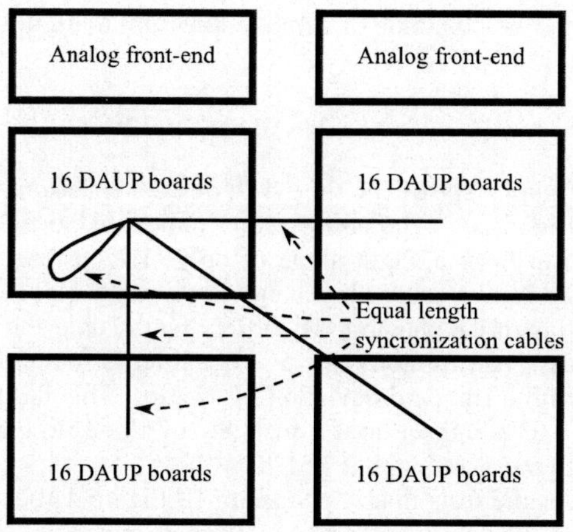

Fig. 1 The architecture of SARUS.

shown in Figure 1. In this way, the system can be configured as one system with 1024 channels, as smaller individual systems with 256 channels each, or in a number of intermediate steps. Multiple racks are connected with purpose-made synchronization cables of equal length.

For testing purposes, it is possible to connect the output of the digital-to-analog converters (DACs) to the input of the analog-to-digital converters (ADCs) using loopback cables. This allows us to sample the emitted signals and measure the time stability of the system.

For high-quality imaging, it is important that all channels are sampled at not only the same rate, but also at the exact same time [12]. The sampling clock signal is a periodic, 70 MHz square-wave signal, whose rising or falling edges determine the sampling instant. To assure that this signal is in phase at all of the ADCs, it is distributed along equal-length synchronization cables from a central point to all of the DAUP boards. These cables also carry control signals for the emitters and the sampling. Achieving such globally synchronicity in a digital system is a non-trivial task.

III. PERFORMACE METRICS

This section discusses some of the performance metrics that are of interest for high-quality imaging and high-accuracy flow estimates. Besides a good signal-to-noise ratio (SNR), the amplification in and the delay through the system should be stable over time. Both the amplification and the delay may be changed by temperature changes in the electronics. Changes in the delay through the system can significantly decrease the accuracy of blood velocity estimations and should therefore be minimized.

It has previously been shown that even small errors in the delay have a significant impact on both conventional beamforming [13] and synthetic aperture imaging [12]. Therefore, the sampling should occur simultaneously in the whole system at any time. Specifically, it should be verified that the sampling clock signal and the emissions and sampling control signals are in phase throughout the system.

The criteria for these metrics is that they must be sufficiently stable to not impact the subsequent data processing significantly whether it be conventional beamforming, synthetic aperture imaging or blood flow estimation.

IV. EXPERIMENTAL SETUP

To evaluate the metrics defined in Section III, two experimental setups are used.

- Setup A: 12 DAUP boards (192 channels) in one rack are connected to an analog front-end with a 192 channel linear array transducer (BK 8804, BK Medical, Herlev, Denmark) attached.
- Setup B: 12 DAUP boards are split over two racks and interconnected using loopback cables. Some transmitters are connected to receivers on the same board, while other transmitter/receivers pairs are connected across the two racks, i.e., the transmitters on a DAUP board in one rack are connected to the receivers on a DAUP board in the other rack.

Using Setup A, the transducer is put on a tissue-mimicking phantom. In order to measure the SNR, a beam focused at a depth of 6 cm is emitted along a line for which only speckle has been observed. This emission is repeated 100 times, and the corresponding A-lines are beamformed. The average of the A-lines are taken as the signal, while an A-line subtracted this average is taken as the noise. The SNR is then the ratio of the powers of these two signals as a function of depth.

Similar data is used to determine the time stability of the system. The 100 beamformed A-lines are ideally identical (less the noise), and by correlating them to each other, the time drift of the sampled signals can be determined.

With Setup B, a short sinusoidal signal is emitted on all channels simultaneously. By correlating the sampled signals where the transmitter and receiver are in the same rack with those where the transmitter and receiver are in different racks, it can be determined if the sampling clock signal is in phase throughout the system.

V. SIGNAL-TO-NOISE RATIO MEASUREMENTS

This section presents the results of the signal-to-noise ratio experiments. Figure 2 shows a conventional B-mode

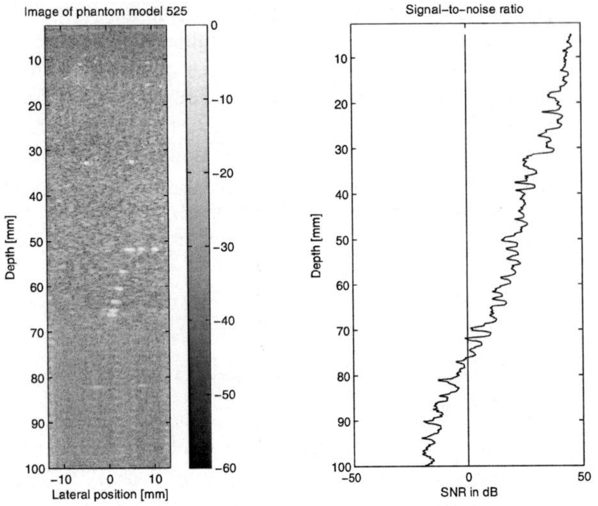

Fig. 2 On the left, a conventional B-mode image. On the right, the SNR is shown as a function of depth with 97 active elements.

image of a tissue mimicking phantom. The B-mode image was created with 64 active elements, a transmit focus in 6 cm, no transmit apodization, and dynamic receive apodization with a constant F-number of 2 and a von Hann apodization. The excitation is 2 periods of a 7 MHz sinusoidal pulse with an amplitude of 50 V. The penetration depth is about 7 cm (SNR of 0 dB), which is comparable to the RASMUS scanner and commercial systems. The FWHM is 0.27 mm (1.25 λ) in the axial direction and 0.62 mm (2.81 λ) in the lateral direction at the focus point. The SNR is plotted as a function of depth on the right in Figure 2.

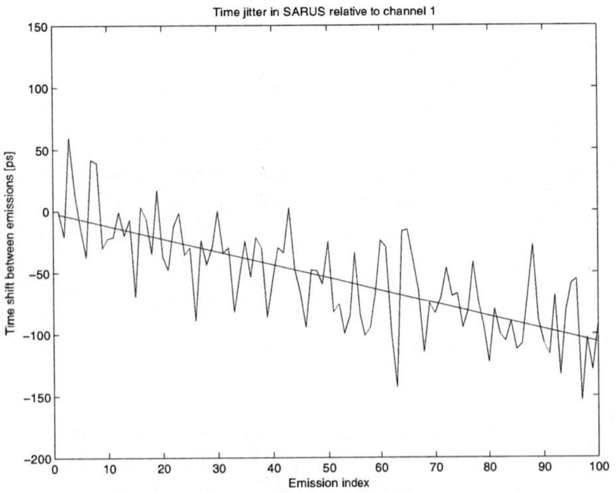

Fig. 3 The time jitter from the first to the 100th emission. The straight line indicates the trend.

VI. System Time Stability

Figure 3 shows the time shift of each emission relative to the first one. The straight line indicates the trend and is seen to have a slight slope of only -1.22 ps per emission, which can be caused by changes in the speed of sound from heating of the phantom. By observing the time shift between successive emissions, the average jitter is found to be 1.41 ps with a standard deviation of 48.3 ps. This has a negligible 0.03% impact on the accuracy of blood flow estimation when measured with a 7 MHz probe.

For the tests made with setup B, Figure 4 shows the average time shift between the channels where the transmitter and receiver are located in the same rack. The time shift is averaged over 100 identical data acquisitions, i.e., identical emissions and sampling setup, and the dashed lines indicate three standard deviations above and below the mean. The vertical line delimits the channel indices in the two racks, i.e., the channels to the left of the line are in one rack, while the channels on the right are in the other rack. The time shift for all channels is given relative to the average shift of all channels relative to the first channel on the left. As can be seen, there is no significant difference between the two racks. Such a difference would be seen as a jump in the curve at the vertical line.

Figure 5 shows the average time shift and three standard deviations between all channels, i.e., both the channels where the transmitter and receiver are located in the same rack and those where they are located in different racks. As can be seen, there are no significant jumps in the curve, which would indicate phase differences in the synchronization signals between the two racks.

The average time shift over all channels is approximately 111 ps with a standard deviation of 86 ps. Considering the time shift as a function of the emission number, it is seen to be highly stable, i.e., varying by around or less than 2 ps per emission. Therefore, the shift for each channel can be considered as being deterministic and can, thus, be compensated for. Likely sources for such deterministic behavior include slight variations in the manufacturing process for the electronic components and the relative length of the cables being used to connect transmitters and receivers.

VII. Conclusions

This paper has presented and evaluated a number of performance metrics for the SARUS research scanner. The system has a signal-to-noise ratio which gives a penetration depth around 7 cm in conventional B-mode imaging. The time stability was measured by acquiring the same B-mode line 100 times and cross-correlating the beamformed lines. This shows a very small jitter between emissions that has negligible impact on blood flow estimations. Finally, it was documented that the emission and sampling on different channels is in phase throughout the system.

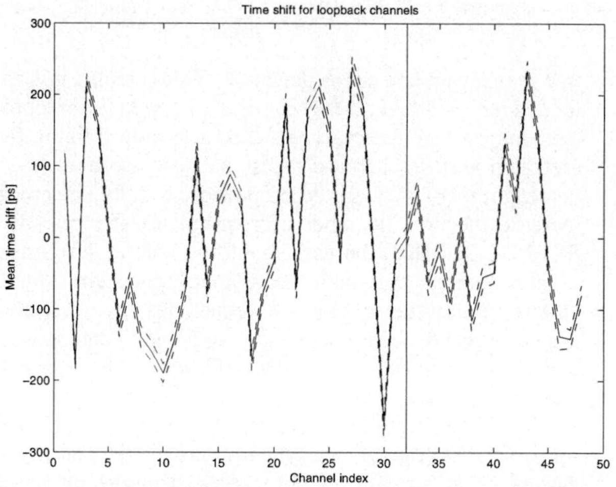

Fig. 4 Test of the synchronization between different racks. The average and three standard deviations of the time shift relative to a fixed point of the channels connected with loopback cable. To the left of the vertical line are the channels in one rack, to the right are those in the other rack.

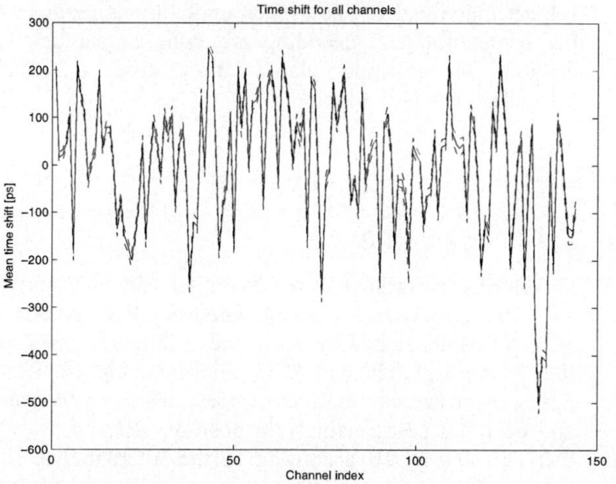

Fig. 5 Test of the synchronization between different racks. The average and three standard deviations of the time shift of all channels relative to a fixed point.

ACKNOWLEDGEMENT

This work was supported by grant 9700883, 9700563 and 26-04-0024 from the Danish Science Foundation and by B-K Medical Aps. This project is also supported by grant 024-2008-3 from the Danish Advanced Technology Foundation.

REFERENCES

1. Jensen J, Holm O, Jensen L, Bendsen H, Pedersen H, Solomonsen K, Hansen J, Nikolov S (1999) Experimental ultrasound system for real-time synthetic imaging, IEEE Proc. vol. 2, Ultrason. Symp., pp 1595–1599
2. Jensen J, Nikolov S (2004) Directional synthetic aperture flow imaging. IEEE Trans. Ultrason., Ferroelec., Freq. Contr., 51:1107–1118
3. Jensen J, Oddershede N (2006) Estimation of velocity vectors in synthetic aperture ultrasound imaging. IEEE Trans. Ultrason., Ferroelec., Freq. Contr., 25:1637–1644
4. Nikolov S, Dufait R, Schoisswohl A, Jensen J (2002) Three-dimensional real-time synthetic aperture imaging using a rotating phased array transducer, IEEE Proc., Ultrason. Symp., pp 1545–1548
5. Nikolov S, Jensen J (2003) In-vivo synthetic aperture flow imaging in medical ultrasound. IEEE Trans. Ultrason., Ferroelec., Freq. Contr., 50(7):848–856
6. Misaridis T, Pedersen M, Jensen J (2000) Clinical use and evaluation of coded excitation in B-mode images, IEEE Proc. vol. 2, Ultrason. Symp., pp 1689–1693
7. Gran F, Hansen K, Jensen J (2006) Preliminary in-vivo results for spatially coded synthetic transmit aperture ultrasound based on frequency division, IEEE Proc., Ultrason. Symp., pp 1087–1090
8. Lu J, Cheng J, Wang J (2006) High frame rate imaging system for limited diffraction array beam imaging with square-wave aperture weightings. IEEE Trans. Ultrason., Ferroelec., Freq. Contr., 53(10):1796–1812
9. Tortoli P, Bassi L, Boni E et al. (2009) ULA-OP: An advanced open platform for ultrasound research. IEEE Trans. Ultrason., Ferroelec., Freq. Contr., 56(10):2207–2216
10. Tortoli P, Jensen J (2006) Introduction to the special issue on novel equipment for ultrasound research. IEEE Trans. Ultrason., Ferroelec., Freq. Contr., 53(10):1705–1706
11. Jensen J, Holm O, Jensen L et al. (2005) Ultrasound research scanner for real-time synthetic aperture image acquisition. IEEE Trans. Ultrason., Ferroelec., Freq. Contr., 52(5):881–891
12. Andresen H, Nikolov S, Jensen J (2009) Precise time-of-flight calculation for 3D synthetic aperture focusing. IEEE Trans. Ultrason., Ferroelec., Freq. Contr., 56(9):1880–1887
13. Holm S, Kristoffersen K (1992) Analysis of worst-case phase quantization sidelobes in focused beamforming. IEEE Trans. Ultrason., Ferroelec., Freq. Contr., 39:593–599.

Author: Matthias Bo Stuart
Institute: Center for Fast Ultrasound Imaging, Department of Electrical Engineering, Technical University of Denmark
Street: Ørsteds Plads, Building 349
City: DK-2800 Lyngby
Country: Denmark
Email: mbs@elektro.dtu.dk

Steady State Visual Evoked Potential (SSVEP) - Based Brain Spelling System with Synchronous and Asynchronous Typing Modes

H. Segers[1], A. Combaz[2], N.V. Manyakov[2], N. Chumerin[2], K. Vanderperren[1], S. Van Huffel[1], and M.M. Van Hulle[2]

[1] ESAT – SCD/SISTA, K.U.Leuven, Kasteelpark Arenberg 10, POBox 2446, 3001 Heverlee, Belgium
[2] Laboratorium voor Neuro- and Psychofysiology, K.U.Leuven, Herestraat 49, POBox 1021, 3000 Leuven, Belgium

Abstract— **The paper presents an EEG-based wireless brain-computer interface (BCI) with which subjects can mind-spell text on a computer screen. The application is based on the detection of steady-state visual evoked potentials (SSVEP) in EEG signals recorded on the scalp of the subject. The performance of the BCI is compared for two different classification paradigms, called synchronous and asynchronous modes.**

Keywords— **brain-computer interface, mind speller, steady-state visual evoked potentials, synchronous and asynchronous spelling.**

I. Introduction

Research on *brain-computer interfaces* (BCIs) has witnessed a tremendous development in recent years, and is now widely considered as one of the most successful applications of neuroscience. BCIs can significantly improve the quality of life of patients suffering from amyotrophic lateral sclerosis, stroke, brain/spinal cord injury, cerebral palsy, muscular dystrophy, etc.

Among the different BCIs, mostly the noninvasive ones (see Fig. 1 for a general overview) received a lot of attention lately, since they mostly employ *electroencephalograms* (EEGs) recorded from the subject's scalp without requiring any surgery. In this paper, we study one such type of BCI, based on the detection of *steady-state visual evoked potential* (SSVEP) responses. This type of BCI relies on the psychophysiological properties of EEG brain responses recorded from the occipital pole during the periodic presentation of a visual stimulus (*i.e.,* flickering stimulus). When the periodic presentation is at a sufficiently high rate (>6 Hz), the individual transient responses overlap, leading to a *steady state* signal: the signal resonates at the stimulus rate and its multipliers [1]. This means that, when the subject is looking at stimuli flickering at frequency f_1, the frequencies f_1, $2f_1$, $3f_1$ ... can be detected in the Fourier transform of the EEG signal recorded form the occipital pole. Since the amplitude of a typical EEG signal decreases as $1/f$ in the spectral domain, the higher harmonics become less prominent. Furthermore, the fundamental harmonic f_1 is embedded into other, ongoing brain activity and (recording) noise. Thus, when considering a small recording interval, it is quite possible to make an erroneous detection. To overcome this problem, averaging over several time intervals [2], recording over longer time intervals [3], and/or system training [4–5] are often used to increase the signal-to-noise ratio and the detectability of the targted responses. Finally, in order to increase the usability and information transfer rate of the SSVEP-based BCI, the user should be able to select one of *several* commands, which means that the system should be able to reliably detect *several* frequencies $f_1,\ldots, f_n$ (one for each command). This makes the frequency detection problem more complex, and requires efficient signal processing and decoding algorithms.

An SSVEP-based BCI could be build as a system operating in a synchronous or asynchronous mode. The first one assumes that the subject observes a stimulus for a fixed, predefined amount of time, after which the classification is performed. This mode requires either a fixing of the stimulation duration for all subjects' or to perform a preliminary training/calibration to adjust the stimulus duration to each subject individnally. The asynchronous mode assumes that the stimulation and decoding are done in parallel, thus, enabling for a proper classification, given a sufficient amount of data is available.

II. Methods

A. EEG data acquisition

The EEG recordings were performed using a prototype of an ultra low-power 8-channel wireless EEG system. This system was developed by *imec*[1] and built around their ultra-low power eight-channel EEG amplifier chip [6]. Recordings were made with eight electrodes located on the parietal and occipital poles, namely in positions P3, Pz, P4, PO9, O1, Oz, O2, PO10 according to the international 10–20 system. The reference electrode and ground were placed on the left and right mastoids respectively.

The raw EEG signals are filtered above 3 Hz, with a fourth order zero-phase digital Butterworth filter, so as to remove the DC component and the low frequency drift. A

[1] Interuniversity Microelectronics Centre, http://www.imec.be

K. Dremstrup, S. Rees, M.Ø. Jensen (Eds.): 15th NBC on Biomedical Engineering & Medical Physics, IFMBE Proceedings 34, pp. 164–167, 2011.
www.springerlink.com

notch filter is also applied to remove the 50 Hz powerline interference.

B. Experiment design

Eight healthy subjects (aged 24–60 with average 35, two female and six male) participated in the two experiments. In the first experiment the subject had to observe a flickered square (white square on the black background) in the center of the screen. For this, eight different frequencies were used, one following the other. Those frequencies were selected as a dividers of the screen refreshing rate (60 Hz) and were 30, 20, 15, 12, 10, 60/7=8.57, 7.5 and 60/9=6.67 Hz. This set of frequencies has been chosen because of the stimulation method: an *intense* (white square) followed by a *non-intense* (black square) stimulus presented for an integer amount of frames (*e.g.*, 1 frame for the intense- and 1 frame for the non-intense square for a stimulation frequency of 30 Hz). The subject's task is to simply concentrate on the flickering stimuli. The data from this experiment is used when we wish to select the best frequencies, for which the SSVEP can be detected, for a given subject.

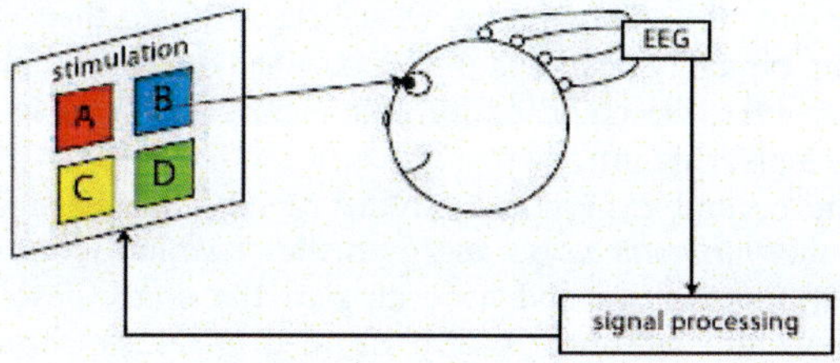

Fig. 1 General overview of the Brain Spelling System.

The second experiment is the actual typing with the spelling device. The subject is presented with a screen with a set of characters arranged in a 8 by 8 matrix. This matrix is divided into four quadrants (sub-matrices of 4 by 4 characters) against a differently colored background. The background of each quadrant is flickering with a particular and unique frequency (selected from the above mentioned set), whereby the subject is allowed to select one group of characters via her/his SSVEP responses while gazing at the correspondent quadrant. After the desired quadrant is selected, it is enlarged in to cover the entire screen and it replaces the initial 8 by 8 matrix. Then, the selection is done and the procedure is repeated: the 4 by 4 matrix is also split into 4 quadrants from which the subject can select only one. Eventually, after three selections, the system detects the character the subject was focusing on. Figure 1 presents the last level in this selection hierarchy.

C. Synchronous and asynchronous modes

For the spelling device two different decoding modes are available: *synchronous* and *asynchronous*. In the synchronous mode the stimulation, signal processing and decoding are sequential: the stimulation lasts for a fixed time Δt, after which the acquired EEG-signals are processed to detect one out of four stimulation frequencies. In the asynchronous mode, all the system components work in parallel: the signal recording, processing and decoding are done during the stimulation phase. For the asynchronous mode, the decoding could be performed on a separate computer, connected to the stimulation computer by mean of a TCP/IP protocol, or on the same computer as one of the concurrent processes. Decoding starts after a short initial pause Δt_p following the beginning of the visual stimulation. During this time the system keeps collecting EEG data. If after Δt_p seconds the collected data allows the classifier to make a "firm" decision, this decision is considered as "final" for this selection stage, and the system goes to the next selection stage. Otherwise, the classifier tries to detect the winner frequency using more data, which have been acquired during a bit longer period $\Delta t_p+\Delta t_c$, where Δt_c is the time needed for the classifier to perform the first classification attempt. The process repeats until the decision is reached or the stimulation time exceeds a threshold Δt_{max} (5 seconds in our case). In the latter case, a most probable classification result is taken (see further).

D. Spatial filtering

For enhancing the decoding accuracy, a feature extraction procedure is applied, which in our case consists of spatial filtering. This means that we search for the optimally weighted linear combination of recording channels, which result in a smaller (or equal) number of signals $S = Y*W$, with an improved signal-to-noise ratio for the frequencies of interest, with W the weight matrix and Y the original EEG signals. To estimate W, we opted for the *minimal noise energy* [7] approach, the method of *maximizing the contrast between SSVEP and the noise energy* [7], and for the methods based on the extraction of the *independent components*. The first two methods utilize irrelevant information Y_{ir} which is defined as the result of the subtraction, from the original recordings Y of all information contained in the projections onto the subspace spanned by the sine and cosine transform of all stimulus frequencies and their harmonics. This irrelevant information, considered as noise, has to be minimized to increase the signal-to-noise ratio. Thus, we could look for the weighted combination $Y_{ir}*W$, which minimizes the variances of the resulting signals. The *minimal noise energy* approach uses the principal component analysis to select the directions that correspond to the smallest eigenvectors of the covariance matrix of Y_{ir}. But the application of the previously determined weights W to the EEG signals can lead also to a decrease in the amplitude of the frequencies of interest. Thus, alternatively, we can look for those weights that increase the signal-to-noise ratio at the frequencies of interest, by maximizing the Rayleigh quotient $\max_w(\|Y*w\|^2/\|Y_{ir}*w\|^2)$. This can be done by computing the generalized eigenvalue-decomposition of the matrices Y^T*Y and $(Y_{ir})^T*Y_{ir}$, leading to a *maximum contrast* solution, by taking the directions corresponding to the highest eigenvectors. The idea behind using an independent component analysis (ICA) is based on the assumption that the recorded EEG signals are linear mixtures of independent sources caused by the visual stimulation, the ongoing brain activity and the recording noise. Hence, the application of ICA can lead to the extraction of relevant information. But here a problem arises: which independent component(s) relate to the stimulus activity?

Table 1 Mean classification accuracy for different spatial filtering methods (minimum energy approach, maximum contrast methods and ICA with different numbers of independent components kept)

Method	min energy	max contrast	ICA-8	ICA-7	ICA-6
accuracy	**64%**	63%	63%	61%	59%

E. Classification

Classification was done with the use of $T(f)$, which is an average of the signal-to-noise ratio in the power spectral density (psd) function for all signals remaining after spatial filtering, and all considered harmonics of frequency f (we used 2 harmonics). This statistic was assessed for all possible stimulation frequencies, leading to the selection of the highest value as a "winner" frequency in the synchronous mode. In the asynchronous mode, the "precise" or "firm" classification is done only if the ratio of the highest $T(f)$ to the second highest is greater than some quality threshold Q. Otherwise, the system will require more data for the classification (see Sec. II.C).

The signal-to-noise ratio for a frequency f is estimated as the ratio of the *psd* amplitude at this frequency, during stimulation, and the *psd,* at the same frequency, but with no stimulation. The latter is computed as an approximation of the *psd* when applying the autoregressive method [8] on the signal obtained after subtracting of all relevant information, as was described in Sec. II.D. But this time, the subtraction is applied after spatial filtering.

III. Results and Discussion

The system is implemented in MATLAB as a client-server application, and can run either in parallel MATLAB mode (as two labs) or on two MATLAB sessions started as separate applications (possibly on different systems).

To assess the potential accuracy of the different spatial filtering techniques, we used the data recorded in the first experiment. We tried to decode the frequency the subject was looking each time 5 seconds at one of the all eight stimulation frequencies used in the experiment. This leads to a chance level of 12.5%. The results in Table 1 show that all three spatial filtering strategies lead to almost similar results, while the *minimum noise energy* method performs slightly better, managing to correctly classify 64% of all collected SSVEP-signals. For the decoding performance, without spatial filtering, only 39% of the signals were correctly classified. Including the spatial filtering thus leads to an about 25% increase in detection performance.

We have also determined the best detectable frequency from those eight among all subjects. Based on the data from the first experiment, we have found a peak at 12 Hz (89%). But when choosing an optimal combination of four frequencies (since our online speller has 4 flickering stimuli), out of eight candidates, it is important to not only look for the best individual frequencies, but also to eliminate those that cause the most *false positives*. 12 Hz seems to yield good results, but its usability should be checked for each subject, since it falls into the alpha-range (8–13 Hz). Since this alpha rhythm typically occurs in the EEG, when the subject closes his/her eyes, we could easily obtain misclassifications due to the alpha rhythm.

The second experiment, online typing, was been done with the *minimum noise energy* method, which is the best spatial filtering method according to the study described above. The experiment was performed in the synchronous mode with 5 seconds per selection stage and with the best decoding frequencies selected based on the first experiment. Averaged among all subjects, the typing accuracy was 81%, with the chance level being 100/64=1.5625%. This result shows the potential of our application for a typing device.

To make a qualitative comparison between the synchronous and asynchronous modes, the data recorded with the previous on-line typing method was also subjected to a classification based on an asynchronous decoding. We would like to mention here that this mode also works online, and that it was applied in a way that mimics online decoding. Table 2 shows the averaged detection percentages for different initial pauses Δt_p and quality thresholds Q. Additionally, Table 3 shows the corresponding averaged detection times. Note that, in some cells, we have a stimulation and detection time larger than $\Delta t_{max} = 5$ sec, since the table shows the time required for a stimulation *with classification*. The results show that the higher Q, the better the classification results become, but the slower the detection time is. This is as expected because the frequency to be classified needs to be more pronounsed. This takes more time to achieve, but once this threshold is reached, it is more plausible that the classified SSVEP-frequency is the correct one. Longer initial pauses also lead to better classification results and slower detection times. A possible explanation is that the SSVEP-response is not prominent enough if the initial pause is too short, due to the latency of the responses, or the time required setting a *steady state* mode.

Table 2 Accuracy for different initial pauses Δt_p and quality threshold Q

		Quality threshold Q				
% detected		1.1	1.3	1.5	1.7	1.9
	0.5	15%	20%	36%	47%	57%
Δt_p [s]	1	37%	47%	58%	60%	65%
	1.5	44%	56%	62%	64%	66%

Table 3 Averaged detection time for different initial pause and threshold

		Quality threshold Q				
Avg time [s]		1.1	1.3	1.5	1.7	1.9
	0.5	0.55	0.97	2.34	3.41	4.35
Δt_p [s]	1	1.12	2.25	3.56	4.41	5.12
	1.5	1.74	3.11	4.38	5.20	5.71

Table 4 contains the typing accuracy per subject in the asynchronous mode (Q=1.5 and Δt_p =1.5 s). The first row gives the detection percentages. All subjects manage to achieve near perfect classification results. The second row gives the average detection times. A rather large inter-subject variability can be found here. The third row gives the detection percentages for the eight frequencies, when taking the data of the first experiment. This can be considered as a measure of the *quality* of the SSVEP-response for that person. There is a correlation of 92% between the average detection time and this measure. This reflects a great advantage for the asynchronous classification: all subjects (in this study) can reach almost perfect detection rates, but the classification times for persons with a strong SSVEP-response are a lot shorter. The stimulation time adapts to the specific needs of the subject.

Table 4 Classification results and time per person for 4 command asynchronous typing together with general detection accuracy

	person							
	A	B	C	D	E	F	G	H
% correct	94	100	100	100	94	100	95	100
avg time [s]	2.04	2.66	2.05	2.65	6.36	2.55	5.12	4.86
% general det	74	78	77	74	58	81	63	64

We also made a comparison between synchronous and asynchronous modes based on the theoretical *information transfer rate* (ITR), which tell us how many bits per minute the system can theoretically communicate. It implies that we assume a zero time for changing from one selected target to the next. The ITR averaged over our subjects was used for the assessment, since we wanted to compare the asynchronous with the synchronous one, when the duration of the stimulation was fixed before the experiment, and does not depend on the subject. Hence, we do not consider specially selected stimulation timings for each particular subject, but rather consider the system without any preliminary calibration/training. We can conclude from Table 5, that, in general, the asynchronous mode (Q=1.5 and Δt_p =1.5 s) yields higher ITR's than the synchronous one. Examining the performance of each subject for asynchronous typing, we see that the theoretical ITR's, which can be achieved with a 4 target system, are between 17,57 and 59,16 bits/min. If we use 8 target stimuli, in the asynchronous mode, the averaged ITR drops to 29.4 bits/min.

Table 5 Averaged ITR [bits/min] for different modes and 4 targets

Synchronous with different stimulation durations					Asynch
1s	2s	3s	4s	5s	
35.7	33.4	28.8	22.9	19.0	38.2

IV. CONCLUSIONS

Due to the large inter-subject variability, no pre-selected combination of parameters could be found that leads to an optimal detection percentage or ITR. As such, either a calibration/training stage to tune the system to a particular subject, or an asynchronous classification is needed. In this way, an average ITR of 38 bits/min can be achieved.

REFERENCES

1. Luck S.J (2005) An introduction to the event-related potential technique. MIT Press Cambridge.
2. Cheng M., Gao X., Gao S., and Xu D. (2002) Design and implementation of a brain-computer interface with high transfer rates, IEEE TBE, 49(10): 1181-1186.
3. Wang Y., Wang R., Gao X., Hong B., and Gao S. (2006) A practical VEP- based brain-computer interface, IEEE TNSRE, 14(2): 234-240.
4. Manyakov N.V., Chumerin N., Combaz A., Robben A., and Van Hulle M.M. (2010) Decoding SSVEP responses using time domain classification, Proceedings of the International Conference on Fuzzy Computation and 2nd International Conference on Neural Computation, pp. 376-380.
5. Sergio P., Luca M., Turconi A.C., and Andreoni G. (2009) A robust and self-paced BCI system based on a four class SSVEP paradigm: algorithms and protocols for a high-transfer-rate direct brain communication, Computational Intelligence and Neuroscience 864564.
6. Yazicioglu R.F., Merken P., Puers R., Van Hoof C. (2006) Low-power lownoise 8-channel EEG front-end ASIC for ambulatory acquisition systems, Proceedings of the 32nd European Solid-State Circuits Conference, pp. 247–250.
7. Friman O., Volosyak I., and Graeser A. (2007) Multiple channel detection of steady-state visual evoked potentials for brain-computer interfaces, IEEE TBE 54(4): 742-750
8. Kay S. (1988) Modern Spectral Estimation: Theory and Application. Upper Saddle River, NJ: Prentice-Hall

Localization of Heart Sounds Based on S-Transform and Radial Basis Function Neural Network

A. Moukadem[1,3], A. Dieterlen[1], N. Hueber[2], and C. Brandt[3]

[1]MIPS Laboratory, University of Haute Alsace, 68093 - MULHOUSE CEDEX FRANCE
[2] ISL: French-German Research Institute of SAINT-LOUIS, 68300 - SAINT-LOUIS FRANCE
[3]University Hospital of Strasbourg, CIC, Inserm, BP 426, 67091 STRASBOURG CEDEX FRANCE

Abstract— **This paper presents an original method for heart sounds localization based on S-Transform and radial basis function neural network (SRBF). The S-Transform is used to extract the features of heart sound. These features are then applied as inputs to RBF classifier. The performance of the localization is evaluated according to a data base of 50 subjects (including 25 cardiac pathologies sounds) which correspond to 1074 S1 and S2 heart sounds, selected from The University Hospital of Strasbourg and the Mars500 project. This study is made under the control of an experienced cardiologist. The SRBF was shown to have 95% sensitivity and 98% positive predictivity value. The proposed solution is compared with other existing methods and the robustness is shown against additive white Gaussian noise.**

Keywords— **Heart sounds localization, S-Transform, neural network, RBF, clinical heart sounds.**

I. INTRODUCTION

Auscultation is a basic part of even the hastiest cardiac exam. It provides a wealth of information about structural and functional cardiac defects, using a simple and efficient medical device: the stethoscope. Invented in the nineteenth century, this acoustic instrument has proved since to be of paramount importance to the physical examination and differential diagnosis of cardiac pathologies. Over the course of the past two centuries, the stethoscope underwent numerous improvements; among reach the development of the electronic stethoscope capable of registering and optimizing the quality of the acoustic signal termed the phonocardiogram (PCG) has been improved.

The PCG signal confirms, and mostly, refines the auscultation data and provides further information about the acoustic activity concerning the chronology of the pathological signs in the cardiac cycle, by locating them with respect to the normal heart sounds.

The cardiac sounds are by definition non-stationary signals, and are located within the low frequency range, approximately between 30 and 450 Hz [1].

The analysis of the cardiac sounds, solely based on the human ear, remains insufficient for a reliable diagnosis of cardiopathologies, and for a clinician to obtain all the qualitative and quantitative information about cardiac activity.

Information, such as the temporal localization of the heart sounds, the number of their internal components, their frequential content, and the significance of diastolic and systolic murmurs, could all be studied directly on the PCG signal. In order to recognize and classify cardiovascular pathologies, advanced methods and techniques of signal processing and artificial intelligence will be used. For that, two approaches could be considered for improve the electronic stethoscope:

- Tool with embedded autonomous analysis, simple for home use by the general public for the purpose of autodiagnosis and warning in case of necessity.
- Tool with sophisticated analysis (coupled to a PC, Bluetooth link) for the use of professionals in order to make an in-depth medical diagnosis.

Whatever the approach, one of the first phases in the analysis of heart sounds, after preprocessing, is the localization and segmentation. The heart sound localization consists to find the first and the second heart sounds, without distinguishing the two from each other. Heart sound segmentation partitions the PCG signals into cardiac cycles and further into S1 (first heart sound), systole, S2 (second heart sound) and diastole.

This phase of autonomous detection, without the help of ECG, is based on signal processing tools such as: time-frequency representation, wavelet transforms, short-time Fourier transform [2, 3], Shannon energy [4, 5], homomorphic filtering [6], Hilbert Transform [5], high order statistics [7], moment analysis [8], hidden Markov model [9] ...

In this article, a method for heart sounds localization based on S-Transform technique and radial basis functions neural network is proposed. Our objective is to find a tool provides acceptable performance for clinical use. The proposed method is compared with other existing methods and its robustness is shown against additive Gaussian noise.

K. Dremstrup, S. Rees, M.Ø. Jensen (Eds.): 15th NBC on Biomedical Engineering & Medical Physics, IFMBE Proceedings 34, pp. 168–171, 2011.
www.springerlink.com

II. METHOD

The dataset contains 50 subjects, including 25 cardiac pathologies sounds which contain different systolic murmurs. Each subject corresponds to one recording sound. The length of each sound is 8 seconds. The total number of S1 and S2 sounds is 1072.

A. Acquisition and Pre-processing of PCG Signals

Several factors affect the quality of the acquired signal, above all, the type of the electronic stethoscope, its mode of use, the patient's position during auscultation, and the surrounding noise [10]. According to the cardiologist's experience, it's preferable that the signals remain unrefined; filtration will only be applied subsequently in the purpose of signal analysis. For this reason we used prototype stethoscopes produced by Infral Corporation, and comprising an acoustic chamber in which a sound sensor is inserted. Electronics of signal conditioning and amplification are inserted in a case along with a Bluetooth standard communication module.

Different cardiologists equipped with a prototype electronic stethoscope have contributed to a campaign of measurements in the Hospital of Strasbourg. In parallel, 2 prototypes have dedicated to the MARS500 project in order to collect sounds form 6 patients (astronauts). The use of prototype electronic stethoscopes by different cardiologists makes the database rich in terms of qualitative diversity of collected sounds, which in turn makes the heart sounds localization more realistic.

The sounds are recorded with 16 bits accuracy and 8000Hz sampling frequency in a wave format, using the software "Stetho" developed under Alcatel-Lucent license. At first the original signal is decimated by factor 4 from 8000 Hz to 2000 Hz sampling frequency and then the signal is filtered by a high-pass filter with cut-off frequency of 30 Hz, 3 dB per octave, to eliminate the noise collected by the prototype stethoscope. The filtered signal is refiltered reverse direction so that there is no time delay in the resulting signal. Then, the Normalization is applied by setting the variance of the signal to a value of 1. The resulting signal is expressed by:

$$x_{norm}(t) = \frac{x(t)}{|\max(x(t))|} \tag{1}$$

B. Localization

The localization algorithms operating on PCG data try to emphasize heart sound occurrences with an initial transformation that can be classified into three main categories: frequency based transformation, morphological transformations and complexity based transformations. [7]

The transformation try to maximize the distance between the heart sounds and the background noise, and the result is smoothed and tresholded in order to apply a peak detector algorithm.

In the same approach, the proposed method (SRBF) tries to extract the envelope of the signal by applying to the radial basis function (RBF) neural network the features extracted from the S-Transform matrix of the heart sound signal. The SRBF module is illustrated in Fig. 1.

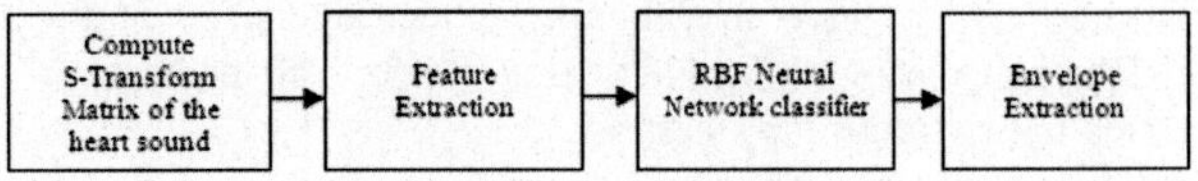

Fig. 1 Block Diagram of SRBF Module

C. S-Transform

S-Transform originates from two advanced signal processing tools, the short time Fourier transform (STFT) and the wavelet transform. It can be viewed as a frequency dependent STFT or a phase corrected wavelet transform. The S-Transform has been proven in [11] to perform better than other time-frequency /scale transforms and it's suitable for heart sounds signal analysis [12].

The generalized S-Transform of a time varying signal $h(t)$ is defined by:

$$S(\tau, f) = \int_{-\infty}^{\infty} h(t) w(\tau - t) e^{-2\pi i f t} dt \tag{2}$$

Where the window function $w(\tau - t)$ is chosen as:

$$w(t, f) = \frac{1}{\sigma(f)\sqrt{2\pi}} e^{\frac{-t}{2\sigma f^2}} \tag{3}$$

And $\sigma(f)$ is a function of frequency as:

$$\sigma(f) = \frac{\alpha}{|f|} \tag{4}$$

The window is normalized as:

$$\int_{-\infty}^{\infty} w_n(t, f) dt = 1 \tag{5}$$

The width of the window varies inversely with frequency so the S-Transform performs multi-resolution analysis on the signal. This gives high time resolution at high frequencies and high frequency resolution at low frequencies.

The discrete version is given by:

$$S(j,n)=\sum_{m=0}^{N-1}H[m+n]G(m,n)e^{\frac{i2\pi mj}{N}} \quad (6)$$

Where $G(m,n)=e^{\frac{-2\pi^2\alpha^2m^2}{n^2}}$ and $H(m,n)$ are obtained by shifting the discrete Fourier transform (DFT). Experimentally in this paper α is set to 3.

The output of S-Transform is an N by M matrix, where rows are belongs to the frequency and columns to the time. In this paper a 0-100 Hz frequency range was used to cover the main frequency band of S1 and S2 and to avoid murmurs which have in general a spectral energy above the frequency of 100 Hz [14]. This matrix will be used in the feature extraction process.

D. Feature Extraction

The absolute value of each element of the S-Transform matrix is calculated. Then the feature extraction is done by applying some standard statistical techniques and transformations. A sliding window of 50 ms (so 100 samples) was operated on the S-Transform matrix and an overlap of 75% was chosen to increase the accuracy of localization. These features have been found to be useful for heart sounds localization:

1. RMS: the root mean square rows value of each column of S-Transform matrix.

$$RMS=\sqrt{\frac{1}{N}\sum_{n=1}^{N}x_j^2(n)} \quad (7)$$

2. Max: Maximum rows value of each column of S-Transform matrix.
3. Avg: Average rows value of each column of S-Transform matrix.

Each array (100 samples) was divided into 5 segments and the mean of calculated features of each segment was calculated and taken as input to the classifier [14]. So for each step we have a 100 by 100 matrix which gives 15 descriptors.

E. RBF neural network

Radial basis function networks are embedded in a two layer neural network, where the hidden layer implements a radial activated function. The output units implement a weighted sum of hidden unit outputs. The input (calculated 15 descriptors) onto an RBF network is nonlinear while the output is linear.

The output of the RBF network is given by:

$$f_r(x)=\lambda_0+\sum_{i=1}^{n}\lambda_i\phi(\|x-c_i\|) \quad (8)$$

Where λ_i are the weights, c_i are known as the RBF centers, and n_r is the number of centers, and the transfer function $\phi(.)$ of a radial basis neuron is the Gaussian function:

$$\phi(n)=e^{-n^2} \quad (9)$$

In this paper, the learning procedure is based on the orthogonal least squares method. Neurons are added to the network until the sum-squared error falls beneath an error goal or a maximum number of neurons have been reached.

The network is trained on two heart sounds samples (S1 and S2) and two no heart sound samples (systole, diastole) selected randomly from the database. The target is fixed to 1 for S1 or S2 and 0 for the other components. A major advantage of the RBF network comparing to others neural networks types is its fast learning speed and the simplicity of its implementation.

III. Results and Comparaison

The performance of the SBRF method was measured as the method's capacity to locate S1 and S2 correctly. It was measured by sensitivity and positive predictive value:

$$Sensitivity=\frac{TP}{TP+FN} \quad (10)$$

And positive predictive value:

$$PPV=\frac{TP}{TP+FP} \quad (11)$$

A sound is true positive (TP) if it is correctly located, all others detected sounds are considered as false positive (FP) and all missed sounds are considered as false negative (FN).

In [15], authors proposed a comparative study of five algorithms of heart sounds localization; radial basis function based method (RBF) [15], Hilbert transform based method [5], Shannon energy based method [4], variance fractal dimension based method [7] and Cardiac sound characteristic waveform (CSCW) method [5]. The robustness of each method is shown against additive white Gaussian noise. Authors found that the RBF and CSCW based methods seem to be the most promising. RBF can be used in low level noise signals and CSCW method can be trusted in the presence of high level of noise. The goal of the SRBF method was to combine the advantages of each method.

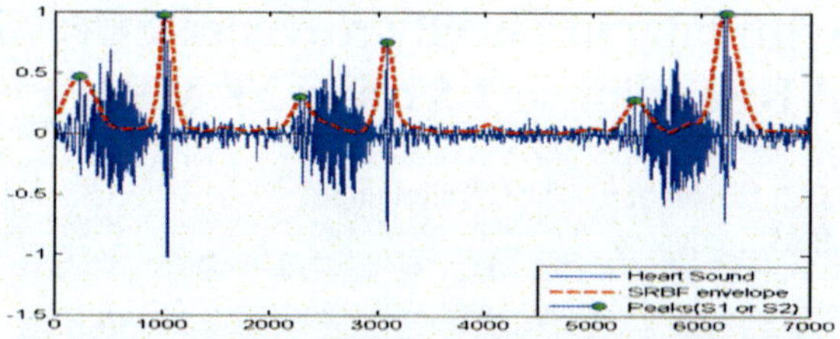

Fig. 2 SRBF method applied on a heart sound with heavy systolic murmur.

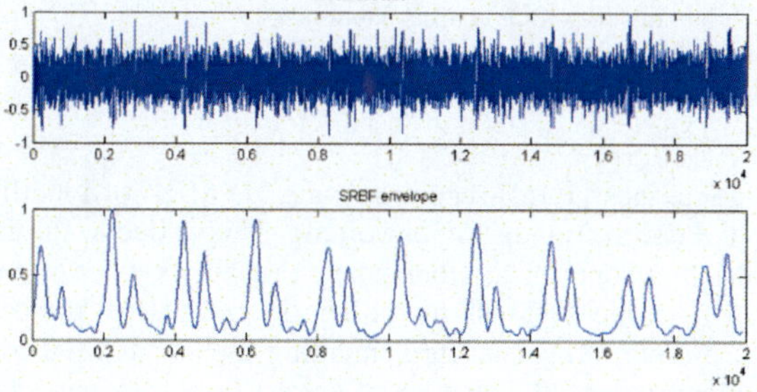

Fig. 3 SRBF method applied on a normal heart sound hardly degraded by additive white Gaussian noise.

Table 1 Sensitivity and Positive Predictive Values for the 3 methods applied on the clinical sounds set without and with additive Gaussian noise.

Method	Sensitivity	PPV	Sensitivity Noise	PPV Noise
SRBF	95%	98%	91%	91%
RBF	94%	98%	79%	89%
CSCW	92%	98%	84%	91%

Results in Table1 show a significant difference between SRBF and other methods, especially in presence of additive noise. RBF based method operates directly on heart sound signal without any previous features extraction, so it's not surprising that the distance between signal and noise was decreasing by the networks in the presence of high level additive noise. CSCW method, it is based on a frequency transformation so it will be less sensitive to noise than other morphological transforms but still more sensitive than the SRBF method.

IV. Conclusion

An original and robust method for heart sound localization based on S-Transform features and radial basis function neural network, named SRBF, is presented in this paper.

The robustness of this method was shown against additive white Gaussian noise and against the diversity of the database. The efficiency of heart sounds localization reached by SRBF method, measured on 1074 heart sounds (S1 and S2), gives 95% sensitivity and 98% PPV.

Compared with other existing methods for heart sounds localization, SRBF shows a significant enhancement in term of sensitivity.

The high localization efficiency, even in the presence of heavy murmurs or high level noise, makes the proposed method suitable for clinical use.

References

1. Z. Dokur, T. Ölmez, Feature determination for heart sounds based on divergence analysis, Digital Signal Process. (2007), doi:10.1016/j.dsp.2007.11.003.
2. P Rakovic, E Sejdic, LJ Stankovic, J Jiang. Time-Frequency Signal Processing Approaches with Applications to Heart Sound Analysis. Computers in Cardiology 2006; 33:197–200.
3. S.M. Debbal, F. Bereksi-Reguig Time-frequency analysis of the first and the second heartbeat sounds. Applied Mathematics and Computation 184 (2007).
4. H Liang, S Lukkarinen, I Hartimo, Heart Sound Segmentation Algorithm Based on Heart Sound Envelogram, Helsinki University of Technology, Espoo, Finland.
5. Samjin Choi, Zhongwei Jiang, Compariason of envelope extraction algorithms for cardiac sound signal segmentation, Micro-Mechatronics Laboratory, Yamaguchi University, 2006, Japan.
6. Cota Navin Gupta, Ramaswamy Palaniappan, Sundaram Swaminathan, Shankar M. Krishnan. Neural network classification of homomorphic segmented heart sounds. Applied Soft Computing 7 (2007) 286–297.
7. Christer Ahlstrom, NonLinear Phonocardiographic Signal Processing thesis, Link¨oping University, SE-581 85 Link¨oping, Sweden, April 2008.
8. Yan, Z., Jiang, Z., Miyamoto, A., Wei, Y. The moment segmentation analysis of heart sound pattern (2010) Computer Methods and Programs in Biomedicine, 98 (2), pp. 140-150.
9. Schmidt, S.E., Holst-Hansen, C., Graff, C., Toft, E., Struijk, J.J.Segmentation of heart sound recordings by a duration-dependent hidden Markov model (2010) Physiological Measurement, 31 (4), pp. 513-529.
10. A. Ch. Stasis, E.N. Loukis, S.A. Pavlopoulos, D. Koutsouris. A multiple decision trees architecture for medical diagnosis: The differentiation of opening snap, second heart sound split and third heart sound. CMS Springer-Verlag 2004.
11. R.G. Stockwell, L. Mansinha, R.P. Lowe, Localization of the complex spectrum: the S-transform, IEEE Trans. Sig. Proc. 44 (4) (1996) 998–1001.
12. G. Livanos, N. Ranganathan, J. Jiang, Heart sound analysis using the S transform, IEEE Computers in Cardiology 2000;27:587-590.
13. Amir A. Sepheri, et al., A nouvel method for pediatric heart sound segmentation without using the ECG, Comput. Methods Programes Biomed. (2009), doi:10.1016/j.cmpb.2009.10.006
14. H. M. Hadi, M. Y. Mashor,Mohd Zubir Suboh, Mohamed Sapawi Mohamed, Classification of heart sound based on S-Transform and neural network, IEEE 10th International Conference on Information Science, Signal Processing and their Applications (ISSPA 2010).
15. A. Moukadem, A. Dieterlen, N. Hueber, C. Brandt, Comparative study of heart sounds localization, Bioelectronics, Biomedical and Bio-inspired Systems SPIE N° 8068A-27, Prague.

Diffusion Weighted MRI (DWI) for Brachytherapy in Locally Advanced Cervical Cancer – Determining the Degree of Distortion at 1.5T and 3T MRI

S. Haack[1], S.N. Jespersen[2], L. Fokdal[3], J.C. Lindegaard[3], J. F. Kallehauge[4], K. Tanderup[3], and E.M. Pedersen[5]

[1] Department of Clinical Engineering, Aarhus University Hospital, Aarhus, Denmark
[2] CFIN, Aarhus University Hospital, Aarhus, Denmark
[3] Department of Oncology, Aarhus University Hospital, Aarhus, Denmark
[4] Department of Medical Physics, Aarhus University Hospital, Aarhus, Denmark
[5] Department of Radiology, Aarhus University Hospital, Aarhus, Denmark

Abstract— **Diffusion Weighted MRI (DWI) enables identification of tissue with high cellular density such as tumors. This makes DWI a potentially valuable tool in oncology imaging for both diagnostic imaging and monitoring of treatment. Locally advanced cervical cancer is usually treated with brachytherapy using an intracavitary applicator (Figure 1). MRI guided brachytherapy can be performed by imaging with the applicator in place prior to dose planning and treatment. This study evaluates the amount of distortion in DW images in vivo at both 1.5T and 3T MRI. DWI was performed in six patients at 1.5T MR and in two patients at 3T MRI. All MRI examinations were performed with the plastic applicator for brachytherapy in place. The cervix and lower uterus was manually contoured on T2 weighted images (T2W) and on DW images with b-value = 0 s/mm^2 (Figure 2). The contours were compared by calculating the Jaccard similarity coefficient (the common area compared to the union area). The center of the applicator tandem was identified and marked on T2W and DW images and the shift was calculated. The Jaccard coefficient (mean±std.dev.) was 73.6±8.3 (1.5T) and 78.5±3.8 (3T). The difference between the location of the tandem center was (mean±std.dev.) 2.2±1.2 mm (1.5T) and 1.5±1.6 mm (3T). If DW images should be used dose planning of brachytherapy the shift and distortion should be corrected to match the morphological images.**

Keywords— **Diffusion, brachytherapy, cervical cancer, distortion.**

I. INTRODUCTION

Brachytherapy (BT) is an important modality for treatment of patients with cervix cancer. BT is following external radiotherapy and chemotherapy to take advantage of the tumor shrinkage. The major advantage of BT is that it allows a very high dose to the cervix while the rapid dose fall-off result in relative sparing of the surrounding organs such as the bladder, bowel and rectum. The steep dose-gradients imply that correct tumor identification and contouring is very important. This is emphasized by the GEC-ESTRO guidelines [1;1;2] recommending 3D MRI for identification and delineation of tumor volume. This is due to the fact that MRI is superior to other image modalities when it comes to soft tissue contrast including differentiating tumor tissue from normal tissue. Implementing the use of MRI for dose-planning of BT has been of great value decreasing the dose to the surrounding organs while still delivering the optimal dose to the tumor [3-5].

Functional imaging such as dynamic contrast enhanced MRI (DCE-MRI) and diffusion weighted MRI (DW-MRI) are both potential valuable tools for identifying the tumor and for monitoring the tumor during therapy [6-9]. Both methods may improve dose-planning by identifying the most therapy resistant part of the tumor allowing a dose-boost to this volume.

DW-MRI measures the random motion of water in tissue [10;11]; hence the measured signal in DW-MRI depends on the density of the tissue. Tumor tissue is more dense than normal tissue and water will experience decreased diffusion in tumor tissue compared to normal tissue. The diffusion can be quantified by the Apparent Diffusion Coefficient (ADC) and this parameter might be used for differentiating tumor tissue and for monitoring therapy response [12-14]. Acquisition of DW-MRI is done using a single shot Echo-Planar Sequence (SS-EPI), that acquires an entire image at a time. The major disadvantages of the EPI sequence is the high sensitivity to susceptibility changes (air/tissue) and magnetic in-homogeneity resulting in shape distortion and spatial displacement. This problem increases further when the applicator for BT is present.

This study evaluates the displacement of the cervix and the BT applicator in EPI based DW-MR images compared to standard morphological T2-weighted Turbo-spin echo (TSE) MR images. This will give an indication of the amount of geometric distortion correction needed to performed to make DW-MRI suitable for dose-planning of locally advanced cervical cancer.

K. Dremstrup, S. Rees, M.Ø. Jensen (Eds.): 15th NBC on Biomedical Engineering & Medical Physics, IFMBE Proceedings 34, pp. 172–175, 2011.
www.springerlink.com

II. MATERIALS AND METHODS

A. Imaging

MR imaging was performed in six patients using a 1.5T MR scanner (Magnetom Symphony, Siemens, Erlangen, Germany) and in two patients using a 3T MR scanner (Achieva 3T-X, Philips, Best, Netherlands). All MR examinations were performed with the plastic BT applicator (Varian Medical Systems) in place. (fig. 1)

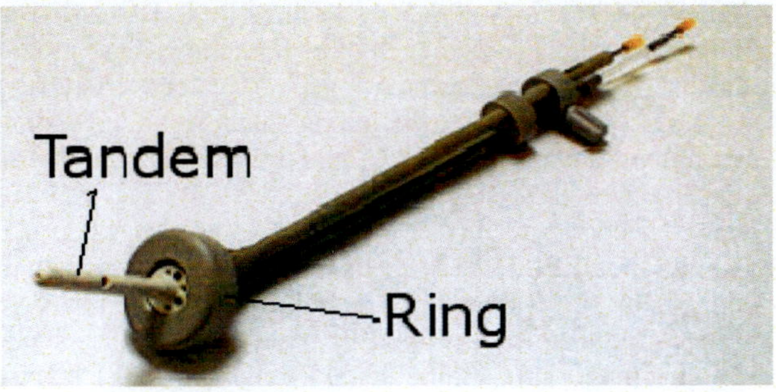

Fig. 1 Tandem applicator for brachytherapy, Varian Medical Systems

Transversal diffusion weighted images and T2 weighted images were acquired at the same slice position using the same FOV and slice thickness to allow geometrical comparison. Parallel imaging was used for DWI acquisition at both 1.5T and 3T. The imaging parameters for both DWI sequences are listed in table 1.

Table 1 DWI sequence parameters

	1.5T	3T
FOV (LR x AP)	280 x 280 mm	280 x 260 mm
Slice thickness	5.5 mm	4 mm
Matrix	128 x 128	112 x 112
Phase direction	AP	AP
Parallel imaging	Grappa 2	Sense 2
TR	2500 ms	1720 ms
TE	82 ms	86 ms
b-values (s/mm^2)	0, 150, 600, 1000	0, 150, 500, 1000

B. Image analysis

Images were analyzed using an in-house Matlab application (Mathworks). T2W images and DW-images with b-value = 0 were analyzed slice by slice. The volume of interest is from the surface of the applicator ring, which is at the beginning of the cervix, to the end of the tandem, which resulted in an average of 6 slices pr. patient at 1.5T and 9 slices pr. patient at 3T. For each slice the cervix/uterus was contoured on both the T2W image and the DWI, b=0 and the common and union areas were calculated as shown in figure 2 and 3 where the two areas are overlaid on SE image. At b-value = 0 the DWI is a T2 weighted EPI image and the contrast is comparable to the one in the T2 weighted TSE image.

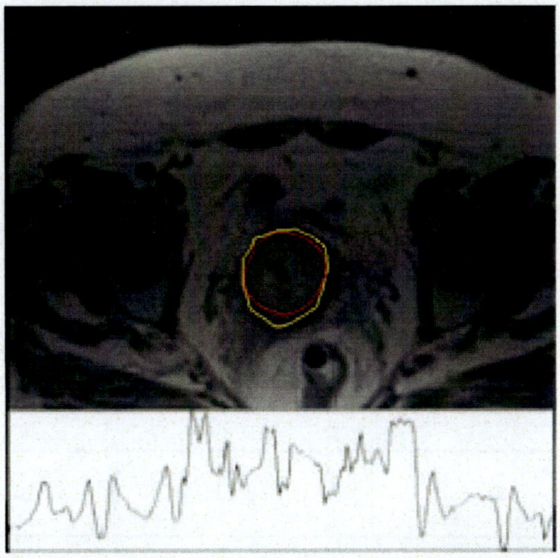

Fig. 2 Union area (orange) and common area (red) at 1.5T of spin-echo image and DWI, b=0 image. Intensity profile through center of tandem shown at bottom.

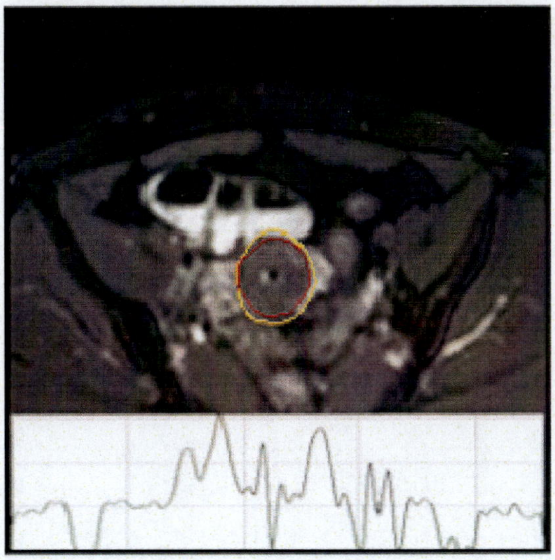

Fig. 3. Union area (orange) and common area (red) at 3T of spin-echo image and DWI, b=0 image. Intensity profile through center of tandem shown at bottom.

The areas were compared by calculating the Jaccard similarity coefficient, J (eq. 1), where X is the contoured area in the T2W SE image and Y is the contoured area in the DWI image.

$$J = \frac{|X \cap Y|}{|X \cup Y|}$$

The center of the tandem was identified on the images and the image coordinates were compared.

III. Results

The results of the Jaccard similarity coefficient for area comparison for both 1.5T and 3T are shown in table 2.

Table 2 The Jaccard similarity coefficient, J for areas contoured on T2W TSE and DWI,b=0 image

	1.5T mean ± sd	3T mean ± sd
J	73.6 ± 8.3	78.5 ± 3.8

The differences of the location of the tandem center in the images are shown in table 3.

Table 3 Difference of tandem center between T2W TSE and DWI, b=0 for both 1.5T and 3T.

	1.5T mean ± sd	3T mean ± sd
Δ Tandem center	2.2 ± 1.2 mm	1.5 ± 1.6 mm

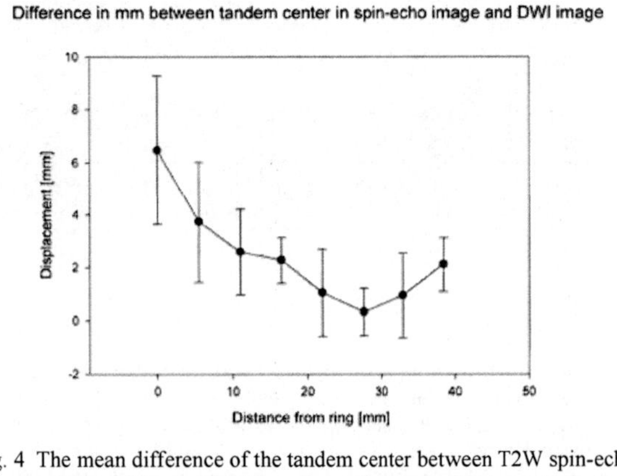

Fig. 4 The mean difference of the tandem center between T2W spin-echo images and the EPI based DW images as a function of distance to the top of the ring for the 1.5T MR-examinations.

The largest tandem center differences were found closest to the ring applicator and the difference decreased when moving away from the ring (fig. 4). This was not the case for the Jaccard coefficient.

IV. Discussion

Several studies have shown that diffusion weighted MRI add valuable information to tumor diagnosis and monitoring [12;13;15-17]. There is an increasing interest to include functional imaging modalities such as PET, DCE-MRI and DW-MRI for dose-planning of radiotherapy. One problem is that these modalities have geometrical uncertainties or low resolution. Before including such techniques in the treatment planning pipeline quantification of geometrical uncertainties is necessary. This study shows that geometrical correction and further validation is needed when using single shot EPI DWI measurements in patients with BT applicator. Several methods have been proposed for correction of the distortion in the EPI based DW images, such as B0-mapping [18], Point Spread Function (SPF) mapping [19] or use of the reverse gradient method [20]. It is challenging to implement a method that corrects for both the geometrical distortions and susceptibility artefacts such that the calculated ADC values can be trusted. The organ motion in the pelvis becomes an issue when one of these approaches are to be used due to the time lapse between the DWI acquisition and the correction sequence (e.g. a B0-map). This time lapse should be kept as small as possible.. The use of a tandem applicator and possibly additional needles for radiation dose delivery in BT further increases the amount of distortion in the DW images. Corrections and further validation of the influence of the calculated ADC value form both distortions and correction methods should be made before DWI can be used for dose planning of brachytherapy in locally advanced cervical cancer.

V. Conclusion

The EPI based DWI sequence induces shape distortions in the images which would introduce erroneous contours if used for dose-planning. The Jaccard coefficients for 1.5T and 3T are comparable. The largest shifts are seen near the ring. This is unfortunate since this is the position where the tumor is most often located.

The study shows that distortion correction is necessary at both 1.5T and 3T if DWI should be used for dose-planning of brachytherapy for locally advanced cervical cancer.

Acknowledgment

The study has been supported by research grants from the Danish Cancer Society, Danish Council for Strategic Research and CIRRO - the Lundbeck Foundation Centre for Interventional Research in Radiation Oncology and the Danish National Research Foundation.

References

1 R. Pötter, C. Haie-Meder, L. E. Van, I. Barillot, B. M. De, J. Dimopoulos, I. Dumas, B. Erickson, S. Lang, A. Nulens, P. Petrow, J. Rownd, and C. Kirisits, "Recommendations from gynaecological (GYN) GEC ESTRO working group (II): concepts and terms in 3D image-based treatment planning in cervix cancer brachytherapy-3D dose volume parameters and aspects of 3D image-based anatomy, radiation physics, radiobiology," *Radiother. Oncol.*, vol. 78, no. 1, pp. 67-77, Jan.2006.

2 C. Haie-Meder, R. Pötter, L. E. Van, E. Briot, B. M. De, J. Dimopoulos, I. Dumas, T. P. Hellebust, C. Kirisits, S. Lang, S. Muschitz, J. Nevinson, A. Nulens, P. Petrow, and N. Wachter-Gerstner, "Recommendations from Gynaecological (GYN) GEC-ESTRO Working Group (I): concepts and terms in 3D image based 3D treatment planning in cervix cancer brachytherapy with emphasis on MRI assessment of GTV and CTV," *Radiother. Oncol.*, vol. 74, no. 3, pp. 235-245, Mar.2005.

3 R. Pötter, J. Dimopoulos, P. Georg, S. Lang, C. Waldhausl, N. Wachter-Gerstner, H. Weitmann, A. Reinthaller, T. H. Knocke, S. Wachter, and C. Kirisits, "Clinical impact of MRI assisted dose volume adaptation and dose escalation in brachytherapy of locally advanced cervix cancer," *Radiother. Oncol.*, vol. 83, no. 2, pp. 148-155, May2007.

4 J. C. Lindegaard, K. Tanderup, S. K. Nielsen, S. Haack, and J. Gelineck, "MRI-Guided 3D Optimization Significantly Improves DVH Parameters of Pulsed-Dose-Rate Brachytherapy in Locally Advanced Cervical Cancer," *Int. J. Radiat. Oncol. Biol. Phys.*, vol. 71, no. 3, pp. 756-764, 2008.

5 K. Tanderup, S. K. Nielsen, G. B. Nyvang, E. M. Pedersen, L. Rohl, T. Aagaard, L. Fokdal, and J. C. Lindegaard, "From point A to the sculpted pear: MR image guidance significantly improves tumour dose and sparing of organs at risk in brachytherapy of cervical cancer," *Radiother. Oncol.*, vol. 94, no. 2, pp. 173-180, Feb.2010.

6 B. A. Moffat, T. L. Chenevert, T. S. Lawrence, C. R. Meyer, T. D. Johnson, Q. Dong, C. Tsien, S. Mukherji, D. J. Quint, S. S. Gebarski, P. L. Robertson, L. R. Junck, A. Rehemtulla, and B. D. Ross, "Functional diffusion map: a noninvasive MRI biomarker for early stratification of clinical brain tumor response," *Proc. Natl. Acad. Sci. U. S. A*, vol. 102, no. 15, pp. 5524-5529, Apr.2005.

7 V. N. Harry, S. I. Semple, F. J. Gilbert, and D. E. Parkin, "Diffusion-weighted magnetic resonance imaging in the early detection of response to chemoradiation in cervical cancer," *Gynecol. Oncol.*, vol. 111, no. 2, pp. 213-220, Nov.2008.

8 G. Groenendaal, C. A. van den Berg, J. G. Korporaal, M. E. Philippens, P. R. Luijten, V. M. van, and U. A. van der Heide, "Simultaneous MRI diffusion and perfusion imaging for tumor delineation in prostate cancer patients," *Radiother. Oncol.*, vol. 95, no. 2, pp. 185-190, May2010.

9 S. Haack, E. M. Pedersen, S. N. Jespersen, J. F. Kallehauge, J. C. Lindegaard, and K. Tanderup, "Apparent diffusion coefficients in GEC ESTRO target volumes for image guided adaptive brachytherapy of locally advanced cervical cancer," *Acta Oncol.*, vol. 49, no. 7, pp. 978-983, Oct.2010.

10 B. D. Le, E. Breton, D. Lallemand, P. Grenier, E. Cabanis, and M. Laval-Jeantet, "MR imaging of intravoxel incoherent motions: application to diffusion and perfusion in neurologic disorders," *Radiology*, vol. 161, no. 2, pp. 401-407, Nov.1986.

11 P. J. Basser, J. Mattiello, and D. LeBihan, "Estimation of the effective self-diffusion tensor from the NMR spin echo," *J Magn Reson. B*, vol. 103, no. 3, pp. 247-254, Mar.1994.

12 S. Naganawa, C. Sato, H. Kumada, T. Ishigaki, S. Miura, and O. Takizawa, "Apparent diffusion coefficient in cervical cancer of the uterus: comparison with the normal uterine cervix," *Eur. Radiol.*, vol. 15, no. 1, pp. 71-78, Jan.2005.

13 P. Z. McVeigh, A. M. Syed, M. Milosevic, A. Fyles, and M. A. Haider, "Diffusion-weighted MRI in cervical cancer," *Eur. Radiol.*, vol. 18, no. 5, pp. 1058-1064, May2008.

14 A. R. Padhani and A. A. Khan, "Diffusion-weighted (DW) and dynamic contrast-enhanced (DCE) magnetic resonance imaging (MRI) for monitoring anticancer therapy," *Target Oncol.*, vol. 5, no. 1, pp. 39-52, Mar.2010.

15 C. K. Kim, B. K. Park, J. J. Han, T. W. Kang, and H. M. Lee, "Diffusion-weighted imaging of the prostate at 3 T for differentiation of malignant and benign tissue in transition and peripheral zones: preliminary results," *J. Comput. Assist. Tomogr.*, vol. 31, no. 3, pp. 449-454, May2007.

16 K. Tamai, T. Koyama, T. Saga, S. Umeoka, Y. Mikami, S. Fujii, and K. Togashi, "Diffusion-weighted MR imaging of uterine endometrial cancer," *J Magn Reson. Imaging*, vol. 26, no. 3, pp. 682-687, Sept.2007.

17 G. Groenendaal, M. R. Moman, J. G. Korporaal, P. J. van Diest, V. M. van, M. E. Philippens, and U. A. van der Heide, "Validation of functional imaging with pathology for tumor delineation in the prostate," *Radiother. Oncol.*, vol. 94, no. 2, pp. 145-150, Feb.2010.

18 P. Jezzard and R. S. Balaban, "Correction for geometric distortion in echo planar images from B0 field variations," *Magn Reson. Med.*, vol. 34, no. 1, pp. 65-73, July1995.

19 H. Zeng and R. T. Constable, "Image distortion correction in EPI: comparison of field mapping with point spread function mapping," *Magn Reson. Med.*, vol. 48, no. 1, pp. 137-146, July2002.

20 R. Bowtell, D. J. O. McIntyre, M.-J. Commandre, P. M. Glover, and P. Mansfield, "Correction of geomtric distortion in echo planar imaging," 1994, p. 411.

Author: Søren Haack
Institute: Dept. of Clinical Engineering, Aarhus University Hospital
Street: c/o AUH Skejby, Brendstrupgaardsvej 100
City: 8200 Aarhus N
Country: Denmark
Email: soeren.haack@stab.rm.dk

A Novel Hierarchical Semi-centralized Telemedicine Network Architecture Proposition for Bangladesh

S.H. Choudhury, C.B. Peterson, S. Kyriazakos, and N.R. Prasad

Center for TeleInFrastruktur (CTIF), Aalborg University, Aalborg, Denmark
{samiulhc,cbp,sk,np}@es.aau.dk

Abstract— **One of the major functions of telemedicine is the prompt delivery of modern healthcare to the remotest areas with reduced cost and efficient use of communication resources. The establishment of a well organized telemedicine system is therefore exigent for the developing countries like Bangladesh where there are extreme paucities of efficient healthcare professionals and equipments, specifically in the rural areas. In this paper a novel, hierarchical and semi-centralized telemedicine network architecture has been proposed holistically focusing on the rural underdeveloped areas of Bangladesh. The model utilizes the existing fiber optic backbone and wireless telecommunication infrastructures to connect the remote healthcare centers with the urban specialized hospitals. The proposed network is of low cost, flexible and faster as well as more concrete than the existing healthcare organogram of Bangladesh. Finally, some features and services associated with the model have also been proposed which are pragmatic and easily implementable.**

Keywords— **Bangladesh, e-Health, Telemedicine, Tele-infrastructure, Tier, Upazila.**

I. Introduction

Telemedicine is one of the most promising branches of Information and Communication Technologies (ICT) and is possibly the most prominent e-business service that can have a major visible effect on the development of healthcare sector. It can reduce mortality, morbidity, expenses and psychological strains by catering healthcare services to the remotest parts of a country. In countries where access to medical services is restricted by distance and poor transportation, and where health care services are inadequate, telemedicine offers a great opportunity and possibilities to distribute medical services by utilizing ICT. Developing countries like Bangladesh are still striving to provide minimal health services to all citizens due to the insufficient number of health care professionals and medical services available, economic regression, geographical hindrance, inadequate transportation system, political instability and improper healthcare management. Telemedicine, therefore, may be an effective solution of the health care predicaments of Bangladesh.

Telemedicine activities were first materialized in Bangladesh in the mid of 1999 [1]. Henceforth different potential projects on telemedicine have been undertaken [2,3]. But none of them is commercially well established due to the lack of efficient workforce, technical limitations, high cost, discrete choices of location and services, illiteracy in technical knowledge and the absence of a concretely defined telemedicine policy. Moreover, the rural areas which consist of about 50,000 villages [4] and accommodate 80% of the total population of Bangladesh [5] get lower priority in healthcare sector due to the excessive healthcare demand of the urban areas.

In this paper, we have proposed a novel tele-infrastructure based integrated telemedicine network architecture that focuses primarily on the extensive but deprived rural parts of Bangladesh and is compatible with the existing telecommunication infrastructure, economic, social and geographic condition.The paper is organized as follows: the next part describes the methodology that has been followed in developing the paper. The succeeding section delineates the prospective areas of telemedicine of Bangladesh. Finally, the proposed telemedicine network architecture has been portrayed with its features and services.

II. Methodology

In this research work, we have studied a considerable number of scientific articles on telemedicine network architecture and applications in the developed and developing world. Necessary data and other information have been accumulated from the reliable official websites, newspaper articles, and up-to-date publications. The undeveloped rural parts of Bangladesh have been chosen as major concern. The applicability of the establishment of an organized telemedicine system in Bangladesh has been scrutinized. Comprehensive discussions with pertinent professionals, such as doctors and IT professionals have also been conducted and a network model has been developed on the basis of the outcomes of research and discussions.

III. Background Study: Health Sectors of Bangladesh and Prospect of Telemedicine

Bangladesh has made significant progress in health sector, specifically in primary healthcare since the Alma Ata

K. Dremstrup, S. Rees, M.Ø. Jensen (Eds.): 15th NBC on Biomedical Engineering & Medical Physics, IFMBE Proceedings 34, pp. 176–179, 2011.
www.springerlink.com

Declaration in 1978 [6]. Infant, maternal, and under-five mortality rates have decreased over the last decades, with a marked increase in life expectancy at birth [6]. But the progress is insufficient compared to the demand. In Bangladesh, there are 7 divisions, 64 districts and 482 upazilas which are the administrative units [7]. Each Upazila has an existing governmental Upazila Health Complex (UHC). According to the monthly statistical bulletin of Bangladesh Bureau of Statistics (BBS, 2008) the statistics of UHC is provided in Table 1. The statistics represents the impoverished healthcare infrastructures of Bangladesh. Moreover, around 75% of the professional physicians work in the urban areas [5]. There is a huge disparity in health care distributions between rural and urban areas. Rural people have to spend a higher proportion of their limited income to receive specialist's advice, because they have to travel long distances to go to the urban areas. It often becomes impossible to transfer a patient with a critical health condition to the urban hospitals due to the poor transportation systems in the rural areas. So, telemedicine may be an easier and cheaper way to disseminate healthcare facilities to the rural areas of Bangladesh. Compared to developed countries, telemedicine activities are still in the primary level in Bangladesh. So there is a huge window open for advanced and organized telemedicine system development in Bangladesh.

Table 1 Statistics of upazila health complex in Bangladesh

Parameters	2002	2003	2004	2005	2006
No. of hospitals	1,362	1,384	1,676	1,676	1,683
No. of hospital beds	45,607	46,125	50,655	50,827	51,044
Person per hospital bed (No.)	4,109	2,801	2,550	2,736	2,732
Person per Physician (No.)	4,043	3,532	3,213	3,317	3,125
Registered Physician (No.)	34,502	36,678	40,210	41,933	44,632

IV. Proposed Telemedicine Network Architecture

In this section, we have presented our proposed telemedicine network architecture. The objective of the proposed telemedicine model architecture is to develop a link between remote healthcare centers and the mainstream centralized urban healthcare centers utilizing the existing telecommunication infrastructures, incorporating government and private organizations in building up the panorama of an effective telemedicine network.

A. *Network Model Description*

The network architecture is shown in Fig. 1. The entire medical infrastructures will be partitioned into *four tiers*. The tier-1 health centers will be some specific, internationally recognized, government and non-government hospitals situated at the Capital and some big cities of Bangladesh. Tier-2 health centers are the district level hospitals while the Upazila Health Complexes (UHC) will lie in tier-3. The remote health care units will be designated as tier-4 medical units. The tier-1 units will communicate with the tier-2 units using the high speed fiber optic backbone network of BTTB covering all the districts of Bangladesh [2] while the communication between the tier-2 units and tier-3 units and also between tier-3 units and tier-4 units will be accomplished by wireless internet using the network covered by the mobile phone operators [8].

The Tier-1 medical units, adorned with the expert professsionals will be responsible for developing and maintaining the online telemedicine network and central server system and will regulate the overall telemedical activities throughout the country. The units will form a *fully connected mesh network* for quicker and safer performance. The logical central server system will be the integration of all the physical servers located at different tier-1 units for data protection, decentralization, and security reasons. The iPath software solution can be used as the online telemedicine platform [9]. The tier-2 units will directly regulate the operations of tier-3 units while providing some additional services such as mobile medical units, post disaster medical service units, etc. A tier-3 unit will be the one where the local people will sign themselves up for the online server system. It will communicate with the tier-4 units and provide the common medical treatments and diagnoses. The tier-4 units will be either rural government healthcare centers or NGO (Non-Government Organization) healthcare units affiliated with the central system. These units will be organized by cheap telemedicine kits and tablet PCs or simple ham radios.

In Bangladesh, the remote healthcare centers (mentioned as tier-4 units) are still not well developed; therefore, the tier-3 units (UHC) have got the primary focus in our model. Each tier-3 unit will be provided sufficient telemedicine equipments (desktop or laptop computer, broadband internet connection, etc.) with trained medical officers, nurses, pharmacists, and IT operators depending on the population density of the respective area.

B. *Storage and Archive Subsystem*

In order to develop a compact telemedicine network, a well organized data storage system is essential. The data base and server system of the proposed telemedicine network architecture is depicted in Fig. 2. Due to the large population, the centralized server system is hierarchically stratified into division level servers and district level servers. Each district level server will contain the information related to every citizen under that district.

In medical system, each patient will be identified by a unique *'National Health Card' (NHC)* number. Recently, Bangladesh Government has successfully implemented the

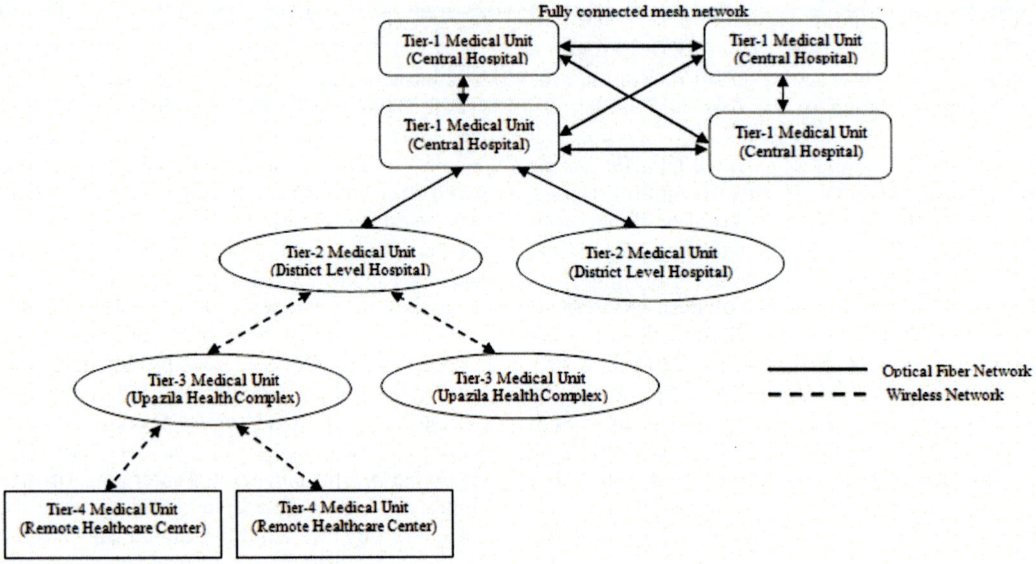

Fig. 1 Proposed Telemedicine network architecture

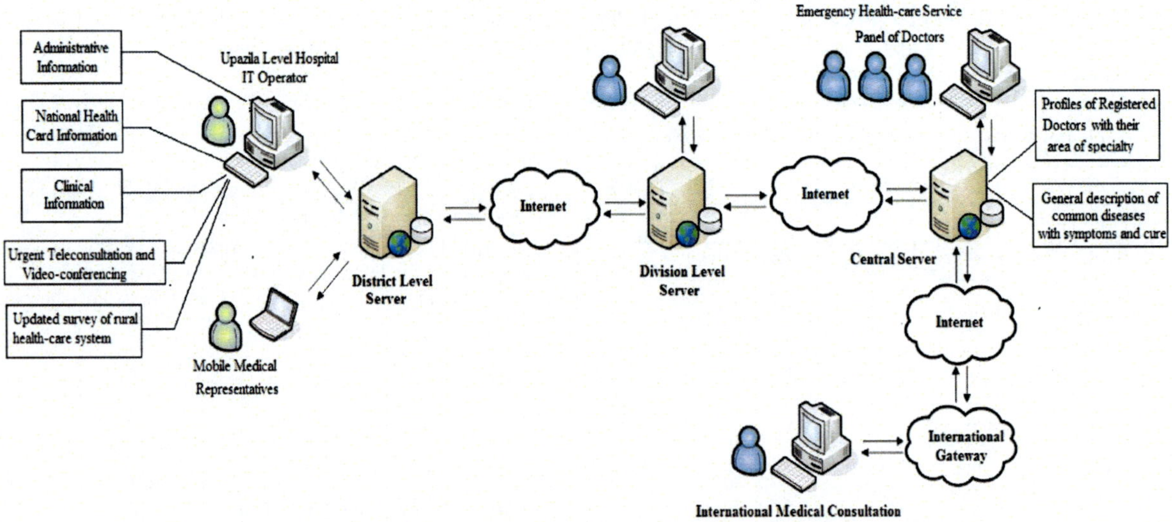

Fig. 2 Storage and Archive system of medical information for the proposed architecture

National ID card consisting of a 13 digit unique number for every single person of the country. The unique number of national ID card can be concatenated with the 'Date of Birth' to form the unique NHC number. The patients will be signed up in the e-health server with their NHC number and the online profile will contain the complete medical records and treatments. On the other hand, the registered doctors all over the country will have their own registration numbers and against these attributes, their profiles will be displayed which will contain their backgrounds, areas of specialty, experiences, and professional affiliations.

Apart from NHC information, the district level server will contain the hospital administrative information, patients' clinical information, and rural periodic health survey information. The central server will contain the NHC information, doctors' profiles, general descriptions of the common and chronic diseases with symptoms, and treatment options and links to medical journals and literatures.

When a patient will be moved to the UHC (tier-3), the local operator will find out the corresponding profile from the server and print a prescription form containing the NHC number and other required information for the patient. Then the patient will be moved to the tier-3 medical officer who will primarily observe the condition of the patient and analyze the information along with the symptoms to prescribe the patient for medicine or further diagnosis. If the physical condition of the patient is complex enough to seek a second opinion or improved consultation from a specialist doctor, the tier-3 medical officer will inform the specialist doctor in tier-2 unit through a mail or any other messaging service with special emergency codes (e.g. 'Red' for critical condition, 'Yellow' for moderate condition, etc.). After scrutinizing medical records and physical condition, the experienced doctor will provide expert opinion. If the tier-2 unit doctor justifies the case as critical, then he/she will suggest the tier-3 medical officer to arrange a video conferencing with the experts in the central tier. In case of rare diseases or epidemic level diseases, international teleconsultation or medical knowledge and information sharing will be accomplished via internet using the international gateway.

V. Features and services offered by the model

The following features and services can be incorporated with the proposed network model.

(1) Mobile medical and diagnostic unit Monthly Diagnosis Day: Mobile medical and diagnostic units consist of expert doctors and portable telemedicine kit will visit each area once in every month. The visiting day will be announced as '*the Monthly Diagnosis Day'* and people will be encouraged to join the program spontaneously with the help of local and national media. The mobile units will use bus or launch where the telemedicine equipments will be deployed. The prime functions will be to provide basic healthcare to the deprived people and collect organized health records from the patients.

(2) Emergency rescue team with telemetry equipment and transportation: The tier-3 medical units will be equipped with 24 hour emergency rescue team and ambulatory units, such as ambulances and specially equipped boats as roads are not available in many of the rural places and waterways are the only means of transportation.

(3) Pre and post disaster medical services: As Bangladesh is a disaster-prone country, pre and post disaster healthcare services are of prime importance. Pre-disaster services include the training and educating general people about disaster management. Post-disaster services include rehabilitation and providing of medicine and emergency treatment for diseases like malaria, diarrhoea, skin diseases, etc. that occur in an epidemic range in the rural areas after the natural disasters.

(4) Immediate Teleconsultation via mobile phone: Leased Helpline: This facility will be provided in the remotest areas where the austere people cannot afford to manage even a mobile phone. The gross networks of the existing mobile phone operators cover proximately the entire area of Bangladesh except the hilly and forest areas. These networks can be utilized here. The village phone program of Grameen Phone is an example of the solution [10]. The people will feel more comfortable to discuss directly with the doctor via cell phone. All of the mobile operators can dedicate some common *leased helpline* for emergency medical calls.

(5) e-Library facility: e-learning concept: E-mail and internet access for urban and rural health-care centers will facilitate the doctors in all tiers to access to the recent medical literatures and update their knowledge.

(6) e-Medical board for critical cases: In case of rare and chronic diseases or accidents occurred at any places throughout the country, special medical board of specialists will be formed in the tier-1 units to provide medical support for the victims.

(7) Special sections for mother, child, and aged people: Mother, child and aged people are vulnerable to diseases and death. Therefore medical units in each tier can contain intensive care sections for these groups of patients.

VI. Conclusion

In this paper, we have proposed a contemporary and pragmatic telemedicine network architecture with some services that are innovative for Bangladesh. The proposed system merges all the active government and non-government organizations and utilizes the existing telecommunication and medical infrastructures and is therefore readily implementable, cost effective and more compact, multifaceted, reliable and faster than the existing telemedicine systems. We have also proposed the organogram of a simple but complete medical data storage system that is compatible with the network architecture. The model will certainly facilitate the promulgation of the modern telemedical application to the remotest regions of Bangladesh.

References

1. Ministry of Health and Family Welfare, http://www.mohfw.gov.bd /health _policy.htm, Last Visited 2008. (2)
2. A. Nessa, M. A. Ameen, S. Ullah, and K. Kwak, "Applicability of Telemedicine in Bangladesh: Current Status and Future Prospects," The International Arab Journal of Information Technology, vol. 7, no. 2, April 2010. (3)
3. M. S. Chowdhury, Md. H. Kabir, K. Ashrafuzzaman, K. S. Kwak, "A Telecommunication Network Architecture for Telemedicine in Bangladesh and Its Applicability," International Journal of Digital Content Technology and its Applications, vol. 3, no. 3, September 2009.(4
4. http://www.reb.gov.bd/about_reb.htm.(5)
5. Encyclopedia of the Nations, http://www.nationsencyclopedia.com/ economies/Asiaand-the-Pacific/Bangladesh.html, Last Visited 2009.
6. http://en.wikipedia.org/wiki/Bangladesh
7. http://en.wikipedia.org/wiki/Upazilas_of_Bangladesh
8. http://www.bdgateway.org/ict_mobile_phone_operator.php?directores=MobilePhoneCom
9. http://sourceforge.net/apps/trac/ipath/
10. http://www.grameenphone.com/index.php?id=79

Masters Program in Biomedical Engineering and Informatics – Research-Based Teaching and Teaching-Based Research

J.J. Struijk, P.B. Elberg, and O.K. Andersen

Aalborg University, Department of Health Science and Technology, Aalborg, Denmark

Abstract— **In typical Master level Biomedical Engineering curricula the focus on research is most clearly visible in the final thesis work, whereas the first year (and a half) focuses on knowledge and engineering skills. The organization of the Aalborg Biomedical Engineering and Informatics program gives a framework for a much more intensive focus on research, throughout the Masters program, an opportunity that is used to involve the students in research activities from the very beginning. With positive results, in terms of study efficiency, grades, employment and research output.**

Keywords— **Education, research, qualification profile.**

I. Introduction

Ever since the start of the five-years curriculum in Biomedical Engineering and Informatics (BMEI) at Aalborg University in the year 2000, the first semester and upwards of the Masters program in the two-cycle curriculum (before 2007 this was the seventh semester in an undivided program) have been strongly focused on research and on the students' participation in research projects with the intent of publishing results of interest in peer-review journals and/or international conferences. In relation to science and research the qualification profile of the education states that the successful graduate

i. has knowledge of scientific communication and … key areas within Biomedical Engineering and Informatics (BMEI), based on the highest international level of research within the areas,
ii. is able to reflect on this knowledge on a scientific basis, and is able to identify scientific problems, either related to clinical research or basic research, within the area,
iii. masters the BMEI's scientific methods and tools,
iv. is able to judge and to choose from the discipline's scientific theories, methods, tools and general skills, and is able, on a scientific basis, to propose new models for analysis and problem solving within BMEI,
v. is able to communicate research based knowledge.

In Denmark the profile [1] is generally subdivided into three levels: 1) knowledge and comprehension, 2) skills, and 3) competences. The list given above (only related to research qualifications) mainly focuses on the first two levels, but aims at research skills that extend beyond the general requirements of research-based MSc programs.

In the following we describe how the requirements i. - v. are fulfilled with respect to the design of the curriculum, the implementation of the curriculum, the requirements for the staff and the research-base and finally, the evaluation and employment of graduates.

II. Implementation

The masters program in BMEI follows the Aalborg problem based learning (PBL) model [2-4], which greatly facilitates student involvement in research projects. The Aalborg PBL model is based on a 50%, nominally, participation in group projects and 50% course work. In the final year of the masters program the project work amounts to 100% of the students' time and the group sizes are limited to one, two or, occasionally, three students. The projects are proposed by a researcher who acts as the group's advisor or, less often, by the students themselves who then find a supervisor among the scientific staff. Under all circumstances the students have to redefine the project and to define an objective for the project.

Throughout the bachelor program the students are required to produce project reports for their semester projects, but the first semester of the master program requires the groups to participate in the so-called semester conference (SEMCON), which includes student groups from other educational programs as well. Therefore, they are asked to write a conference abstract, to give a 10-minutes oral presentation, to present a poster at the conference and to produce a paper in the format of a journal paper. In order to support those efforts, the students follow a mandatory 5-ECTS course in Scientific Methods and Communication with focus on scientific processes, hypothesis-driven research, proper conduct, ethics, and principles for scientific communication.

Also in the projects of the second, third and fourth semesters the students are allowed to use the journal paper as

K. Dremstrup, S. Rees, M.Ø. Jensen (Eds.): 15th NBC on Biomedical Engineering & Medical Physics, IFMBE Proceedings 34, pp. 180 – 182, 2011.
www.springerlink.com

an alternative way of reporting instead of using the traditional project report. The third and the fourth semester projects can be merged to allow for a single "long" master thesis project of 10 months duration.

In order to ensure research based teaching, the teaching staff at the master level consists of active researchers only and their project proposals are within their own research areas, sometimes with collaborators from academia, the health care sector or industry. For the final project, only active senior researchers (associate professors and full professors) are responsible for the supervision of the students. At least two of the projects are examined in the presence of an external, accredited, examiner (censor) in accordance with the Danish national regulations.

III. Results

The BMEI master program allows for 75% project work and 25% course work. Since all four (three in the case of a long final project) projects must have a scientific focus, the students are well trained in research and also contribute to the research of the department.

The course in scientific methods and communication together with the project work and the SEMCON during the first semester fulfill qualifications i. and v. This is further trained in the following semesters. It turns out that approximately half of the students continue in the 2nd to 4th semesters to use the format of the journal paper, together with edited worksheets as appendices, as their way of reporting the work accomplished in the projects. Importantly, this format also gives an excellent basis for the exams, which for competent students often takes the form of a scientific discussion.

At the same time the students are led to think in terms of research, how to design a study, to think of a clear scientific problem, to search and to read the literature as well as to report on a scientific basis. This greatly improves the students' thinking in relation to objective ii.

In general, projects are an excellent way of teaching in depth knowledge and skills, if only because the students have to participate actively and during a prolonged period of time. With the focus on scientific questions in those projects, objectives iii. and iv. are well met.

Student projects also yield a some research production. In contrast to projects at the bachelor level, which often are design projects, the master level projects are based on scientific problems, and therefore they often give scientific answers as is exemplified by an average of almost 3 journal papers or conference papers with students as (co)authors each year.

Group based projects that are perceived by the students as meaningful have a strong tendency to improve the students' involvement in their education, thus also improving the efficiency (relative number of ECTS awarded per student per year) of the program. Among the Danish universities, Aalborg University has the highest student efficiency, and also the BMEI masters program has a high efficiency of 90% on average.

More than 90% of the projects are indeed within the research areas of the department of Health Science and Technology, with the only exceptions occurring at the third semester, where the students are allowed to spend one semester abroad or to perform a research project in a company. Approximately 70% of the students take this opportunity, most of them to perform a full semester project in foreign academic institutions where already collaborations exist with the department's own researchers. So, also in this case the students typically learn about the department's own research areas, although sometimes projects are chosen that are less closely related to the home department.

The grades for the final project, based on the exam with an external examiner are relatively high. The grades for projects are higher with external examination in comparison with projects where the examination is internal only, probably, because the students from Aalborg University are familiar with project work and with project exams, and therefore perform relatively well in this type of educational element.

Since 2005 an average of 29 Master students have graduated each year (171 in total from 2005 to 2010). With the exception of the year 2009, where it was difficult for engineering graduates to find jobs, the employment rate is higher than 70% within 3 months from graduation. Within 6 months employment is near 100% [5]. Approximately 80% of the students find their first job within biomedical engineering areas, be it academic, within the health care sector or in industry.

IV. Conclusions

The Master of Science program in BMEI at Aalborg University graduates qualified engineers who soon find jobs within academia or within industry. The graduates are brought up with interdisciplinary scientific thinking from day one and they master communication with both technically oriented and medical oriented professionals.

The strong focus on project work is on the expense of many hours of course work. One implication may be a lesser depth in theoretical knowledge, but on the other hand substantial experience with design, implementation, and testing solutions and a well-trained ability of independent professional development is obtained.

The format of the journal paper is an excellent way of reporting if the projects are based on real research. The strict framework of Introduction, Methods, Results and Discussion structures the thinking of the students and encourages them to search the literature, to reflect about their methods, to be clear about their results and to put them into perspective. Moreover, using the project work together with real research assignments makes it a small step from research-based teaching to teaching-based research.

REFERENCES

1. Danish qualifications framework for higher education http://en.iu.dk/transparency/qualifications-frameworks/danish-qualifications-framework-for-higher-education
2. Kjærsdam F, Enemark S. (1994), "The Aalborg Experiment", Aalborg University Press.
3. Kolmos A, Fink FK, Krogh L, (eds), (2004), "The Aalborg PBL Model", Aalborg University Press.
4. Barge S (2010), "Principles of Problem and Project Based Learning – The Aalborg PBL Model" Aalborg University.
5. Karrierecenteret, (2009), 'Kandidat- og aftagerundersøgelsen 2009 delrapport for sundhed og teknologi (SSN)'

Author: Johannes Jan Struijk
Institute: Aalborg University
Street: Fredrik Bajersvej 7D1
City: Aalborg
Country: Denmark
Email: jjs@hst.aau.dk

Real-Time Photoplethysmography Imaging System

U. Rubins, V. Upmalis, O. Rubenis, D. Jakovels, and J. Spigulis

[1] Institute of Atomic Physics and Spectroscopy, University of Latvia, Riga, Latvia

Abstract— **Real-time non-contact photoplethysmography imaging (PPGI) system for high-resolution blood perfusion mapping in human skin has been proposed. The PPGI system comprises of LED lamp, webcam and computer with video processing software. The purpose of this study is to evaluate the reliability of the PPGI system when measuring blood perfusion. The validation study of PPGI and laser-Doppler perfusion imager (LDPI) was performed during local warming of palm skin. Results showed that the amplitude of PPGI increases immediately after warming and well correlated with the mean LDPI amplitude (R=0.92+-0.03, p<0.0001). We found that PPGI technique has good potential for non-contact monitoring of blood perfusion changes.**

Keywords— **Imaging photoplethysmography, blood perfusion, blood pulsations, optophysiological imaging, non-contact technique.**

I. INTRODUCTION

Assessment of blood perfusion can provide information about cardiovascular processes, such as the pulse amplitude, oxygen saturation, and heart and respiration rate [1]. Photoplethysmography imaging (PPGI) is a non-contact optical diagnostic technique for detection of blood perfusion in skin using backscattered radiation [2-11]. In contrast to conventional photoplethysmography the PPGI technique is suitable in situations when free movement is required or skin is seriously damaged. Similarly to Laser Doppler perfusion imaging (LDPI) technique, PPGI is able to highlight blood perfusion in every point of the skin surface in real-time.

Evaluation of high resolution maps of blood perfusion takes a lot of computational power. This is the main problem for real-time imaging of blood perfusion. The problem could be solved by reducing the image resolution [6-8] or reducing the video frame rate, which is not acceptable in cases of fast blood perfusion changes over large areas of skin.

In this work, a system for non-contact real-time monitoring of skin blood pulsations is presented. Previously we showed that the amplitude of PPGI correlates with skin temperature under local anesthetic input [11]. The current study confirmed that the amplitude of blood pulsations is in good correlation with the mean value of blood perfusion obtained by LDPI.

II. METHODS

A. The principle of imaging PPG

The basic concept of the PPGI technique is shown in Fig.1. Optical radiation after penetration into skin is partially absorbed in tissue and modulated by skin blood pulsations. The backscattered radiation containing important information about cardiovascular processes can be detected by video recording device as extremely weak light pulsations. Light intensity varies from frame to frame due to blood pulsations. Every frame of video can be represented as 2-D intensity distribution of red/green/blue (RGB) pixels. Blood hemoglobin is strongly absorbed in the green spectral band, so the green component of RGB contains valuable information on the skin blood volume changes [8].

B. The PPGI system

The developed PPGI system consists of light source (4W bulb of 80 white LEDs), webcam (Logitech QuickCam) and portable computer (dual-core, 2.5GHz, 2GB RAM) with custom designed Matlab software. The computer, webcam, lamp, portable tripod and power supply can be packed in a portable case. During the measurements the lamp and webcam were mounted on tripod, and the webcam was connected to computer by USB interface. The system is able to acquire video frames with resolution of 640x480 with rate 15 frames/second and to calculate blood perfusion changes in real time.

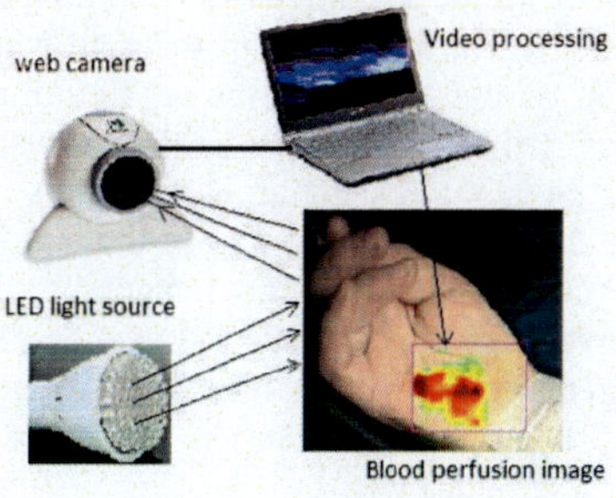

Fig. 1 The principle of imaging photoplethysmography

K. Dremstrup, S. Rees, M.Ø. Jensen (Eds.): 15th NBC on Biomedical Engineering & Medical Physics, IFMBE Proceedings 34, pp. 183–186, 2011.
www.springerlink.com

C. The algorithm of video processing

The algorithm of PPGI processing comprises the following steps (Fig. 2):

- Acquiring video frame from the webcam and conversion to single precision floating point pixel values (640x480x3, values in range 0-255);
- Extracting the region of interest (RoI) of video frame;
- Extracting the green channel of video frame (in Truecolor Red-Green-Blue space);
- Motion compensation for video stabilization;
- Blurring of frame by applying of spatial Gaussian filter with kernel size of 30x30 pixels;
- Thresholding of object illumination by replacing the pixels with zero values where the areas of object are too dark and too bright;
- Evaluation of the slow varying intensity (DC) in every pixel of frame;
- Evaluation of the fast pulsatile intensity (AC) in every pixel of frame;
- Buffering of AC and evaluation of 2-D distribution of PPG amplitudes;
- Evaluating of averaged AC and bandpass filtering by 5th order Butterworth filter (frequency range 55-90 beats per minute);
- Buffering of averaged AC, calculation of amplitude and frequency;
- Displaying of PPGI map, RoI averaged PPG amplitude and pulse rate in real-time window;
- Repeating of all steps from start of the algorithm.

The algorithm can be performed in real-time with frame rate 15 Hz with the maximum resolution of RoI area 50 000 pixels. Spatial filtering (blurring) increases the signal/noise ratio of PPG signal in every pixel of PPGI [7]. The motion compensation algorithm helps to avoid slow movements of skin [12]. The skin-reflected light thresholding helps to avoid background from the scene, high contrast borders and areas with under/over-exposure.

The interface of PPGI software is shown in Fig. 3. The current video frame and the PPGI map are shown in the left part of window. In the right part there are video parameters: the amount of free memory, the resolution and frame rate, the size of RoI and the size of PPGI areas, as well as the evaluated physiological parameters – pulse amplitude and pulse rate. In the lower right part of the window four graphs show the pulse rate, PPG signal shape and PPG amplitude changes over time.

During the video acquisition, the software saves pulse rate and PPG amplitude data to hard disk, which can be useful for later analysis.

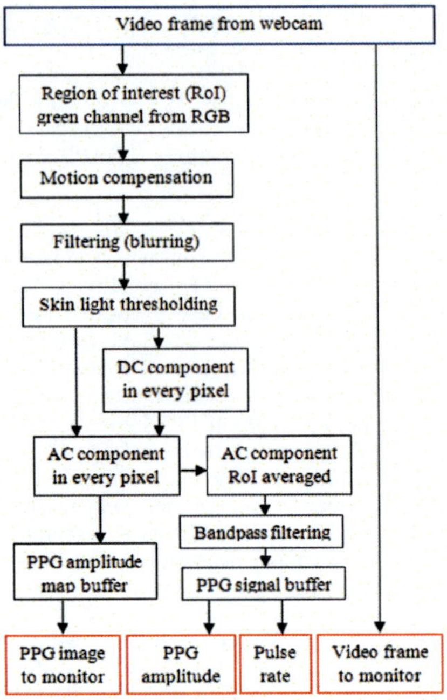

Fig. 2 The algorithm block set

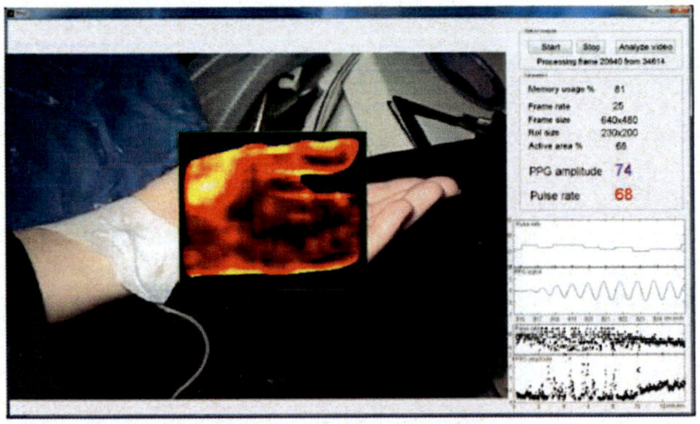

Fig. 3 The interface of PPG imaging software

D. The experimental setup

The PPGI system was tested in clinical conditions. 12 healthy subjects (2 males, 10 females) were involved in this study. All subjects were asked to lie in supine position. The thermostat cylinder filled with water was placed on the skin of the right palm dorsal side. The temperature in thermostat was 23°C first 3 minutes, and then the temperature was increased to 35°C and maintained constant for the next 20 minutes. Light source and camera were placed 40 cm apart from skin surface. The camera video settings (white balance, brightness, gain, exposure) were set in manual mode.

For achieving full dynamic range the exposure was set to 80% of maximum of luminance from RoI area. The PPGI system was taking video simultaneously with the LDPI measurements that were taken with sampling rate of 1 frame in 30 seconds and 30 second pauses (reserved for the analysis of PPG dynamics). The mean value of PPG and LDPI recorded in the first 3 minutes were assumed as the reference.

III. Results

Fig. 4 shows the video frame before and 5 minutes after the warming of skin surface. Warming of skin provokes the increase of blood supply into the skin tissue, and the skin color becomes reddish. This indicates to increased amplitude of blood pulsations. Fig. 5 shows PPG image maps before and after the warming. It is seen that the amplitude of PPG increases over the skin surface after skin warming.

Fig. 6 shows the dynamics of PPG amplitude, averaged across the RoI, and the averaged LDPI values 3 minutes before and during the warming of the skin of 24-years old female subject. In result of the analysis, significant correlation between the amplitudes of PPG and the LDPI values during the thermal provocation of skin has been found (statistical values R=0.97, p<0.0001).

Fig. 7 shows R-values of correlation between the averaged PPG and averaged LDPI in group of 12 subjects. Results showed statistically significant correlation between the amplitude of PPG and LDPI signals (statistical values R=0.92+-0.03, p<0.0001).

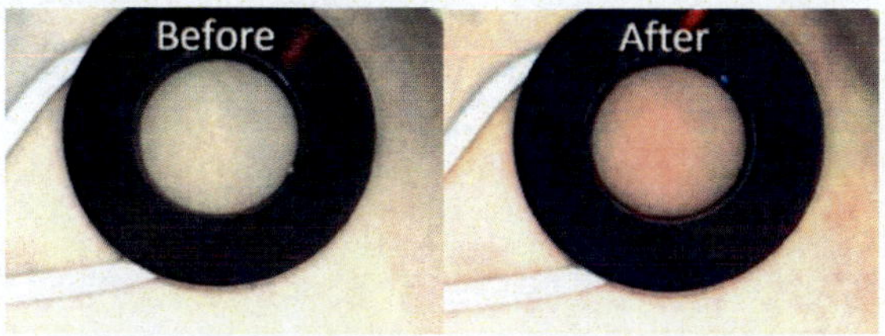

Fig. 4 The hand skin before and after the 5 minutes warming

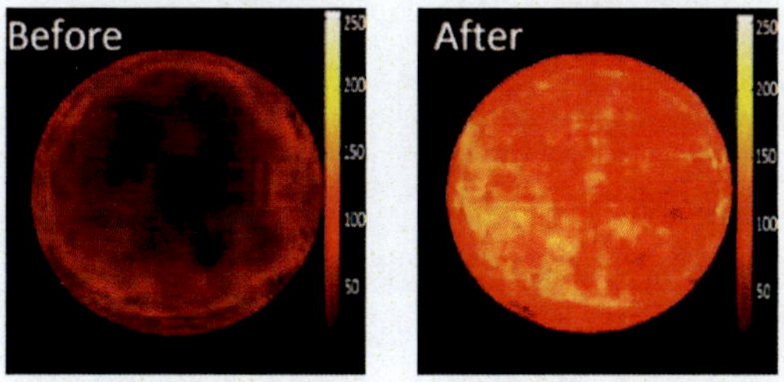

Fig. 5 The PPGI map of hand skin surface inside thermostat before and after warming 5 minutes later

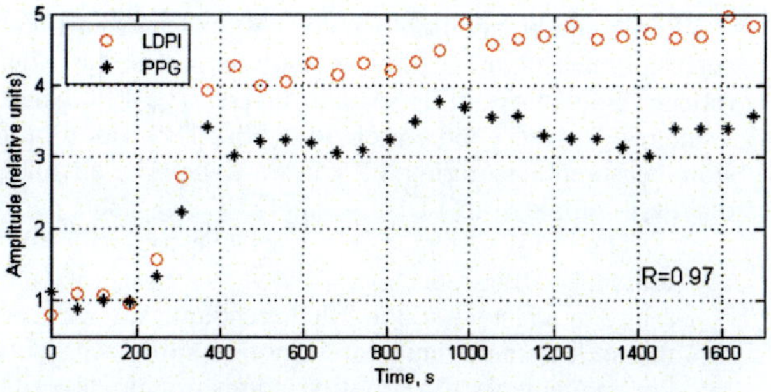

Fig. 6 Amplitude changes of RoI-averaged PPG and RoI-averaged LDPI during the skin warming (24-years old female). Mean value of first three measurements assumed as the reference value.

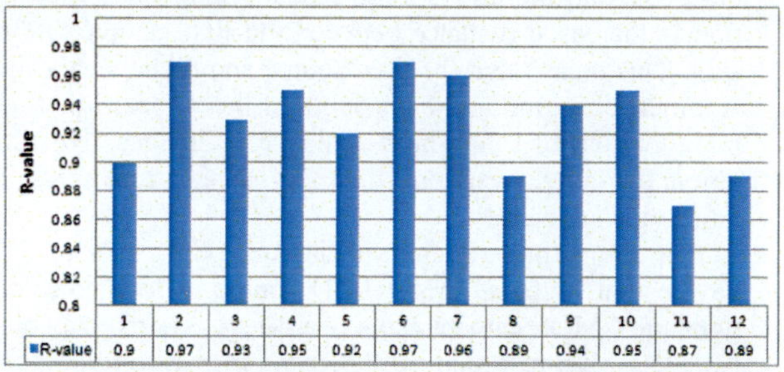

Fig. 7 The R-values of correlation between RoI-averaged PPG and RoI-averaged LDPI in group of 12 subjects. Mean value of R=0.92+-0.03

IV. Discussion

The main advantages of PPGI technique are non-invasiveness and simplicity. In contrast to the traditional contact photoplethysmography used for continuous monitoring of temporal changes in dermal perfusion at a single site [1], the PPGI technique allows to get information about skin perfusion and blood pulsatile processes over a certain skin area.

Our results showed that PPG signals could be remotely accessed through the imaging PPG system in real-time. It was possible to get high resolution PPGI maps by using a consumer type webcam and LED light source. In previous studies [6-8] the resolution of video should be downsampled (averaging pixels in small groups) to achieve better signal-to-noise ratio of PPGI. Our pixel averaging method used spatial filtering that provides PPG images of improved quality and higher resolution.

The calculation of high resolution PPGI takes a lot of computational power. We developed optimized real-time video processing algorithm that enables calculation of PPG amplitudes of a limited 50000 pixels/ frame. Such resolution is large enough for real-time monitoring of skin blood perfusion in the micro-vascular bed.

PPGI could be applicable in clinical diagnostics of blood perfusion measurements. The pulsatile component of signal reflects blood flow in upper skin layers. The PPGI measurements showed good correlation with LDPI blood perfusion measurements during the skin warming. The advantage of PPGI compared to LDPI is simplicity and faster performance. The PPGI system is less expensive compared to the conventional LDPI systems.

Despite to advantages, the PPGI technique still has some limitations. The main limitation is motion artifacts produced by skin movements. The motion compensation algorithm and skin illumination thresholding helps to reduce these artifacts but they cannot be fully avoided. The second limitation is non-homogeneity of illumination. Skin surface is not even, and the only way to reach homogenous illumination is increased distance between the light source and the skin. This means that the light source should be very bright. Another limitation is the bit-depth of the webcam that does not allow use of sharp, high quality PPG images. This limitation could be overcome by using a high resolution/low noise video camera.

Our future plans include designing of a portable PPGI device that includes bright LED source, a low noise/high dynamic CMOS camera and a board-level system capable to calculate high resolution PPGI maps in real-time. A future study with more subjects and clinical states would better characterize the performance of the PPGI blood perfusion mapping.

V. CONCLUSIONS

Real-time PPG imaging system consisting of webcam, white LED bulb and computer with original video processing software was developed in this study. The PPGI system was able to detect skin blood pulsations and visualize the blood perfusion with high resolution. The system was verified in comparison with LDPI measurements, and the results showed high correlation between the PPGI and LDPI. The developed PPGI system could be useful for real-time non-contact monitoring of skin blood perfusion in clinical conditions.

ACKNOWLEDGMENT

Financial support from European Social fund, project number 2009/0211/1DP/1.1.1.2.0/09/APIA/VIAA/077, is highly appreciated. The authors are grateful for the clinical support provided by Institute of Experimental and Clinical Medicine, University of Latvia.

REFERENCES

1. Allen, J, (2007) Photoplethysmography and its application in clinical physiological measurement. Physiol. Meas., 28, R1-39
2. Wu T, Blazek V, Schmitt H.J (2000) Photoplethysmography imaging: a new noninvasive and noncontact method for mapping of the dermal perfusion changes. Proc. SPIE, vol. 4163, pp. 62-70
3. Wu T (2003) PPGI: New Development in Noninvasive and Contactless Diagnosis of Dermal Perfusion Using Near InfraRed Light. J. GCPD e.V., vol. 7(1), pp. 17-24
4. Humphreys K, Markham C, Ward T.E (2005) A CMOS camera-based system for clinical photoplethysmographic applications. Proc. SPIE, vol. 5823, pp. 88-95
5. Wieringa F.P, Mastik F, van der Steen A.F (2005) Contactless multiple wavelength photoplethysmographic imaging: a first step toward "SpO2 camera" technology. Ann Biomed Eng, vol.33(8), pp. 1034-41
6. Zheng J, Hu S (2007) The preliminary investigation of imaging photoplethysmographic system. J.Phys.: Conf. S. vol. 85, pp. 012031
7. Zheng J, Hu S et al (2008) Remote simultaneous dual wavelength imaging photoplethysmography: a further step towards 3-D mapping of skin blood microcirculation, Proc. SPIE 6850, pp 68500S
8. Verkruysse W, Svaasand L.O, Nelson J.S (2008) Remote plethysmographic imaging using ambient light. Opt. Express, vol. 16(26), pp. 21434–21445
9. Erts R, Rubins U, Spigulis J (2009) Monitoring of blood pulsation using non-contact technique. Proc. IFMBE 25/VII, pp. 754–756
10. Rubins U, Erts R, Nikiforovs V (2010) The blood perfusion mapping in the human skin by photoplethysmography imaging. Proc. IFMBE, vol. 29, pp. 304-306
11. Rubins U, Miscuks A et al (2010) The analysis of blood flow changes under local anesthetic input using non-contact technique. IEEE conf. proc. Vol 2, pp. 601–604, 2010
12. Poh, M.Z, McDuff, D.J, Picard, R.W (2010) Non-contact, automated cardiac pulse measurements using video imaging and blind source separation. Opt. Express, 18, pp. 10762-10774

Author: Uldis Rubins
Institute: Institute of Atomic physics and spectroscopy
Street: Raina Bulv. 19
City: Riga
Country: Latvia
Email: uldisrubins@yahoo.com

Study of the Muscular Force/HOS Parameters Relationship from the Surface Electromyogram

F. Ayachi[1], S. Boudaoud[1], J.F. Grosset[2], and C.Marque[1]

[1] University of Technology of Compiegne, BMBI-CNRS UMR 6600, Compiegne, France
[2] University Paris 13, Departement of sport science, Bobigny, France

Abstract— **The aim of the present study is to investigate a possible relationship between High Order Statistic (HOS) parameters and muscle force. In fact, it is guessed that Motor Unit (MU) recruitment during contraction has an influence on surface EMG (sEMG) amplitude distribution shape. For this purpose, skewness and kurtosis are used to monitor variation of monopolar sEMG data according to contraction level. First, a simulation was performed to evaluate the sensitivity of both proposed parameters to physiological and instrumental parameters. Then, 3 healthy young males took part to an experimental protocol on the biceps brachii muscle. The sEMG and force data were recorded and analyzed for different voluntary contraction levels. According to the results obtained, a relationship between HOS parameters and muscle force appears to exist. However, HOS parameters are sensitive to the tested parameters.**

Keywords — **Motor unit recruitment; Surface EMG; Twitch/Force; HOS.**

I. INTRODUCTION

Extracting information about Motor Unit (MU) recruitment schemes during muscle contraction from the analysis of surface EMG data is a challenging task [1], [2]. In fact, temporal and spatial recruitment of the MUs, driven by the Central Nervous System (CNS), are fundamental mechanisms for generating sEMG signal and force by the studied muscle [1], [2]. The evaluation of the relationship between neural command and sEMG/Force data is essential for the comprehension of complex processes underlying muscle contraction [1],[2]. Several studies, based on simulation and human experiments demonstrate this relationship [1], [2]. From the sEMG signal, typical descriptors usually investigated are amplitude and frequency information [1], [2]. However, it has been reported a poor discriminative ability of these descriptors in the evaluation of MUs recruitment scheme [1]. Furthermore, these descriptors are sensitive to clinical variability (anatomical, instrumentation, etc...) [1]. In the present study, we propose to investigate the relationship between Force and High Order Statistics (HOS) parameters extracted from monopolar sEMG signal. Indeed, HOS parameters are sensitive to shape variation of the sEMG amplitude distribution. Moreover, these shape variations are supposed to occur following temporal and spatial MUs recruitment. A recent study [3], using a shape modeling formalism based on simulation, has shown the ability of this formalism to separate several contraction force levels, according to sEMG amplitude distribution shape. Following this idea, two parameters, skewness and kurtosis, will be tested in the present study by using simulation as well as human experimentation. Previous studies [4], [5] have already used skewness to detect MU firing synchrony during fatigue.

For this purpose, a first study based on simulation is reported using a sEMG/force model. This simulation focused on the evaluation of the force/HOS parameters relationship for three contraction levels. Influence of the MUs recruitment strategy, fat thickness, and electrode distance from the innervations zone, is investigated. The second step is to perform an experimental study on three young healthy subjects to extract HOS parameters according to the same three contraction levels of the biceps brachii muscle. Finally, the obtained results are discussed.

II. MATERIAL AND METHODS

A. sEMG/Force model

The simulations were based on a commonly used recruitment model of a population of MUs [2]. In this model, each MU activation is tuned by a motoneuron firing rate which increases linearly with the force level, from recruitment threshold toward the peak firing rate level as in [6]. In fact, each MU has its own firing/force slope. To provide a better physiological realism, different firing rate ranges were imposed according to MU type (see Table. 1). The InterPulse Interval (IPI) probability is supposed to have normal distribution (SD=15%) [1]. A layered volume conductor model proposed by [7] is considered. It includes muscle, fat and skin tissues, electrodes (shape and configuration), finite length fibers, etc... Mathematically, it consists in applying two-dimensional filters to the input current density function, derived from the intracellular action potential [7]. The muscle is assumed to have a circular section (20 mm diameter). The mean length of the

K. Dremstrup, S. Rees, M.Ø. Jensen (Eds.): 15th NBC on Biomedical Engineering & Medical Physics, IFMBE Proceedings 34, pp. 187–190, 2011.
www.springerlink.com

muscle fibers is 120 mm (L1=55mm, L2=65mm), as in human biceps brachii [5]. Ends of the fibers were scattered uniformly within a 18 mm range. The overall width of the innervation region was 30 mm [5]. For each individual MU, the end plate region width was set to 10 mm as in [5]. Other anatomical properties of MU and fibers are shown in Table.1. Signal was recorded from one monopolar circular electrode (10 mm diameter) located above the longer semi-length, 20 mm away from the middle of the innervation zone. Calculation of the conduction Velocity (CV) for each fiber belonging to a given MU was performed following the same equation (linking CV to the fiber diameter) and using a reference CV=4 m/s as reported in [5].

Table 1 Main MU and anatomical parameters (mean±maximal deviation)

MU Type	Number of MUs	MU diameter (mm)	Number of fibers	Fiber diameter (μm)	Firing Rate (Hz) Fmin - Fmax
S	50	5 ± 1	100± 10	45±5	8 - 20
FR	25	5.5± 1	150± 15	50±5	10 - 25
FIN	25	6± 1	200± 20	55±5	15 - 30
FF	50	6.5± 1	250± 25	60±5	20 - 35

When an action potential arrives at muscle fibers, the contractile response appears and is named "twitch." It is the mechanical response of the fiber. The MU twitch (the mechanical response of a MU to a single neural impulse) has a specific bell shaped that is dependent on the MU type: fast twitch fatigable (FF), fast-twitch fatigue resistant (FR), intermediate (FIN), slow-twitch (S), and is analytically described in [8]. In all simulations, the total muscle forces are obtained from the summation of individual MU forces computed as in [8] and depending on the recruitment strategy (see Figure.1).

B. HOS analysis

In this work the HOS descriptors (Skewness, Kurtosis) were used to evaluate the shape variation of the sEMG amplitude distribution according to contraction level. The SEMG signal can be considered as a stochastic process whose amplitude variation is related essentially to summation of MUAPs (MU Action Potentials) of a large number of active MUs according to the level of muscle activation. The activation of the corresponding MUs follows stochastic discrete processes [4]. It is guessed that sEMG amplitude distribution shape is strongly dependent on the activation of the MUs during contraction [8]. The asymmetry of the distribution can be described by using the skewness statistic (normalized 3rd order central moment), which is defined for a random variable X as:

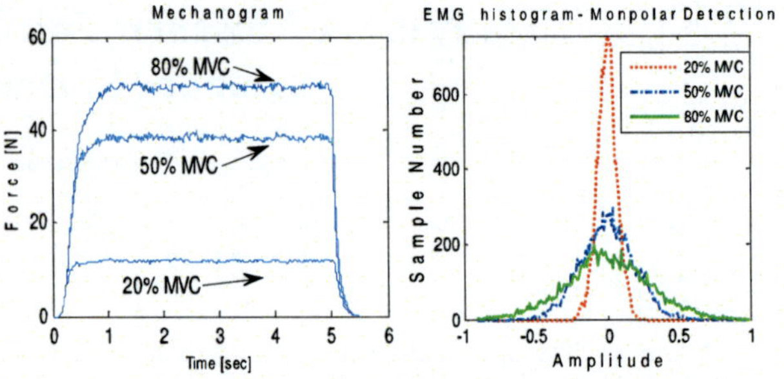

Fig.1 Simulated Force/SEMG Amplitude histograms for 3 contraction levels

$$\gamma_1 = \frac{E[(X-\mu)^3]}{\sigma^3} \quad (1)$$

A normal distribution has a skewness equal to 0. A positive skewness corresponds to an important right tail. A negative value is due to an important left tail of the distribution. Elsewhere, another parameter is used for the quantification: the kurtosis of the distribution of a random variable X, which is defined as:

$$\gamma_2 = \frac{E[(X-\mu)^4]}{\sigma^4} \quad (2)$$

Kurtosis statistic (normalized 4th order central moment) is linked to the degree of peakedness of a distribution. We substract 3 (normal distribution value) from the computed value, so that the normal distribution has a kurtosis of 0. However, a positive value corresponds to a peaked distribution and a negative one to a flattened distribution.

C. Experimental Protocol

Three healthy males participated to this study (age=27±5.0 year, height=173.6±8 cm, weight=81±12 kg). After preparing the skin of the subject, the sEMG activity was recorded (sampling frequency 3 kHz) by TeleMyo 2400 G2 PC interface of Noraxon telemetry system; with monopolar electrode (10 mm diameter) placed along the longitudinal axis of the Biceps Brachii of their right arm, at a distance of 1cm away from the middle of the belly toward the elbow. A reference electrode was placed on the elbow. The sEMG activity and the associated force developed were simultaneously measured for the different contraction levels investigated. Testing was performed on the right arm. The elbow was maintained at 90 degrees with the palm up. A maximal isometric voluntary contraction (MVC) of the Biceps Brachii was developed in isometric conditions as fast as possible and maintained for three seconds. Two MVCs were performed with two minutes rest between trials. The best attempt was recorded as the participant's

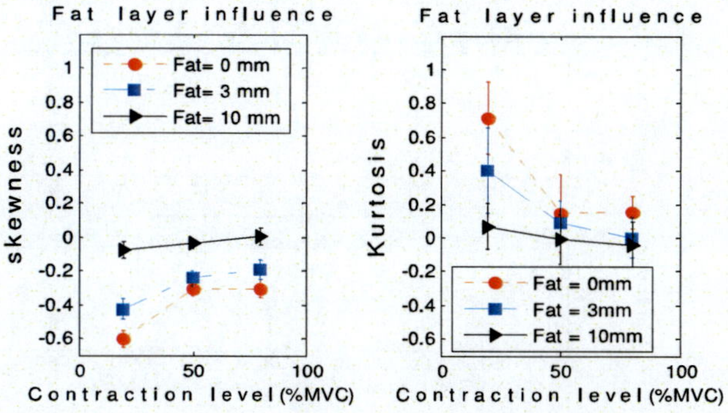

Fig.2 Fat layer influence with (R2) Recruitment strategy

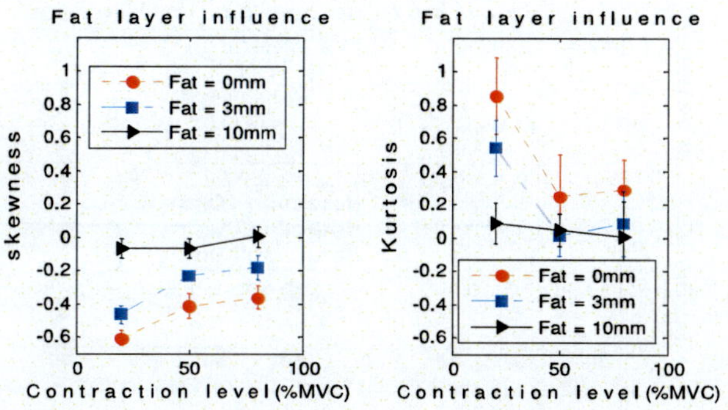

Fig.3 Fat layer influence with (R1) Recruitment strategy

MVC score for the day. Then subject were instructed to maintain a torque corresponding to 20, 50 and 80% MVC for five seconds with a visual feedback of their voluntary torque developed. A minimum of 90 sec of rest was given between contractions. Two trials were done for each contraction level. To avoid changes due to circadian rhythm, for each subject 3 testing sessions were conducted at the same hour for 3 different days, with 1 to 3 days rest in between testing sessions, to assess reproducibility of the used method.

D. Simulation protocol

To evaluate the sensitivity of the proposed approach to clinical variability, different simulation sets (according to the same random location of different MUs in the muscle) were generated (see Figure 1), by using the model described in Section II.A, with a Gaussian noise (20 dB, SNR). We used each combination of the following parameters:

(1) The same muscle anatomy with three different fat layer dimensions (0, 3, and 10 mm) using R2 recruitment strategy (see below).

(2) Two different recruitment strategies, (R1) as depicted in [2]. The MU increased its firing rate with force in a linear way, with the same slope for the whole MUs, min and max firing rate for each MU are used as Fmin=8 Hz, Fmax=35 Hz. The (R2) strategy is an improvement of [6] with different MU firing/force slope according to Table.1 (see Section II.A).

(3) Three electrode positions (10, 20, 40mm) from the innervation zone, with fat thickness equals to 3 and 10 mm. For each simulation, 5 contractions of 5 seconds for each force level (20, 50, 80 %MVC) are modeled.

The average skewness and kurtosis were calculated by segmenting the sEMG signal into 4 epochs of 1 sec. duration.

III. RESULTS

A. Simulation study

On Figure 2 and 3, we can observe clear tendencies concerning the HOS parameters with force level: an increase of the skewness from negative value toward zero and a decrease of the kurtosis from a positive value toward zero. Indeed, whatever the recruitment strategy considered (R1 or R2), both parameters are sensitive to fat layer with less dynamic according to thickness increase. Concerning force relationship, this can be explained by two phenomena occurring during contraction. The first phenomenon is more visible at low contraction level and consists on the influence of the MUAP amplitude distribution on the overall signal. In fact, monopolar MUAP recording is asymmetric with high negative peaks. At low contraction level, few MUs are recruited and MU superposition is not important. For this reason, skewness is negative. In the other side, the kurtosis is positive due to an important number of small values in the signal. The second phenomenon is the increase in MU superposition with force level. In fact, and following the central limit theorem, superposition of random variables tends to normal distribution. For this reason, an increase and decrease of the skewness and kurtosis respectively, toward zero value (normal distribution). Both recruitment strategies (R1, R2) give almost the same tendencies for skewness and kurtosis. Concerning the strong influence of the fat layer thickness on the HOS parameters, this is mainly due to the spatial filtering induced by increasing thickness. For a large thickness of the fat layer, the HOS parameters are less sensitive to different levels of contraction especially the skewness.

Concerning the electrode position, for a fat layer of 3mm, the influence on the HOS parameter tendencies is small (see Figure 4) for the three positions (10, 20, 40 mm). For a larger thickness (10 mm), the influence is greater especially for the distance of 40 mm as presented on Figure 5.

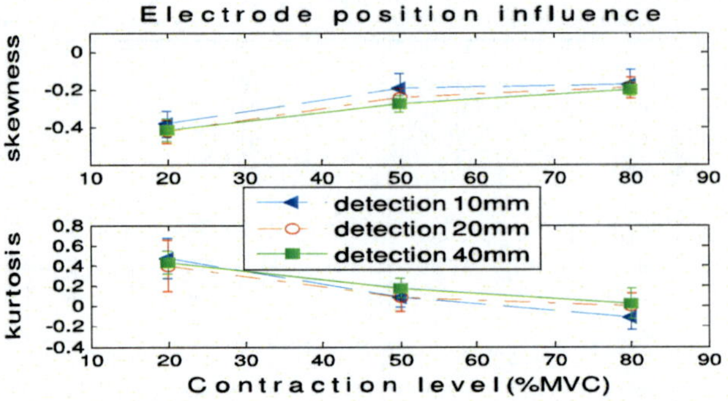

Fig.4 influence of detection position with fat layer = 3mm

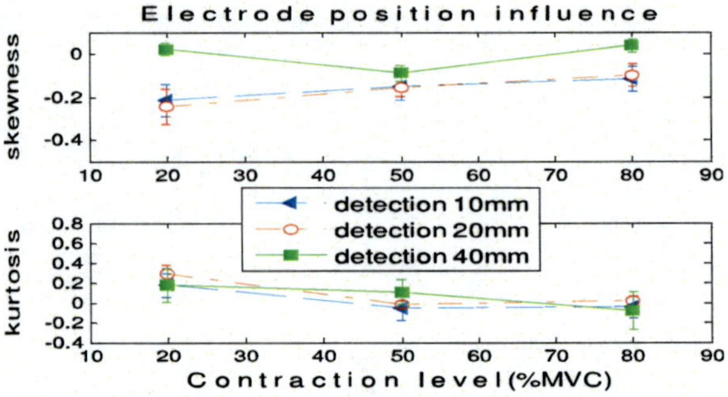

Fig.5 influence of detection position with fat layer = 10mm

B. Experimental study

In experimentation, we can observe interesting results on Figure 6. As for simulation, we show an increase and decrease of the skewness and kurtosis respectively. But, the tendencies are not so clear as in simulated results. In fact, an increase toward positive skewness values is observed. This is mainly due to errors in the placement of the electrode according to the innervations zone, and to the influence of the fat thickness and muscle anatomy. The subcutaneous fat tissue influence is clearly shown in subject 1. In fact, his BMI was 34 instead of 23 for subjects 2 and 3. For subject 1, it becomes difficult to extract any tendency for skewness and the range of variation is small. In the other side, the tendencies are more pronounced and the variation range is larger for subjects 2 and 3.

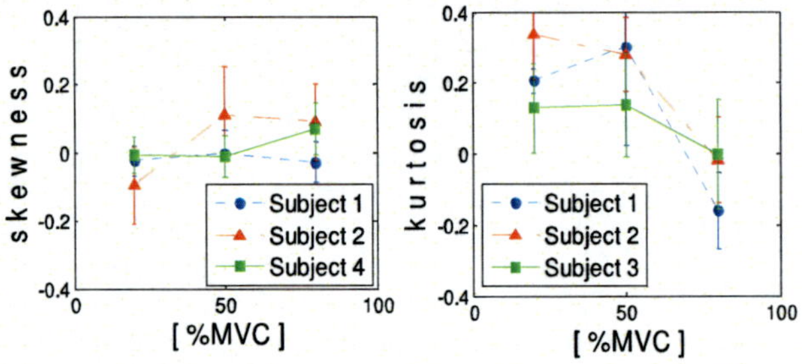

Fig. 6 experimental results for 3 subjects.

IV. CONCLUSIONS

The present study investigated High Order Statistic (HOS) parameters to evaluate muscle force. HOS parameters (skewness and kurtosis) were used to study possible relation between variations of the sEMG amplitude distribution shape, due to MU recruitment, and isometric contraction levels. First, a study by simulation has been proposed to evaluate HOS parameter tendencies and robustness against clinical and instrumental variability. It appears that a relationship with force seems to exist. But it shows a great sensitivity to fat thickness and electrode position. For experimental results, the same tendency seems to occur in the evolution of HOS parameters, but with less clarity. This is mainly due to errors in electrode placement according to innervation zone, and also to the presence of muscle anatomy variability. Finally, this study is part of a preliminary study and further investigations are needed to precisely demonstrate the capability of HOS parameters to track force variations despite the presence of several sources of variability.

REFERENCES

[1] D. Farina, R. Merletti, and R.M. Enoka. The extraction of neural strategies from the surface emg. J Appl Physiol, 96:1486–1495, 2004.
[2] A. J. Fuglevand, D. A. Winter, A. E. Patla, “Models of recruitment and rate coding organization in motor-unit pools,” J Neurophysiol, vol. 70, pp. 2470-2488, 1993.
[3] S. Boudaoud, F.Ayachi, C.Marque. Shape analysis and clustering of surface EMG data. 32th IEEE EMBS .Int. Conf., pages 4703 - 4706, August 2010.
[4] A. Holtermann, C. Grolund, J.S. Karlsson, and K. Roeleveld. Motor unit synchronization during fatigue: Described with a novel semg method based on large motor unit samples. J Electromyo and Kinesiol, 19:232–241, 2009.
[5] T.I. Arabadzhiev, V.G. Dimitrov, N.A. Dimitrova, and G.V. Dimitrov. Influence of motor unit synchronization on amplitude characteristics of surface and intramuscularly recorded emg signals. Eur J Appl Physiol, 108:227–237, 2010., **DOI 10.1007/s00421-009-1206-3.**
[6] H. Cao, F. Marin, S. Boudaoud, C. Marque, “Muscle force simulation by using two motor-unit recruitment strategies,” Computer Methods in Biomechanics and Biomedical Engineering, vol. 11 (Supplement 1), pp. 51-52, 2008.
[7] Farina D, Merletti R. A novel approach for precise simulation of the EMG signal detected by surface electrodes. IEEE Trans Biomed Eng 2001;48:637–46.
[8]R. Raikova, H. Aladjov, The infuence of the way the muscle force is modeled on the predicted results obtained by solving indeterminate problems for a fast elbow flexion, Comput. Methods Biomech. Biomed. Eng. 6 (2003) 181–196.

Fuzzy Inference System for Analog Joystick Emulation with an Inductive Tongue-Computer Interface

H.A. Caltenco, E.R. Lontis, and L.N.S. Andreasen Struijk

Center for Sensory-Motor Interaction, Dept. of Health Science and Technology, Aalborg University, Denmark

Abstract— **This paper describes the development of a fuzzy inference system (FIS) for emulating an analog joystick using an inductive tongue-computer interface. The principle of operation of the interface and the inductive sensors signals are described. The FIS receives sensor signals and output the Cartesian position of the virtual joystick, which can be used to control the mouse pointer in a personal computer, wheelchairs or other joystick enabled applications at varying magnitude and directions proportional to the tongue position over the palatal plate. This provides a significant advantage to individuals with tetraplegia using this computer interface.**

Keywords— **Fuzzy inference system, fuzzy control, machine learning, computer-interfacing, assistive device.**

I. INTRODUCTION

Individuals with spinal cord injury and similar disabilities that lead to tetraplegia, may not have the ability to efficiently control standard computer input devices, such as a keyboard or a mouse. These individuals may need a computer interface with a minimum number of physical operations or movements. However a fair amount of operations can still be obtained using specialized computer interfaces for individuals that still have complete mobility and control of mouth and eyes, as is the case for individuals with spinal cord injury.

The tongue can perform sophisticated motor control for vocalization and mastication, which suggest a potential for computer input. Moreover tongue interfaces might be intra-oral and invisible to other people. In a study comparing three input interfaces [1]: the *Tongue-Touch-Keypad* (TTK®), the *HeadMaster*® and the Mouthstick, the TTK® from New Abilities [2] was preferred by users due to its discretion and low exertion rate, even though it was not the most efficient interface.

The aforementioned tongue capabilities have resulted in the development of a few tongue-computer interface devices for individuals with tetraplegia. The TTK® [2], the palatal tongue controller [3] and the tongue-operated switch array [4] are intra-oral interfaces that can be fixed in the roof of the mouth and have buttons or switches that are pressed with the tongue. These intra-oral interfaces are invisible to other persons and often preferred by users [1]. However these devices do not take advantage of the fine motor control of the tongue, as they only use four to nine sensors, while the tongue can easily pick out all of our 32 teeth.

The *Tongue drive system* (TDS) [5] is a wearable wireless headset that detects the position of a small magnetic tracer attached to the users' tip of the tongue. Any specific tongue movements in 3D space can be translated into user-defined commands with high information transfer rate. This interface has the advantage of detecting the position of the tongue inside the mouth at any time. However is highly dependent on the intra-oral environment (humidity and temperature) for accurate detection. It is also visible to other persons.

An inductive tongue-computer interface (ITCI) for individuals with tetraplegia, developed at Aalborg University [6], consists of inductive sensors (coils) that change their inductance if a ferromagnetic material is placed nearby. The ITCI (as other intra-oral interfaces) is invisible to other persons but has the ability to detect the position of the tongue without the need of pressing buttons (as the TDS). It has higher number of sensors than other intra-oral interfaces, and can combine and interpolate signals from 2 or more sensors to create a combined activation signal that accurately determines the position of the tip of the tongue inside the mouth.

This paper describes the development of a virtual analog joystick using the inductive sensor coils of the ITCI. Analog joystick emulation was performed with a fuzzy inference system that takes signals from the ITCI sensors as inputs and outputs the virtual configuration (magnitude and direction) of an analog joystick, which can be used to control the mouse-pointer in personal computers, wheelchairs, assistive-technology robots, etc.

II. MATERIALS AND METHODS

A. Intra-oral Inductive Sensors

The inductive tongue computer interface (ITCI) contains a palatal plate, fastened to the teeth by dental retainers. The palatal plate contains inductive sensors (coils) that change their inductance, according to Faraday's Law, if a ferromagnetic material is placed nearby. Inductive sensors are grouped in two different printed circuit boards (Fig. 1),

K. Dremstrup, S. Rees, M.Ø. Jensen (Eds.): 15th NBC on Biomedical Engineering & Medical Physics, IFMBE Proceedings 34, pp. 191–194, 2011.
www.springerlink.com

one with 8 sensors as a tongue mousepad area (TMP), and another with 10 sensors as a tongue keypad area (TKP).

The ferromagnetic activation unit (Fig. 2) is a cylinder made of biocompatible stainless steel, which has a 4 mm diameter and 2 mm height. The cylinder is fixed (e.g. pierced or glued) 7 to 10 mm posterior to the users' tongue tip. Sensors can be activated by appropriate positioning of the activation unit over the palatal plate surface.

A battery-driven 50 kHz sine wave current with an amplitude of 30 μA provides power to the coils. The induced voltage (ε) is rectified and amplified by hardware, giving in result an activation signal, which is sampled with a resolution of 1 byte per sensor. From Faradays law the induced voltage is:

$$\varepsilon = -L\frac{di}{dt} = -\mu_0 \mu_r N^2 \frac{A}{l}\frac{di}{dt} \quad (1)$$

Where L is the inductance, μ_0 is the vacuum permeability, μ_r is the relative magnetic permeability of the core material, N is the number of turns, A is cross section area, and l is the length of the coil.

B. Signal Processing

Sampled activation signal is transmitted wirelessly to the computer or other hardware designed to process this signal. Signal processing software monitors signals coming from the ITCI, normalizes sensor signals, and calibrates signal baseline. A sensor is considered active when the difference exceeds a 15% (TMP sensors) and 25-50% (TKP sensors) threshold, relative to the maximum activation signal for that specific sensor. In case two or more TKP sensor signals exceed the threshold, the sensor with the greatest signal amplitude is chosen. However, signals from TMP sensors are treated as an input vector to a fuzzy inference system to emulate the position of a joystick and move the pointer.

C. Sensor Geometry and Signals

There are two types of sensors in the TMP: round and

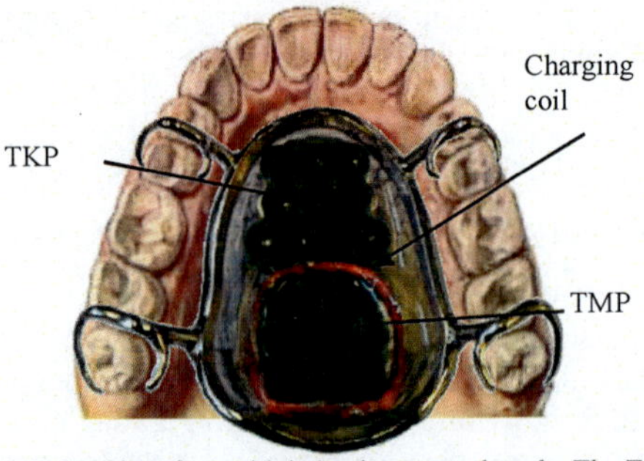

Fig. 2. Palatal interface with inductive sensor boards; The Tongue mousepad (TMP) area contains 8 sensors and the tongue keypad area (TKP) contains 10 round sensors. TMP is fenced by a charging coil.

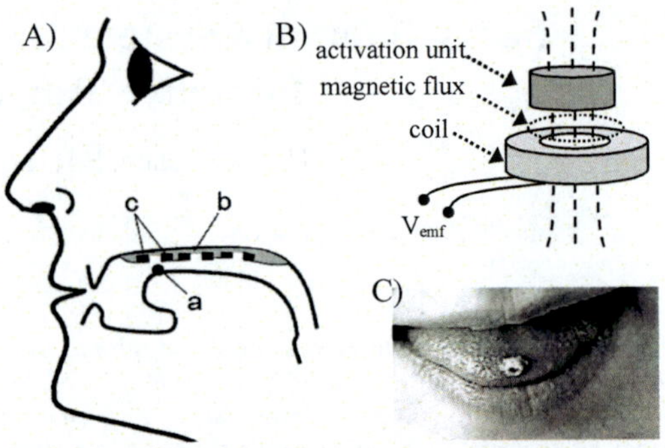

Fig. 1. The activation unit: A) placement of sensors [c] in the palate [b] and activation unit [a], B) principle of activation, C) activation unit. Modified from [6] with permission, © 2006 IEEE.

oval. Each type of coil provides signals with variable amplitude dependent on the coil's geometry and on where the center of the activation unit is positioned over the coil. The geometry of the coil determines the strength of maximal activation, i.e. the maximal influence of the activation unit (blue circle in Fig. 3) on the magnetic flux generated by the coil. Placing the center of the activation unit outside the maximal activation point (blue star in Fig. 3) determines a decreased activation. The round coil is the most efficient in concentrating a generated magnetic field and provides the greatest strength of maximal activation. The oval coil generates a more dispersed magnetic field with lower maximal activation, but with increased transit area (red lines in Fig. 3).

Fig. 4 shows the activation signal dependent on the activation unit position, using a stainless steel activation unit 4 mm (diameter) x 2 mm (height) placed 0.3 mm above the surface of the coil. The center of the activation unit, relative to the maximal activation point of each coil defines its placement. The activation signal can be interpreted by the signal processing software that takes signals from individual sensors or the interpolation of sensor signals to perform actions, e.g. mouse movement of character typing.

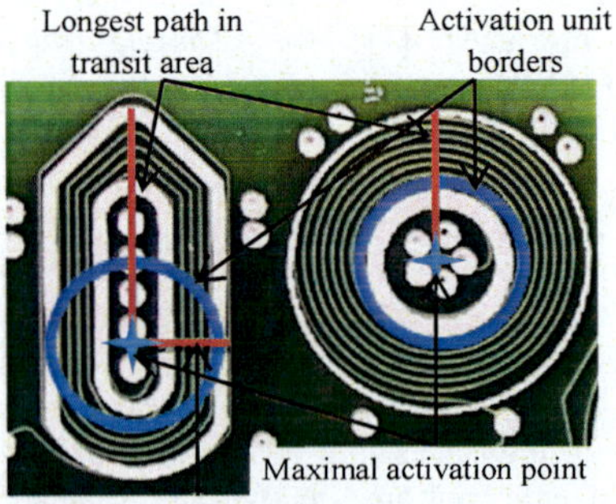

Fig. 3. Examples of coil geometry that provides different maximal activation strength and transit areas relative to the maximal activation point for each coil.

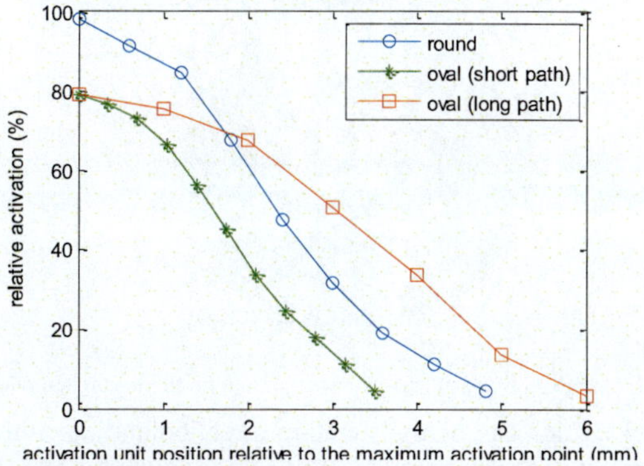

Fig. 4. Activation signal of round coils and oval coils relative to the maximum activation signal of both coils. The signal is dependent on the activation unit position relative to the maximal activation point of each coil.

III. Fuzzy inference system design

A Sugeno- type FIS for mouse movement ("x" and "y" axis) was designed to give a proportional relation between the position of the activation unit over the TMP and the "joystick position" output. Training was performed with 25 activation points within the TMP. The boundaries of the TMP are marked by a copper coil (Fig. 1), in order to help the users to maintain the activation unit within the TMP area. The coil is also used to wirelessly provide power to the ITCS by induction.

The Sugeno FIS lends itself to the use of adaptive techniques for constructing fuzzy models [7]. These adaptive techniques can be used to customize membership functions so that the FIS best models the data. Therefore a Sugeno-type FIS with two linear outputs was used to interpolate input signals and determine the position of the activation unit over the sensor board. The aggregation function used is the maximum of the output membership functions and a weighted average defuzzifier is used.

Inputs to the FIS are processed signals from the eight TMP sensors ($s_1 \ldots s_8$), with values ranging between zero and

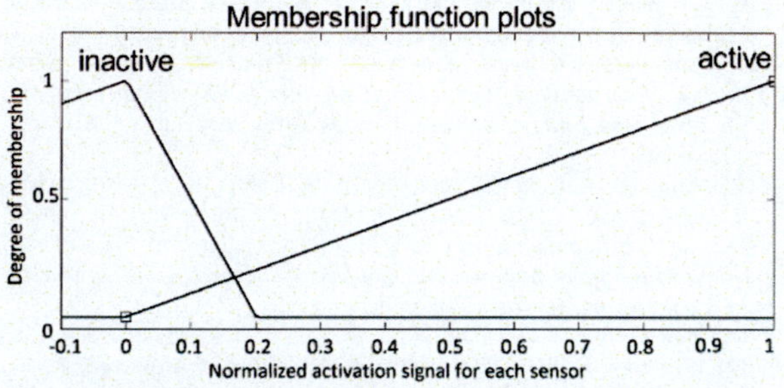

Fig. 5. Membership functions for input variables s_1 to s_8. Sensor signals range from no activation ($s_i = 0$) to maximum activation ($s_i = 1$), relative to the maximum activation signal for each sensor. Negative values exist due to noise.

one. The outputs from the FIS emulate an analog joystick position in Cartesian coordinates (X,Y) with values ranging between zero and one. Each input has 2 membership functions (*active* and *inactive*), and the outputs have 9 membership functions, one per each rule. Membership functions for the inputs are illustrated in Fig. 5. The fuzzy rules are in the form:

$$Ru^{(1)}: \text{if } s_1 \text{ is } active_1 \text{ then } (X \text{ is } fx^1) \text{ and } (Y \text{ is } fy^1)$$

$$\vdots$$

$$Ru^{(8)}: \text{if } s_8 \text{ is } active_8 \text{ then } (X \text{ is } fx^8) \text{ and } (Y \text{ is } fy^8) \quad (2)$$

$$Ru^{(9)}: \text{if } (s_1 \text{ is } inactive_1) \text{ and ... and } (s_8 \text{ is } inactive_8) \text{ then } (X \text{ is } zero) \text{ and } (Y \text{ is } zero)$$

Where, *fx* and *fy* are in the form:

$$\begin{aligned} fx^l &= a_1^l * s_1 + \ldots + a_8^l * s_8 + a_0^l, \quad l = [1,\ldots,9] \\ fy^l &= b_1^l * s_1 + \ldots + b_8^l * s_8 + b_0^l, \quad l = [1,\ldots,9] \end{aligned} \quad (3)$$

Training sets for designing the FIS are presented as I/O data pairs. Three datasets (training, testing and checking) are recorded for each palatal plate. Each dataset consists of 25 activation points within the TMP (Fig. 6) and a null-activation point. The null-activation point consists of data from sensors without the activation unit being present. The output coordinate (*fx*, *fy*) of the datasets corresponds to the physical position of the center of the activation unit in relation to the TMP sensors, e.g. (-1, 0) for the leftmost, (0, 1) for the backmost, and (0.71, -0.71) for the rightmost-frontmost activation points (Fig. 6).

Heuristic values for the output membership function constants (a_i and b_i) were initially defined based on the activation point coordinates. Automatic tuning of the output membership functions was performed using an adaptive neuro-fuzzy inference system training (ANFIS). The ANFIS change input and output membership function parameters through the optimization process similar to a neural network optimization. The adjustment of membership function parameters is facilitated by a gradient vector, which provides a measure of how well the FIS is modeling the data for a given set of parameters.

Data from the "training set" was used for the optimization process, while data from the "checking set" was used to control the possibility for the model to over-fit the data, i.e. when the "checking error" stopped diminishing; the training was stopped even though "training error" was still diminishing. Data from the testing set was used to compare the generalization capabilities of the trained FIS (as plotted in Fig. 7). The Sugeno-type output membership function parameters were significantly modified by the optimization. In Fig. 7, it can be observed

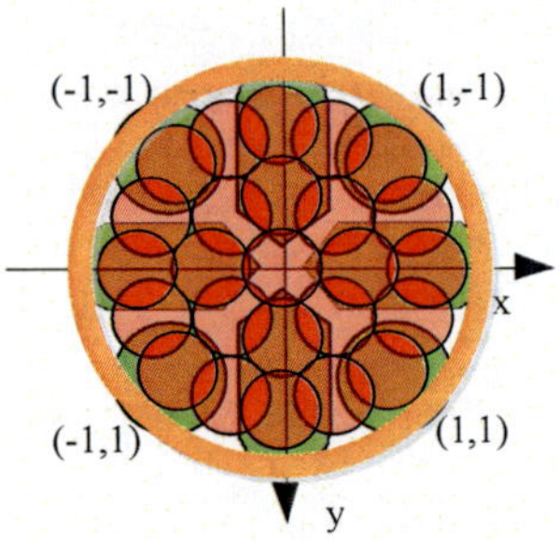

Fig. 6. Activation points for training the fuzzy inference system. The position of the activation unit for each point is marked with a semi-transparent red circle. The output value ranges from -1 to 1. The TMP area is fenced by a charging coil (orange circle) in order to help the users to maintain the activation unit within the TMP area.

that there is a clustering of output points corresponding to the 25 activation points.

IV. Analog Joystick Emulation

Emulation of analog joystick position using the fuzzy inference system allows continuous and proportional control of the virtual joystick position by placing the activation unit over the TMP sensors. One application of the joystick is to control the mouse-pointer in personal computers. Mouse-pointer speed (4) is calculated by multiplying the magnitude of the output vector by the sampling rate (30 Hz) and by a constant (c), which represents the amount of pixels the mouse will move. The mouse movement direction (5) is calculated simply by the angle of the output vector.

$$r = 30c\sqrt{X^2 + Y^2} \tag{4}$$

$$\theta = \tan^{-1}\frac{Y}{X} \tag{5}$$

For example, if the output of the fuzzy system is (0.2, -0.5), and the constant multiplying the output is 3 pixels/sample, then the mouse pointer speed will be approximately 50 pixels per second towards 292°:

$$\begin{aligned} r &= 30*3\sqrt{0.2^2+(-0.5)^2} &= 48.47\tfrac{\text{pixels}}{\text{s}} \\ \theta &= \tan^{-1}\left(-0.5/0.2\right) &= 291.80^\circ \end{aligned} \tag{6}$$

Placing the activation unit closer to the center of the TMP will produce low speed mouse movement in the corresponding direction, while placing the activation unit farther from the center of the TMP will produce mouse movement with higher speed. This was achieved by using a fuzzy inference system to receive calibrated and normalized signals and output the Cartesian position of the virtual joystick, which can be translated to direction and magnitude of the joystick command. With this development, the inductive tongue-computer interface were be used to control the mouse pointer in a personal computer [8, 9], and a wheelchair [10] at varying speed and multiple directions with respect to the tongue position over the palatal plate. This provides a significant advantage to individuals with tetraplegia using this computer interface.

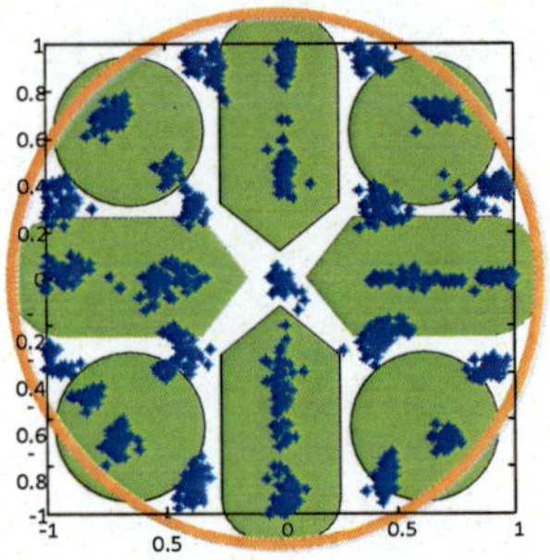

Fig. 7. Untrained fuzzy inference system output for the testing dataset.

References

[1] C. Lau and S. O'Leary, "Comparison of computer interface devices for persons with severe physical disabilities," *Am J Occup Ther,* vol. 47, pp. 1022-1030, 1993.

[2] D. Fortune, J. E. Ortiz and J. N. Tran, "Tongue Activated Communications Controller,", 1993.

[3] C. Clayton, R. G. S. Platts, M. Steinberg and J. R. Hennequin, "Palatal tongue controller," *J. of Microcomputer Applications,* vol. 15, pp. 9-12, 1992.

[4] D. Kim, M. E. Tyler and D. J. Beebe, "Development of a tongue-operated switch array as an alternative input device," *Int. J. Hum. -Comput. Interact.,* vol. 18, pp. 19-38, 2005.

[5] X. Huo, J. Wang and M. Ghovanloo, "A magnetic wireless tongue-computer interface," in *3rd International IEEE EMBS Conference on Neural Engineering,* Kohala Coast, HI, 2007, pp. 322-326.

[6] L. N. S. Andreasen Struijk, "An inductive tongue computer interface for control of computers and assistive devices," *IEEE Trans. Biomed. Eng.,* vol. 53, pp. 2594-2597, DEC, 2006.

[7] L. Wang, *A Course in Fuzzy Systems and Control.* Upper Saddle River, N.J.: Prentice Hall, 1997.

[8] E. R. Lontis, M. E. Lund, H. V. Christensen, B. Bentsen, M. Gaihede, H. A. Caltenco Arciniega and L. N. S. Andreasen Struijk, "Clinical evaluation of wireless inductive tongue computer interface for control of computers and assistive devices," in *Proceedings of the 32nd Annual International Conference of the IEEE Engineering in Medicine and Biology Society,* Buenos Aires, Argentina, 2010, pp. 3365-3368.

[9] M. E. Lund, H. A. Caltenco Arciniega, E. R. Lontis, H. V. Christensen, B. Bentsen and L. N. S. Andreasen Struijk, "A framework for mouse and keyboard emulation in a tongue control system," in *Proceedings of the 31st Annual International Conference of the IEEE Engineering in Medicine and Biology Society,* Minneapolis, MN, USA., 2009, pp. 815-818.

[10] M. E. Lund, H. V. Christensen, H. A. Caltenco Arciniega, E. R. Lontis, B. Bentsen and L. N. S. Andreasen Struijk, "Inductive tongue control of powered wheelchairs," in *Proceedings of the 32nd Annual International Conference of the IEEE Engineering in Medicine and Biology Society,* Buenos Aires, Argentina, 2010, pp. 3361-3364.

Investigation of In-Vivo Hinge Knee Behavior Using a Quasi-Static Finite Element Model of the Lower Limb

L. Zach, S. Konvickova, and P. Ruzicka

[1] CTU in Prague – Faculty of Mechanical Engineering/Department of Mechanics, Biomechanics and Mechatronics, Laboratory of Biomechanics, Prague, Czech Republic

Abstract— **A key goal of joint endoprosthesis is to become a full-featured functional and anatomical replacement. The joint damage may occur for several reasons - primarily a disease of different nature and magnitude, resulting in gradual and irreversible changes and in an extreme solution in the implantation of artificial joints. However, there should be also mentioned accidents leading to joint destruction, which are often "trigger mechanism" of the disease. This work therefore presents a quasi-static computational finite element analysis of a hinge-type knee replacement, which aim to streamline and accelerate the development of knee endoprosthesis. It tackles a question of the overall strength of the implant and detects sites of elevated concentrations of stresses that may be potential sources of implant damages. It also studies the behavior of the endoprosthesis under quasi static loads with emphasis on the study of the shape and size of the contact surfaces, which are closely related to the size of the contact pressure and material wear. Aside the hinged knee replacement, the computational model consisted of femur, fibula, tibia, patella and 25 most important muscles of the lower limb. Due to realistic definition of the boundary conditions, this model is suitable for investigation of in-vivo knee joint replacement behavior.**

Keywords— **Knee, knee replacement, finite element method, lower limb.**

I. INTRODUCTION

Finite element method (FEM) is a common and an effective tool used in mechanics for a development or a verification of various components or mechanisms. In biomechanics, using FEM means to undergo many compromises and simplifications. All these simplifications have to be reasonable and must take into account as many tissue characteristics as possible.

With reference to this fact, there are two groups of FEA of lower extremity models. The first ones are used to simulate a behavior of a healthy knee joint in-vivo [1, 2, 3, 4] and the second group which deals with a knee joint after a total knee endoprosthesis (TKE) implantation [5,6].

Since our laboratory participates on development of total knee endoprosthesis (TKE), the aim of this paper is to present the assembled complex model of the lower limb, consisting of all bones of the knee and 25 main muscles of the lower limb and 8 ligament units of the knee. This complex lower limb model simulates behavior of the hinged knee endoprosthesis and predicts contact pressure and stress distribution for the TKE.

II. MATERIALS AND METHODS

A. Geometric model

For the presented nonlinear quasi-static analysis solved in Abaqus CAE, a universal size of the hinged knee endoprosthesis by ProSpon [7] was chosen.

The modular hinged knee ProSpon is made up of several components to cover individual operation demands. For the presented model, all its main components were modeled, i.e. femoral component, femoral stabilizing rod, tibial component, tibial stabilizing rod, meniscal component (tibial plateau), hinge post, hinge pin, hinge lock and two hinge bushings (see Fig. 1). Position of the TKE on the corresponding bones respected the formerly designed mechanical axis and producer's recommendations to a surgeon concerning an endoprosthesis implantation.

A bone anatomy was reconstructed based on the male cadaver CT scans of the Visible Human Project [8] provided by the National Library of Medicine. A pelvic bone, necessary for muscles origins definition, was adopted from a model library of the BEL Repository, managed by the Istituti Ortopedici Rizzoli, Bologna, Italy [9].

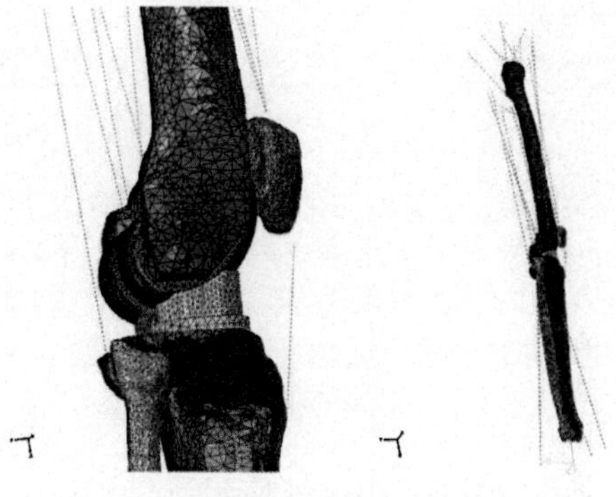

Fig. 1 Lower limb geometric model

K. Dremstrup, S. Rees, M.Ø. Jensen (Eds.): 15th NBC on Biomedical Engineering & Medical Physics, IFMBE Proceedings 34, pp. 195–198, 2011.
www.springerlink.com

Having already the replacement well positioned on the corresponding bones, the data provided by White [10] and Brand [11] were used to locate muscles origins and attachments.

Fig. 1 illustrates the detailed view of the complete geometric model of the lower limb.

B. Material properties and mesh generation

For material definitions, only isotropic homogenous material models were used (see Table 1).

Table 1 Material properties

Entity (Material)	Young's modulus [MPa]	Poisson's ratio [-]
Bones	14 000	0.36
Femoral component (TiAl6V4)	113 800	0,34
Femoral stabilizing rod (TiAl6V4)	113 800	0,34
Tibial component (TiAl6V4)	113 800	0,34
Tibial stabilizing rod (TiAl6V4)	113 800	0,34
Meniscal component (UHMWPE)	820	0.44
Hinge post (TiAl6V4)	113 800	0,34
Hinge pin (TiAl6V4)	113 800	0,34
Hinge lock (TiAl6V4)	113 800	0,34
Hinge bushings (PEEK)	3 650	0.44

All materials excluding UHMWPE were supposed to behave according to Hook's low; the tibial plateau formed from UHMWPE has been defined as an elasto-plastic material (see Fig. 2). All muscles were represented by lines of actions with no material properties definitions. The patellar ligament was modeled as a spring with stiffness of 1000 N/mm.

A mash of elements was created semi-automatically using mixture of hexahedral, tetrahedral and wedge elements. The assembly consisted totally of 338 003 elements.

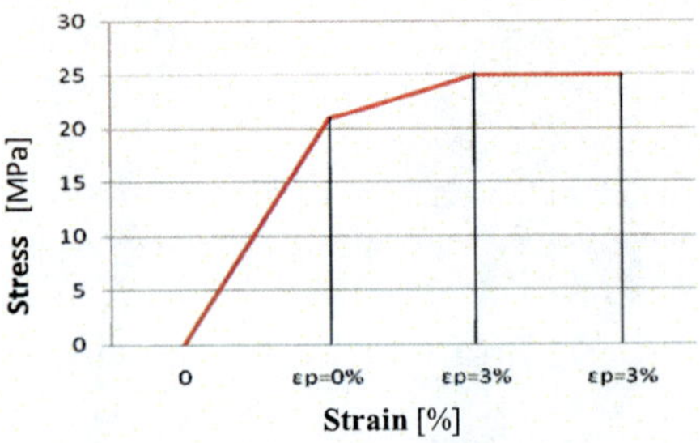

Fig. 2 UHMWPE elasto-plastic material model

C. Boundary conditions

Magnitudes of muscle forces were adopted from Vilímek [12] who calculated by a static optimalisation muscle forces for a group of 31 musculotendon actuators. Following 25 muscles took part of the presented FEA: two parts of gluteus medius (GLMED), two parts of gluteus minimus (GLMIN), semimembranosus (SM), semitendinosus (ST), biceps femoris long head (BFL), biceps femoris short head (BFS), sartorius (SR), adductor longus (ADL), adductor breve (ADB), tensor fascia lata (TFL), pectineus (PCT), gracilis (GRC), gluteus maximus (GLMAX), ilio-psoas (ILPS), rectus femoris (RF), vastus medalis (VM), vastus intermedius (VI), vastus lateralis (VL), medial gastrocnemius (MG), lateral gastrocnemius (LG), soleus (SOL), tibialis anterior (TA) and tibialis posterior (TP).

The resulting ground reaction force was also adopted from Vilimek [12].

All shifts and rotations were constrained in case of femur. For the tibia, only the varus-valgus and internal-external rotations were allowed as well as the proximal-distal shift. The patella was allowed to move only in anterior-posterior axis direction which simulated a simplified articular capsule.

There were defined surface-to-surface hard-contacts only between the corresponding master and slave pairs of the hinge. All other contacts were considered as tie contacts which agreed with the acceptable simplification of the model or the ideal fixation of the TKE to the bone tissue.

The static FEA was run for the following flexions in the hip joint (ankle joint respectively): 18.3 ° (9.46 °), 69.4 ° (28.3 °) and 93.1 ° (31.3 °).

III. CONCLUSIONS

Primary indications for a hinge knee include medial or lateral collateral loss, massive bone loss, and metaphysis and cortical shell, which includes collateral origins or insertions, and severe flexion gap imbalance requiring a link system for stability. Indications for a hinge in primary TKA include patients with neuromuscular deficits such as polio or flail knee, who require the hyperextension stop.

Main advantages of the presented FE model are obvious. Since there are already main muscles of a lower limb and some knee ligaments in the model, and a femur and a tibia bones can move independently, a very realistic analysis can be made. It could be used not only for static or quasi-static analysis and simulations but also for dynamic ones. Unfortunately, no relevant result comparison of this finite element analysis (FEA) with other authors could be done since no published paper on FEA of the hinge knee endoprosthesis has been found.

Nevertheless, our models are based on experimentally evaluated simplified models (knee joint simulator, MTS 885 MiniBionix).

Fig. 3 illustrates the deformed assembly at 18.3 ° (9.46 °), 69.4 ° (28.3 °) and 93.1 ° (31.3 °) of the hip joint flexion (ankle joint flexion respectively).

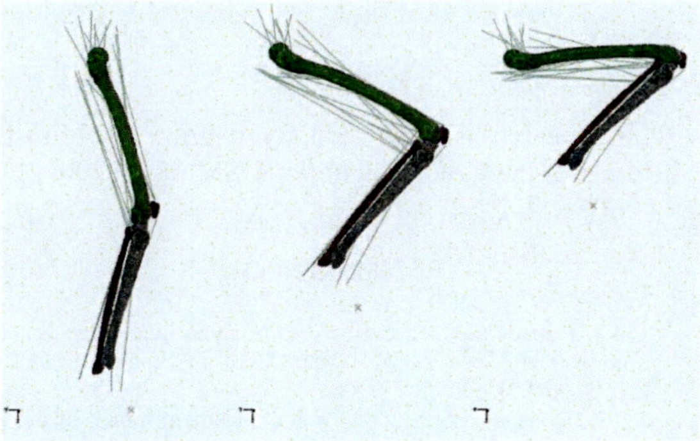

Fig. 3 Deformed assembly at 18.3 °, 69.4 ° and 93.1 ° of the hip joint flexion

Magnitudes of contact pressures, reduced stresses and also their distributions can significantly influence a lifespan of an endoprosthesis. Since the weakest part of the hinge knee endoprosthesis are the PEEK formed hinge bushings, only results of contact pressures, HMH stresses and logarithmic strains for the bushings are published.

Contact pressures on inner contact surface of the hinge bushings (articulating with the hinge pin) are illustrated on Figs. 4 - 6. As for magnitudes, the maximal value of approx. 50 MPa was calculated at 18.3 ° of the hip joint flexion (Fig. 4).

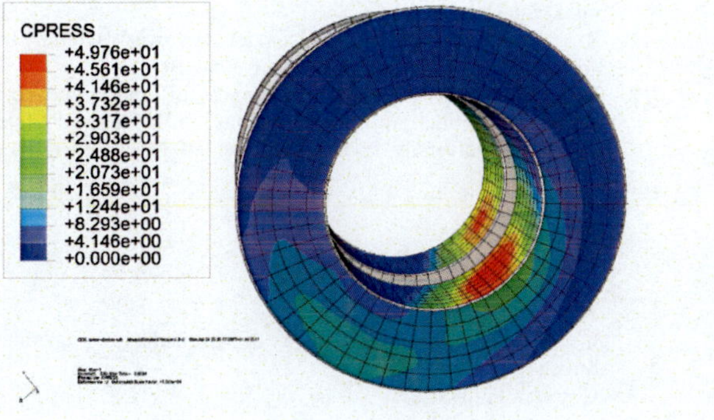

Fig. 4 Contact pressures on hinge bushings [MPa] at 18.3 ° of the hip joint flexion

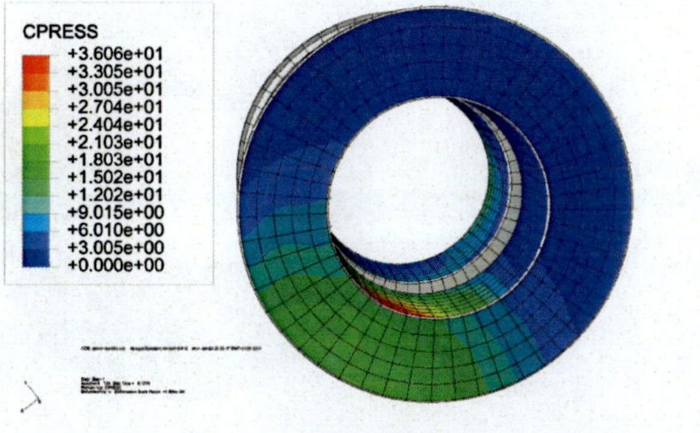

Fig. 5 Contact pressures on hinge bushings [MPa] at 69.4 ° of the hip joint flexion

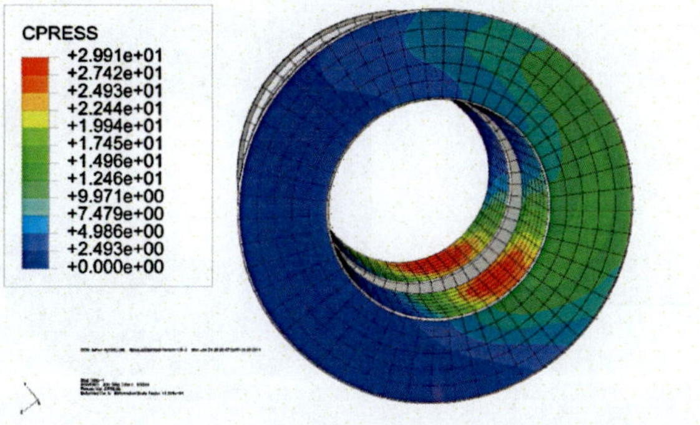

Fig. 6 Contact pressures on hinge bushings [MPa] at 93.1 ° of the hip joint flexion

In the matter of HMH stresses (e.g. Fig. 7-9), for the hinge bushings the maximal value of 45 MPa was calculated at 93.1° of the hip joint flexion (Fig. 9).

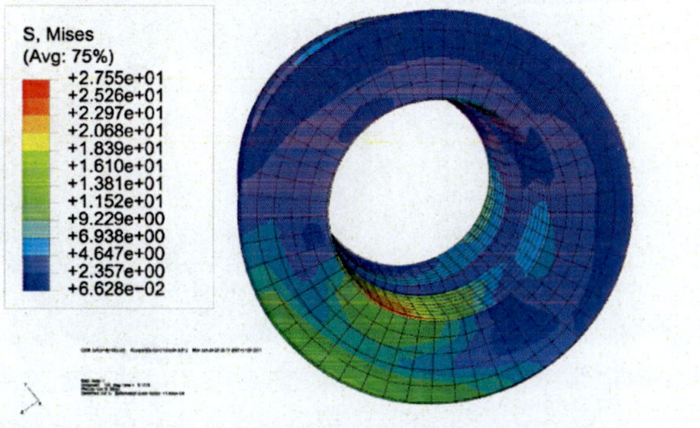

Fig. 7 HMH stress distribution on hinge bushings [MPa] at 18.3 ° of the hip joint flexion

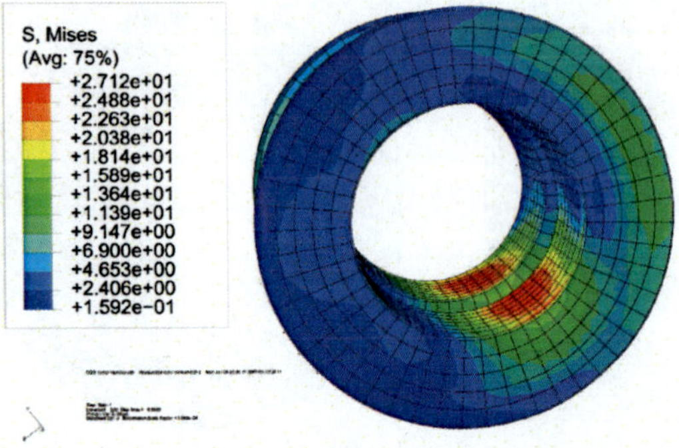

Fig. 8 HMH stress distribution on hinge bushings [MPa] at 18.3 ° of the hip joint flexion

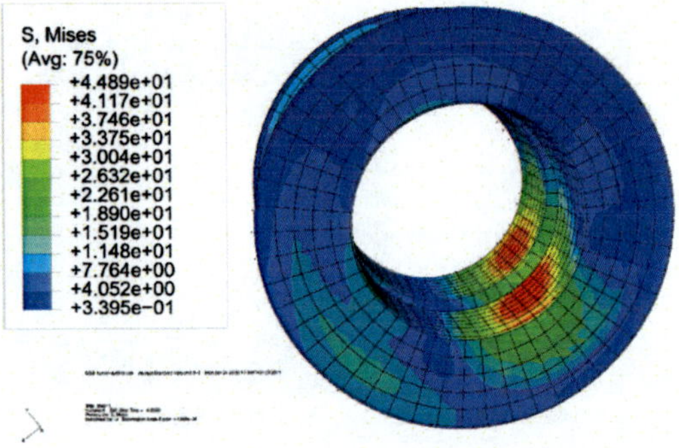

Fig. 9 HMH stress distribution on hinge bushings [MPa] at 18.3 ° of the hip joint flexion

No plastic strain was found for the PEEK bushings. The maximal value of a logarithmic strain of 0.0073 can be expected at 93.1 ° of the hip joint flexion (Fig. 10).

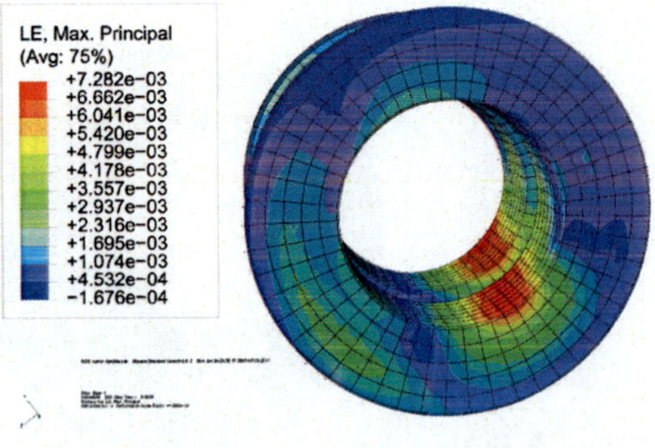

Fig. 10 Logaritmic strain distribution on hinge bushings surface

The presented FEA was used for the development of hinge knee replacement design. All the results of the realistic quasi-static simulation predict a good outcome of the endoprosthesis during the clinical tests. Since no other validation and comparison of the presented FE model could be done, therefore the clinical tests will be useful for its further development.

Acknowledgment

This research is supported by a grant of Ministry of Education of the Czech Republic: MSM 6840770012.

References

1. J.M.T. Penrose (2002) Development of an accurate three dimensional finite element knee model. Comp. Meth. in Biomech. and Biomed. Eng. 5, pp. 291-300
2. T.L.H. Donahue, et al (2002) A finite element model of the Human knee joint for the study of tibio-femoral contact. J. Biomech. Eng. 124, pp. 279-280
3. J.A. Heegaard (2001) A computer model to simulate patellar biomechanics following total kneee replacemnet: the effects of femoral component alignment. Clinical Biomech. 16, pp. 415-423
4. P. Beillas, et al (2004) A new method to investigate in vivo knee behavior using a finite element model of the lower limb. J Biomech. 37, pp. 1019-1030
5. A.C. Godest (2002) Simulation of a knee joint replacement during a gait cycle using explicit finite element analysis. J. Biomech. 35, pp. 267-275
6. J.P. Halloran (2004) Explicit finite element modeling of total knee replacement mechanics. J. Biomech. 38, pp. 323-331
7. ProSpon,s.r.o. at http://www.prospon.cz
8. National Library of Medcine, Visible Human Project at http://www.nlm.nih.gov/research/visible/visible_human.html
9. Viceconti, Visible Human Male - Bone surfaces, From The BEL Repository at http://www.tecno.ior.it/VRLAB/
10. S.C. White, et al (1989) A Three Dimensional Musculoskeletal Model for Gait Analysis. Anatomical Variability Estimates, J. Biomech. 22, pp. 885-893
11. R.A. Brand, et al. (1982) A model of lower extremity muscular anatomy, J. Biomech. Eng., 104, pp. 304-310
12. M. Vilimek (2005) The challenges of musculotendon forces estimation in multiple muscle systems, PhD Thesis. Prague, Czech Technical University in Prague - Fac. of Mechanical Engineering, 2005

Author: Lukas Zach
Institute: CTU in Prague– Faculty of Mechanical Engineering
Street: Technická 4
City: Praha 6
Country: Czech Republic
Email: Lukas.zach@fs.cvut.cz

Reliability of Hemodynamic Parameters Measured by a Novel Photoplethysmography Device

A. Grabovskis[1], E. Kviesis-Kipge[1], Z. Marcinkevics[2], V. Lusa[2], K. Volceka[2], and M. Greve[2]

[1] Institute of Atomic Physics and Spectroscopy, University of Latvia, Riga, Latvia
[2] Department of Human and Animal Physiology, Faculty of Biology, University of Latvia, Riga, Latvia

Abstract— **Three channel photoplethysmography (PPG) signal pulse wave studies of the leg's conduit arteries during rest conditions were performed. The obtained data of each channel showed similar values, proving arterial PPG as a reliable and repeatable method to assess arterial waveform parameters. A validation experiment was carried out by acquiring signals from three identical IR PPG sensors, which were placed on different sites over the leg's conduit arteries during rest conditions. Coefficients of variation (CV) were calculated at a 95% confidence interval by comparing results of each subject during multiple attempts. This data processing leads us to certain criteria of improvements in our methodology. Results show that the arterial PPG technique can give trusted and accurate information about the changes in hemodynamics, and therefore, makes it promising for early diagnostics of vascular disease.**

Keywords — **conduit arteries, photoplethysmography, pulse wave velocity, second derivative.**

I. INTRODUCTION

Photoplethysmography (PPG) is a well known, non-invasive, optical method used for detecting blood pulsations. Currently PPG measurements are typically performed by recording signals from defused vascular beds, such as fingertips and ear lobes, thus providing information about their microcirculation and tone of the small arteries [1]. There were attempts to obtain PPG signals from the large conduit arteries since 1971, when Weinman and Sapoznikov first described continuous measurement of the arterial pulse wave velocity (PWV) [2]. Later studies were focused on methodological and PPG device validation [3], and comparison of PWV obtained by PPG in healthy subjects and patients [4]. Despite the potential clinical value, arterial PPG has not become a widely used method among researchers and clinicians. The reasons are the lack of available, commercial versions of arterial PPG sensors and the protocol for receiving correct measurements. Also, there are methodological and technical difficulties in its application which requires highly skilled personnel.

Our studies are related to the arterial health assessment using PPG waveform analysis.

To our knowledge, the PWV and wave form parameter assessment using three site arterial PPG, has not been previously performed.

Therefore, the aim of this study was to verify the usability of the arterial PPG technique for measuring multiple pulse wave parameters which refer to local and regional arterial stiffness which is an independent predictor of cardiovascular events [5].

II. MATERIALS AND METHODS

A. Subjects

Youthful (18 to 26 years old) volunteers (2 males, 4 females) with a healthy lifestyle, body mass index from 16.2 to 25.9 kg/m^2 and no signs or symptoms of cardiovascular diseases were enrolled in this study. All of the subjects gave their informed consent to participate. The Scientific Research Ethics Committee of the University of Latvia, Institute of Experimental and Clinical Medicine approved the research protocol.

B. Equipment and Experimental Design

Every volunteer was subjected to three attempt series, each five minutes long. During the experiment, a subject was held in a supine position on an ergonomic pedestal at room temperature (24°C) in quiet and comfortable conditions. Prior to data recording, the PPG sensors were placed while the subject adapted. The volunteers were cautioned not to intake any caffeine or eat a meal within 2 hours before the experiment.

Physiological measurements require an accurate and easy to use PPG signal measuring device, which ensures low signal noise, and high spatial and temporal resolution. One of the novelties of this study is the usage of identical PPG sensors and custom fastening strips for their attachment to multiple sites of the leg arteries, instead of measuring the

K. Dremstrup, S. Rees, M.Ø. Jensen (Eds.): 15th NBC on Biomedical Engineering & Medical Physics, IFMBE Proceedings 34, pp. 199–202, 2011.
www.springerlink.com

pulse transit time between these sites with different methods – ultrasound, ECG, sphygmogram etc. In physiological measurements, we are looking to achieve an accuracy higher than 8 %, therefore same type of sensor applications with common device clock is required, instead of using different types of pulsation detection methods with unknown hardware delay times.

Such PPG devices are not produced commercially, hence we designed a laboratory-made prototype. This prototype is a custom made 3-channel digital PPG device (sampling rate: 1 kHz per channel; 875 nm LED; photodiodes with visible light filter and peak spectral response wavelength of 880nm). Originally this device was designed for scientific purposes, which require an analog signal output to one common data acquisition system. Therefore, the digital PPG signals were converted using a 12-bit eight-channel DAC. The device has an integrated LED driver that provides stable power throughout the battery discharge range. LEDs were typically driven by 55 ± 15 mA (5 mA stepping), so that no perceptible warming of the upper tissue layer was produced. A special screening barrier for the photodiode was made within the sensor to lower the influence of ambient light. The design of the scheme is based on a photodiode discharge time measurement using a 32-bit timer built into a microcontroller [6].

The only filter integrated in the design is a second order Butterworth low pass filter with the cut off frequency of 42Hz at 3dB. This filter did not distort the signal shape and phase because the typical bandwidth of the PPG signal is 0.05 – 40Hz. The measured noise level of the device is -30 to -40dB compared to the PPG signal level.

Three custom-made, reflection type PPG probes were developed and adapted to meet criteria necessary for the measurements of arterial blood pulsation from the skin over the conduit artery. This provided the ability to take contact PPG measurements virtually from any site of superficial arterial tree. The PPG method is very sensitive to tissue motion and to the sensor-to-tissue contact force [7]. Therefore, to prevent signal artifacts, the arterial PPG sensors were fastened with custom-made holders.

Sensors were placed as follows: S1 over femoral a. near the groin, S2 over popliteal a. in the popliteal fossa, and S3 over posterior tibial a. near the ankle (Fig. 1.). Distances between sensors were noted to derive the pulse wave velocities from the pulse transit time (PTT) – delay time of the pulse wave reached all three PPG sensors consecutively (S1-S2 for thigh, S2-S3 for calf, S1-S3 for both thigh and calf). The analog signals from the PPG device were captured by a 12-bit ADC USB data acquisition module at 1 kHz per channel, and stored in a PC.

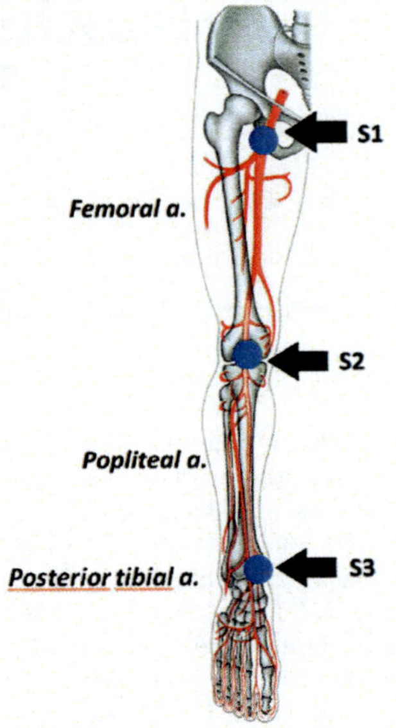

Fig. 1 Placement of PPG sensors during experiment

Later, the PPG signals were processed offline with custom developed Matlab software (signal smoothing with wavelet and Savitsky-Golay filters to reduce signal artifacts and ADC stepping noise.); foot-to-foot PTT was computed in a beat-per-beat manner, and second derivative waveform parameter *b/a* of PPG signal was calculated by using a well-known method of a normalized amplitude ratio *b/a* of the second derivative shape [8] (Fig. 2.).

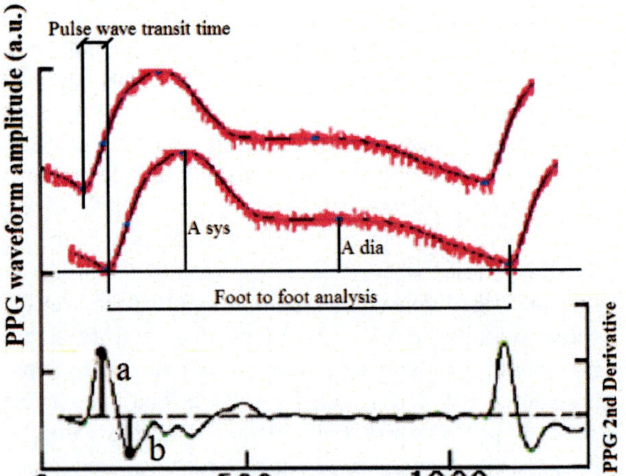

Fig. 2 PPG Waveform analysis parameters: pulse wave transit time (PTT), 2nd derivative parameter *b/a*, and diastolic to systolic peak amplitude ratio A_{dia}/A_{sys}

Every single beat of the PPG waveform was also analyzed by a pulse peak amplitude ratio parameter – a ratio of amplitudes of systolic and diastolic components of pulse wave.

The measure of variability of PPG parameters between different measurement times and other factors were assessed by computing coefficients of variation (CV) in %.

III. RESULTS AND DISCUSSION

High quality PPG recordings were acquired from each subject and from each of the arterial sites (Fig. 3.).

However, in some cases it was difficult to obtain the low noise signals from the S2 site. Partly, that could be explained by the anatomical peculiarities of the subject and the ability to achieve the correct position over the artery with the PPG probe.

Different subjects showed diverse parameter values, reflecting acceptable individual inter-subject variability, yet, there was a good agreement within all three attempts for each subject, thus proving ability to obtain repeatable signal (Table 1).

The most reliable results were obtained for PWV measured between the sites S1 and S3 (coefficient of variation 4±1%) which represents the total limb PWV and could be explained by more precise sensor location on these sites. While the PWV data obtained for calf and thigh were less reliable, 11±7% and 10±3%, respectively. The precision of the thigh and calf PWV largely depended on sensor positioning on the polpiteal fossa, over the artery.

The average variability between different measurements (three times) for parameter *b/a* is 7±4%, which is acceptable, considering that its calculation is sensitive to signal-noise ratio and consequently on processing software settings (smoothing level, filter settings).

The variability of A_{dia}/A_{sys} ratio (8±3%) was similar to that of parameter *b/a*.

Table 1 Summary of parameters acquired during all three measurements: the coefficient of variation in % of PWV, second derivative waveform parameter *b/a*, and pulse peak amplitude ratio parameter A_{dia}/A_{sys}

Subject	Sex	PWV (S1-S2)	PWV (S2-S3)	PWV (S1-S3)	*b/a*	A_{dia}/A_{sys}
1	F	9%	15%	5%	10%	8%
2	F	14%	5%	3%	8%	11%
3	F	12%	6%	2%	12%	12%
4	M	11%	23%	5%	2%	5%
5	F	7%	5%	4%	5%	5%
6	M	5%	11%	4%	7%	8%

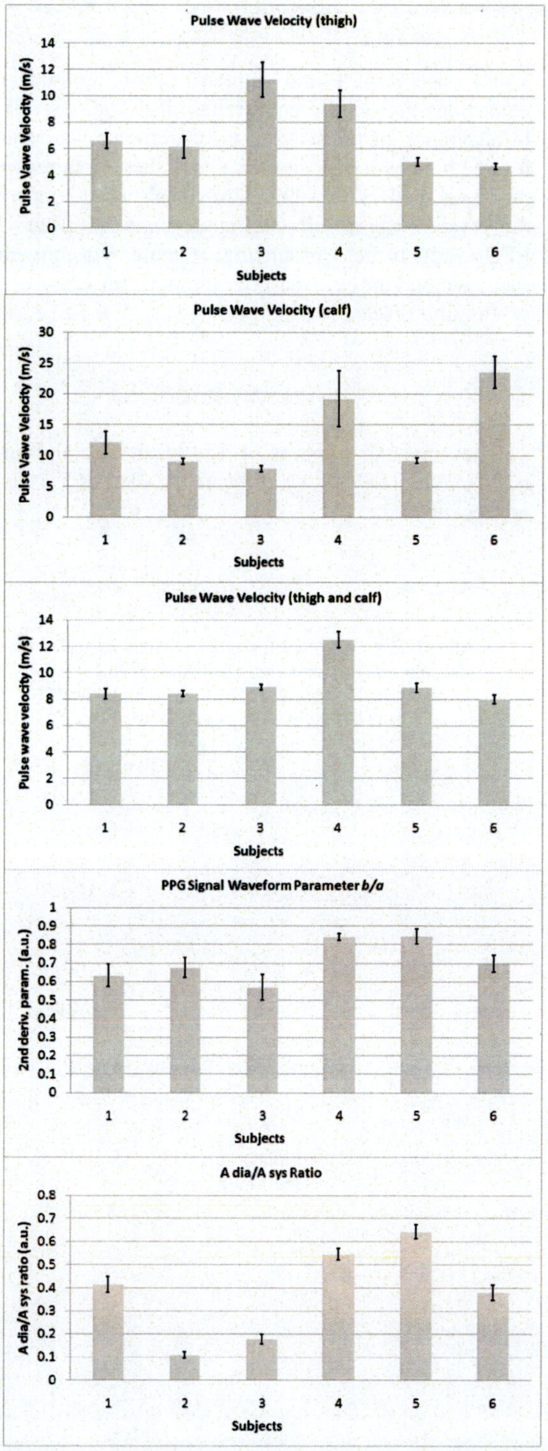

Fig. 3 Obtained mean values and standard deviations for six subjects of all measured parameters for three attempts.

IV. Conclusions

The measurements and standardisation protocol for PPG recordings provided a good reliability. The possible sources in dispersion of pulse wave measurements were the contact force, the precise positioning and the orientation of PPG probe, as well as the settings in signal processing software – the level of signal smoothing and applied filters. Arterial PPG seems to be a promising, reliable and convenient method, which can be used in equipment for early diagnostics of vascular disease.

Acknowledgment

The financial support of European Social Fund (grant #2009/0211/1DP/1.1.1.2.0/09/APIA/VIAA/077) is highly appreciated.

References

1. Allen J. (2007) Photoplethysmography and its application in clinical physiological measurement. Physiol Meas 28:1-39
2. Weinman J, Sapoznikov D. (1971) Equipment for continuous measurements of pulse wave velocities. Med Biol Eng 9:125–38
3. Loukogeorgakis S, Dawson R, Phillips N, Martyn C N and Greenwald S E. (2002) Validation of a device to measure arterial pulse wave velocity by a photoplethysmographic method. Physiol Meas 23:581–596
4. Eliakim M, Sapoznikov D, Weinman J. (1971) Pulse wave velocity in healthy subjects and in patients with various disease states. Am Heart J 82:448–57
5. Laurent S, Cockcroft J, Van Bortel L, Boutouyrie P, Giannattasio C, Hayoz D, Pannier B, Vlachopoulos C, Wilkinson I, and Struijker-Boudier H. (2006) Expert consensus document on arterial stiffness: methodological issues and clinical applications. Eur Heart Journal 27:2588–2605
6. Stojanovic R, Karadaglic D. (2007) A LED–LED-based photoplethysmography sensor. Physiol Meas 28:19-27
7. Teng X F, Zhang Y T. (2006) The effect of applied sensor contact force on pulse transit time. Physiol Meas 27:675-684
8. Takazawa K, Tanaka N, Fujita M, Matsuoka O, Saiki T, Aikawa M, Tamura S, Ibukiyama C. (2003) Assessment of vasoactive agents and vascular aiging by the second derivative of photopletysmogram waveform. J of the Neurological Sciences 216:17–21

Author: Andris Grabovskis
Institute: Institute of Atomic Physics and Spectroscopy, University of Latvia
Street: Raina Blvd. 19, LV-1586
City: Riga
Country: Latvia
Email: andris.grabovskis@gmail.com

Development of a Test Rig for MEMS-Based Gyroscopic Motion Sensors in Human Applications

C. Gerdtman[1], Y. Bäcklund[2], and M. Lindén[1]

[1] Department of Computer Science and Electronics, Mälardalen University, Västerås, Sweden
[2] Office for Science and Technology, Uppsala University, Uppsala, Sweden

Abstract— **This paper describes the development of a test rig for MEMS gyroscopes. The purpose of the test rig is testing and verification of various gyroscopes that are intended for human motion analysis. The test rig will be a tool to test functionality and help in the selection process of appropriate MEMS-gyroscopes. Human movement pattern differs from mechanical motion and thus puts specific demands on the test equipment and verification procedures.**

The main function of the test rig is to rotate the gyroscope and measure the precision in the sensor signal response in different situations. This includes detection of different movement patterns and performances in different environment conditions (e.g. temperature, vibrations, etc).

Several components can be tested at the same time in the test rig. Among the things that can be evaluated is the performance of the components, comparisons between different individual components or batches, aging processes of components and verification of the component performance for comparison of the specifications from the manufacturer. There are several different pre-programmed test-programs available but the test rig can also be manually operated. The data from the tests are stored and can be analyzed and processed afterwards.

Keywords— **test rig, MEMS, gyroscope.**

I. Introduction

The market of MEMS gyroscopes is very volatile in that sense that new gyroscopes are coming out on the market quite often. Some manufacturers change their models of existing gyroscopes every year, some manufacturers put down their production and new manufacturers are established on the market. It is therefore important not only to look at the performance, but also consider the price, production lifetime and replacements possibility, especially for smaller companies that cannot build up a large stock of components. Developing a product is costly and it is important not to be forced to redesign the whole construction just because a single component has to be replaced. So to be able to replace or compare different gyroscopes, a test rig for MEMS gyroscopes would be a good tool for the job.

When using MEMS-based components such as gyroscopes or accelerometers in different applications it is of high importance to know the performance of the component. A good source of information of the component is the datasheet [1-6] from the manufacturer. The datasheet gives a vast amount of information, but not all. A regular datasheet contains specification about the most common parameters and performance in the working range that the component is intended for.

A regular datasheet usually does not specify what happens with the performance if the component is operated outside the specified range, such as operating temperature. Also the manufacture can use different units to describe the same parameters, for example nonlinearity [2] [5]. When studying a datasheet, one issue is how the manufacturer has measured the parameters specified [1-6]. For example, which variability does a certain parameter exhibit between different individuals of the same type of gyroscopes from one manufacturer? It is also of interest to compare the variation between different production batches.

Another issue is to know from the datasheet which behavior a sensor will have in an application. The manufacturer cannot foresee all specific applications demands a certain user will have on the sensor, and thus it is of great importance for the user to be able to specify a test scheme for motion sensors within the scope of his specific application.

When developing sensor systems for human motion analysis, it is of great importance to choose the right sensor for the application. Human movement pattern differs from mechanical motion and thus puts specific demands on the test equipment and verification procedures. In a previous study, an alternative computer mouse for disabled, based on a gyroscopic sensor, was developed [7]. In this process, it was a problem to find a sensor with the desired behavior. Thus, the aim of the present study is to develop a test rig that can mimic human motion and verify critical parameters of a gyroscopic sensor, and also assist the designer to choose the best sensor for his application.

II. Materials and Methods

When choosing a suitable MEMS-gyroscope for a certain application, the information from the manufacturers' data-

K. Dremstrup, S. Rees, M.Ø. Jensen (Eds.): 15th NBC on Biomedical Engineering & Medical Physics, IFMBE Proceedings 34, pp. 203–206, 2011.
www.springerlink.com

sheets often is the first way to sort out good candidates, and it is essential to know which parameters that is most important for that application. In the case of detection of human motion in order to control a computer cursor, i.e. an alternative computer mouse for disabled [7], the identified important parameters and theirs preferred values were:

- Small size & low weight
- Low noise
- Good long-time stability (low drift)
- Low power consumption
- High sensitivity
- Short start-up time

These parameters are not the same as those of a typical industrial application. In a vibrating machine, for example, the shock and vibration durability would be of higher importance [8] than in a consumer situation [9]. After identification of these parameters, comparison of data-specification of different MEMS-gyroscopes was undertaken. From this investigation, several gyroscopes that were likely to fulfill the demands of performance were identified as candidates for the application. These gyroscopes (six different models, see table 1) were selected for further investigation and tests in the test rig that was to be developed. Further, an investigation on common test procedures of gyroscopes was performed. Different manufacturers have different test devices and routines to test their sensors [10] [11], which not always are in line with standardized test procedures [12].

Often these tests are more a quality control to check that the sensors fulfill the performance and tolerance levels specified in the datasheets. Manufacturing tests are designed to test as many MEMS gyros as possible in the shortest possible time to reduce the total test cost [13] [14]. The test equipment is both big and expensive and, as mentioned before, focused on the production test, for example wafer test [15], and therefore not appropriate to use when testing and evaluating various sensors from different manufacturers for specific applications. With a small portable test rig, it is possible to test the sensor in its intended use environment.

Quality performance of sensors is important but for a specific application, functionality tests are just as important. Several of the investigated test procedures used by manufacturers were developed for industrial applications, which in many aspects differ from human motion analysis.

A specification of requirements for the test rig was made based on the identified important parameters for this specific application and the common practice to test such sensors. The main function of the test rig was to ***rotate the sensors***, since MEMS-gyroscopes measure angular rotational velocity. It is also important that it is possible to trace the measurements and to repeat the measurements in an identical way. Therefore it is necessary keep track on which sensors that has been tested and during what circumstances (temperature, humidity, etc.), which test programs, date, and other relevant information that can effect the measurement.

In order to allow data from the measurements to be stored for further comparison and analyze after the measurements, a ***log function*** is needed. This is achieved with computer software especially developed for the test rig, which automatically stores all information. Another identified requirement was that the ***interface to the computer should be standardized and easy to use***. This was achieved by an USB-bus. It should be possible to test ***several sensors at the same time***. The computer software keeps track on how many sensors and their mutual position and stores this information. A common problem with MEMS-gyroscopes is the ***long-time stability***. Another is the ***power-up period***, how long time it takes from power-on until the gyroscope signal is stabilized. Also ***tests for specific situations*** should be possible to perform. Therefore different test-programs were developed, allowing the software to handle speed, time of rotation, direction of rotation and how many repetition of each movement the test program should run. The basic test programs implemented and tested in this study were:

- Measurement during start-up (cold components)
- Long-time drift (no motion)
- Noise level (motion and no motion)
- Signal response
- How well the gyro signal correlates to the actual motion of the test rig

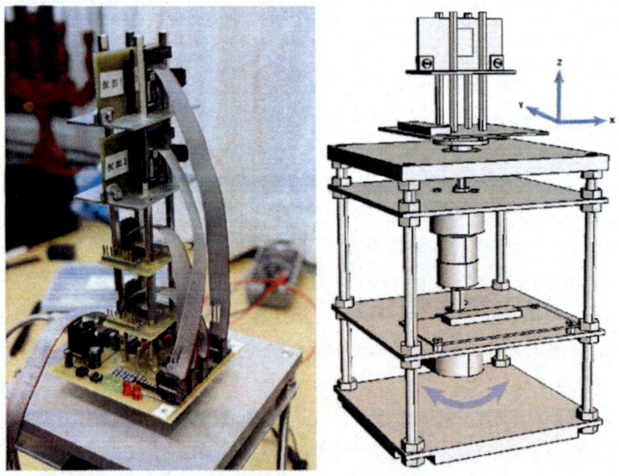

Fig. 1 The Test Rig. (A) Top view with 4 gyros. 2 X/Y at top and 2 Z below. (B) Bottom view with motor and pulse decoder and a Z gyro at top.

Thus, the test rig should be able to handle ***X/Y and Z-axis rotation***. This was solved by various orientations of the circuit boards that the gyros were placed on in the test rig, se figure 1. The test rig is placed on a desk, and the X-axis is in the desk's long side. The Y-axis is turned 90 degrees to the right from the X-axis, still in the tabletop plane (desk's short side). The Z-axis is turned 90 degrees from the Y-axis up from the desktop's plane. Therefore the Z-axis gyros were mounted on a PCB in the same plane as the desk and gyros with X/Y-axis was mounted on a PCB that was turned 90 degrees up from the desk, se figure 1B, so they also could be able to measure the rotation.

In total, 18 sensors were tested; six different models and three sensors of each model (see table 1). Sensors from the same manufacturer batch were used. Thus, it was easy to detect if one sensor had a failure. All sensors were tested in the test rig by same test programs. A totally of four test programs were implemented and tested. The test rig was on same place for all measurements, in the same environment. The temperature was about 20 degrees Celsius and the humidity was around 85%.

Table 1 Tested gyroscopes, single-axis

Manufacture	Model	Structure	Materials	Axis
Analog Devices	ADIS16100	Comb	Micromachined	Z
Analog Devices	ADXRS150	Comb	Micromachined	Z
Bosch	SMG040	Comb	Micromachined	X/Y
Murata	ENC05S	Triangular rod	Piezoceramic	X/Y
Nec-Tokin	CG-L43	Rod	Piezoceramic	X/Y
SensoNor	SAR10	Butterfly	Micromachined	X/Y

The test programs were short and long term measurement to determine the drift and noise on cold respective warm sensors. Both these tests were performed on stationary sensors. For cold sensors measurement, the sensors had been turned off for at least 4 hours and then powered up at the same moment the measurement was started. For warm sensors the sensors hade been powered up at least 10 minutes before the measurement started.

The test rig has been designed so it can be placed in a test chamber for measure the signal response of the sensor to humidity and temperature changes. This option was not utilized in this study. The other two test programs were a step response and a hysteresis test. The step response is simulating a single movement with a pause before next movement. The hysteresis is a back and forth motions, with an abrupt turn, no pause before next movement. Both movements was smoothed and performed in a fast (90°/s) and a slow mode (15°/s). The fast mode is to simulate a regular human movement and the slow mode to simulate a movement done by a motion disabled person.

The step response is a similar, but not so challenging test, as the back and forward test. Therefore the back and forth test is presented here. For the back and forward test, the ideal output signal from the gyroscope sensor should be a square wave followed by an identical square wave in opposite value. So doing this test, differences in rise and fall time, levels, hysteresis and other non-uniform slope changes will be captured.

III. Results

A test rig that can measure different MEMS-gyroscopes has been developed. By using the same test-program it is possible to get a repeatable test procedure.

The logged data that is collected during the measurement contains information about the test-objects, -programs and -equipment. Also test environmental parameters such as temperature are stored. This gives traceability in the measurements. It is very easy to run the same test procedure to get comparable measurements of different sensors, even if the measurements are made on different times.

It is also possible to evaluate the same sensor in the same way after long time to see if the sensor has aged or got impaired in some other way over time.

The cold sensor test showed that during the first minutes after power-up of the sensor, the output signal differs without moving the sensor. This is because the sensor is heated up internally. When the sensors have been heated up, the drift decreases. For the warm sensor tests, the behavior is very similar and stable after 30 minutes operation.

The test rig is not ideal due to minor mechanical errors, tolerances, mass inertia and other construction problems that affect the measurement. Therefore the gyro signal is compared to a measured signal from a pulse decoder with high resolution. The test was made simultaneous on different MEMS-gyros. The ENC05S from Murata, it is an old component (from 1993) with high noise. The measured signal from the sensors (green) and compared to the output signal from the test rig (blue), as shown in figure 2.

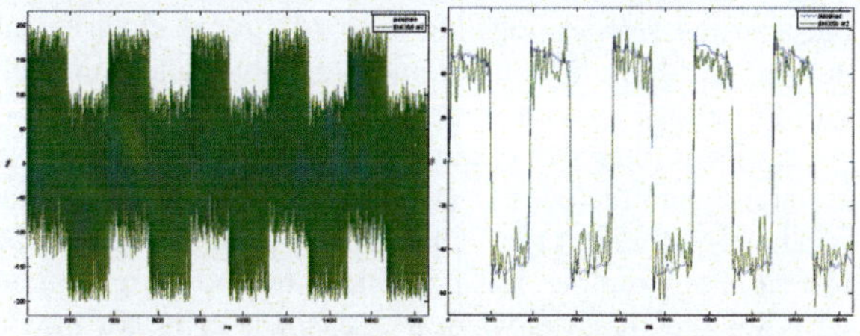

Fig. 2 (A) Measurement of ENC05S. (B) ENC05S with low pass filter.

SMG040 from Bosch is a rather big sensor. Compared to the other tested MEMS-gyros it has low noise and follows the test-signal quite well, as shown in figure 3.

SAR10 from SensoNor is an interesting gyro. It is an angular rate sensor designed for automotive detection with a butterfly structure. As seen in figure 4 it did very well.

Before deciding which sensor to use in a construction, it is possible to test different solutions that can improve the signal quality in the test rig. Figure 2 shows the signal from ENC05S tested both with and without a low pass filter. The low pass filter reduces the noise and clearly improves the quality of the signal.

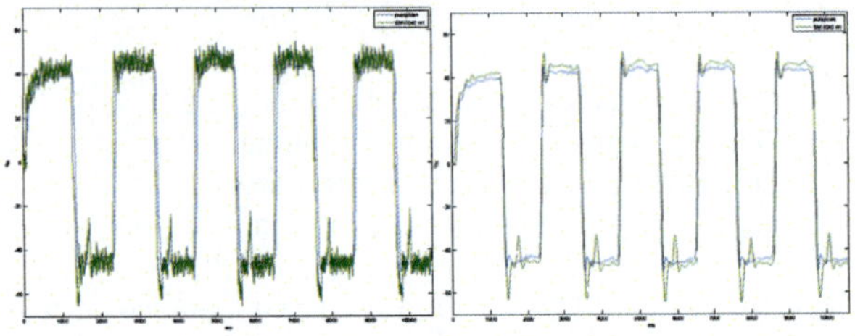

Fig. 3 (A) Measurement of SMG040. (B) SMG040 with low pass filter.

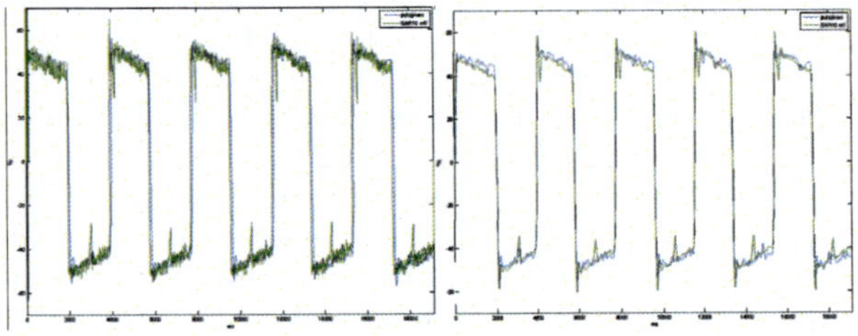

Fig. 4 (A) Measurement of SAR10. (B) SAR10 with low pass filter.

IV. Discussion

The test rig has proven to be very usefully when it is desirable to compare different sensors to each other. It is also possible to test the components performance to verify the sensors performance as specified by the manufacturer. It is possible to test the sensors outside of their specified operating range to see how they will behave in these situations.

The test rig is a good tool to evaluate the sensor alone but also the sensors together with additional electronically designs as filters. It is favorable to be able to test the behavior of parts of the constructions.

An ideal filter will eliminate all errors and leave the sensor signal unaffected. But analogue signal processing is not ideal and will affect the signal. This is possible to test in the test rig. For example, it is possible to test how big the latency is for the signal is when the sensor is or is not moving. The test rig can help the engineer to see if some unwanted behavior occurs when adding filters to the sensor.

In applications for human motion analysis, especially if you want to control a device by the movement, it is not acceptable with a big delay in the signal. This can be tested by the test rig and used to decide which sensor to choose.

The tests show that a general rule is that newer sensors are better than older once. For sensors from the same date a general rule is that the bigger the sensor is, the better is the performance. Also a trend is that the manufacturers want to shrink the size. If the choice is between getting a smaller sensor or one with better performance, the choice usually is to shrink the package.

V. Conclusions

The test rig is a very useful tool to verify that a sensor meets the demands for a certain application and provides a tool to compare sensor to each other. It is very difficult for a human to repeat the exact same movement. Thus, the test rig can be programmed to mimic a human motion pattern and test a various sensors in the same way as a tool to choose the best component. The test rig highlights the differences between different sensors and shows the weakness or strengths each sensor has. So for different applications (different test programs) it is possible to come to different conclusion on what sensor that is best.

References

1. Datasheet ADXRS150 at http://www.analog.com/static/imported-files/data_sheets/ADXRS150.pdf
2. Datasheet ADIS16100 at http://www.analog.com/static/imported-files/data_sheets/ADIS16100.pdf
3. Datasheet CG-L43 at http://www.nec-tokin.com/english/product/pdf_new_pro/Ceramic_Gyro.pdf
4. Datasheet Murata ENC-05S, Specification and data sheet of gyrostar, Piezoelectric Vibratory Gyroscope, Murata
5. Datasheet SAR10 at http://www.sensonor.com/media/39650/datasheet%20sar10(h).pdf
6. Datasheet SMG040 at http://www1.futureelectronics.com/doc/BOSCH/SMG040-0273102002.pdf
7. Gerdtman C, Lindén M (2005) Development of a gyro sensor based computer mouse, with USB interface as technical aid for disabled persons, IEEE, 3rd European Med. and Bio. Eng. Conf. (EMBEC'05), Prague, Czech Republic, 2005, ISSN: 1727-1983, 2456F.pdf.
8. Azevedo R G, Jones D G, Jog A V et al. (2007) A SiC MEMS resonant strain sensor for harsh environment applications. IEEE Sensors Journal, 7(3-4):568–576
9. Song I, Lee B (2004) MEMS-based angular rate sensors, IEEE Proc. vol. 2. The 3rd Conf. of Sensors, Vienna, Austria, 2004, pp 650-653
10. Rodjegard H, Sandstrom D, Pelin P et al (2004) A novel architecture for digital control of MEMS gyros, IEEE Proc. vol. 3, The 3rd IEEE Conference on Sensors, Vienna, Austria, 2004, pp 1403-1406
11. Skvortzov V, Cho Y C, Lee B-L et al (2004) Development of a gyro test system at Samsung Advanced Inst. of Tech., PLANS Position Location and Nav. Symp., Monterey, California, 2004, pp 133-142
12. IEEE Std 1431-2004(R2010) IEEE Standard Specification Format Guide and Test Procedure for Coriolis Vibratory Gyros
13. FocusTest MXP-2 at http://www.focustestinc.com/benefits.htm
14. Motion Dynamics TES-3 at http://www.motiondynamic.com/v5/
15. Simon I, Billat S, Link T et al. (2007) In-situ pressure measurements of encapsulted gyroscopes. The 14th Int. Conf. on Solid-State Sensors, Actuators and Microsystems, Lyon, France, 2007, pp 1175-1178

Author: Christer Gerdtman
Institute: Department of Computer Science and Electronics
Street: Högskoleplan 1
City: Västerås
Country: Sweden
Email: christer.gerdtman@mdh.se

Photoplethysmographic Measurements of Finger/Toe Arterial Pulse Waveforms and Their Compound Time Domain Analysis

Matti Huotari, Kari Määttä, and Juha Kostamovaara

[2] Department of Electrical and Information Engineering, Electronics Laboratory, University of Oulu, Oulu, Finland

Abstract— **Optical methods, especially, photoplethysmography (PPG) is an interesting and valuable method in studying human circulatory physiology. This novel model makes it possible to estimate very accurately mechanical properties of arteries. This estimation is based on the analysis of the arterial pulse waveforms and their compound decomposition analysis derived from photoplethysmographic measurements.**

Measured PPG signal can contain information on autonomic neural system, gastric mobility, and stress states of the human under study. However, quantification of the mental stress level is not easy to confirm stressed state and also the validity of the index. It would be important to compare the stress indexes between the stress and resting stages.

We see that the PPG signal information is not yet fully understood. After analyzing the PPG signals in time domain it would be possible to uncover the characteristic features. These features will describe arterial properties, autonomic waving, human stress state, and also gastric mobility which will be shown in this paper. For example, high arterial stiffness is a symptom and increases the risk of cardiovascular diseases. Arteries stiffen normally as a consequence of age, but also because of arteriosclerosis. Age related stiffness occurs when the elastic fibers within the arterial walls begin to weaken due to age, but diseases as arteriosclerosis accelerate this process.

Keywords— **photoplethysmography, pulse wave analysis, wave decomposition, arterial stiffness.**

I. Introduction

The difference between elasticity of young and elderly arteries implies that their mechanical properties change strongly during aging. Thus, monitoring measurements of in lifetime from proper locations during cardiac cycles could be necessary to characterize properly the elastic properties of an artery. We propose a novel model of estimating these mechanical properties based on the analysis of the arterial pulse waveforms and their compound decomposition analysis derived from photoplethysmographic measurements. The NIR (over 900 nm) based photoplethysmograph is continuously recorded from the finger and toe arteries. Because of the exact pulse waveform, the separation between its intrinsic compounds as it appears in the recorded PPGs curve can be analyzed on different planes.

In principle, one heart beat cycle in finger and toe PPG is one at a time analyzed in each analysis procedure. In the second derivative PPG (SDNPPG), each wave consists of four systolic waves and one diastolic wave namely *a*-wave (early systolic positive wave), *b*-wave (early systolic negative wave), *c*-wave (late systolic reincreasing wave), *d*-wave (late systolic redecreasing wave) and *e*-wave (early diastolic positive wave). The amplitude of each wave is measured from the baseline=0. Over the years, various characteristic PPG points have been proposed, especially, the characteristic points of the finger PPG are numerous.

Specific features in the photoplethysmographic waveform could be used to identify normal and abnormal arterial stiffness. For example, cardiovascular diseases or even arterial stiffness does not cause any symptoms, but after the person exercises the symptoms appear. During the first symptoms 60% of the affected persons die. These persons are and have been in danger for long time. But measuring blood pressure is not enough, because it do not see the arterial stiffness at all. That's why we have been developing an optical device for arterial stiffness measurement and software for analysis of the measurement results.

The pulse waveform decomposition analysis, each wave consists of at least five compound waves, namely percussion wave, which is caused by the left ventricular contraction, tidal wave caused by the aorta elasticity release after the ventricular contraction, dicrotic wave caused by the reflection from the low body (bifurcation), and the pair of pre-ejection waves. The amplitude and area of each wave is measured from the baseline=0, with the values above the baseline being positive.

The use of multi-lognormal functions as a model can be justified in this case, because they well represent the vascular network with many asymmetric arterial double-branching and lognormal distribution of the length of capillary arteries [1].

Compliance, distensibility and can then be calculated based on pulse waveforms. Thus, the mechanical behaviour of peripheral human arteries can now be characterized non-invasively over the whole cardiac cycle [2]. The results obtained in vivo on human radial and brachial arteries show that an exact analysis of the pulse wave curves and their modifications is necessary to uncover two different vessels in a meaningful way [3]. Here we have been developing a photoplethysmograph for arterial stiffness measurement and its software for analysis of the measurement results.

K. Dremstrup, S. Rees, M.Ø. Jensen (Eds.): 15th NBC on Biomedical Engineering & Medical Physics, IFMBE Proceedings 34, pp. 207–210, 2011.
www.springerlink.com

II. MATERIALS AND METHODS

A. Instrumentation

The new PPG system consists of two optical measurement probes, one for a finger and the other for a toe, and a compound electronics unit for handling the optically measured signals based on phase sensitive detection (PSD). The measurement head consists of two LEDs and one large area semiconductor photo detector for collecting light emitted by the LEDs through the finger or toe. The compound electronic unit contains electronics for driving the LEDs, two preamplifiers for signals, four PSD channels, an analog-to-digital converter and an USB-interface for transferring the digitized results onto a laptop. In addition, parallel methods of electrocardiogram (ECG) and phonocardiogram (PCG) have been measured simultaneously to support the later PPG analyses. The subjects were measured a.m. in supine position without coffee or tobacco in the morning. Each measurement took about five minutes to obtain consecutive 300 pulses, of which parallel 10 to 20 most stable were selected for pulse wave decomposition analysis. In this research, the index finger and index toe were always under measurement..

B. Data handling

After measurements we draw the toe PPG as a function of the finger PPG which describe complex non-harmonic motion in all cases. The phase shifts were so that delay semantics can be difficult to define and used with a causal system relations. In the pulse wave decomposition analysis, each pulse wave was divided into four lognormal wave components. The compound decomposed wave-forms were after computation and fitting visually compared to the original waves to make sure the best fitting. This comparison and the four lognormal functions can be used to obtain a residual error curve and its chi-square value will describe the goodness of the fit. The use of multi-lognormal functions can be justified as they well represent the vascular network with many asymmetric arterial double-branching and lognormal distribution of the length of capillary arteries.

The Origin 7.5 (OriginLab®) lognormal procedure was utilized for analyzing the pulse waves in time domain to obtain best mathematical fitting with minimal residual error. In this procedure, the Levenberg-Marquart algorithm (LMA) is a very popular curve-fitting algorithm used in many applications for solving non-linear curve-fitting problems, e.g. logarithmic normal function curves. LMA provides a numerical solution to the problem of minimizing a function which can be nonlinear, over a space of parameters of the function. In our case we have four lognormal components which have 4x3 parameters and the correlation coefficients (R^2) are 0.995 or over.

III. RESULTS AND DISCUSSIONS

In Figure 1 it is shown PPG waveforms of the pulse wave signal at 940 nm measured through the finger tip (straight) pulse wave and the toe tip (dash). Signals are normalized for the amplitude.

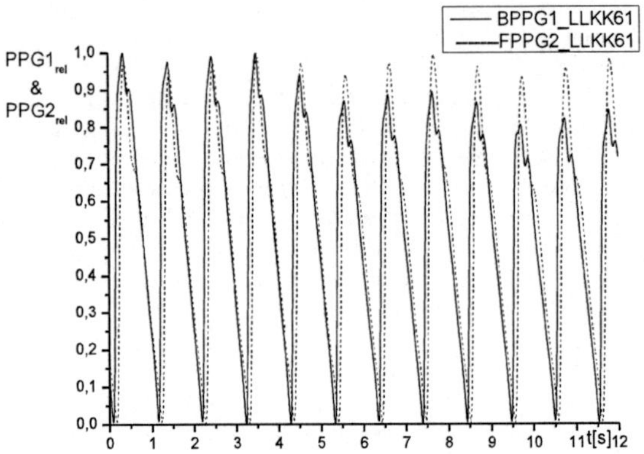

Figure 1. PPG waveforms through the finger & toe tip pulse waves for 12 s of a 61 y male person.

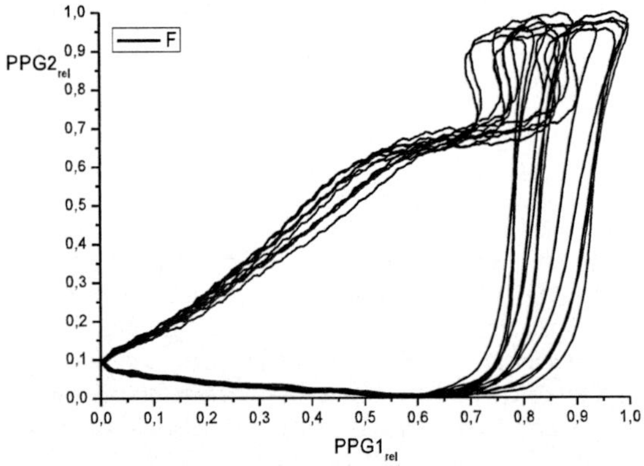

Figure 2. The toe PPG waveforms as a function of the finger PPG of the above male person. Compare Lissajous-figure.

Figure 3 shows the causal relation between the PPG1 and PPG2 in a complex way. In Figure 2 it is shown an analyzed compound finger PPG waveform. It contains the typical PPG components. In this case the PPG waveform analyses are covering the following four pulse components in each pulse wave: percussion (P), tidal (T), dichrotic (D), and peripheral reflection component.

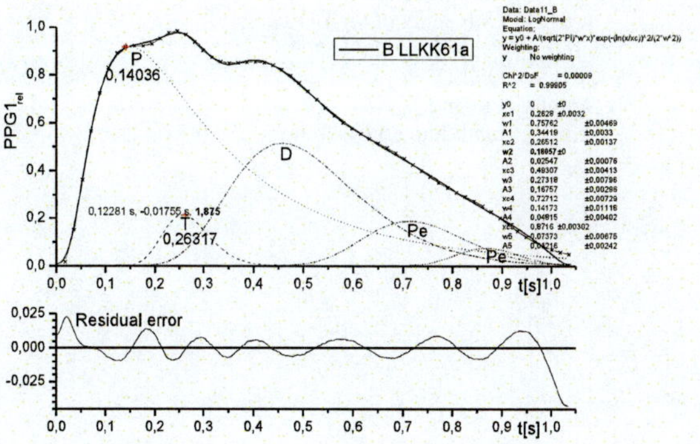

Figure 3. An analyzed finger PPG waveform for the Figure 1 which contains the percussion component (peak value 0.14036), the tidal component (0.26345), the dicrotic component, and the peripheral reflection components. In the lower parts of the figure there is shown the residual error. See text.

Percussion is caused by the contraction of the heart left ventricular muscle. The second component is the tidal wave, occurring during the later part of the systole, caused by the elastic properties of aorta. The dichtrotic component is the reflected pulse from lower periphery elasticity and vessel branching [2]. Peripheral reflection is another reflection component occurring towards the end of each pulse wave. All these components appear in the finger pulse wave.

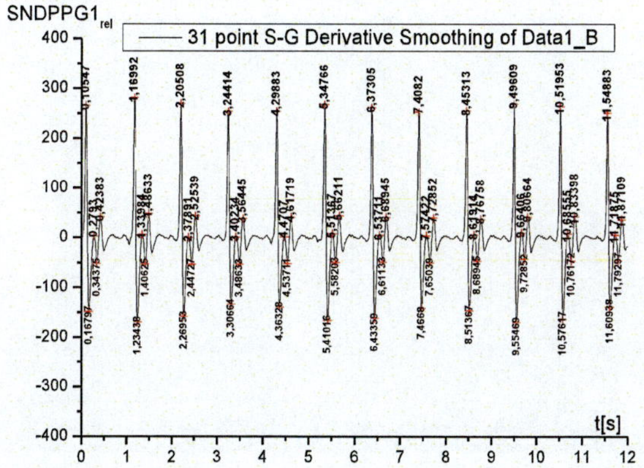

Figure 4. The 2nd PPG1 waveform for the PPG1 in Figure 1. The 2nd derivative of PPG1 contains the parameters **a, b, c, d,** and **e** which were found by the search window (SW).

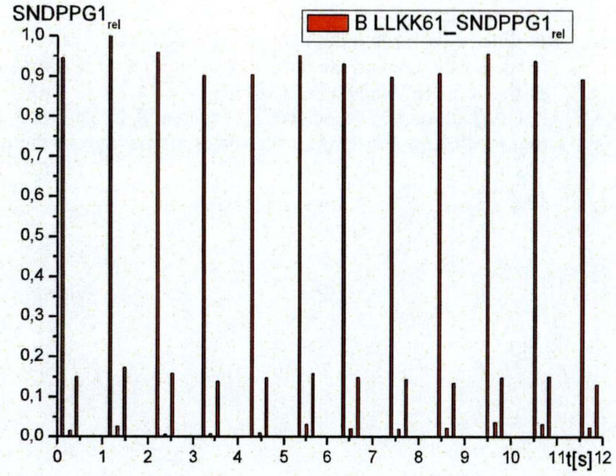

Figure 5. The SNDPPG1 waves as bars for the PPG1 in Figure 1. The 2nd derivative of PPG1 contains the parameters a, c, and e, or shortly ACE-figure.

In Figure 5, the results of the PPG1 measurement can be in the form of the ACE-figure. For example, in this case the bar C is always positive. This research studies the potential of PPG for early diagnosis of arterial stiffness. PPG technology is widely available at the pulse oxygen saturation measurements and is relatively cheap and does not require special expertise. PPG can be utilized for detecting pulse wave-forms. In the blood circulatory system, the arterial pulse wave reflections depend on the arterial wall stiffness. This study includes also creating a mathematical model for pulse wave-forms for analyzing the four wave components of the human pulse.

Photoletysmographic equipment is cheap and relatively simple to use. The PPG will be a potential method for analyzing arterial stiffness. It is possible to mathematically decompose the measured PPG compound wave into five components, namely percussion, tidal, dichrotic, and peripheral reflection. The tidal component proved to be the most interesting in this study. The division into the four components is easier for younger than older persons. However, older persons with good physical fit (like in this case, the male 61 y person) resemble younger persons more than those with less physical exercise. PPG measurements taken from a finger are easier to analyze and divide into the four components than those taken from a toe.

Acknowledgment

Acknowledgment is given for the Finnish Cultural Foundation for the support to one of the authors (MH).

References

1. H Qian, J B Bassingthwaighte (2000) A Class of Flow Bifurcation Models with Lognormal Distribution and Fractal Dispersion, Journal of theoretical Biology 205, 261-268.
2. A G Scandurra, G J Meschino, L I Passoni, A L Dai Pra, A R Introzzi and F M Clara (2007) Optimization of arterial age prediction models based in pulse wave, Journal of Physics: Conference Series **90** 012080
3. A Qasem, A Avolio (2008) Determination of aortic pulse wave velocity from waveform decomposition of the central aortic pressure pulse, Hypertension 51;188-195

Author: Matti Huotari
Institute: University of OULU
Street: Pentti Kaiteran katu 1
City: OULU
Country: Finland
Email: matti.huotari@ee.oulu.fi

Quasi-stability Theory: Explaining the Inevitability of the Magic Numbers at Various Stages from Subatomic to Biological

K. Naitoh

Waseda University, Faculty of Science and Engineering, Tokyo, Japan

Abstract— **A statistic fluid-dynamical model derived based on a quasi-stability concept extended from our previous reports (Naitoh, JJIAM, 2001, Artificial Life Robotics, 2010) reveals the magic numbers observed in various systems including living beings and non-living systems. First, this model explains the reason why particles such as biological cells, nitrogenous bases, liquid droplets, and child atoms resulting from the fission of uranium 235 have bimodal size ratios of 1:1 and about 2:3 between the golden and silver ratios. Next, a higher order of analysis also clarifies the other asymmetric ratios, i.e., the super-magic number of about 1:3.5, 1:2.5, 1:2.1, 1:1.78, 1:1.35, and 1:1.27 in various systems including amino acids, proteins, atomic systems, and atoms appearing at the cold fusion. This paper also shows that the same theory holds true for several levels of parcels from baryons to stars in the cosmos: specifically, at the levels of nuclear force, van der Waals force, surface tension, and the force of gravity.**

Keywords— **magic number, statistic fluid dynamics, asymmetry.**

I. Introduction

Let us consider the silver and golden ratios [the silver ratio of $1:\sqrt{2}$ (about 1.41) and golden ratio of $1:[1+\sqrt{5}]/2$ (about 1.62)]. Purines and pyrimidines in biological base pairs, biological cells after divisions, liquid droplets, and stars in the cosmos also have a fusion of symmetry of 1:1 and asymmetry of around 2:3. [1-4] (Fig. 1) A neutron impacting uranium 235 produces smaller child atoms that often have an asymmetric weight ratio of about 2:3 between the silver and golden ratios. In contrast, varying the impact speed of neutrons results in a nearly symmetric division of uranium 235. [5] There are also mesons with "two" quarks and baryons with "three". [6] A model we developed previously based on fluid dynamics has qualitatively revealed the reason for the fusion of symmetry of 1:1 and asymmetry of around 2:3 in the fractals found in nature. [1-4] The fusion of symmetry and asymmetry appears in various systems from atoms to stars, because each system commonly stems from "breakup of flexible particles deformed".

The present report gives a further explanation of the inevitability of the fusion of asymmetric and symmetric ratios.

Then, a higher order of analysis clarifies the reason why size ratios over 2.0 are also seen in other biological molecules such as amino acids as well as liquid droplets.

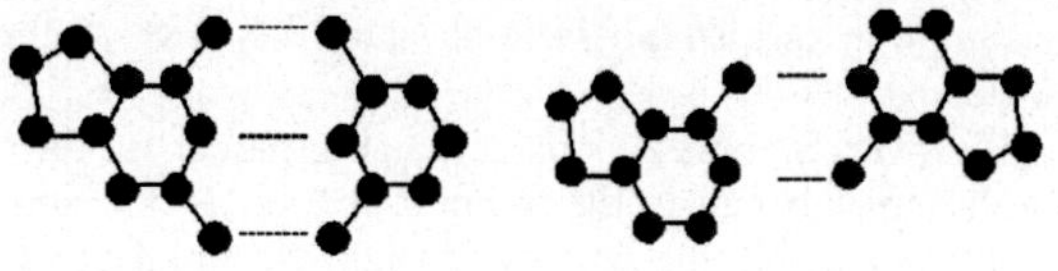

(a) Asymmetric Watson-Crick type (b) Symmetric type

Fig. 1. Asymmetric and symmetric base pairs.

II. Model [1-4]

Here, we define a parcel as a flexible spheroid having two long and short radii of $a(t)$ and $b(t)$ dependent on time *t*, for a nitrogenous base in biological base-pairs of nucleic acids hydrated with a lot of water molecules, a biological cell, the aggregation of neutrons and protons in each child atom resulting from the fission of a uranium 235 atom, and a star at breakup in the cosmos. The parcel becomes a sphere of the radius $r_d(=[ab^2]^{1/3})$ under an equilibrium condition. The deformation rate $\gamma(t)$ is defined as $a(t)/b(t)$, while a sphere without deformation corresponds to $\gamma = 1$.

Then, we consider the form of two spheroid parcels connected at the replication stage of biological base-pair, at cell division, at the time of the breakup processes of uranium 235, and at division of star. We derive a theory for describing the deformation and motions of the two connected spheroid parcels having two radii of r_{d1} and r_{d2} under equilibrium conditions and two deformation rates of γ_k $[k=1, 2]$, while the size ratio of two parcels is defined by $\varepsilon = r_{d1}/r_{d2}$.

We model the relative motion between the two parcels, nonlinear convections inside the parcels, and the interfacial force at the parcel surface. The interfacial force is evaluated in the form of σ / r^m where *m* and σ are constants and *r* is

K. Dremstrup, S. Rees, M.Ø. Jensen (Eds.): 15th NBC on Biomedical Engineering & Medical Physics, IFMBE Proceedings 34, pp. 211–214, 2011.
www.springerlink.com

the curvature of parcel surface. Several types of forces such as nuclear force, van der Waals force, surface tension, coulomb force, and gravity can be explained by varying m. The relation $m = 1$ implies the surface tension of liquid. The mean density of the parcels is ρ_L.

We assume that the convection flow inside parcel is irrotational, i.e., potential one. There are random collisions of water molecules and electrons with the parcel such as biological molecules and cells. It is stressed that these random collisions from the outer region induce potential flow inside the flexible continuum particle, i.e., irrotational flow, because the fluctuations of impulsive starts and stops generate potential flow. This potential flow is also applicable, because the fluctuations are close to those of thermal fluctuations, less dissipative for atoms and molecules. (Fluctuation dissipation theorem) Then, this assumption is also possible to stars, because of the large scale.

Moreover, we must consider that a parcel is not often a continuum, because the number of nucleons and water molecules inside the parcels for atom and nitrogenous base will be less than the order of 1,000. The scale for averaging, i.e., the minimum scale representing the phenomenon, will be smaller than that in continuum mechanics. Thus, this small averaging window leads to a weak indeterminacy of physical quantities such as parcel shape, deformation and density because of discontinuity of nucleons and molecules.

It is stressed that analyses with the indeterminacy clarify the dynamical processes of two parcels connected, which also have various shapes except for spheroid.

Here, we derive the relation between dimensionless deformation rate $\gamma_k (\equiv a_k / b_k \; [k = 1, 2])$ of each parcel dependent on dimensionless time

$$\bar{t}_k = \sqrt{\frac{8\sigma}{\rho_L r_{dk}^{\;2+m}}}\, t \quad [k = 1, 2]$$

and the size ratio of the two parcels of $\varepsilon = r_{d1} / r_{d2}$.

The stochastic governing equation having indeterminacy can be described as

$$\frac{d^2}{d\bar{t}_i^{\,2}}\gamma_i = \{m_{ci}(\frac{d}{d\bar{t}_i}\gamma_i)^2 + m_{cj}(\frac{d}{d\bar{t}_j}\gamma_j)^2 + m_{si}\,\gamma_i^{\frac{5}{3}-\frac{2}{3}m} + m_{sj}\,\gamma_j^{\frac{5}{3}-\frac{2}{3}m}\} / Det + \delta_{st}$$

$$[\text{for } i = 1, 2.\; j = 1, 2.\, i \neq j] \tag{1}$$

with $m_{ci}(\gamma_i, \gamma_j, m), m_{cj}(\gamma_i, \gamma_j, m), m_{si}(\gamma_i, \gamma_j, m), m_{sj}(\gamma_i, \gamma_j, m)$
where the parameter δ_{st} denotes random fluctuation. [1-4]

It is stressed that this system is not the simple two-body problem of rigid body, because of flexible nonlinear deformations of the parcels. [Equation 1 is derived only by the above assumptions and also purely mathematical transformation. The long derivation of Eq. (1) in Ref. 2 is confirmed by the referees, although only the stochastic term δ_{st} is not in Ref. 2.]

We then define the deviation from a sphere as y_i, which is equal to $\gamma_i - 1$. Taking the first order of approximation in the Taylor series leads to

$$\frac{d^2 y_i}{d\bar{t}_i^{\,2}} = [-\frac{2}{3}(3 - \varepsilon^3 - 2\varepsilon^{2+m})\left(\frac{dy_i}{d\bar{t}_i}\right)^2 + 3(3 - \varepsilon^3) m y_i - 4\varepsilon^{1+m}\left(\frac{dy_j}{d\bar{t}_j}\right)^2 + 12\varepsilon^{1+m} m y_j]$$
$$/[3(\varepsilon^3 + 1)] + \delta'_{st}, \tag{2}$$

where the parameter δ'_{st} denotes random fluctuation.

Equation 2 shows that a symmetric ratio of 1.0 ($\varepsilon = 1$) makes the first term on the right-hand side of the equation zero, while an asymmetric ratio of $\sqrt[3]{3}$ around 1.5 ($\varepsilon^3 = 3$) makes the second term zero for each m. [The size ratios of 1.00 and approximately 1.50 can be described by the unified number of the n-th root of n.]

We define a system as being quasi-stable when only one term on the right-hand side of the differential equation system governing the phenomenon is zero. Then, the system of two parcels connected is relatively quasi-stable because d^2x/dt^2 becomes smaller when the size ratio of connected parcels takes the values of $\varepsilon = 1$ or $\varepsilon^3 = 3$.

III. First-order of analysis

Here, we can classify the five bases of adenines (A), guanines (G), cytosines (C), thymines (T), and uracils (U) into two groups: purines and pyrimidines. Purines, i.e., A and G, have a relatively large size, while pyrimidines, i.e., C, T, and U, are small. Asymmetric base pairs such as the Watson-Crick type of about 1: $\sqrt[3]{3}$ are used in living beings. This grouping specifically refers to the asymmetric size ratio of purines and pyrimidines of around 1.50 in their hydrogen bonds within DNA and RNA, although a symmetric size ratio of 1.00 is often observed in RNA. [1-4] Symmetric and asymmetric size ratios are also observed at the cell level of microorganisms such as yeast. [1]

These ratios of 1:1 and about 1:1.5 are also similar to those of child atoms generated by the breakup of uranium 235.

As Eqs. 1 and 2 show a slightly vague solution for the phenomenon, this indeterminacy also implies that size variations of ε are possible in a limited range. This indeterminacy permits the weak possibility of sizes around 1:1 and also around 1: $\sqrt[3]{3}$, i.e., between 1:1 and about 2:3.

The concept of quasi-stability is weaker than neutral stability. The quasi-stability concept is necessary for living beings, because stronger stability cannot bring variations, i.e., adaption for environmental change and evolution.

The quasi-stable ratios of 1: $\sqrt[n]{n}$ for n=1 and 3 appear for each m (see Eq. 2). This universality also leads to the possibility that the present model can be applied for several levels of parcels from baryons to stars in the cosmos [8]: specifically, at the level of nuclear force, coulomb force, van der Waals force, surface tension, and force of gravity.

It is also well known that several atoms in nature have the "number" ratios of protons and neutrons between 1:1 and 2:3. (Our previous reports also clarify the reason why larger atoms have larger "number" ratios close to 2:3. This is because of mass conservation law, i.e., because heavier particles such as purines and protons can be generated with less numbers. [2, 4])

IV. THIIRD-ORDER OF ANALYSIS

Then, the third order of approximation in the Taylor series for Eq. 1 brings

$$\frac{d^2 y_1}{d\tau_1^2} = \frac{1}{m_2}\left\{ a_{31}y_1 + a_{32}y_1^2 + a_{36}y_1^3 + \left(b_{31} + b_{32}y_1 + b_{35}y_1^2\right)\left(\frac{dy_1}{d\tau_1}\right)^2 \right\}$$

with

$$\begin{aligned}
a_{31} &= -9\left(\varepsilon^3 - 3\right)\left(\varepsilon^2 + 1\right)^2 \\
a_{32} &= \frac{3}{4}\left(\varepsilon^3 - 3\right)\left(\varepsilon^6 + 34\varepsilon^3 + 33\right) \\
a_{36} &= \frac{1}{2}\left(\varepsilon^3 - 3\right)\left(\varepsilon^6 - 42\varepsilon^3 - 171\right) \\
b_{31} &= 6\left(\varepsilon^3 - 1\right)\left(\varepsilon^3 + 1\right)^2 \\
b_{32} &= -2\left(5\varepsilon^9 + 5\varepsilon^6 - 21\varepsilon^3 - 21\right) \\
b_{35} &= \frac{8}{3}\left(5\varepsilon^9 + 9\varepsilon^6 - 27\varepsilon^3 - 63\right)
\end{aligned} \tag{3}$$

, when the terms related to parcel 2 are eliminated. [7] [We consider only the disturbance for parcel 1, because the system of two parcels is mathematically symmetric.]

Equation 3 shows the quasi-stable ratios: 1:1 and about 1:1.27, 1:1.35, 1:1.44, and 1:3.50. (Table 1) The quasi-stable ratios of about 1:1, 1:2.7, 1:1.35, and 1:1.44 correspond to those for identical base pairs such as U-U and A-A, A-T, G-C, and A-U among the base pairs in nucleic acids, respectively. (Fig. 2)

Let us think about 1:3.50 in Table 1. It is also stressed that liquid fuel droplets generated by injectors and child atoms broken up from the fission of uranium 235 also have a threefold variation of sizes at the maximum [9] and also that the molecular weights of the twenty types of amino acids show a threefold variation between 240 of cysteine as the maximum and 75 of glycine as the minimum. The higher-order of analysis clarifies the ratios over 2:3 in several systems.

First order of approximation in the Taylor series mathematically shows the quasi-stable ratios for small disturbances of deformation and deformation speed, while higher order of analyses bring the ratios for larger disturbances. Thus, some amino acids, which are more flexible than nitrogenous bases and can largely deform due to disturbances, can be explained by the higher order of the Taylor series.

Table 1. Size ratios of parcels (The third order of approximation)

Quasi-stability	surface tension term	convective term
1st order	1:1.44	1:1
2nd order	1:1.44	1.27:1
3rd order	1:1.44 1:3.50	1.35:1

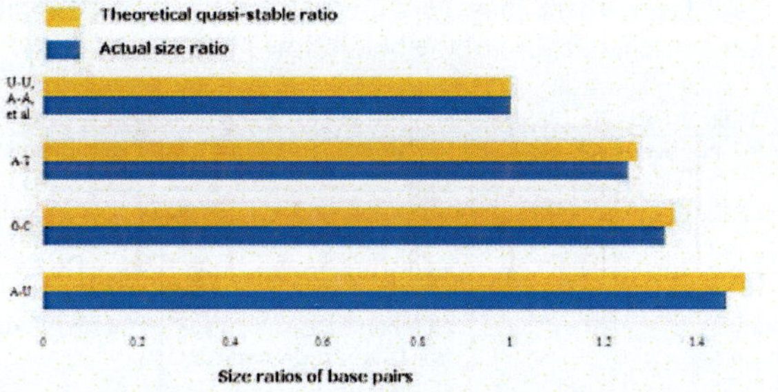

Fig.2. Four types of base pairs clarified by the higher-order of Taylor series.

V. HIGHER-ODER OF ANALYSIS

Next, we take a higher order of the Taylor series for Eq. 1. Odd-numbered terms result in other quasi-stable ratios. The third term in the Taylor series results in a quasi-stable ratio of about 3.5, the fifth term in ratios of about 2.5 and 2.1, and the seventh in a ratio of about 1.78. Table 2 demonstrates the biological molecules corresponding to the quasi-stable size ratios derived. Moreover, the quasi-stable ratios shown in Table 2 also explain the inevitability of various size ratios of amino acids. (Fig. 3) [4]

An important point is that the present analysis (Table 2) also reveals the magic numbers appearing in the frequency ratios of protons and neutrons in various atoms with halo structures and the atom groups generated by the cold fusion in Fig.4. [9]

VI. CONCLUSION AND OUTLOOK

The statistic fluid dynamics reveals the inevitability of various biological molecules and atomic systems. The inevitability on the frequency ratios are shown in reference [3]. There will also be some proteins close to unstable conditions, which have the sizes larger than 1:3.5 in Table 2, because of the flexibility.

The super-string theories (Types I, IIA, IIB, et al.) and the M theory [6] will not be universal, because of various shapes of super-strings. This is similar to the fact that a biological cell includes various shapes of molecules having strings and rings such as DNA and RNA with nitrogenous bases, proteins with amino acids, sugars, and lipids, because these biological molecules are also with various types of symmetries. This similarity between subatomic system and biological cell can be explained by repeats of strings and particles for various systems from subatomic to cosmic. [4, 8] Thus, the super-strings may be constructed with the smallest spheroid-like particles having the ratios such as 1:1, about 1.1.44, and 1:3.5, which may be the Planck length. Moreover, there may be super-universes having the above size ratios, because of the black flow [10]

Table 2 Quasi-stable ratios observed in atoms and biological systems.

n-th term in Taylor series	Quasi-stable ratio	Atoms	Bio-molecules
1	1:1 & about 1.1.5	Main size ratios of child atoms broken from uranium 235. Number ratios of neutrons and protons in atoms	Nitrogenous bases in DNAs and RNAs (Purine:Pyrimidine)
3	about 1:3.5	Higher limit of size ratios of child atoms broken from uranium.	Amino acids
5	about 1:2.5 & 1:2.1	The intermediate size ratios of child atoms broken from uranium.	rRNA (23S:16S)
7	about 1:1.78	The intermediate size ratios of child atoms broken from uranium.	Ribosome (50S:30S)
9, 11	unstable	-	-

REFERENCES

1. Naitoh K Oil & Gas Science and Technology 54 205. 1999
2. Naitoh K Japan Journal of Industrial and Applied Mathematics 18-1 75. 2001
3. Naitoh K Artificial Life and Robotics 13 10. 2008
4. Naitoh K Artificial Life and Robotics 15 117. 2010. (also presented at the International and Interdisciplinary Workshop on Novel Phenomena in Integrated Complex Sciences, 2010 and submitted to a Journal.)
5. El-Wakil M Nuclear Heat Transport, DoubunShoin. 1971
6. Henley E M and Garcia A 2007 Subatomic physics. World Scientific.
7. Naitoh K, Hashimoto K, and Inoue. Proceedings of 6th WCB, Singapore. 2010. (Also accepted for Artificial Life and Robotics.)
8. Naitoh K J. of Cosmology 5, 999. 2010.
9. Miley GH and Patterson JA. J. of New Energy. 1-3. 5, 1996.
10. Kashlinsky A, Atrio-Barandela F, Ebeling H, Edge A and Kocevski D. The Astrophysical Journal Letters. 712-1. 2010.

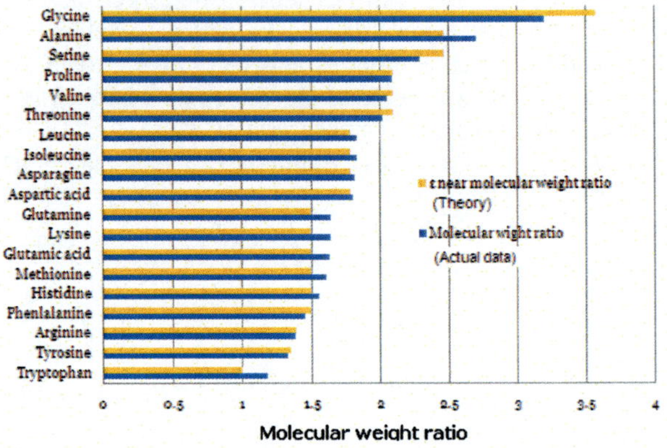

Fig.3 Molecular weight ratios of amino acids (Cystine / the other amino acid) and the corresponding theoretical quasi-stable ratios.

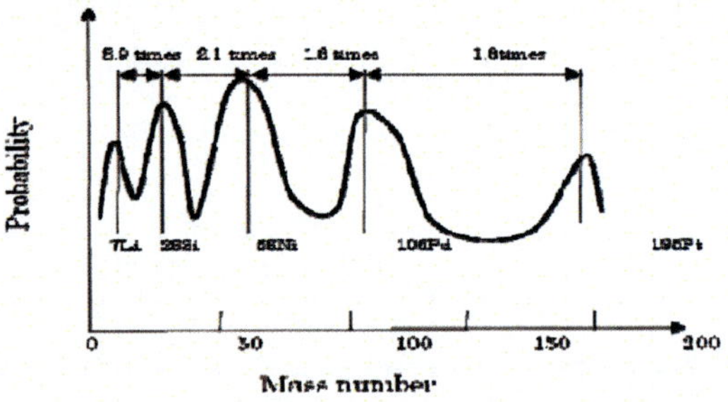

Fig.4. Quasi-stable atoms generated by cold fusion.

ACKNOWLEDGMENT

This article is part of the outcome of research performed under a Waseda university Grant for special research project (2009B-206).

Author: Ken Naitoh
Institute: Waseda University, Faculty of Science and Engineering
Street: 3-4-1 Ookubo
City: Shinjuku, Tokyo
Country: Japan
Email: k-naito@waseda.jp

The Engine: Inducing the Ontogenesis

K. Naitoh

Waseda University, Faculty of science and engineering, Tokyo, Japan

Abstract— **The stochastic Navier-Stokes equation essentially solves the mysteries underlying the morphogenetic processes of human beings having arms, legs, and inner organs, because seventy percent of living things are filled with water. First, a computational fluid dynamics code used on super-computers reconstructs the three-dimensional structures of systems such as the human body with complex convexoconcave shapes. Second, the theory also explains the reason why inner organs such as heart and liver are left-right asymmetric at the later stage of the developmental process. Analysis based on the present fluid-dynamics on space and the molecular theory on time proposed in our previous reports will bring a new insight on the spatiotemporal structure of ontogeny.**

Keywords— **Ontogenesis, organs, brain, asymmetry.**

I. Introduction

Researches involving the use of continuum mechanics [1, 2] have revealed some aspects of large structures such as trees, shells, and cells. After an informatics study[3] showed the three-dimensional bio-structures of trees and cells, a lot of research has been done to model some aspects of developmental processes. However, a design diagram of natural morphogenetic process of human beings remains a mystery.

Understanding of the spatiotemporal structure controlling the living beings is very important in order to reveal the mechanism of cancer, which is related to apoptosis. Let us think about the mystery by classifying into spatial and temporal aspects.

One of the main spatial mysteries is the bipolar order of symmetry: the left-right asymmetric Watson-Crick base pairs in DNA and symmetric ones in RNA, the asymmetric and symmetric divisions of microorganisms and stem cells, and the left-right asymmetric liver and symmetric kidneys. [4] We fundamentally revealed the reason of the bipolar order of symmetry. [5-7] However, we do not know how the three-dimensional structures of systems such as embryos and the brain are generated in detail.

The main temporal mystery concerns the basic molecular instrument regulating the biological rhythm common to the cell cycle, stem cell cycle, circadian clock [8, 9], the neural network circuits, the heart beat, and so on.

Our other reports on living beings and artifacts presented some recent clues to clarify these mysterious processes. [10-18]

By synthesizing and extending our previous studies, we will outline here the fluid-dynamic mechanism controlling the spatiotemporal network of molecules and cells.

II. Brain

Observations of the development of the human brain reveal that the bones of the skull above the neck become increasingly larger and a lot of soup-like fluid for generating flexible cells, including nerves and synapses, enters the skull from the body. This process is qualitatively similar to the intake process of an internal combustion engine, because the increase in cylinder volume as the piston descends corresponds geometrically to the development of the skull and also because the human neck resembles the intake port that functions as the throat of an engine (Fig. 1). [16]

The flow of air during the intake process of a piston engine and fluid flow in the brain, including water as the main component, can be approximated as being incompressible. [The intake process of an engine at low engine speeds is incompressible, although the compression process shows a strong density variation with time in the zero-Mach number regime.]

We employ the incompressible stochastic Navier-Stokes equation with a random force [19, 20] as the basic governing equation that describes the unsteady flows in the intake process of engines as well as the flexible dynamic motion of the soup-like fluid for generating brain cells. The equation can be written as

$$\vec{F} = \begin{bmatrix} \sum_i \dfrac{\partial u_i}{\partial x_i} \\ \dfrac{\partial u_i}{\partial t} + \sum_j u_j \dfrac{\partial u_i}{\partial x_j} + \dfrac{\partial p}{\partial x_i} - \dfrac{1}{\mathrm{Re}} \sum_j \dfrac{\partial^2 u_i}{\partial x_j^{\,2}} \end{bmatrix} = \begin{bmatrix} \varepsilon_1 \\ \varepsilon_2 \end{bmatrix}$$

where u_i, p, ε_i, t, and Re denote the dimensionless quantities of velocity components in the i-direction,

K. Dremstrup, S. Rees, M.Ø. Jensen (Eds.): 15th NBC on Biomedical Engineering & Medical Physics, IFMBE Proceedings 34, pp. 215–218, 2011.
www.springerlink.com

pressure, random force, time, and Reynolds number, respectively. Details on the boundary and initial conditions and numerical methods are depicted in Refs. 10 and 19.

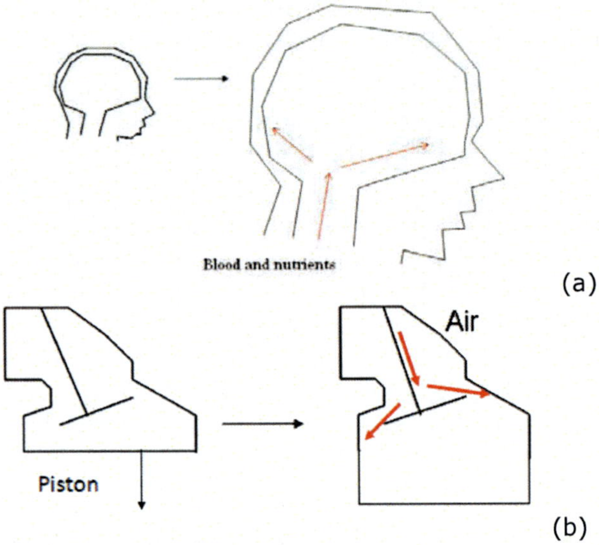

Fig. 1 Topological similarity of the brain and a piston engine.
(a) Brain volume increasing during cerebral development (b) Cylinder volume increasing during the intake process of a piston engine

In our previous report [16], we qualitatively revealed the similarity between the brain shape and engine flow, by solving the incompressible stochastic Navier-Stokes equation on a supercomputer using a finite difference method with a higher order of accuracy. In the computational results we can see two spheres representing the eyeballs and a cylindrical form representing the nose.

Figure 2(a) shows an image of the human brain obtained with magnetic resonance imaging (MRI) at a horizontal cross section which includes the two eyes; Fig. 2(b) is the result of the corresponding computation performed with the stochastic Navier-Stokes equation. The computational figure was visualized at a crank angle of 72 degrees after top dead center (ATDC) in the intake process, because at that point the cylinder volume of the piston engine considered here is about the same as that of the skull.

The configurations around points A and B in Fig. 2(a) are similar to those in Fig. 2(b). The triangular region around point C in Fig. 2(a) can also be seen in Fig. 2(b). It should be noted that the two circular regions at the top in Fig. 2(a), which are the two human eyeballs, can also be seen in Fig. 2(b). The MRI tomogram agrees fairly well with the computational results.

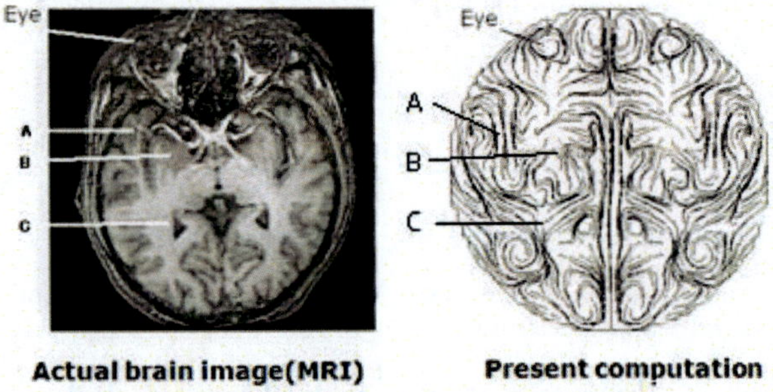

(a) MRI image [23]. (b) Computational results
Fig. 2. MRI tomogram and computational results for the brain.
(a) http://riodb.ibase.aist.go.jp/brain/welcomej.html [23] (b) Present computation based on the Navier-Stokes equation

It is stressed that the brain stem is similar to the intake-valve stem in Fig. 1. Blood and nutrients, including water, enter the skull through the area around the brain stem. It should also be noticed that the expansion angle of the brain near the neck is geometrically very similar to that near the intake port of the engine. [16] The expansion angle of the brain is optimized during the natural evolutionary process, while engine designers optimize the pentroof angle of a piston engine.

Our previous report [21] also shows that four main blood vessels in the brain are simulated well by the numerical code. The blood vessels in brain correspond to the path lines in engine computed.

III. Organs

Next, let us think about the developmental process of human beings, which include various organs, arms, and legs. Figure 3 demonstrates the shapes of the brain and human embryo. It is stressed that both the brain and embryo have the valley in the lower middle area and also a pipe for picking the blood flow coming from the body. Then, both brain and embryo gradually expand, while blood flows with a lot of water enter into the tanks. Thus, the shapes of the brain and embryo will topologically be similar, although qualitatively. (Fig. 3) It is stressed that the weight of unborn baby is almost same as that of the brain in adult human being.

Therefore, the similarities between engine flow and brain shape shown in Fig. 2 and between engine flow and embryo in Fig. 3 lead to that between engine flow and embryo. (Fig. 4)

The valleys can be generated between the vortices in Fig. 2(b), because the regions having large deformation between vortices are destroyed due to the strong stress, which induces apoptosis. Thus, the principle of "no cells between vortices" can also be applied to the embryo in Fig. 4, which brings separations such as those between the main body and arms.

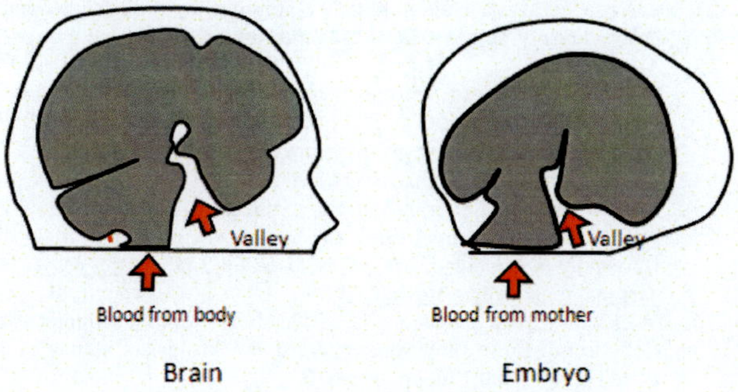

Fig. 3 Topological similarity between brain and embryo.

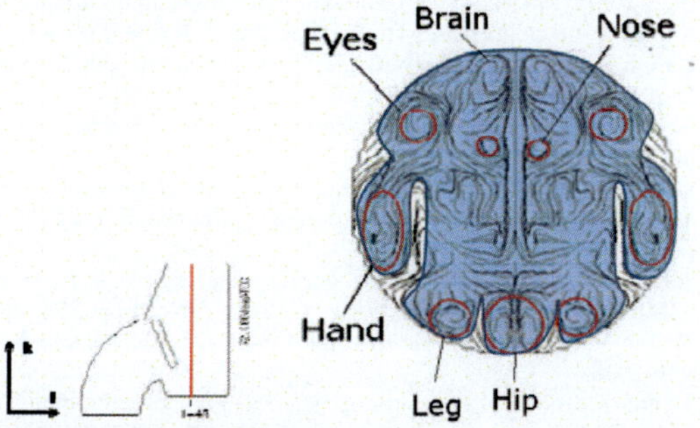

(a)

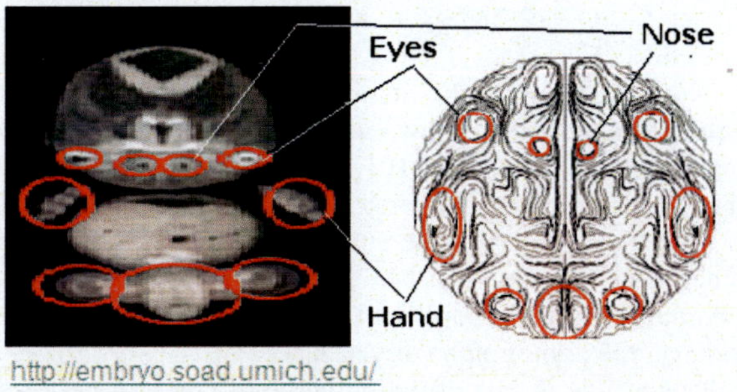

(b)

Fig. 4 Similarity between embryo [25] and the flow field in engine computed. Eyes, nose holes, hands, and legs computed are shown by red circles on the cross section of I= 45 close to the front surface of the human body. (a) Present computation (b) Comparison with actual image and our computation.

Next, let us think about the reason why asymmetric organs are generated at the later stage of the morphogenetic process, whereas the outer geometries of the piston engine and uterus are left-right symmetric and also the early stage of the intake process induces left-right symmetric distributions of flow patterns. An important point is that the later stage of the intake process in the piston engine demonstrates some left-right asymmetric vortices in the relatively inner area, although the other vortices around the outer surface are symmetric. (Fig. 5) The principle of "inner asymmetry and outer symmetry" is observed in both engine vortices and biological organs.

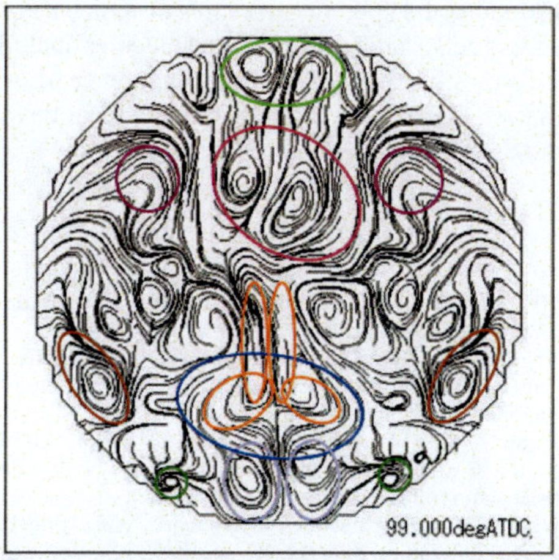

Fig. 5 Inner asymmetry and outer symmetry observed in the inner ross section of I = 56 for the engine flow, which is close to the back in the human body. (Orange and pink regions are left-right asymmetric, while brown, green, and purple parts are symmetric.)

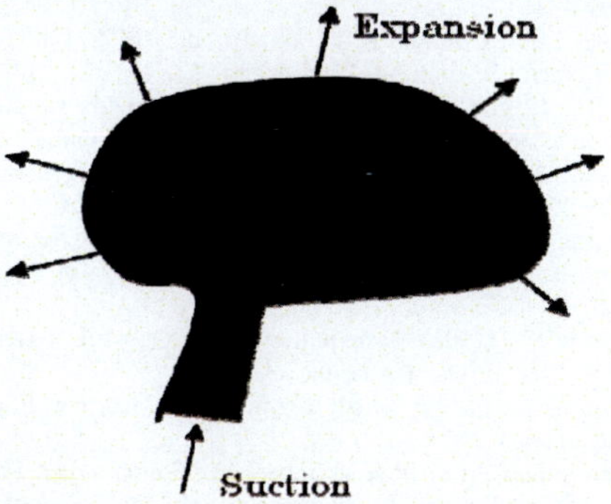

Fig. 6 Fundamental topology common to engine, embryo, and brain: expanding tank and suction port.

IV. CONCLUSIONS

This report proposes a model for revealing some aspects underlying the developmental processes of living things, including human beings. The topological structure of convexoconcave repeats with a fractal nature as the forms such as corrugation in the human brain and the human body with arms and legs. (See Fig. 6) It can also be understood from the fact that plantar energy centers for acupressure are connected with various organs such as stomach and liver: the other fractal of various organs in the planta pedis.

Moreover, an attempt was made to clarify the further developmental mechanics on the temporal feature. [28-33]

A bio-grammar is that, after natural thermo-fluid dynamic forces induce the specific structures of molecules, molecular networks, cells, and organs, the molecules including DNA and enzymes fix the timing of the emergence of organs.

However, so many mysteries related to developmental processes still remain unsolved.

References

1. Thomson D. (1961) On Growth and Form, Cambridge University Press. Pp.1-359
2. Ingber D.E. (1988) The Architecture of Life. Scientific American. 52.
3. Prusinkiewicz P. and Lindenmayer A. (1990) The Algorithmic Beauty of Plants: Springer-Verlag. Pp. 1-228
4. Scott G. (2005) Developmental biology, 8th edition. SINAUER. 1-750
5. Naitoh K. (2001) Cyto-fluid Dynamic Theory. Japan Journal of Industrial and Applied Mathematics **18**-1, 75-105.
6. Naitoh K, (2008) Stochastic determinism underlying life. Artificial Life and Robotics, Springer, **13,** pp. 10-17
7. Naitoh K (2010) Onto-biology. Artificial Life and Robotics, 15: pp. 117-127 .
8. Lowrey P and Takahashi J. (2004) MAMMALIAN CIRCADIAN BIOLOGY: Elucidating Genome-Wide Levels of Temporal Organization. Annual Review of Genomics and Human Genetics, **5**, 407-441
9. Bear M.F., Connors B.W., and Paradiso M.A. (2007) Neuroscience, Lippincott Williams & Wiklins Inc.
10. Naitoh K. and Kuwahara K. (1992) Large eddy simulation and direct simulation of compressible turbulence and combusting flows in engines based on the BI-SCALES method. Fluid Dynamics Research 10, 299-325.
11. Naitoh K. (1999) Cyto-fluid dynamic theory for atomization processses. Oil & Gas Science and Technology, Vol. 54, No. 2. Pp. 205-210
12. Naitoh K. (2005) Basic pattern underlying life. NIKKEI SCIENCE, Vol. 6, pp. 58-65. (in Japanese)
13. Naitoh K. (2006) Gene Engine and Machine Engine, Springer-Japan,pp. 1-251
14. Naitoh K. (2008) Physics underlying Topobiology, Proc. of 13th Int. Conf. on BioMedial Engineering, (ICBME), Singapore. Springer-Verlag.
15. Naitoh K. (2008) Onto-biology: Proc. of 13th Int. Conf. on BioMedial Engineering, (ICBME), Singapore. Springer-Verlag
16. Naitoh K. (2008) Engine for cerebral development. Artificial Life Robotics, Springer, **13, pp.** 18-21
17. Naitoh K. (2008) Inevitability of nTP: information-energy carriers. Artificial Life and Robotics, Springer, 13,pp. 81-83
18. Naitoh K. (2008) Fluid Dynamics underlying Morphogenesis. Proceedings of 13th International Symposium on Artificial Life and Robotics.
19. Naitoh K. and Shimiya H. (2011) Stochastic determinism for capturing the transition point from laminar flow to turbulence. Japan J. of Industrial and Applied Mathematics. **28**-1, pp.3-14.
20. Naitoh K, Noda A., Kimura S., Shimiya H., and Maeguchi H. (2010) Transition to turbulence and laminarization clarified by stochastic determinism. Proceedings of ETMM8. Marseille, pp. 775-780.
21. Kawanobe H. and Naitoh K (2011) Engine for brain development. Proceedings of 16th Artificial Life and Robotics.
22. Matsuda K., Oishi T., Higo N. (2007) S128, P1-k22 Web-based MRI brain image database system Neuroscience Research Volume 58, Supplement 1, 2007(ISSN 0618-0102). (http://riodb.ibase.aist.go.jp/brain/welcomej.html)
23. Proceedings of 5th WBC 2010 (2010).
24. http://www.otsuka-elibrary.jp/library/support/101/252
25. http://embryo.soad.umich.edu/ (in the website of Prof. Brad Smith, University of Michigan, NICHD NO1-HD-6-3257: http://www-personal.umich.edu/~brdsmith/)
26. Takahashi K. and Yamanaka S. (2006) Induction of Pluripotent Stem Cells from Mouse Embryonic and Adult Fibroblast Cultures by Defined Factors. Cell **126**, pp. 663-676
27. Jin Y. (2008) Identification of L4 as a Novel Downstream Gene of Oct4 and Its role in Promoting Differentiation of Mouse Embryonic Stem Cells. Proceedings of international symposium on induced pluripotent stem cell research frontier and future, Kyoto. Pp.14-15.
28. Naitoh K. (2009) Onto-biology. Proceedings of 2009 Chemical, Biological, Environmental Engineering, (CBEE2009).
29. Naitoh K. (2011) Morphogenic economics. Japan Journal of Industrial and Applied Mathematics Vo. 28, No.1.
30. Naitoh K. (1998) Macroscopic kinetic equation of a genetic algorithm. Japan Journal of Industrial and Applied Mathematics Vol.15, No.1, pp. 87-133
31. Naitoh K. (1998) Introns for accelerating quasi-macroevolutions. JSME International Journal, Series C, Vol.41, No.3, pp. 398-405.
32. Naitoh K. (2010) Onto-biology. the Cosmos. J. of Cosmology, 5: pp. 999-1008.
33. Naitoh K, Inoue H, Hashimoto H. (2010) Topo-embryology: WCB 2010. IFMBE Proceedings, Springer-Verlag. Pp.1163-1166.

Acknowledgment

The author sincerely thanks Dr. Keiji Matsuda of the Neuroscience Research Institute, AIST, Tsukuba, Japan, for permission to reuse the MRI images of the brain and also Prof. Brad Smith of University of Michigan for permission to reuse the images on the embryo. Sincere thanks are also due Mr. A. Suzuki, and Mr. H. Kawanobe in the author's laboratory for their help with visualization of engine flow and also the photographs of yeast.

This article is part of the outcome of research performed under a Waseda university Grant for special research project (2009B-206).

Author: Ken Naitoh
Institute: Waseda University, Faculty of Science and Engineering
Street: 3-4-1 Ookubo
City: Shinjuku, Tokyo,
Country: Japan
Email: k-naito@waseda.jp

Electrical Characterization of Screen Printed Electrodes for ECG Measurements

L. Rattfält[1], F. Björefors[2], X. Wang[3], D. Nilsson[3], P. Norberg[3], and P. Ask[1]

[1]Dept of Biomedical engineering, Linköping University, Sweden
[2]Dept of Materials Chemistry, Uppsala Universitet, Sweden
[3]Acreo, Norrköping, Sweden

Abstract— **Screen printed electrodes with conductive ink made of Carbon and Ag/AgCl were tested for polarization potentials and electrode impedances. In 30 minutes the mean decrease of polarization potential was 2 mV. The electrode impedances at 10 Hz were between 670 and 250 Ohms. These characteristics seem adequate for personalized health care applications.**

Keywords— **Screen printed electrodes, ECG, impedance spectroscopy, polarization potentials.**

I. INTRODUCTION

With the increasing number of elderly there is a health care driven need to develop new biomedical engineering solutions for home health care. For example, cardiovascular disease in this population is highly prevalent. Among other methods, the electrocardiogram (ECG) serves as an important diagnostic method which traditionally only is used in primary and specialized care for this group of patients. However, it is a clear need to obtain such information from the patients wherever they are located, e.g. in the home. This demands reliable acquisition systems that are easy to use, reliable and robust. The scope of this article is to evaluate the electrical properties of screen printed electrodes as a potential part of such a system. It was done by investigating the polarization potentials and electrode impedance in an electrochemical cell with a physiological saline electrolyte.

II. THEORY

A common modality to record bioelectric events is to measure the difference in potential between two locations on the body surface with a differential amplifier. Ideally, the small but unavoidable indifferences between the electrodes and their connections with the skin could be neglected. However, a poor choice of electrode material and electrolyte could lead to unnecessary polarization. Polarization potentials tend to cause a baseline drift which in many cases are several orders of magnitude larger than the desired signal. The signal at hand in our study is the ECG which typically has an amplitude of a few mV. The polarization potentials should therefore preferably be held below this value in order to not corrupt the ECG.

Furthermore, at the interface between the electrode and electrolyte, a charged layer will appear. This causes a phase shift due to a reactive behavior. It is important to keep the impedance as small as possible in order to maintain an adequate common mode rejection ratio. If not, disturbances from the mains, for example, might corrupt the measured signal.

III. METHODS

A. Screen Printed Electrodes

The electrodes were screen printed on a plastic foil substrate (Polyfoil Bias, 125 Mic). A first layer of conductive ink was applied as a conduction line. It was either a 2 cm wide carbon based ink (C7102 from Dupont with 10% Dupont 3610 thinner) or a 0.5 mm wide Ag based ink (Ag5000, Dupont), see Fig. 1. A second layer consisted of a Ag/AgCl ink (Creative Materials, 113-09) circle which served as the active electrode area with a diameter of 14 mm. A third layer of insulating lacquer (Sericol Uvivid Screen CN-622 Tactile Varnish) was applied resulting in an electrode diameter of 10 mm.

In order to verify the importance of layer thickness, a second layer of Ag/AgCl and lacquer were reprinted in some of the samples. The different specimens tested are summarized in Table 1.

B. Measurement setup

The electrodes were tested for polarization potentials and impedance in an electrochemical cell consisting of physiological saline solution. (0.9% NaCl) The electrodes were mounted on a rigid fixture and submerged in the electrolyte, A potentiostat, Autolab PGSTAT30, Metrohm was used for the measurements along with a Ag/AgCl reference electrode and a palladium auxiliary electrode. Polarization potentials were measured every minute for 30 minutes. Impedance measurements were done in the range of 0.05 Hz to 2 KHz

K. Dremstrup, S. Rees, M.Ø. Jensen (Eds.): 15th NBC on Biomedical Engineering & Medical Physics, IFMBE Proceedings 34, pp. 219–221, 2011.
www.springerlink.com

(50 samples, logarithmically distributed) with an rms amplitude of 10mV at open circuit potential.

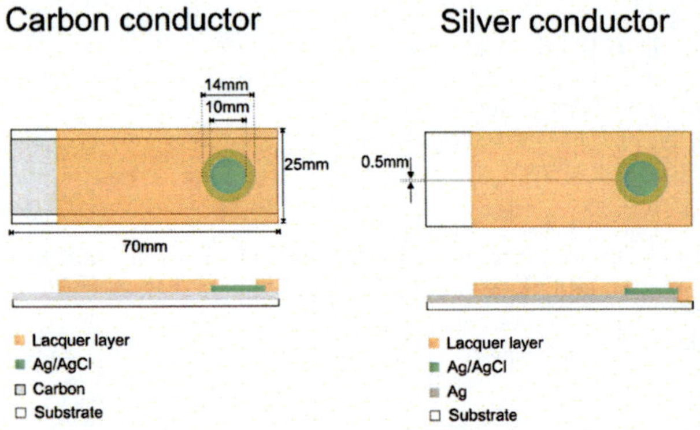

Fig. 1 Schematic setup of the two types of electrodes. The main differences are the conduction line widths and conductive components in the ink.

Table 1 Printed electrodes setups

Denotation	Material of conduction line	Layers of Ag/AgCl	Layers of Lacquer
A: Ag	Ag	1	1
A: Carbon	Carbon	1	1
B: Ag	Ag	1	2
B: Carbon	Carbon	1	2
C: Ag	Ag	2	1
D: Ag	Ag	2	2

IV. Results

A. Polarization potentials

Results show that the polarization potential in average decreased 2 mV during 30 minutes where the largest decrease was during the first 5 minutes, see Fig. 2. This is coherent with the individual measurements for each electrode, which in general showed the same behaviour. Electrode series B: Carbon diverged from the rest by having a 1.5 mV smaller polarization potential.

B. Impedance measurements

In Fig. 3, results from the impedance measurements are displayed. Both the real part and the negative imaginary part are shown. The two electrodes with carbon conductor lines show a slightly higher resistive value than the Ag conductor lines. At f = 10 Hz the resistive values are approximately, 670 Ohms (A: Carbon and B: Carbon), 330 Ohms (A:Ag and B:Ag) and 250 Ohms (C:Ag and D: Ag).

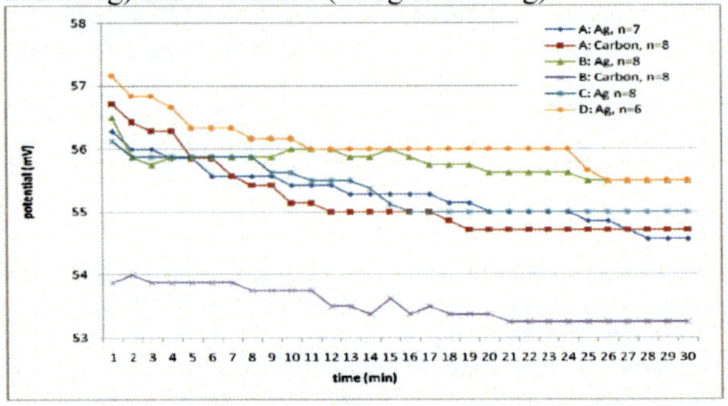

Fig. 2. Mean of the polarization potentials for the six electrode setups.

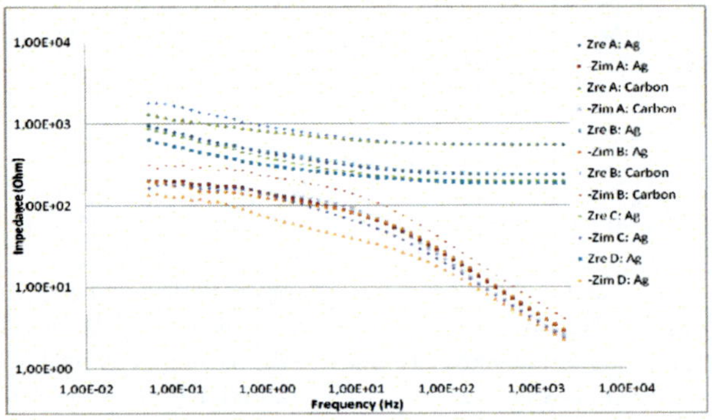

Fig. 3 Mean impedance for the six lectrode setups. Both the real part and the negative imaginary part are plotted.

V. Discussion

The polarization potential results are promising, since they are rather stable. Keeping in mind that it is not the absolute magnitude of the potential that is crucial, but how stable it is. However, further tests on a skin model would be interesting.

In Fig. 3, results from electrode 1 in each batch have been discarded, due to how the measurements were performed. It showed that the first electrode in each batch generally had increased impedance (both real and imaginary part) with respect to the other electrodes. Probably due to that impedance measurements were performed right after polarization potential measurements and that electrode 1 consequently had been submerged in the electrolyte for 30 minutes already. However, repeated impedance measurements for the same electrode showed no difference at all even though the submersion time was comparable with that of the first experiment.

In a prior study a conductive ink was used on a textile substrate and with stainless steel powder.[1] However, it is uncertain how the measurements were conducted or how the skin was pretreated. Their results show a magnitude difference of a factor of 100 to 1000 Ohms higher than the results presented here. A way to decrease the skin/electrode impedance is by adding an electrolyte. However, since it is preferable to have a sensor without pretreatment, a hydrogel might be interesting to test together with the electrode. Previous studies have tested hydrogels, both together with textile ECG electrodes [2] and EEG electrodes [3]. In a future study it would be interesting to incorporate a hydrogel in the printed electrode.

VI. Conclusions

Six screen printed electrode setups have been tested for polarization potentials and electrode impedance. Overall, the electrodes show stable polarization potentials. The carbon conduction lines had the highest resistance of the electrodes. Results also indicate that there might be a difference in resistance due to the number of Ag/AgCl layers.

Acknowledgment

We would like to thank the Swedish Research Council and NovaMedTech for funding our research.

References

1. Yan L, Yoo J, Kim B, Yoo HJ: A 0.5-mu V-rms 12-mu W Wirelessly Powered Patch-Type Healthcare Sensor for Wearable Body Sensor Network. Ieee Journal of Solid-State Circuits 2010;45:2356-2365.
2. Paradiso R, Loriga G, Taccini N: A wearable health care system based on knitted integrated sensors. Ieee Transactions on Information Technology in Biomedicine 2005;9:337-344.
3. Alba NA, Sclabassi RJ, Sun MG, Cui XT: Novel Hydrogel-Based Preparation-Free EEG Electrode. Ieee Transactions on Neural Systems and Rehabilitation Engineering 2010;18:415-423.

Author: Linda Rattfält
Institute: Department of Biomedical Engineering
Street: Linköping University
City: Linköping
Country: Sweden
Email: Linda.rattfalt@liu.se

Temporal Characteristics of Cervical Muscle Activation Patterns before, during and after the Completion of a Repetitive Arm Task

P. Blummer, K. Emery, and J.N. Côté

Department of Kinesiology and Physical Education, McGill University, Montreal, Canada;
Michael Feil and Ted Oberfeld / CRIR Research Centre, Jewish Rehabilitation Hospital, Laval, Canada

Abstract— **Cervical pain has previously been linked with abnormal activation patterns at rest and during movement; however, few studies have assessed how recovery patterns immediately following an arm task may be associated with symptoms. As a first step, the objective of this study was to quantitatively describe the activation and recovery patterns of the cervical musculature before, during and after a repetitive arm task in asymptomatic people. We recorded bilateral surface electromyography (sEMG) from the Sternocleidomastoid (SCM), Scalenus Anterior (SA), Upper Trapezius (UT) and Cervical Erector Spinae (CES) muscles at rest before, at three time points during a 2.5min repetitive unilateral arm task, immediately after and every minute for five minutes at rest after the task. ANOVA analyses with within-subject factors of Time and Muscle were performed on each subject's average and standard deviation values of each muscle's EMG root-mean squared (RMS) amplitude calculate across 1s samples. Results show that there was a significant interaction effect (p<0.0001), showing that muscles responded differently through time. There were main effects of Time on the mean peak amplitude for all muscles, with the majority of the activity peaking during the middle of the task. There were main Time effects on amplitude variability for SCA and UT muscles that also showed increases during the task but that occurred at the same time or earlier than when peak activity was reached. Findings suggest that pre-task sEMG amplitude levels were reached immediately upon completion of the task in asymptomatic subjects and that this could have been accomplished by modulating amplitude variability early on in the task.**

Keywords— **EMG, fatigue, recovery, variability, biomechanics.**

I. Introduction

About 14% of the population suffers from chronic neck pain in the cervical region [1], with higher prevalence in women [2]. Pain in the cervical region is associated with reduced force and mobility, discomfort, stiffness as well as other signs of impaired function [3-4]. Moreover, signs of cervical functional impairment involving cervical muscular patterns, such as the diminishment in the capacity to relax muscles in the painful region, have been used as a means of diagnosis for pathologies of the neck. Szeto et al. [5] found that people with chronic neck pain had higher EMG activity in posture stabilizing muscles when enduring physically stressful conditions. In addition, patients experiencing high discomfort in the cervical region showed increased trapezius activity at baseline resting postures [5-7], and in a series of studies, Elert et al. [8-9], workers with myalgia of the superior trapezius muscle presented a reduced capacity to relax the painful superior trapezius muscle in between repetitive movements.

Although these studies have documented elevated activity with pain at rest and during activities, very few studies focused on the ability of cervical muscles to relax after a task, which could be a very important characteristic of muscle behavior in association with workplace health. Johnston et al. [7] found that 10s after completion of a motor task, the upper trapezius, cervical extensors, and anterior scalene muscles were unable to relax in subjects with reported neck pain. In a study by Nederhand et al. [10] people diagnosed with cervical sprain, as well as a group of healthy controls, were given a task of repetitive movements of the arm for 2.5 minutes. The activity of bilateral superior trapezius muscles was measured before and immediately after the task. The authors found a greater ipsilateral activation in subjects with cervical pain, and trapezius muscles from both sides had difficulty relaxing before and immediately after the task. However, the authors did not measure the duration of this hyper activation following task completion, although they do point to this as a recommendation for future studies.

Another characteristic of interest to describe motor behavior of an individual is the amount of variability of their biomechanical signals, calculated through consecutive epochs or small time samples. Recent studies have documented increased kinematic and kinetic variability with fatigue and with acute experimental pain [11-12]. In a recent study, we showed increasing electromyographic variability of neck-shoulder muscles with fatigue induced by a repetitive task in healthy subjects and, conversely, lower and decreasing variability throughout the same repetitive task in people with chronic neck-shoulder pain [13]. Despite these recent advances, very few studies have investigated the amount of variability of electromyographical signals, and to our knowledge, none has investigated how these characteristics progress through time before, during and after the completion of a repetitive task. This knowledge could shed more light into muscular control mechanisms that are triggered by repetitive tasks.

K. Dremstrup, S. Rees, M.Ø. Jensen (Eds.): 15th NBC on Biomedical Engineering & Medical Physics, IFMBE Proceedings 34, pp. 222–225, 2011.
www.springerlink.com

To address these knowledge gaps and as a first step towards designing an intervention study involving patients with cervical symptoms, this project aimed to quantitatively describe the time progression of cervical activity patterns before, during and after the completion of a repetitive arm task. We hypothesized that task performance would be associated with a significant increase in activity amplitude and variability across all muscles and that these data would return to pre-task, baseline values within 5min post-task.

II. Methods

A. Participants

Healthy female adults (N = 16) were recruited from the University and Research Centre community and agreed to participate in this study. They could not have been diagnosed with a cervical injury or disorder, have reported pain or have displayed reduced range of motion in the neck and/or shoulder area during the year prior to their participation. At arrival, they all provided informed consent by signing forms approved by the Research Ethics Board of the Centre for Interdisciplinary Research in Rehabilitation (CRIR) of Greater Montreal.

B. Procedure

Upon arrival, subjects were outfitted with electrodes on the cervical muscle sites of interest. The electromyographic (EMG) activity of 8 muscles was measured (sampling rate: 1080 Hz) using the Telemyo 900 system (Noraxon, USA). Bipolar, Ag-AgCl surface electrodes with a 10-mm diameter circular conductive area (Ambu, ON, Canada) were positioned with a 3-cm centre-to-centre distance, parallel to the muscle fibers. The skin overlying the muscle sites was lightly abraded, shaved and cleaned before positioning the electrodes [14]. Electrodes were placed bilaterally on the moving, or Dominant side (D) and the stationary, or Nondominant side (N) over the following muscles: sternocleidomastoid (SCM), anterior scalene (SCA), upper trapezius (UT), and cervical erector spinae (CES). Then, subjects performed two reference tasks while we recorded sEMG. In the first one, they stood, abducted their shoulders 90 degrees, elbows extended. In the second one, they laid supine, with the head in neutral posture, then lifted their head until the occiput was not in contact with the support surface. In each case, when the position was reached, it was maintained 5s, and 3 trials of each task were performed. Both tasks have been previously used in experiments on cervical muscle activity [11,15].

Then subjects were placed in front of a table, seated feet flat on the floor. The table was adjusted 5cm below elbow height. Before the experimental task, 10s of baseline sEMG was recorded with subjects passively resting both forearms on the table. Then, they were to perform the experimental task designed by Nederhand et al. [11] in which they continuously moved the dominant arm/hand between three target areas by putting marks with a pencil in circles with a diameter of 70 mm, moving the arm in a clockwise motion. During this task the nondominant arm rested passively on the table. The pace of 88 marks/min was kept constant with the help of a metronome. The task was performed for 2.5 minutes, and data was collected for 10s each time, starting 10s after the start of the task, then at 1min, and 2min into the task. Immediately when the task was completed, subjects assumed the baseline position for 5min, while we recorded 10s of sEMG immediately at task completion and every minute thereafter.

C. Analysis

As part of preliminary analyses (N = 8), heartbeats were removed from each EMG signal by detecting a heartbeat reference signal for each trial over a selected muscle and applying cross-correlation filters over the other muscle signals of the corresponding trial. Signals were then filtered (zero-lag second order 20-500 Hz band-pass) and full-wave rectified. Normalization values for each muscle were obtained from the reference trials by finding the peak EMG value across all trials and calculating the RMS value of the time interval beginning 50ms before and ending 50ms after the peak. The 10s-EMG samples were partitioned into 10-1s samples, and root mean square (RMS) values were calculated for each sub-sample and were normalized to the corresponding muscle's normalization value. The 10 normalized sub-samples were averaged and their standard deviation, representing the amount of variability, was computed.

Repeated measures ANOVA with within-subject factors of Muscle (8) and Time (10) were computed, with the 10 Time conditions being (see Fig. 1):

- rest, baseline
- 10s into the repetitive task
- 1min into the repetitive task
- 2min into the repetitive task
- rest, immediately after the repetitive task
- rest, 1min after the end the repetitive task
- rest, 2min after the end of the repetitive task
- rest, 3min after the end of the repetitive task
- rest, 4min after the end of the repetitive task
- rest, 5min after the end of the repetitive task

In cases of significant effects, Tukey post-hoc tests were applied, and significance was set at alpha = 0.05.

III. Results

A. EMG RMS amplitude

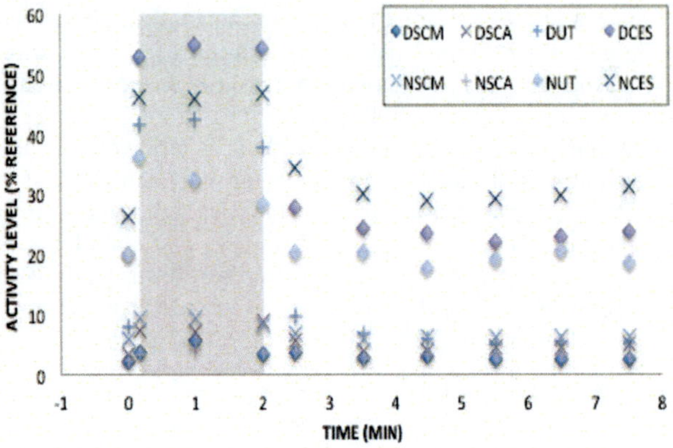

Fig. 1 . Time course of EMG RMS at 8 cervical muscle sites before, during and after the repetitive task (outlined by grey zone).

There was a Muscle x Time interaction (p<0.00001), showing that muscles responded differently from each other to time during the protocol (Fig. 1). All muscles on the moving (Dominant, D) side showed significant main effects of Time but only the UT and the CES did so on the Non-moving (N) side. Post-hoc analyses revealed that the activity of all muscles showing main Time effects significantly increased during the task (p<0.05). The onset of significant increase from baseline occurred 10s into the repetitive task for DUT, DCES and NCES, 1min into the repetitive task for DSCM, and NUT, and 2min into the repetitive task for DSCA (Table 1). Peak activity for muscles showing a time effect was observed at 10s into the repetitive task for NCES (23% above baseline), 1min into the repetitive task for DSCM (6% above baseline) and NUT (16% above baseline), and 2min into the repetitive task for DSCA (9% above baseline), DUT (32% above baseline) and DCES (29% above baseline). All muscles displayed activity levels not significantly different from those at baseline immediately following task completion (Time 5), which remained unchanged until 5min post-task. During the 5min post-task, all muscles showed activity levels < 2.5% different from those at baseline; interestingly, the largest difference between baseline and post-task levels was a decrease of 2.24% observed at the DUT.

Table 1 Summary of results

Muscle	Onset of change from baseline	Time to peak	Peak value (difference from baseline value in %ref)	Onset of increased variability	Offset of increased variability
DUT	10s	2'	+32%	10s	End of task
DCES	10s	2'	+29%		
NCES	10s	10s	+23%		
NUT	1'	1'	+16%	10s	1'
DSCM	1'	1'	+6%		
DSCA	2'	2'	+9%	NS	NS
NSCM					
NSCA					

Empty cells: no main Time effects

B. EMG RMS amplitude variability

ANOVA analysis for the amount of EMG RMS variability, calculated within 10s sEMG data samples, also showed a significant Muscle x Time interaction (p<0.0001). However, only three muscles, DSCA, DUT and NUT, showed significant Time main effects. Post-hoc analyses revealed that variability of both DUT and NUT significantly increased from baseline as soon as 10s into the repetitive task. For the NUT, variability returned to baseline values as soon as 1min into the protocol, whereas for DUT, variability remained significantly higher than at baseline until the end of the repetitive task. No significant post-hoc comparisons were observed for DSCA variability.

IV. Discussion and Conclusions

Results show that in healthy female subjects performing a repetitive task for 2.5min, the activity of the main cervical muscles increases and returns to baseline immediately upon completion of the task. This will provide useful information against which to compare activity patterns of individuals with cervical pain. Indeed, Nederhand et al. [11] published a study using the same experimental task and observed higher activity in people with pain, compared to healthy controls. However, authors did not record activity beyond immediately after task completion, and concluded by recommending that other studies address the possibility that group differences remained minutes after task completion. Interestingly, our data shows a slight depression of DUT activity during the minutes that follow task completion. If confirmed by additional analyses, these healthy patterns may be interpreted as facilitatory recovery responses and/or warm-up effects. In addition, these results support the use of frequent breaks during work involving load on the cervical muscles since muscles may not only quickly return to baseline values but slightly below, which may represent a supplementary strategy to facilitate recovery. Finally, our results confirm those of others in showing that the posterior musculature is more heavily involved in upper limb movements performed in a plane anterior to the body, and that both sides are involved in stabilizing the cervical region during such a task [11].

No studies had previously documented the amount of variability at the cervical muscle sites before, during and after a repetitive task. We support findings from previous studies that with increased time performing an upper limb task, the amount of biomechanical (in this case, electromyographical) variability increases. More specifically, we confirm findings of our previous study, which showed increased variability of UT and of Supraspinatus following the completion of a repetitive arm task to fatigue [13]. Authors have previously interpreted such findings to reflect healthy exploratory strategies to find load-minimizing patterns [12,13]. Moreover, our findings show that this occurs as soon as 10s from the beginning of a repetitive task. This suggests that such strategies may precede the actual development of muscle fatigue since variability increase occurs around the same time as the onset of activity amplitude change from baseline for the DUT, and actually earlier than the NUT amplitude increase. These observations are reminiscent of experimental findings associated to the development of the Cinderella Hypothesis [16], suggesting that healthy muscles benefit from mechanisms of rotation and load sharing among motor units (MUs). Increased variability could be indicative of such changes in load sharing among MUs of different sizes and/or among synergistic muscles that could in turn affect the amplitude of the sEMG signal. Whether these changes are operated via feedforward mechanisms or using feedback from some muscle characteristics is not known at this point. Moreover, the impact of experimental and chronic pain on biomechanical variability has been documented before, with equivocal findings [12,13]. Future studies are needed to document the time course of variability changes during a repetitive task in asymptomatic as well as in symptomatic individuals. In turn, this knowledge could be implemented into the design of more efficient myofeedback-assisted training and rehabilitation protocols to treat and prevent cervical pain.

Acknowledgment

We thank Karine Rivest and Jean-Pierre Dumas for their insight into the design of this protocol. Equipment and infrastructure used in this project were provided though funds obtained from the Canadian Foundation for Innovation.

References

1. Cote P, Cassidy JD, Carroll LJ et al. (2004) The annual incidence and course of neck pain in the general population: a population-based cohort study. Pain 112(3):267–273
2. Carroll LJ, Holm LW, Hogg-Johnson S et al. (2008) Course and prognostic factors for neck pain in whiplash-associated disorders (WAD): results of the Bone and Joint Decade 2000-2010 Task Force on Neck Pain and Its Associated Disorders. Spine 33(4 Suppl):S83–92
3. Nordin M, Carragee EJ, Hogg-Johnson S et al. (2008) Assessment of neck pain and its associated disorders. Eur Spine J 17(Suppl 1):S101–S122
4. Larsman P, Sandsjo L, Kadefors R et al. (2009) Prognostic factors for intervention effect on neck/shoulder symptom intensity and disability among female computer workers. J Occup Rehabil 19:300–311
5. Szeto GPY, Straker LM, O'Sullivan PB (2009) Neck-shoulder muscle activity in general and task-specific resting postures of symptomatic computer users with chronic neck pain. Man Ther 14:338–345
6. Sjors A, Larsson B, Dahlman J et al. (2009) Physiological responses to low-force work and psychosocial stress in women with chronic trapezius myalgia. BMC Musculoskelet Disord 10:63–79
7. Johnston V, Jull G, Darnell R et al. (2008) Neck movement and muscle activity characteristics in female office workers with neck pain. Spine 33(5):555–563
8. Elert J, Gerdle B (1989) The relationship between contraction and relaxation during fatiguing isokinetic shoulder flexions. An electromyographic study. Eur J Appl Physiol Occup Physiol 59(4):303–309
9. Elert J, Rantapaa-Dahlqvist SB, Henriksson-Larsen K et al. (1992) Muscle performance, electromyography and fibre type composition in fibromyalgia and work-related myalgia. Scand J Rheumatol 21(1):28–34
10. Nederhand MJ, IJzerman MJ, Hermens HJ et al. (2000) Cervical muscle dysfunction in the chronic whiplash associated disorder grade II (WAD-II). Spine 25(15):1938–1943
11. Madeleine P, Mathiassen SE, Arendt-Nielsen L (2008) Changes in the degree of motor variability associated with experimental and chronic neck-shoulder pain during a standardized repetitive arm movement. Exp Brain Res 185:689–698
12. Cignetti F, Schena F, Rouard A (2009) Effects of fatigue on intercycle variability in cross-country skiing. J Biomech 42:1452–1459
13. Lomond KV, Cote JN (2010) Movement timing and reach to reach variability during a repetitive reaching task in persons with chronic neck/shoulder pain and healthy subjects. Exp Brain Res 206(3):271–282
14. Basmajian J, Blumenstein R (1980) Electrode placement in EMG biofeedback. Williams & Wilkins, Baltimore
15. Sterling M, Jull G, Vicenzino B et al. (2003) Development of motor system dysfunction following whiplash injury. Pain 103(1-2):65–73
16. Sjogaard G, Sogaard K (1998) Muscle injury in repetitive motion disorders. Clin Orthop Relat Res 351:21–31

Julie N. Cote, PhD
Department of Kinesiology & Physical Education, McGill University
475 Pine Avenue West
Montreal
Canada
julie.cote2@mcgill.ca

Data Mining Techniques for Analyzing Demographic Factors in Relation to Chronic Pain Patients

G.P. Nguyen[1], J.A. Biurrun Manresa[1], M. Curatolo[2], and O.K. Andersen[1]

[1] Integrative Neuroscience group, Center for Sensory-Motor Interaction,
Department of Healthy Science and Technology, Aalborg University, Denmark
[2] University Department of Anaesthesiology and Pain Therapy, Inselspital, Bern, Switzerland

Abstract— **The purpose of this study was to analyze the influence of demographic factors on the nociceptive withdrawal reflex (NWR) with chronic pain patients. Different from existing researches, which are mainly concentrated on differences in reflex thresholds, a new approach was proposed. The approach uses original reflex electromyography (EMG) signals as "features" of patients. Data mining techniques, including statistical measurement and pattern recognition, were applied to these "features" to evaluate relations among different groups of patients with respect to gender and age. Experiments revealed that these "features" allowed to separate those groups of patients. Results proved the strong influence of those demographic factors on NWR of chronic pain patients.**

Keywords— **Nociceptive withdrawal reflex, chronic pain, demographic factors.**

I. Introduction

Understanding the influence of demographic and physiologic factors in chronic pain patients plays an important role in clinical evaluation. The nociceptive withdrawal reflex (NWR) is considered as a reliable tool in pain assessment [1,2,3]. Several attempts to find gender and age differences on NWR were based on reflex thresholds [1,3,4,5]. It is pointed out that most thresholds are determined by expert observers i.e. subjective evaluation. Other objective evaluations are dependent on a variety of criteria such as reflex interval mean, reflex interval peak z-score or number of samples above certain amplitude. However, none of these methods used the full reflex electromyography (EMG) signal itself as a representative "feature" of a pain patient.

The purpose of this study was to provide a new objective way of analyzing the influence of different demographic factors on the NWR. When EMG signal was recorded for each patient, it was used as a "description" or "feature" of that patient. In that sense, data mining provides a number of powerful methods to analyze relations among data elements in their "feature space". This field includes both statistical evaluation and pattern recognition techniques. When applying these techniques, the issue of small data samples is often faced because experiments usually carried out with a limited number of subjects (healthy volunteers or pain patients). For classifying small data samples, especially when the number of data samples much less than the "feature space" dimensionality, a common approach is to reduce the dimensionality of the feature space.

There are several existing techniques to deal with dimensional reduction, including linear and nonlinear approaches. Some of the well-known techniques in this field are multi dimensional scaling (MDS), principal component analysis (PCA), Isomap, Stochastic neighbor embedding (SNE) and locally linear embedding (LLE). In [6,7] existing techniques were compared based on two main criteria. The first one was preserving local structure of the high dimensional data after reducing the number of dimensions. The second criteria was to reveal any global structure such as the presence of clusters. A technique called t-SNE, an improved version of SNE, was proven to be the best one satisfying both criteria. This technique was employed in this study.

In order to analyze how gender and age interact based on this feature, four different cases were considered:

- *Case 1*: Gender difference in a young group.
- *Case 2*: Gender difference in an old group.
- *Case 3*: Age difference in female group.
- *Case 4*: Age difference in male group.

Each case can be considered as a two-classes classification problem. For instance, if male and female patients can be grouped into two separable classes, this indicates the influencing of gender on pain. The more separation the two classes, the higher the influence the factor. For that purpose, a classification technique can be employed to analyze how each class is separable from another.

In this study, supervised classification techniques were used. Supervised classification uses a set of training samples, i.e. samples with known classes, to predict classes of unknown samples. Comparing available classification techniques in pattern recognition did not show any outstanding method [8,9]. Regarding the accuracy of the classification, more advanced techniques such as support vector machine and neural networks in general give better performance.

K. Dremstrup, S. Rees, M.Ø. Jensen (Eds.): 15th NBC on Biomedical Engineering & Medical Physics, IFMBE Proceedings 34, pp. 226–229, 2011.
www.springerlink.com

However, these techniques were not suitable in this study as they usually require a large number of training samples to achieve high classification rate. The k-nearest neighbors (kNN) was selected to perform the classification task as it works well in case of small training samples.

The paper is organized as follows. Section II presents the material used for analyzing the demographic factors. Selected data mining techniques, namely t-SNE and kNN are described in section III. Discussions of experimental results are followed in section IV.

II. Materials

A. Subjects

A group of 144 patients with chronic pain were recruited at the University Hospital of Bern, Inselspital. These subjects experienced pain for a period of at least 6 months duration (present pain score higher than 3). Written informed consent was obtained from all subjects prior to participation and the Declaration of Helsinki was respected. The study was approved by the local ethical committee.

There were 72 male and 72 female subjects with age ranged from 20 to 80 years. With a mean age of 50 years, those subjects were also grouped into a young group (age < 50 years) and an old group (age >= 50 years), which divided 72 subjects for each group.

Therefore, each of the four cases contained of 72 subjects, namely:

- *Case 1*: 36 young female and 36 young male patients.
- *Case 2*: 36 old female and 36 old male patients.
- *Case 3*: 36 young female and 36 old female patients.
- *Case 4*: 36 young male and 36 old male patients.

B. Setup

a) Electrical stimulation:

Ten surface stimulation electrodes (15 x 15 mm, type Neuroline 700, Ambu A/S, Denmark) were non-uniformly mounted on the sole-of-the-foot (SOF) and a common anode (10 x 7 cm) was placed on the dorsum of the foot (see figure 1).

A computer-controlled stimulator delivered a stimulus to one electrode at a time in a randomized order, with a random inter-stimulus interval ranging from 10 to 15s. Each stimulus consisted of a constant-current pulse train of 5 individual 1 ms pulses delivered at 200 Hz. For each electrode position, the lowest stimulus intensity that evoked pain (i.e., the pain threshold) was assessed, and a stimulation intensity of 1.5 times the pain threshold was selected.

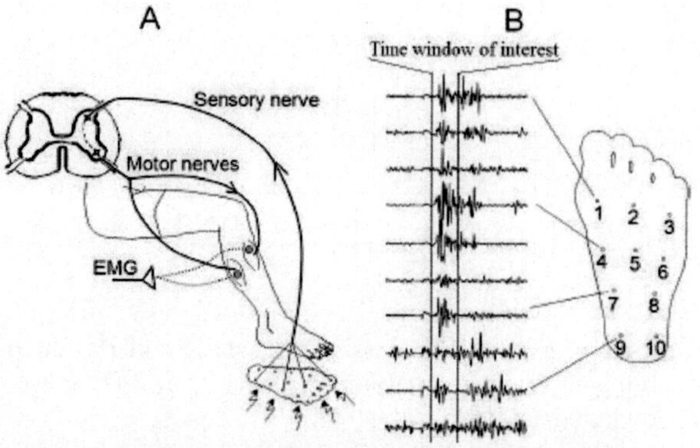

Fig. 1 Methodology for NWR stimulation and recording in humans. (A) Reflex responses evoked by distributed electrical stimulation on the foot sole are recorded by surface EMG at selected muscles. (B) The reflex size is quantified in the time windows of interest (usually 60-180 ms after stimulation). The RRF is the spatial variation in reflex excitability across stimulation sites.

b) EMG recording:

Activity in the tibialis anterior muscle was measured using surface EMG. Initially the skin was lightly abraded, and then two surface electrodes (30 x 22 mm, type Neuroline 720, Ambu A/S, Denmark) were placed along the muscle fiber direction over the muscle with an inter-electrode distance of 20 mm. The signal was amplified (up to 20000 times), filtered (5-500 Hz, 2nd order), sampled (2000 Hz) and stored (1000 ms window including 200 ms of prestimulation activity). The NWR was quantified within the 60-180 ms post stimulation interval.

III. Methods

The data was given as a set of 144 patients x 240-samples EMG signal. For each case, a subset of data contained of 72 patients x 240-samples EMG signals. In pattern recognition terms, this dataset can be described as 72 data points in a 240-dimensional feature space.

A. Two-dimensional representation

t-SNE (t-Distributed Stochastic Neighbor Embedding) [6] was employed in this study to convert the high dimensional data to 2-D ones. The working principle of the technique can be briefly described as follows. First, it computed the probability that two data points X_i and X_j take each other as neighbors, in both the high and the 2-D space. If X_i and X_j were nearby then the probability is relatively high.

Under the Gaussian distribution assumption, the probability in the high dimensional space was given by:

$$P_{ij} = \frac{\exp\left(\frac{-\|X_i - X_j\|^2}{2\sigma^2}\right)}{\sum_{k \neq l} \exp\left(\frac{-\|X_k - X_l\|^2}{2\sigma^2}\right)} \quad (1)$$

where σ denotes the Gaussian variance. Similarly, the probability of mapped data point x_i and x_j in 2-Dl space was computed as:

$$Q_{ij} = \frac{\exp\left(-\|x_i - x_j\|^2\right)}{\sum_{k \neq l} \exp\left(-\|x_k - x_l\|^2\right)} \quad (2)$$

The technique then tried to match the two probability distributions. In other words, if the mapped data points correctly model, the distribution in the high dimensional space then P_{ij} and Q_{ij} will be equal.

To measure the similarity between these two distributions, the Kullback-Leibler distance was used as a cost function:

$$C = \sum_i \sum_j P_{ij} \log \frac{P_{ij}}{Q_{ij}} \quad (3)$$

Finally, the technique searched for optimal positions of mapped data points by minimizing Eq.3 using gradient descent method.

B. Data classification

kNN [10] is based on the principle that data points sharing similar properties should be in close proximity. Therefore, given a set of data points with classification labels, i.e. training data, a label assigned to an unclassified data point was determined by observing labels of its nearest neighbors. The kNN located the k nearest training data points and defined common label of the majority among those data points. Euclidean distance is often used to measure the closeness between two data points x_i and x_j.

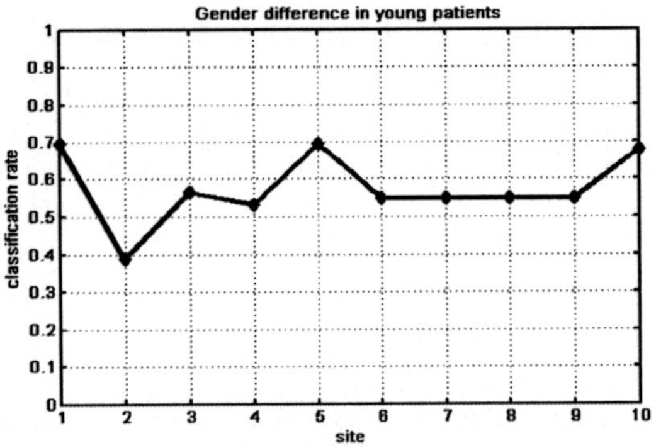

Fig. 2 Gender differences in young group

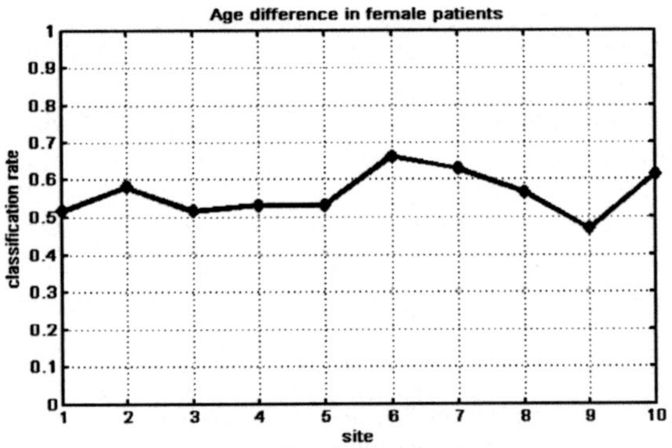

Fig.4 Age differences in female group

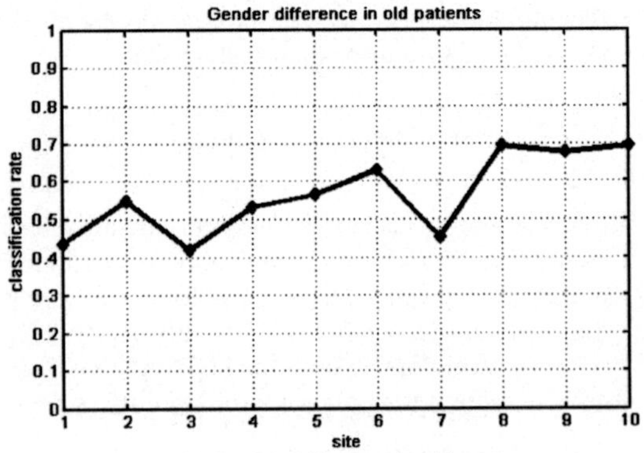

Fig. 3 Gender differences in old group

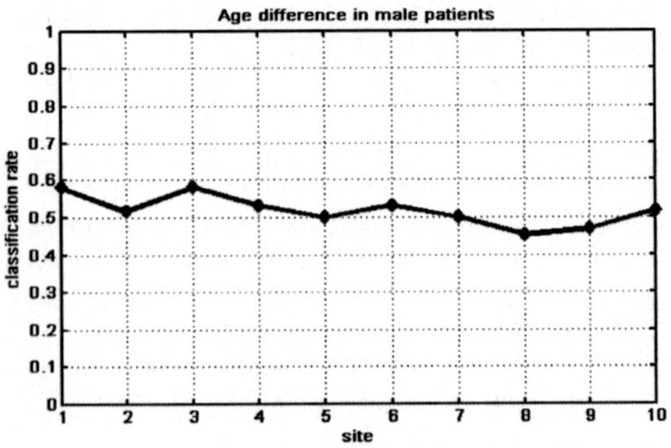

Fig. 5 Age differences in male group

To evaluate the performance of the classification, the following formula was used:

$$r = \frac{TP + TN}{TP + FP + FN + TN} \quad (5)$$

where TP, TN, FP, FN is the number of true positive, true negative, false positive and false negative, respectively.

Regarding the value of k, there is no specific approach for selecting an optimal value as this strongly depends on the data structure. Given our dataset, k was set to 3.

IV. Results and Discussion

Applying the described techniques to the data, t-SNE was first used to project 240-D feature space to a 2-D representation space. For each of the four cases, 5 data points were selected from each group as training samples. For instance, in case 1, 5 young male and 5 young female patients were chosen into a training set. With the kNN classification technique, predicted labels were assigned for 31 remaining male and 31 remaining female patients. Eq. 5 was applied to compute the classification rate for each case. Figures 2, 3, 4, 5 display result obtained for case 1, 2, 3 and 4, respectively.

It should be reminded here that EMG signals were recorded at 10 different electrode sites (see figure 1). Therefore, each site received a separate classification rate. First view of the four figures indicates varying classification rates over 10 electrode sites in all four cases. This result goes along with what discussed in [1] that NWR vary with stimulation sites depending on the skin area and the receptive field.

At some sites, the classification rates were rather low 0.4 (i.e. 40% correctly classified). This means there were no significant different between the two groups. For example, site 2 (figure 2) indicates no significant gender differences in group of young patients. However, at electrode sites 1, 5 and 10, the two groups are highly separable with classification rates up to 0.7 (i.e. 70% correctly classified). A similar observation is obtained in case of gender differences in the older group (figure 3). These results can be used to explain the conflict in existing researches where some pointed out significant differences between gender [4], while others disagreed with that summation [3].

Figure 4 and 5 illustrate that age factor had influence in the group of female and the group of male patients [1,3]. Separation in case 3 (figure 4) reaches maximum of 67%. There are less significant differences between young male and old male patients. Classification rate in this case is ~60%.

In summary, both the gender and age tors are influential on the NWR. These effects vary depending on the different positions of the electrodes.

V. Conclusion

In this paper, we present a new objective approach for analyzing the influence of different demographic factors to nociceptive withdrawal reflex in chronic pain patients. Data mining techniques, including statistical measurement and pattern recognition, were applied to evaluate the influences. Results support existing theory that age and gender have strong influence in patients with chronic pain.

Acknowledgment

The study was supported by the Danish Research Council for Technology and Production Science (FTP).

References

1. Skljarevski V, Ramadan N (2002) The nociceptive flexion reflex in humans – review article. Pain 96: 3-8.
2. Neziri A.Y, Haesler S, Petersen-Felix S, et. al (2010) Generalized expansion of nociceptive reflex receptive fields in chronic pain patients. Pain 151: 798-805.
3. Neziri A.Y, Andersen O.K, Petersen-Felix S, et. al. (2010) The nociceptive withdrawal reflex: normative values of thresholds and reflex receptive fields. European Journal of Pain 14: 134-141.
4. Mylius V, Kunz M, Schepelmann K, Lautenbacher S (2005) Sex differences in nociceptive withdrawal reflex and pain perception. Somatosensory and Motor Research 22(3): 207-211.
5. Rhudy J, France C (2007) Defining the nociceptive flexion reflex (NFR) threshold in human participants: A comparison of different scoring criteria. Pain 128: 244-253.
6. Maaten van der L, Hinton G (2008) Visualizing data using t-SNE. Journal of Machine Learning Research 9: 2579-2605.
7. Maaten van der L, Postma E, Herik van den H (2009) Dimensionality Reduction: A comparative review. Tilburg University Technical Report, TiCC-TR 2009-005.
8. Kulkami S, Lugosi G, Venkatest S (1998) Learning pattern classification – A survey. IEEE Transactions on Information Theory 44(6): 2178-2206.
9. Kotsianstis S, Saharakis I, Pintelas P (2006) Machine learning: a review of classification and combining techniques. Artificial Intelligence Review 26(3): 159-190.
10. Cover T, Hart P (1967) Nearest neighbor pattern recognition. IEEE Transactions on Information Theory 13(1): 21-27.

Author: Giang Phuong Nguyen
Institute: Aalborg University
Street: Niels Jernes Vej 14
City: Aalborg
Country: Denmark
Email: gnp@create.aau.dk

Withdrawal Reflex-Based Gait Training in the Subacute Post-Stroke Phase: Preliminary Results

E.G. Spaich[1], N. Svaneborg[2], and O.K. Andersen[1]

[1] Center for Sensory-Motor Interaction (SMI), Aalborg University, Aalborg, Denmark
[2] Brønderslev Rehabilitation Center, Vendsyssel Hospital, Denmark

Abstract— **The aim of this study was to investigate the feasibility of using nociceptive withdrawal reflexes (NWR) to support gait training in the subacute post-stroke phase. Thirty hemiparetic individuals were randomly divided into: 1) a group that received intensive physiotherapy-based gait training supported by electrical stimulations that triggered the NWR during late stance, and 2) a control group that received intensive physiotherapy-based gait training alone. Both groups received 4 weeks of gait therapy. Electrical stimuli were delivered at the arch of the foot at heel-off with the purpose of eliciting the NWR and thereby support the initiation and execution of the swing phase. Gait was assessed by the Functional Ambulation Category (FAC) test and by the duration of the gait cycle and the stance phase in the hemiparetic side before treatment, immediately after, and one month after finishing treatment. Subjects in both groups showed an improvement in their walking ability, though those who received physiotherapy-based gait training supported by NWR stimulation had a tendency to score better in the FAC-test. Individuals who at inclusion presented severe walking impairments (FAC scores 0 and 1) showed the best improvement as evidenced by a longer duration of the stance phase in the hemiparetic side and a shorter duration of the gait cycle resulting in a pattern closer to normal gait. Intensive physiotherapy training combined with electrical stimulation to evoke a NWR that supports the initiation and production of the swing phase seemed to improve the general walking ability of subacute hemiparetic patients. Strong sensory stimulation seems to be useful in the rehabilitation of the hemiparetic gait.**

Keywords— **nociceptive withdrawal reflex, hemiparetic gait, locomotion, stroke, reflex modulation.**

I. Introduction

People who suffered a stroke often present a compromised gait pattern with diverse spatio-temporal and kinematic deviations from normal gait such as reduced gait speed and longer stance phases, reduced hip, knee, and ankle flexion during swing, and reduced knee extension during early-stance, among others [1]. To rehabilitate the hemiparetic gait, physiotherapy-based gait training is normally used. Combining intensive voluntary exercising with synchronized functional electrical stimulation of involved paretic muscles has shown to be of great benefit for the recovery of upper-limb functions in the acute post-stroke phase [2] and to improve gait performance [3, 4]. Instead of stimulating relevant paretic muscles during gait therapy, movements such as ankle dorsiflexion, and in particular hip flexion can be achieved by eliciting the withdrawal reflex [5, 6]. Electrical stimulation of the sole of the foot has been shown to elicit the nociceptive withdrawal reflex and produce gait phase-, stimulation site-, frequency-, and intensity-dependent muscle and kinematic responses in the lower limb of healthy and hemiparetic individuals [7-10]. In the present study, the therapeutic use of nociceptive withdrawal reflexes to support gait training in the subacute post-stroke phase was investigated.

II Methods

A. Subjects

Thirty subacute stroke patients (36-83 years old) participated in a randomized single-blinded experiment to study the feasibility of using NWR to support gait therapy. The protocol of the study was approved by the local ethical committee (approval number VN-2004/65) and experiments were carried out in accordance with The Declaration of Helsinki. All volunteers provided written informed consent before participating in the study. Inclusion criteria included: first ever cerebrovascular accident (CVA) or second CVA with asymptomatic first CVA, at most 6 weeks since stroke, and ability to walk a maximum of 10 meters without help from a physiotherapist but with eventual help of a supporting aid.

B. Training Protocol

Individuals were allocated to one of two groups: a treatment group that received gait therapy supported by activation of the NWR and a control group that received gait therapy alone. Both groups received 20 days of

K. Dremstrup, S. Rees, M.Ø. Jensen (Eds.): 15th NBC on Biomedical Engineering & Medical Physics, IFMBE Proceedings 34, pp. 230–232, 2011.
www.springerlink.com

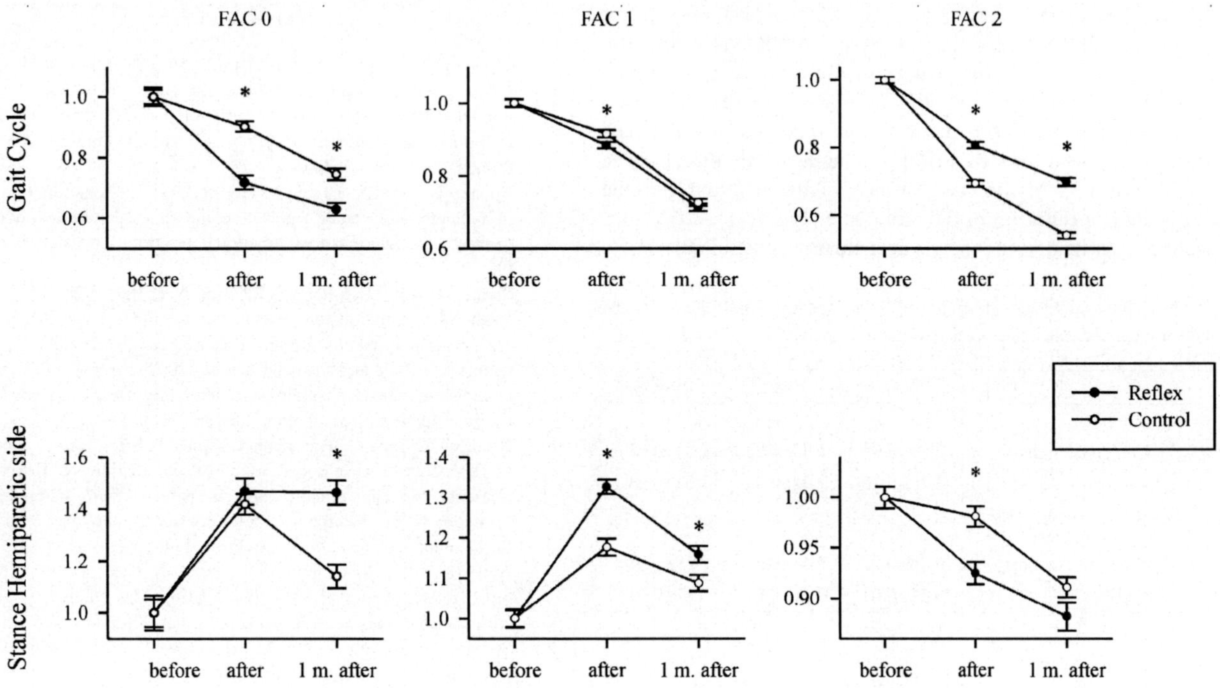

Fig. 1 Duration of the gait cycle and the stance phase in the hemiparetic side. Values are normalized to the measurements at inclusion time and presented separately for the three different walking levels at inclusion time as mean ± SEM. FAC 0: patient needs help from two or more physiotherapists to be able to walk; FAC 1: patient needs constant help from one therapist for weight support and balance; FAC 2: patient needs constant or intermittent help from a therapist to keep balance and coordination

intensive physiotherapy-based gait training. During the 30 minutes long daily sessions, patients had to walk at least 15 minutes allowing for resting intervals. In the treatment group, gait was supported by electrical stimulation of the sole of the foot to activate the NWR. The control group received no electrical stimulation.

The NWR was elicited by transcutaneous electric stimulation delivered at the arch of the foot (2.63 cm2 surface area electrode, AMBU, Denmark) with reference to an anode placed on the dorsum of foot (7x10 cm electrode, Pals, Axelgaard Ltd., USA). Each stimulus consisted of a constant current square pulse train of five 1 ms unipolar pulses delivered at 200 Hz, repeated 4 times at 15 Hz (See details in [10]. Stimulation was delivered at heel-off.

C. *Outcome Measures*

Gait was evaluated at inclusion, immediately after, and one month after completion of training. Quality of walking was assessed by the Functional Ambulation Category Test [11]. Furthermore, the duration of the gait cycle and the stance phase in the hemiparetic side were measured. These temporal parameters might have clinical relevance since they are related to an improvement of walking and a better weight bearing in the hemiparetic side.

Two-way repeated measures analysis of variance (ANOVA) was used to analyze the effect of evaluation time (3 levels) and groups of volunteers (2 levels) on the outcome variables. $P<0.05$ was considered statistically significant.

III Results

At inclusion time, the subjects scored either 0, 1, or 2 in the FAC-test, i.e. patient needed either help from two or more physiotherapists to be able to walk, needed constant help from one therapist for weight support and balance, or needed constant or intermittent help from a therapist to keep balance and coordination. At inclusion, the distribution of individuals with different qualities of gait, as assessed by the FAC-test, was similar in the two groups. One subject was not able to perform the last evaluation due to reasons unrelated to this project.

Subjects in both groups showed an improvement in their walking ability, as assessed by the three outcome measures (Repeated Measures ANOVA, main effect, $p<0.001$ and

Newman-Keuls for post-hoc comparisons, $p<0.001$). After treatment, the subjects who received NWR-based therapy had a tendency to score better on the FAC-test compared with the control group. Subjects who at inclusion scored either 0 or 1 in the FAC test and received NWR-based therapy showed the best improvement as assessed by a longer duration of the stance phase in the hemiparetic side and a shorter duration of the gait cycle (Figure 1). All values were normalized to the means at inclusion time. (Table 1).

Table 1 Duration of the gait cycle and the stance phase in the hemiparetic side at inclusion time (average across subjects ± SEM)

	FAC: 0	FAC: 1	FAC: 2
Gait Cycle [s]			
NWR-based therapy	2.59±0.08	1.95±0.03	1.81±0.03
Control	3.25±0.10	2.32±0.03	2.51±0.03
Stance phase (hemiparetic side) [% of gait cycle]			
NWR-based therapy	0.56±0.01	0.61±0.01	0.69±0.01
Control	0.51±0.01	0.58±0.01	0.60±0.01

IV. Conclusions

In subacute hemiplegic individuals, intensive physiotherapy-based gait training combined with activation of the NWR to initiate and support the swing phase resulted in improved walking, in particular for individuals with severe walking impairment at inclusion time. Withdrawal reflex-based therapy might be useful in the rehabilitation of the hemiparetic gait.

Acknowledgment

This work was supported by The Svend Andersen Fond, the Danish Research Council for Technology and Production Sciences, and The Obel Family Fond. The authors wish to thank physiotherapists Bodil Ottossen, Jørgen Larsen, and Helle R. M. Jørgensen from Brønderslev Neurorehabilitation Center, Vendsyssel Hospital, for their work during the training and evaluation sessions.

References

1. Olney SJ, Richards C (1996) Hemiparetic gait following stroke. Part I: Characteristics. Gait & Posture 4: 136-148
2. Popovic MB, Popovic DB, Sinkjaer T et al. (2003) Clinical evaluation of Functional Electrical Therapy in acute hemiplegic subjects. J Rehabil Res.Dev. 40: 443-453
3. Bogataj U, Gros N, Kljajic M et al. (1995) The rehabilitation of gait in patients with hemiplegia: a comparison between conventional therapy and multichannel functional electrical stimulation therapy. Phys.Ther. 75: 490-502
4. Stanic U, Acimovic-Janezic R, Gros N et al. (1978) Multichannel electrical stimulation for correction of hemiplegic gait. Methodology and preliminary results. Scand.J Rehabil.Med. 10: 75-92
5. Quintern J, Bisle G, Hartmann E et al. (2002) Flexion reflex in spastic hemiparesis: neurophysiological and therpeutic use. 3rd world congress in Neurological rehabilitation 371-372
6. Quintern J, Bisle G, Hartmann E, Maier-Weiterschan C, Bauduin G (2003) Controlled clinical study of stimulation of flexor reflex afferents for gait rehabilitation in patients with hemiplegia. In: Gantchev N (ed) From basic motor control to functional recovery III. St. Kliment Ohridski University Press, Sofia, Bulgaria, pp 240-246
7. Andersen OK, Sonnenborg FA, Arendt-Nielsen L (1999) Modular organization of human leg withdrawal reflexes elicited by electrical stimulation of the foot sole. Muscle Nerve 22: 1520-1530
8. Spaich EG, Arendt-Nielsen L, Andersen OK (2004) Modulation of lower limb withdrawal reflexes during gait: a topographical study. J.Neurophysiol. 91: 258-266
9. Spaich EG, Hinge HH, Arendt-Nielsen L et al. (2006) Modulation of the withdrawal reflex during hemiplegic gait: effect of stimulation site and gait phase. Clin.Neurophysiol. 117: 2482-2495
10. Spaich EG, Emborg J, Collet T et al. (2009) Withdrawal reflex responses evoked by repetitive painful stimulation delivered on the sole of the foot during late stance: site, phase, and frequency modulation. Exp.Brain Res.
11. Holden MK, Gill KM, Magliozzi MR et al. (1984) Clinical gait assessment in the neurologically impaired. Reliability and meaningfulness. Phys.Ther. 64: 35-40

Author: Erika G. Spaich
Institute: Center for Sensory-Motor Interaction (SMI), Integrative Neuroscience Group, Aalborg University
Street: Fredrik Bajers Vej 7-D3
City: 9220 - Aalborg
Country: Denmark
Email: espaich@hst.aau.dk

Biomechanics of Pointing to a Perceived Target: Effects of Fatigue and Gender

J.N. Côté, T. Hsieh, and K. Emery

Department of Kinesiology and Physical Education, McGill University, Montreal, Canada;
Michael Feil and Ted Oberfeld / CRIR Research Centre, Jewish Rehabilitation Hospital, Laval, Canada

Abstract— **Neck/shoulder pain, which is more prevalent in women, has previously been linked to repetitive work and muscle fatigue. We have shown that asymptomatic people performing repetitive upper limb tasks display shoulder fatigue and whole-body compensatory strategies. However, the role of proprioception in controlling these patterns is unclear. The perception of muscular effort has been implicated in the proprioceptive aspects of the fatigue response, although few studies have estimated its impact. In this study, a group of asymptomatic adults (9 women, 9 men) performed a repetitive pointing task (RPT) to fatigue. Before and after the RPT, they performed a shoulder position (SPS) task where they abducted their shoulder to the perceived horizontal, and an endpoint position (EPS) task, where they moved their finger to a perceived target location in front of them. In the SPS, subjects made larger errors after fatigue by raising their elbow higher (~ +1.3cm). In a follow-up study, subjects replicating the same task made smaller errors when actively moving their arm to the target, compared to that when the arm was passively moved. As for the EPS, subjects' finger position accuracy was not affected by fatigue. There were no gender effects on accuracy pre- and post-RPT; however, there were gender differences in the perceived finger target location and in the temporal characteristics of movement towards the target. Results suggest that healthy individuals are able to develop strategies to compensate for fatigue-induced deficits at one joint to maintain the endpoint accuracy of a multijoint task constant, possibly by using feedback from muscle output. Movement strategies and perception of endpoint location may play parts in gender differences in work-related neck/shoulder symptoms.**

Keywords— **Fatigue, proprioception, neck/shoulder, gender.**

I. INTRODUCTION

Repetitive arm movements can lead to the development of neck/shoulder muscle fatigue, which is a risk factor for the development of musculoskeletal disorders [1]. In low force tasks, fatigue gradually develops and leads to a reduction in the muscle's maximal force generating capacity [2]. Moreover, higher-level adaptations have been observed as fatigue develops during multi-joint tasks. In experiments using repetitive tasks that fatigued the shoulder musculature, we showed that reduced elbow motion amplitude was compensated by increased trunk motion [3-4]. Fatigue was also associated to increased electromyographical (EMG) amplitude in the shoulder musculature and in areas distant to the site of fatigue [5]. We also showed that fatigue-related changes also affect temporal characteristics of movement [4]. In these studies, despite the many observed biomechanical changes, inherent task goals (e.g. endpoint movement amplitude) were preserved even when fatigue was reached. These findings suggest that when performing repetitive multijoint movements, the system can modify the relative involvement of individual degrees-of-freedom and even recruit additional ones, possibly to preserve some important global task characteristics constant. However, several elements of the control mechanisms that may underlie these adaptations remain poorly understood.

As an important element of the motor control loop, proprioception is a set of sensory information detected from mechanoreceptors and sent to the central nervous system where it is integrated, with other sensory modalities, as a basis for appropriate motor response [6]. Previous studies have shown that fatigue affects some proprioceptive submodalities including movement acuity [7] and position sense [8]. While fatigue, by definition, is thought to be linked to an increased sense of effort [9], the mechanisms by which the sense of effort may lead to changes in other proprioceptive submodalities is not well understood. Allen and Proske [10] reported that after fatigue, increased effort was necessary to compensate for force decrements to maintain a given elbow position against gravity. It is thought that the observed decline in the effectiveness of muscle outputs following fatigue are compensated centrally by increasing the activity of motor neurons, resulting in increased EMG amplitude and perceived effort [11]. This suggests that proprioceptive mechanisms rely on muscle feedback to obtain proprioceptive information about the corresponding joint, which is supported by recent studies [12].

Several studies have studied the effects of fatigue on shoulder position sense using single-joint, unidirectional experimental tasks [7-8]. However, few studied how position sense is affected during unsupported, multijoint, fatiguing functional activities. Björklund et al. [13] studied the effects of repetitive arm movements on shoulder position sense and found greater errors after fatigue. Angyan et al. [14] studied movement accuracy in athletes after repeated throws and found that they overestimated the target distance more after fatigue. A study of baseball players reported a significant change in their ability to reproduce the same throwing movement after fatigue [15], and the observed reductions occurred in parallel with changes in shoulder and

K. Dremstrup, S. Rees, M.Ø. Jensen (Eds.): 15th NBC on Biomedical Engineering & Medical Physics, IFMBE Proceedings 34, pp. 233–236, 2011.
www.springerlink.com

elbow but not wrist position sense. Thus, it can be hypothesized that fatigue would differently affect not only motor but proprioceptive characteristics of various joints involved in a multijoint task. If confirmed, this would target the role of proprioceptive integration mechanisms as an important one in the control of the whole-body changes that we have previously observed to occur with fatigue.

While women report more neck/shoulder symptoms, several authors have reported that they have greater fatigue resistance than men [16]. However, only few studies have compared the effects fatigue on proprioceptive characteristics between men and women. Of the few that have, one showed no gender differences in upper limb proprioception [14], with the other showing conflicting results [13,17].

The main goal of this study was to assess the effects of repetitive motion-induced fatigue on shoulder joint and upper limb's endpoint position sense. We hypothesized that fatigue would significantly affect accuracy of the shoulder, but not of the limb's endpoint.

II. METHODS

A. Participants

Two experiments were conducted in this study, using two groups of healthy adults (9 men, 9 women in each) of similar ages. Subjects could not have been diagnosed with a neck/shoulder injury, have reported pain or have reduced cervical range of motion during the past year. They all provided informed consent by signing forms approved by the Research Ethics Board of the Centre for Interdisciplinary Research in Rehabilitation (CRIR) of Greater Montreal.

B. Procedure

In the first experiment, subjects performed a shoulder position sense task (SPS) and an upper limb endpoint position sense task (EPS), assigned in random order. In the EPS, subjects stood and we placed a target (a light-reflective marker) at 60% of their dominant arm's length, at shoulder height, in front of their midline. In the starting position, subjects held the non-dominant arm passively to the side, the dominant elbow flexed and the dominant index finger touching the manubrium of their sternum. Subjects were shown the target, then closed their eyes while the target was removed and at the "go" signal, moved the arm smoothly in one ballistic movement at their preferred speed, immobilized the finger at the perceived target location for 1s, moved the arm back to the starting position, and opened their eyes. In the SPS, subjects were seated on a chair, both arms in the same position as in the starting position for the EPS. They closed their eyes and in one smooth ballistic movement at their preferred speed, abducted the shoulder until they perceived that their upper arm was horizontal, maintained this position for 1s and returned to the starting position. In each task, subjects performed 6 trials, with 30s of rest between trials and 1min of rest between tasks.

Once the two position sense tasks were completed, subjects performed a repetitive pointing task (RPT) with the dominant arm alternating touches to two cylindrical targets (length: 6 cm, radius: 0.5 cm) placed at 30% and 100 % of arm's length, in front of their midline, at shoulder height. A mesh barrier was placed immediately under the moving elbow's horizontal movement trajectory and subjects were instructed to not touch it. The RPT was performed at a 1 Hz pace, dictated by a metronome. Subjects were instructed to "perform the task as comfortably and naturally as possible, for as long as possible". During the RPT, they were asked to rate their perceived neck/shoulder exertion on the Borg CR-10 scale [18] at 1min intervals. They continued the RPT until at least one of the following criteria was met: their Borg rating was 8 or higher, the 1-Hz rhythm could not be maintained, they touched the mesh barrier, or they performed the task for more than 30min. Subjects were unaware of these criteria. After the fatiguing protocol, subjects immediately performed the two position sense tasks again in the same order. To prevent recovery, they maintained the shoulder abducted 90° and the index finger on the sternum between trials and tasks until the end of the protocol.

In the second experiment, before and after the same RPT, subjects performed the same SPS task and six trials of a passive version of the SPS, in random order. Their forearm was placed on a horizontal plate, which began at hip height and was moved upward at a constant rate of 5cm/s by a motor-driven apparatus (BTE-Tech). Subjects were instructed to push a button when they perceived their arm to be horizontal, at shoulder height.

EMG of 6 muscles was measured (1080 Hz) using the Telemyo 900 system (Noraxon). Bipolar, Ag-AgCl surface electrodes were placed bilaterally on the dominant (experimental) side over the upper trapezius (UTR), anterior and middle deltoid (AD, MD), bicepts and triceps brachii (BI, TRI) and lumbar erector spinae (LES). Kinematics was recorded in three dimensions using passive, reflective markers. Landmarks of interest were the spinous process of C7, the lateral epicondyle of the dominant elbow and the distal phalange of the dominant index finger. Marker coordinates were captured (120 Hz) using a high-resolution, six-camera motion capture system (Vicon-Peak).

C. Analysis

Marker trajectories were low-pass filtered (zero-lag second order Butterworth, 6 Hz). We analyzed the movement phase starting from the finger leaving the sternum and ending when the finger (for EPS trials) or the elbow (for SPS trials) marker coordinates changed less than 2SD for 0.1s during the 1s when subjects held the perceived target position. For the EPS trials, the difference between the true target position and the final finger marker position was calculated in tridimensional space. For the SPS trials, the C7 marker was set as the reference height and the vertical distance between C7 and the final elbow marker position was calculated as the vertical error. Average and peak velocities of the finger and elbow markers were calculated, and peak velocity occurrence was expressed as a percentage of the movement duration. Heartbeats were removed from each EMG, and signals were filtered (zero-lag second order 20-500 Hz band-pass) and full-wave rectified. Root mean square (RMS) values for each muscle were calculated over 100-ms windows taken at finger or elbow final position.

Analyses of finger position accuracy, average velocity, peak velocity and peak velocity occurrence were performed using repeated measures 3-way ANOVA with factors of Time (pre- post-RPT), Direction (vertical, anteroposterior, mediolateral) and Gender. Accuracy characteristics of the SPS task and movement duration for both tasks were analyzed using 2-way ANOVA with factors of Time and Gender. EMG variables were analyzed using repeated measures 3-way ANOVA with factors of Time, Task and Gender.

III. Results

A. Evidence of fatigue

In the main experiment, there were significant main Time effects for the AD ($p<0.001$), MD ($p<0.001$), TRI ($p<0.005$) and LES ($p<0.01$) muscles, which showed increased activity after the RPT. In the follow-up experiment, preliminary analyses ($N = 5$) show no significant effect of the RPT on EMG amplitude for either the passive or the active SPS.

B. Endpoint and shoulder position sense

There was a significant Gender main effect on final finger position (EPS) ($p<0.024$), with post-hoc analysis showing that women stopped the finger in a less anterior and lower position than men. No other effects were found for accuracy in the EPS task. There was a significant main effect of Time on elbow marker position (SPS) ($p<0.015$) with increased elbow height after fatigue (Fig.1). Before the RPT, subjects raised their elbow 0.9 (±2.4) cm above C7 and after, they raised it even higher, 2.2 (±3.1) cm above C7. In the follow-up experiment, there were no differences in accuracy between the pre- and post-RPT trials. However preliminary analyses show that subjects displayed significantly higher accuracy during the active SPS, both before ($p<0.008$) and after ($p<0.0003$) the repetitive task.

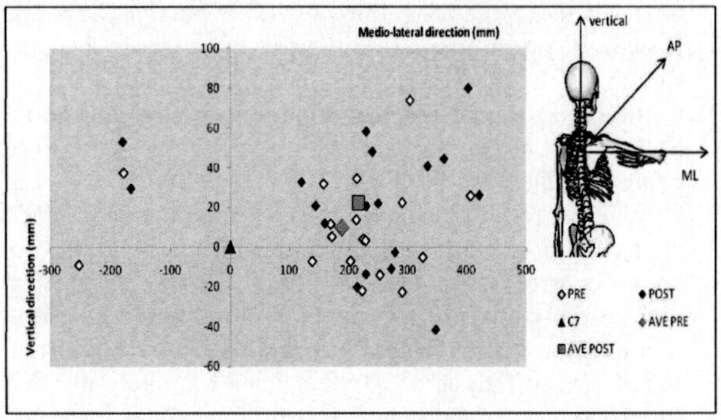

Fig. 1 Vertical (y) and medio-lateral (x) coordinates of the elbow marker position (mm) for the active SPS task, before and after the RPT. Black triangle: position of the C7 marker, set at the (0, 0) coordinate. Big grey diamond: average elbow position pre-fatigue. Big grey square: average elbow position post-fatigue. Elbow height is higher after fatigue, indicating that subjects over-estimated the horizontal, even more so when fatigued.

C. Movement patterns towards the target, EPS task

There was a significant Time x Gender interaction effect on average finger velocity ($p<0.05$), which was higher after the RPT in men and slightly lower in women. A significant Direction x Gender interaction was also found, with men showing higher velocity in the anteroposterior direction than women (343.8 (±106.7) mm/s for men vs. 264.9 (±60.77) mm/s for women; $p<0.02$). There was a significant Time main effect on movement duration, with post-RPT trials being shorter by about 0.18s ($p<0.017$). There was a significant Time x Gender interaction in the peak velocity occurrence of EPS trials, with peak finger velocity occurring 8.8% later in the post-RPT trials in men ($p<0.01$) while no significant timing difference was found for women.

IV. Discussion and Conclusions

Our main hypothesis was confirmed, in that position sense was affected by fatigue only during the SPS and not during the EPS. This suggests that participants were able to benefit from redundancy of the proprioceptive system to compensate for impaired shoulder proprioception, such that they maintained the finger position accuracy constant during the EPS task despite the presence of fatigue. This finding confirms those of Allen and Proske [10]: subjects in our study could have relied on feedback of higher EMG amplitude after the RPT to estimate that the horizontal was actually higher than in reality. However, preliminary analyses of the follow-up experiment preclude us from inferring on the role of EMG feedback in the fatigue mechanisms since in this latter experiment, the repetitive task seems to have failed to induce fatigue, likely due to the opportunity for muscles to rest during the post-RPT passive SPS trials. However, preliminary results still offer some insight in showing better accuracy when the arm was actively moving to the target, supporting the important role of muscle feedback as part of the mechanisms underlying the sense of limb position [12].

We also found that subjects performed the post-RPT trials in less time, with EMG results suggesting that this could have occurred via increased biceps and triceps activity in more rapidly moving and stopping the elbow. The fact that this did not affect their EPS accuracy further supports that the effects of fatigue on SPS accuracy are likely not due to fatigue-related impairments in the motor output. Moreover, this finding reinforces the notion that fatigue adaptation strategies are selected in task-specific ways: given a requirement for position sense performance and no constraint on the temporal characteristics of the task, subjects prioritized the maintenance of the position sense task objective and, independently or not, reduced effort time.

Our results show no gender difference in accuracy. However, when considering the directionality of errors, women systematically made errors that were more anterior and inferior to the target. Interestingly, no gender differences were found in the accuracy of the SPS task, although average velocity was higher and peak velocity occurred later in men, supporting the notion that gender differences in the fatigue response implicate higher levels of control [19]. These findings may have implications on the identification of pathways underlying gender differences in work-related neck/shoulder injuries. For instance women may unknowingly change the perceived location of their workspace to a more proximal one with fatigue, such that the same workspace may be perceived as further away with fatigue in women, compared to in men. Since we did not assess the effects of fatigue on perceived difficulty in attaining the real targets, further studies are needed to address this hypothesis.

In conclusion, our findings support our main hypothesis that healthy subjects can compensate for local proprioception deficits in maintaining the overall sense of the limb's endpoint preserved with fatigue. However, further studies are needed to identify the muscular and gender-specific mechanisms that may underlie these findings.

Acknowledgment

Funds to support this study were obtained from the Canadian Foundation for Innovation and the National Science and Engineering Research Council.

References

1. van der Windt D, Thomas E, Pope DP et al. (2000) Occupational risk factors for shoul- der pain: a systematic review. Occup Environ Med 57:433–442
2. Vollestad NK (1997) Measurement of human muscle fatigue. J Neurosci Methods 74:219–227
3. Côté JN, Mathieu P, Levin MF et al. (2002) Movement reorganization to compensate for fatigue during sawing. Exp Brain Res 146:394–398
4. Côté JN, Raymond D, Mathieu PA et al. (2005) Differences in multijoint kinematic patterns of repetitive hammering in healthy, fatigued and shoulder injured individuals. Clin Biomech 20(6):581–90
5. Côté JN, Feldman AG, Mathieu P et al. (2008) Effects of fatigue on intermuscular coordination during repetitive hammering. Motor Control 12(2):79–92
6. Myers JB, Lephart SM (2000) The Role of the Sensorimotor System in the Athletic Shoulder. J Athl Train 35(3): 351–363
7. Pedersen J, Lönn J, Hellström F et al. (1999) Localized muscle fatigue decreases the acuity of the movement sense in the human shoulder. Med Sci Sports Exerc 31(7):1047–1052
8. Walsh LD, Allen TJ, Gandevia SC et al. (2006) Effect of eccentric exercise on position sense at the human forearm in different postures. J Appl Physiol 100(4):1109–1116
9. Jones LA, Hunter W (1983) Effects of fatigue on force sensation. Exp Neurol 81(3):640–650
10. Allen TJ, Proske U (2006) Effect of muscle fatigue on the sense of limb position and movement. Exp Brain Res 170(1):30–38
11. M Proske U (2005) What is the role of muscle receptors in proprioception? Muscle Nerve 31(6):780–787
12. Fuentes CT, Bastian AJ (2010) Where is your arm? Variations in proprioception across space and tasks. J Neurophysiol 103:164 – 171
13. Björklund M, Crenshaw AG, Djupsjöbacka M et al. (2000) Position sense acuity is diminished following repetitive low-intensity work to fatigue in a simulated occupational setting. Eur J Appl Physiol 81:361–367
14. Angyan L, Antall C, Angyan Z (2007) Reproduction of reaching movements to memorized targets in the lack of visual control. Acta Physiol Hung 94(3):179–182
15. Tripp BL, Yochem EM, Uhl TL (2007) Functional fatigue and upper extremity sensorimotor system acuity in Baseball Athletes. J Athl Train 42(1):90–98
16. Hunter SK, Critchlow A, Enoka RM (2004) Influence of aging on sex differences in muscle fatigability. J Appl Physiol 97:1723–1732
17. Björklund M, Crenshaw AG, Djupsjöbacka M et al. (2003) Position sense acuity is diminished following repetitive low-intensity work to fatigue in a simulated occupational setting. A critical comment. Eur J Appl Physiol 88(4-5):485–486
18. Borg G (1982) A category scale with ratio properties for intermodal and interindividual comparisons. In: Geisler HG, Petzold B, editors. Psychophysical Judgement and the Process of Perception. Berlin: VEB Deutscher Verlag der Wissenschaft
19. Martin PG, Rattey J (2007) Central fatigue explains sex differences in muscle fatigue and contralateral cross-over effects of maximal contractions. Pflugers Arch 454(6):957–969

Julie N. Côté, PhD
Department of Kinesiology & Physical Education, McGill University
475 Pine Avenue West
Montreal
Canada
julie.cote2@mcgill.ca

Biomechanics of Human Movement

P. Madeleine, A. Samani, M. de Zee, and U. Kersting

Physical activity and Human Performance group, Center for Sensory-Motor Interaction (SMI),
Dept. of Health Science and Technology, Aalborg University, Denmark

Abstract— **Biomechanical methods are frequently used to assess human performance in sports and ergonomics in both laboratory and field settings. Musculoskeletal disorders (MSD) are often related to physical activity. Such disorders affect muscles, tendons, cartilage and ligaments and are often accompanied by pain. They are extremely difficult to diagnose/treat and preventing them is still considered as the best treatment. From this perspective, assessments of human performance and related biomechanical loads are primordial as these contribute to assess the physical risk factors in relation to motor activity.**

Human performance and biomechanical load are typically measured by means of physiological, kinetic and kinematic recordings. Physiological recordings can consist of surface electromyography or mechanomyography used to estimate the physical and muscular load. Force transducers, force-platform and pressure sensors are key elements for assessing reaction forces and pressure distribution profiles. Accelerometers, gyroscopes, flexible angular sensors and electromagnetic tracking systems are most popular solutions for kinematic assessments. 3D kinetic and kinematic measurements are often combined to estimate joint load using an inverse dynamics approach.

The three types of recordings are reviewed in relation to sports and ergonomics focusing on the possibilities of each method. Further, computer simulation and modeling approaches are also presented in relation to sports and ergonomics.

Keywords— **Physical activity, musculoskeletal disorders, surface electromyography, mechanomyography, kinetic, kinematics.**

I. Introduction

The study of human movement requires the use of biomechanical assessment, i.e. the use of the laws of mechanics to gain insight in human performance. More importantly, such analyses can provide important information regarding the improvement of performance without increasing the risk of musculoskeletal injuries. Musculoskeletal disorders (MSD) often occur in relation to physical activity in sports or at work. MSD are often accompanied by pain located in muscles, tendons, cartilage and ligaments [1]. Sports related overload injuries are relatively easy to diagnose while an important portion of work-related MSD remains often undiagnosed. Thus, sports and work-related MSD are in general difficult to treat.

Thus, biomechanical assessments of human performance contribute to delineate harmful load pattern to the musculoskeletal system in relation to physical activity. The biomechanics of human movement can be assessed in laboratory or field settings or in a simulated or real environment. Simulating a motor task under laboratory conditions implies that factors like the physical environment and some psychosocial factors (e.g. perceived time pressure) affecting motor behavior can be controlled. Advanced measurements including 3D kinematics are easier to perform in a laboratory. Thus, laboratory studies usually provide data of high reliability and validity.

In the present paper existing assessment methods within physiological, kinetic and kinematics are reviewed. New trends in biomechanical data analysis like e.g. non-linear dynamics and computer simulation and modeling are also within the scope of the present paper.

II. Physiological recordings

A. *Surface Electromyography (SEMG)*

SEMG is usually recorded using classic bipolar single channels. More recently, two dimensional electrode arrays have been used to record muscle activation pattern from e.g. the trapezius muscle (Fig. 1). For classic SEMG assessment, bipolar SEMG electrodes are aligned (inter-electrodes distance 2 cm) on abraded ethanol-cleaned skin along the direction of the muscle fibers. Electrodes are placed in relation to anatomical points [2]. A reference electrode is also used. SEMG signals are usually pre-amplified/amplified (close to the recording site), band-pass filtered e.g. 10-400 Hz and sampled at e.g. 1 kHz. Classic bipolar SEMG recordings are used to show loading pattern during movement as well as level of activation and changes in the power spectral density function.

Nowadays, 2D electrode arrays are becoming more common. The technique consists in the detection of SEMG

K. Dremstrup, S. Rees, M.Ø. Jensen (Eds.): 15th NBC on Biomedical Engineering & Medical Physics, IFMBE Proceedings 34, pp. 237–240, 2011.
www.springerlink.com

signals from a number of closely located points over the same muscle, providing a high-density spatial sampling of the skin surface electric potential due to muscle activation. 2D SEMG assesses heterogeneities in muscle activation during muscle contractions [3]. This can be used to delineate spatial adaptation of activation within individual muscles. The SEMG signals are amplified, band-pass filtered and sampled similarly to classical SEMG recording.

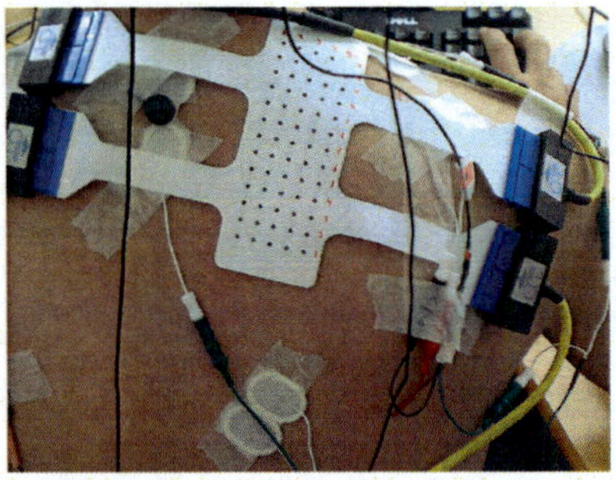

Fig. 1 Example of classic bipolar and 2D electrode arrays SEMG recording from the trapezius muscle [3]

SEMG traces during cyclic movement are rectified, low-pass filtered ($F_{cut-off}$: 6-12 Hz) and averaged over the number of recorded cycles. This enables the detection of muscle contraction onset and offset. Root mean square (RMS) or average rectified values (ARV) are the most common temporal estimators while median frequency (MF) or mean power frequency (MNF) are used to estimate compression of the power spectral density function in relation to e.g. development of localized muscle fatigue. For 2D SEMG recordings, the centroid of the map can be computed. Further, the entropy of SEMG amplitude map is also providing important information about the complexity of muscle activation.

B. *Mechanomyography (MMG)*

MMG reflects the mechanical component of a muscle activation pattern and reflect motor unit recruitment, discharge pattern, synchronization as well as factors affecting the muscle physical milieu (e.g. intra-muscular pressure, stiffness, and osmotic pressure) [4]. MMG signals can be recorded by means of piezoelectric contact sensor, microphones, accelerometers, and laser distance sensors. Accelerometers are currently the most applied sensors for MMG recording due to their small weight and size, easy attachment, and high reliability [5]. Light accelerometers (weight < 5 g) do not interfere with muscle surface dynamics [5]. Furthermore, accelerometers record MMG signal in physical units ($m.s^{-2}$) enabling comparison between different studies.

Like SEMG, single channels and multi-channels surface MMG recordings are possible. A single or several accelerometers are used for this purpose. The physical dimension, sensitivity and linear transmission in the low frequency range [DC-100 Hz] are important. MMG are amplified, band-pass filtered and sampled at e.g. 1 kHz. Accelerometers can be then be arranged in complete/incomplete grid in relation to anatomical points to cover the muscle of interest. MMG signals are then amplified and converted in digital form.

Then, RMS/ARV and MF/MNF values are computed. Moreover, the power spectral variance (2nd order moment) and skewness (3rd order moment) can also be calculated to recover complex modifications of the shape of the power spectrum [6]. The entropy of the MMG amplitude maps can be computed to depict the level of uniformity of the MMG maps which represents the diversity of involved contractile elements [6].

III. Kinetic recordings

A. *Force Sensor Types*

Strain gauges are the most common type of force transducers used to measure force. Strain gauge measures a change in length with respect to the original length. Then, the stress in the material and finally the external load can be registered. Other types of transducers include e.g. piezoelectric, piezoresistive, capacitive sensor. The advantages of strain gauges are numerous including: sensitivity, inexpensive, portable and attachable to devices (knife, pedals) or living tissues (bones, ligaments). The drawbacks to be mentioned are: need for calibration, limited range of measure, easily damageable, cross-talk as well as temperature and pressure sensitivity.

B. *Force Platform*

Force platforms are devices embedded in a biomechanics laboratory integrated in walkway and/or handles. A resultant force is provided by the device. The vertical component describes the change in momentum of the center of mass of the subject in the vertical direction. The anterior-posterior and medial-lateral components correspond to the two horizontal directions. Force signals are in general sampled at frequencies $\geq$ 100 Hz. The measurements of the force sensors in the four corners of a force plate can be used to determine the center of pressure and the free moments of rotation about a vertical axis through the center of pressure. Both piezoelectric and strain gauge force platforms can be used but the latter one are more suitable for postural studies, cheaper and can be custom-built. Ground reaction forces are expressed in absolute values and/or with respect to body weight. Recently, the combination of linear and non-linear analyses has provided new insight in the dynamics of force control by underlining important differences for example gender [7].

C. *Pressure Sensors*

Pressure sensors measure the distribution of reaction forces between two surfaces like foot and shoe. Pressure distribution mats or insoles on the basis of capacitor elements consist of a matrix of m by n stripes of conducting material. Multiplexing techniques are usually applied to assess the force acting on each element. Pressure distribution measurements are often made in sports (shoe insole, ski-boot shaft) and ergonomics (handgrip, seat comfort). These devices are in general rather costly, not very flexible and only measure normal forces. Such devices enable an isobar, path or 3D quantification of loading pattern during human movement.

IV. KINEMATICS RECORDINGS

A. *Accelerometer and Gyroscope*

Accelerometers are typically small, light and can be mounted to equipment or the human body itself. Data from skin-mounted accelerometers are, despite their inherent precision, sometimes difficult to interpret as their fixation is critical due to soft tissue movement. Some research groups have mounted accelerometers directly to the bone [8]. However, such solutions are not generally applicable and may only serve to validate other measurement approaches [8]. Still, it was repeatedly shown that accelerometers attached to the tibia of a runner gives reasonable estimates of the impact peak of the ground reaction force and therefore easily applicable method to evaluate the effects of footwear or running style. There are unidirectional or multidirectional accelerometers available. It is often problematic to separate the source of the acceleration when conduction measurements using uni- or multidimensional devices. Accelerometers can also be used as a trigger signal in high velocity movements where the amplitude is of minor importance [9]. They also find numerous applications in the field of activity monitoring as accelerometer signals can be utilized to infer of body position in static or slow movements as well as identify counts and frequencies of repetitive movements.

Gyroscopes are sensors which measure angular velocity. They are often used in combination with accelerometers and other sensors and provide good estimates for whole body movements [10]. Despite very promising developments in this area, there are still limitations to the precision of such devices.

B. *Flexible Angular Sensor*

Angular movement sensors are often realized as uniaxial goniometers which allow for a direct measure of a joints excursion. Limitations are that only hinge joints can be assessed with these devices. To overcome this limitation flexible two-dimensional angular sensors have been introduced.

C. *Optical imaging system*

Optical systems are probably the most commonly used equipment to assess movement analysis. A single camera system can be used to analyze the movements which are mainly executed in one plane. Walking and running are the classical examples of two-dimensional movement analysis [11]. The extension of this approach is a multi-camera setup which allows to record movement in 3D. By reference to a calibration recording, the method of direct linear transformation can be applied to reconstruct the position of a point in space from minimum 2 cameras [12]. Such setups are nowadays realized using several synchronized video cameras which can be sampled by up to 10000 Hz at a spatial resolution of up to 4 Megapixel. Each camera is typically equipped with its own light source and an optic filter to suppress reflections from items other than a number of retro-reflective markers. The above mentioned advances in camera technology allow for imprecision of less than 0.5 mm in a typical gait laboratory setting.

Other approaches are active marker systems where each marker is a made up out of a small light. In some cases, the higher demand in applying the markers due to cable connections may be overcome by less manual editing work during the tracking process. It is up to the individual to decide which system is most suitable for which application.

Finally, from the area of computer graphics more and more marker-less tracking systems sprung off [13,14] which have partly been validated with respect to high-resolution laboratory based systems.

V. MODELING MUSCULOSKELETAL LOADS

As already mentioned the quantification of the mechanical loads on the human body in e.g. working situations and sports situations is of interest. However, in experimental sessions it is normally only possible to assess external loads on the body, while the internal loads on muscles and joint remains unknown. Joint moments are regularly calculated using motion capture and measured external forces some input based on inverse dynamics. This will though not give information on individual muscle forces and joint reaction forces. The only feasible way to obtain these parameters is to make use of advanced musculoskeletal models based on the laws of physics. Traditionally the methods behind musculoskeletal modeling fall into two categories, inverse dynamics and forward dynamics.

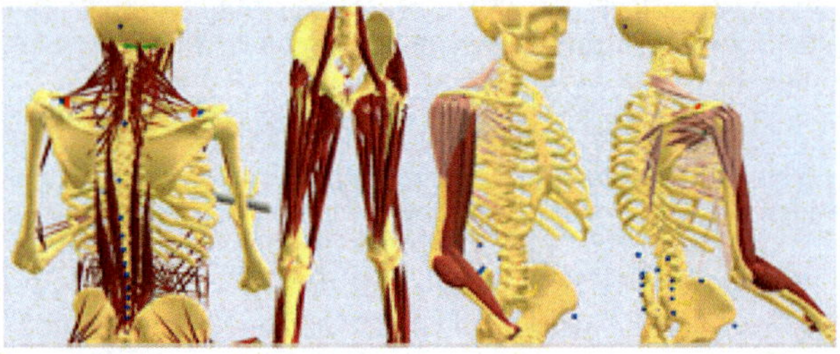

Fig. 2 Example of musculoskeletal models built in the AnyBody Modeling System

In inverse dynamics solutions the movement and external forces are input into the musculoskeletal model. The system has normally many more muscles than strictly necessary to balance the joint degrees of freedom. The solution of the muscle recruitment problem is therefore subject to the so-called redundancy problem. One of the common ways is that the muscles in the model are recruited by an optimality criterion minimizing fatigue. This is for example implemented in the AnyBody Modeling System [15], which was originally developed at Aalborg University. Several musculoskeletal models (Fig 2) are available in the public domain AnyScript Model Repository (www.anyscript.org). An application within ergonomics shows the influence of seat pan inclination and friction on muscle activity and spinal joint forces using a full-body model of the musculoskeletal system [16].

VI. Conclusions

The present paper presents in a concise manner the state of the art within biomechanical assessments of human movement. The sole combination of physiological, kinetic and kinematics recordings provide a full picture of the musculoskeletal loads in relation to physical activity. Experimental and computational approaches are extremely valuable to improve human performances without increasing the risk of MSD.

Acknowledgments

The authors are grateful to Det Obelske Familiefond for supporting the present work.

References

1. Madeleine P (2010) On functional motor adaptations: From the quantification of motor strategies to the prevention of musculoskeletal disorders in the neck-shoulder region. Acta Physiol 679:1-46
2. Hermens HJ, Freriks B, Disselhorst-Klug C, Rau G (2000) Development of recommendations for SEMG sensors and sensor placement procedures. J Electromyogr Kinesiol 10:361-374
3. Samani A, Holtermann A, Søgaard K, Madeleine P (2010) Advanced biofeedback changes the spatial distribution of upper trapezius muscle activity during computer task. Eur J Appl Physiol 10:415-423
4. Orizio C 1993 Muscle sound: bases for the introduction of a mechanomyographic signal in muscle studies. Crit Rev Biomed Eng 21:201-243
5. Watakabe M, Mita K, Akataki K, Ito K 2003 Reliability of the mechanomyogram detected with an accelerometer during voluntary contractions Med Biol Eng Comput 41:198-202
6. Madeleine P, Tuker L, Arendt-Nielsen L, Farina D (2007) Heterogeneous mechanomyographic activation of paraspinal muscles assessed by a two-dimensional array during short and sustained contractions. J Biomechanics 40:2663-2671
7. Svendsen JH, Madeleine P (2010) Amount and structure of force variability during short, ramp and sustained contractions in males and females. Hum Mov Sci 29:35-47
8. Lafortune MA, Henning E, Valiant GA (1995) Tibial shock measured with bone and skin mounted transducers. J Biomechanics 28:989-993.
9. Kersting UG, Janshen L, Bohm H, Morey-Klapsing GM, Bruggemann GP (2005) Modulation of mechanical and muscular load by footwear during catering. Ergonomics 48:380-398.
10. Brodie M, Walmsley A, Page W (2007) Fusion integration: COM trajectory from a force platform. J Appl Biomech 23:309-314.
11. Winter DA (1989) Biomechanics of normal and pathological gait: implications for understanding human locomotor control. J Mot Behav 21:337-355.
12. Abdel-Aziz YI, Karara HM (1971). Direct linear transformation from comparator coordinates into object space coordinates in close-range photogrammetry. Proceedings of the Symposium on Close-Range Photogrammetry. Falls Church, VA, USA, pp 1-18.
13. Rosenhahn B, Brox T, Kersting UG, Smith AW, Gurney JK, Klette R (2006). A system for marker-less motion capture. Kuenstliche Intelligenz 20:46-52.
14. Corazza S, Mundermann L, Chaudhari AM, Demattio T, Cobelli C, Andriacchi TP (2006). A markerless motion capture system to study musculoskeletal biomechanics: visual hull and simulated annealing approach. Ann Biomed Eng 34:1019-1029.
15. Damsgaard M, Rasmussen J, Christensen S T, Surma E, de Zee M (2006) Analysis of musculoskeletal systems in the AnyBody modeling system. Simulat Model Pract Theor 14:1100:1111
16. Rasmussen J, Torholm S, de Zee M (2009) Computational analysis of the influence of seat pan inclination and friction on muscle activity and spinal joint forces, Int J Ind Ergon 39:52-57

Author: Pascal Madeleine
Institute: Dept. of Health Science and Technology, Aalborg University
Street: Fredrik Bajers vej 7
City: Aalborg
Country: Denmark
Email: pm@hst.aau.dk

Stenosis Detection Algorithm for Screening of Arteriovenous Fistulae

Mikkel Gram[1], Jens Tranholm Olesen[1], Hans Christian Riis[1], Maiuri Selvaratnam[1], Helmut Meyer-Hofmann[2], Birgitte Bang Pedersen[2], Jeppe Hagstrup Christensen[2], Johannes Struijk[1], and Samuel Emil Schmidt[1]

[1] Department of Health Science and Technology, Aalborg University, Denmark
[2] Department of Nephrology, Aalborg Hospital, Aarhus University Hospital, Denmark

Abstract— **The aim of the study was to develop an algorithm that can detect stenosis formation in arteriovenous fistulae based on audio recordings. 34 patients with a mature arteriovenous fistula were examined with use of an electronic stethoscope and subsequently by ultrasound. 27 patients had a patent fistula, while the other group consisted of 5 patients with stenosis and 2 with artificial narrowing of the fistula. Feature extraction was carried out using wavelet packet decomposition at depth 4. For each recording the scale energies SE_i and the percentage of scale energy versus total energy SEp_i, were calculated. The two most discriminative features with low correlation were found to be SE_8 and SEp_8. These features were evaluated using leave-one-out cross-validation with a quadratic discriminant function. Cross-validation using SE_8 and SEp_8 yielded a sensitivity of 100% and a specificity of 94%. The algorithm developed using the features obtained by wavelet analysis is reliable for detecting stenosis in a vein segment of an arteriovenous fistula. Based on these results, the prospects of developing an accurate, low-cost screening method for patients undergoing hemodialysis, are promising.**

Keywords— **Arteriovenous fistulae, Stenosis detection, Auscultation, Wavelet packet decomposition, Screening method.**

I. INTRODUCTION

In order to perform hemodialysis, it is necessary to establish a vascular access that allows for easy cannulation and provides a high blood flow. The most common method is the arteriovenous fistula (AVF), where a surgical anastomosis between the vein and the adjacent artery is created [1].

Vascular access dysfunction mostly caused by initial failure of the fistula to mature, as well as later venous stenosis, have proven to be some of the most important reasons for morbidity. Furthermore, access dysfunction is responsible for 20% of the hospitalizations that hemodialysis patients undergo, costing an estimated $1 billion annually in the US [2, 3].

Since venous stenosis is the most common cause of thrombosis, it is necessary for the patient and health staff to monitor the AVF for signs of stenosis formation. This is typically done using color duplex imaging (CDI, also known as ultrasound), venography or fistulography, which are all well tested methods. However, these methods are all expensive and/or invasive, making them unsuitable for screening [4].

Another way of monitoring the vascular access is by use of auscultation, which has been a long-standing, well-proven method of diagnosing heart diseases. Using auscultation, changes in vascular sounds caused by stenosis formation have been observed [5].

Previous studies have shown the presence of increased energy content in different frequency bands ranging from 20 *Hz* to 1000 *Hz* for patients with stenosis formation [6, 7].

This study aims to characterize sound from patent fistulae and distinguish between normal and abnormal sound activity caused by stenosis, making the method usable as a routine screening method.

II. METHOD

A. Subjects

The study included 35 adults of whom seven (20.59%) of the patients had a stenosis formation or unnatural narrowing of the fistula. 24 had a radio-cephalic fistula and 11 a brachiocephalic fistula. Eight suffered from hypertension, seven had diagnosed diabetes, one had an autoimmune disease and one patient from each group had a Gore-Tex shunt. Three patients from the patent fistula group were moved to the stenosed fistula group. Two of them because of an artificial narrowing of the fistula due to banding, and one due to beginning stenosis, which was revealed by the ultrasound. This patient was therefore included in the stenosis group. One patent patient was excluded from the data analysis due to an immature fistula. Patient characteristics are summarized in Table 1.

Table 1 Patient characteristics

Mean characteristics	Patent fistula	Stenosed fistula
N	27	7
Male:female	22:5	4:3
Age, years (±SD)	60.74 (±13.20)	59.14 (±12.33)
BMI (±SD)	23.93 (±3.96)	28.05 (±6.82)

K. Dremstrup, S. Rees, M.Ø. Jensen (Eds.): 15th NBC on Biomedical Engineering & Medical Physics, IFMBE Proceedings 34, pp. 241–244, 2011.
www.springerlink.com

B. Procedure

Over a period of three months, 35 patients with an AVF were evaluated by ultrasound and subsequently by auscultation with use of a 3M Littmann Electronic Stethoscope Model 4000 with a sample rate of 4000 Hz. Basic personal information such as age, sex, height and weight were collected along with information about type of fistula, date of fistula creation, history of fistula problems and other illnesses that may influence the study.

Auscultations were performed with the stethoscope on two or three recording sites according to allocation group, and fistula diameter was measured at the same points by the use of ultrasound (Fig. 1). a) For patent fistulae, 2 x 8 seconds were recorded on top of the anastomosis and 2 x 8 seconds at a reference point, which is a normal vein segment (5-10 cm) proximal to the anastomosis. b) For stenosed fistulae, 2 x 8 seconds were recorded on top of the anastomosis, 2 x 8 seconds at the stenosis and 2 x 8 seconds at a reference point (5-10 cm) proximal to the stenosis. Recording sites were marked by pen for further examination with ultrasound. The recording sites for anastomosis and stenosis were pointed out by a doctor, who confirmed the correctness of the chosen sites using ultrasound. Recordings were then made at the stenosis where the characteristic sound had the highest intensity. Measuring of lumen diameter of the vessels was performed using ultrasound at the marked recording sites.

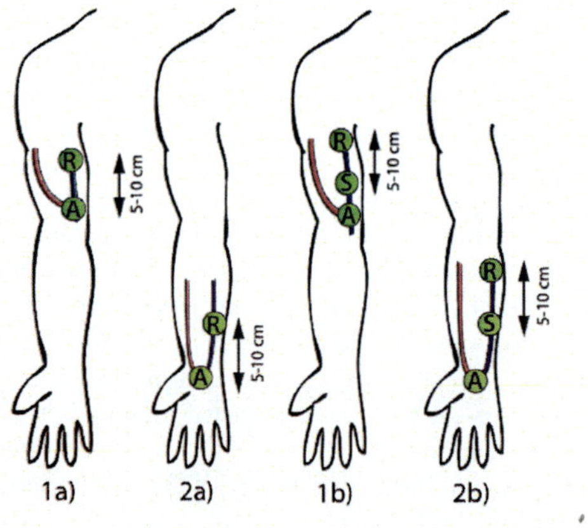

Fig. 1 Recording sites for a) patent fistulae b) stenosed fistulae. A denotes anastomosis, R for reference and S for stenosis 1) An end-to-side brachiocephalic fistulae 2) An end-to-end radio-cephalic fistula.

The procedure was approved by the Danish ethical committee (N-20090011).

C. Pre-processing

A whitening filter using the linear predictor coefficients obtained from an all-pole model of the recordings at reference points, was applied to all the recordings. The recordings at the reference point were thereby given a flat frequency response, while also adjusting the frequency response for the anastomosis and stenosis recordings. Finally, the sound recordings were filtered and preconditioned with a zero-phase 2nd order Butterworth high-pass filter with $\omega_c = 25$ *Hz* and after that a zero-phase 2nd order Butterworth low-pass filter with $\omega_c = 1000$ *Hz*.

After filtering, the recordings were segmented in order to analyze sections of interest, characterized by high intensity. A signal envelope was created by rectifying the signal and filtering it using a zero-phase 5th order Butterworth low-pass filter with $\omega_c = 5$ *Hz*. Peaks on the envelope were then determined using the peak detection algorithm in Matlab, with a specified threshold value for the minimum distance between peaks (Fig. 2). Peaks detected within a time period of half a second at the beginning and end of the signal were removed. Murmur epochs spanning 75 *ms* before and 175 *ms* after each peak were cut out for further investigation.

The segmentation algorithm was tested by manual inspection of each individual signal in order to validate the sensitivity of the algorithm.

D. Feature extraction

The frequency content for each murmur was analyzed by use of the wavelet packet decomposition at depth 4, resulting in 16 different frequency bands to be analyzed. Different families of wavelets were tested (Daubechies 1, 2, 4, 6 and Coiflets 1, 4) but Daubechies 4 was selected since no notable differences on the result appeared. The frequency bands from 8 to 16 were ignored, since content above 1000 *Hz* was removed by filtration (Fig. 3).

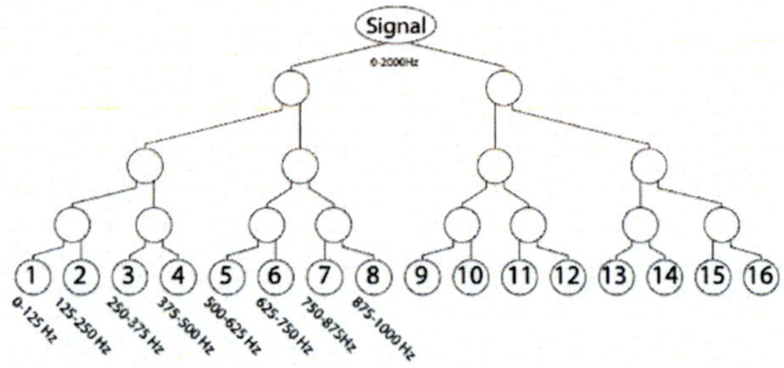

Fig. 2 The wavelet packet decomposition at depth 4 with the frequency bands numerated from 1 to 16.

From each remaining frequency band two values, SE_i and SEp_i, were calculated, where i^{th} denotes the frequency band. SE_i is the scale energies, which is the sum of the squared wavelet coefficients. SEp_i is the percentage of total energy within the ith frequency band. After calculating wavelet features for all murmurs in a recording, the mean was calculated and used as overall features for this recording.

After extraction, all features were log-transformed using the natural logarithm, to achieve normal distribution.

E. Feature selection

A separability and correlation (SEPCOR) analysis was applied to all extracted features from the reference site of patent patients and stenosis site from the stenosis group, in order to find the two most discriminative features. A total of 54 recordings were included from the reference site, while 14 were recorded on stenotic vessels.

Each feature was first described using a measure of variability, which is defined in equation 1.

$$V(x_i) = \frac{\text{variance of the means of the classes}}{\text{mean of the variance of each of the classes}} \quad (1)$$

By calculating the variability, the separability between classes and the compactness of each class is found. Following this, the correlation between the variability of each feature was calculated and the feature is thereby given a rating. A maximum correlation between features was chosen to be 0.80.

F. Classification

The data was classified using a discriminant analysis with a quadratic discriminant function, which provided the best decision boundary in the feature space. Cross-validation was applied in order to determine the effectiveness of the classifier. Leave-one-out was chosen for cross-validation, due to the relatively small number of recordings from stenosed vessels.

III. Results

A. Segmentation

The test of the segmentation algorithm showed that the algorithm correctly detected 1279 murmurs, while 20 murmurs were incorrectly detected and 17 were completely missed. The sensitivity of the algorithm was 98.5% and the positive predictive value of the validation was 98.7%.

B. Feature extraction

In total, 16 features (SE_{1-8} and SEp_{1-8}) were extracted from each recording. The analysis is focused on the 54 recordings from the reference site of the patent patients, and the 14 recordings originating from stenosed vein segments.

Initial investigation of the features using t-test, showed significant differences in all SE_i features ($P \leq 3.6e\text{-}5$ for SE_{1-8}), indicating a significant change in signal energy. SEp_{1-8} which were normalized in relation to intensity, showed significant drops in SEp_2 ($P = 0.004$) and SEp_4 ($P = 0.012$), while increases were seen in SEp_6 ($P = 0.027$) and SEp_8 ($P = 0.002$). These results showed a redistribution of energy content within the signal, with a larger percentage of energy within the higher frequencies. More specifically, for sound recordings at the stenosis site, the frequency ranges 625-750 *Hz* and 875-1000 *Hz* contain a significantly higher percentage of the energy, when compared to the reference sites.

C. Feature selection

SEPCOR analysis revealed SE_8 (V-value: 3.12) as the best feature for differentiating between the stenosis sites and reference sites. Several other SE_i features also had elevated V-values, but were excluded due to close correlation with SE_8. After exclusion of all features with high correlation, SEp_8 was the best remaining feature with a V-value of 0.58.

In order for the analysis using discriminant function to be valid, the data must be normal distributed. To ensure this, normal probability plots where created for the features. Since only minor deviations from normal distribution were observed, analysis using discriminant function was considered valid.

D. Classification

Leave-one-out cross-validation of the 68 recordings from stenosed vessels and reference points, using the features SE_8 and SEp_8, yielded 100% sensitivity and 94% specificity (Fig. 4). The anastomosis recordings from patent fistulae were classified using the same discriminant function. This resulted in a 56% of these recordings being classified as stenotic.

IV. Discussion

The algorithm developed in this study proved to be very reliable for detection of stenosis in a normal vein segment. The algorithm detected all stenosis patients, while only a few patent patients were misclassified, which is important if the algorithm is to be used for screening purposes.

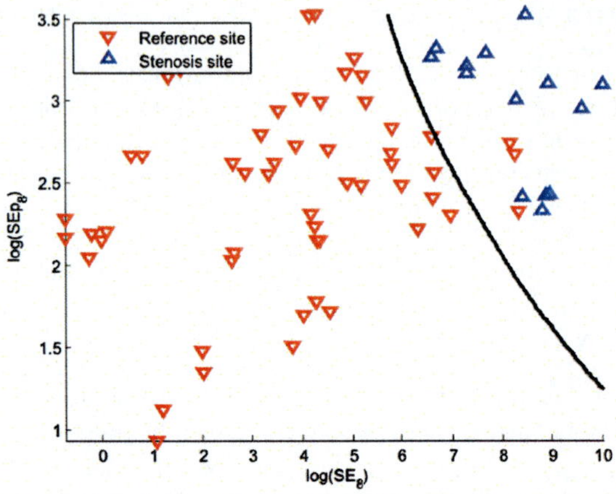

Fig. 3 Scatterplot of the sound recordings based on the two features SE_8 and SEp_8. The decision boundary from the quadratic discriminant function, trained using the 68 recordings is also shown. Red triangles represent recordings from the reference site on normal patients, while blue triangles represent recordings from stenosis sites.

A high sensitivity is a priority for this algorithm since misclassification of patent patients quickly can be resolved using ultrasound. Further studies with more stenosis patients are needed to finally confirm the algorithm and possibly reduce misclassification rate.

The results show a significant increase in sound intensity throughout all frequency bands. Furthermore, the results show redistribution of the frequency content, with more emphasis on higher frequencies (625-750 *Hz* and 875-1000 *Hz*). This is in accordance with the results in previous studies, which indicate that turbulence caused by stenosis generates high frequency sounds [7].

The SEPCOR analysis revealed SE_8 and SEp_8 as the most information carrying features. Differences between stenosis and reference sites were mainly found in frequencies between 875-1000 *Hz*, since this is the frequency band related to SE_8 and SEp_8. This is reasonable, since changes in higher frequencies have been found in previous studies [7].

However, detection of stenosis is complicated by the turbulent sounds in the anastomosis, which are similar to sounds from a stenosed vein segment. 56% of the recordings on top of the anastomosis were classified as stenotic using the developed algorithm. This indicates that the turbulence occurring naturally at the anastomosis can be interpreted by the algorithm as turbulence occurring due to stenosis. This should be taken into account when examining patients, since there may be high risk of false detection within a 5 *cm* distance of the anastomosis. Further studies of stenosis within the anastomosis are needed, in order to improve this. In addition to this, a study focusing on the distance between anastomosis and the reference point should be made, in order to properly evaluate the effects of the distance on the sound characteristics.

The classification relies on recordings on top of the stenosis for best results. The doctors should therefore perform continuous measurements along the vein segment of fistula.

Further development of the algorithm should focus on a conversion to real-time processing, thereby making it possible for the doctor to quickly and effectively evaluate the degree of stenosis formation. The algorithm could be incorporated into a medical device, which uses a visual scale to indicate the level of turbulent flow caused by stenosis.

V. Conclusions

The study confirms previous findings regarding increased high-frequency content in the vascular sound due to stenosis. The algorithm developed using the features obtained by wavelet analysis is reliable for detecting stenosis in a vein segment of an AVF. Based on these results, the prospects of developing an accurate, low-cost screening method for patients undergoing hemodialysis, are promising.

References

Gilpin V, Nichols W. (2010) Vascular access for hemodialysis: Thrills and thrombosis. J Vasc Nurs 2010; 28:78-83

Roy-Chaudhury P, Sukhatme V, Cheung A (2006) Hemodialysis vascular access dysfunction: A cellular and molecular viewpoint. J Am Soc Nephrol, 17: 1112–1127 DOI: 10.1681/ASN.2005050615

Robbin M, Chamberlain N, Lockhart M et al (2002) Hemodialysis arteriovenous fistula maturity: Us evaluation. Radiology 225(1):59–64 DOI: 10.1148/radiol.2251011367

Whittier W (2009) Surveillance of hemodialysis vascular access. Semin Intervent Radiol 26:130–138 DOI 10.1055/s-0029-1222457

Mansy H, Hoxie S, Patel N, Sandler R (2005) Computerised analysis of auscultatory sounds associated with vascular patency of haemodialysis access. Med. Biol. Eng. Comput. 43, pp 56-62

Ask P, Hök B, Loyd D, Teriö H (1995) Bio-acoustic signals from stenotic tube flow: state of the art and perspectives for future methodological development. Med. & Biol. Eng. & Comput. 33, pp 669-675

Vásquez P, Munguía M, Mandersson B (2009) Arteriovenous fistula stenosis detection using wavelets and support vector machines. Conf Proc IEEE Eng Med Biol Soc. pp 1298-1301

Corresponding author:

Author: Mikkel Gram
Institute: Department of Health Science and Technology, Aalborg University
Street: Fredrik Bajers Vej 7 D
City: Aalborg Øst
Country: Denmark
Email: mgram07@student.aau.dk

Quantifying the Effect of Aging on the Autonomic Control of Heart Rate Using Sequential Trend Analysis Plot

L. Ram Gopal Reddy and Srinivas Kuntamalla

Department of Physics, National Institute of Technology, Warangal, India-506004

Abstract— **Heart rate variability (HRV) is gaining acceptance as a non-invasive tool to analyze the influence of the autonomic nervous system on the heart. Age has a strong influence on heart rate variability, which should be considered in the interpretation of HRV data comparing diseased and normal populations. A nonlinear scatter plot technique called sequential trend analysis is used to quantify the sympathetic and parasympathetic nervous system activities on the human heart rate. HRV data obtained from an online, publicly available and widely used database of 19 young (21 - 34 years) and 19 elderly (68 - 85 years) rigorously-screened healthy subjects is used for investigation. In this study, it is observed that there is a decrease in both the sympathetic and vagal components in elder subjects. Further, it is observed that the sympathetic and vagal tonic levels are having the same values, representing the healthy condition of the subjects. There is a significant difference between the ANS activity of younger and elder subjects (p value < 0.001).**

Keywords— **Sympathetic, Vagal, Heart rate variability, Sequential trend analysis.**

I Introduction

Heart rate variability (HRV) is gaining acceptance as a non-invasive tool to analyze the influence of the autonomic nervous system on the heart [1]. Usually HRV data is obtained from R peaks of electrocardiogram (ECG). The beat instants are taken at these points and consequently the beat to beat intervals are determined as the time interval from one R peak to the next one. Therefore, these intervals are called R-R intervals and are plotted against their beat number, which is called a tachogram. In 1996, the Taskforce of the ESC/NASPE published standards in HRV analysis proposing several time and frequency parameters based on short-term (5-min) and long-term (24-h) HRV data [1]. The HRV can be analyzed using several methods which are broadly classified as time domain and frequency domain methods. Time domain measures are simple statistical operations on R-R intervals, such as standard deviation of normal R-R intervals (SDNN), root mean square of successive R-R interval differences (RMSSD) and the percentage change of normal R-R intervals that differ by > 50 ms (PNN50) etc. Frequency domain analysis includes FFT or AR based power spectral density measures which provide information on how variance distributes as a function of frequency. HRV has been categorized into high-frequency (HF) (0.15 – 0.4 Hz), low-frequency (LF) (0.04 - 0.15 Hz), and very-low frequency power (VLF) ($\leq$0.04 Hz) ranges according to frequency domain analysis. HF is an indicator of well-known respiratory sinus arrhythmia and is considered to represent the vagal control of heart rate. LF is attributed to the contribution of both vagal and sympathetic activities. The ratio LF/HF is considered by some investigators as a measure of sympathovagal balance.

Frequency-domain analysis of HRV has gained popularity as a functional indicator of the autonomic nervous system (ANS). Although 24-h analysis of HRV is helpful in increasing the frequency resolution, especially for the lower frequency power, its application is difficult to accomplish. For example, changes in the physical or mental states of the study subjects, changes in environments and even noises in the ambulatory recordings may severely influence the results of HRV frequency analysis. For this purpose, 5-min recording is more practical than 24-h recording.

Since the original publication of the HRV Task Force standards in 1996, a variety of new techniques has been proposed to quantify HRV, borrowing concepts from information theory and nonlinear systems theory such as entropy, scale invariance and symbolic dynamics. These novel indices quantify the structure/patterns embedded in HRV, although, the physiological meaning of these measures is not well understood. Yet, there is clinical evidence that such techniques are superior in capturing HRV features predictive of sudden cardiac death and therefore might carry information on sympathetic and vagal outflow to the heart [2].

Age has a strong influence on short-term heart rate variability, which should be considered in the interpretation of HRV data comparing diseased and normal populations [3]. The aging process reduces the parasympathetic activity on the heart and consequently, decreases the HRV indices [4]. Increased predominance of the sympathetic activity over the parasympathetic is observed at rest in elderly subjects [5].

K. Dremstrup, S. Rees, M.Ø. Jensen (Eds.): 15th NBC on Biomedical Engineering & Medical Physics, IFMBE Proceedings 34, pp. 245–248, 2011.
www.springerlink.com

However, Pagani et al suggests that the sympathovagal balance appears to be unchanged with aging due to decrease observed in both sympathetic and parasympathetic components [6].

In this background, a nonlinear scatter plot technique called sequential trend analysis is used to quantify the sympathetic and parasympathetic nervous system activities on the human heart rate. This study analyzes the effect of age on the autonomic control of heart rate.

II. METHODS

A. Data Group

The data base used in this study is publicly available FANTASIA data base from Physionet website [7]. This data base contains twenty young (21 - 34 years old) and twenty elderly (68 - 85 years old) rigorously-screened healthy subjects electrocardiographic (ECG), and respiration signals of 120 minutes duration. In half of each group, the recordings also include an uncalibrated continuous non-invasive blood pressure signal. Each subgroup of subjects includes equal numbers of men and women.

To maintain wakefulness all the subjects are made to watch the movie fantasia (Disney, 1940) in resting condition. All the signals were digitized at 250 Hz. The database also contains heart beat instances annotated using an automated arrhythmia detection algorithm, and each beat annotation was verified by visual inspection. As per the HRV Task Force standards, the short term records for HRV analysis are of 5 min duration, which normally consists of 300 to 600 R-R intervals. Although, 600 data points are sufficient, in this study, a record segment of 10-20 min duration, which corresponds to 1000 R-R intervals is randomly taken from each long term record for the analysis.

From this database 19 records from young group and 19 records from elderly group are considered for study. The two records f2o08 and f2y09 are not used because of presence of too many ectopic beats.

B. Sequential Trend Analysis

Sequential trend analysis (STA) of HRV is a nonlinear scatter plot technique, in which the differences among successive R-R intervals (Δt) are plotted with Δt_n on x-axis and Δt_{n+1} on y-axis (Fig. 1).

Where $\Delta t_n = t_n - t_{n-1}$

$\Delta t_{n+1} = t_{n+1} - t_n$

and t_n is the nth R-R interval.

This STA plot is similar to the Poincare plot in which R-R intervals are directly plotted with t_n on X-axis and t_{n+1} on Y-axis, whereas for STA plot the differences among successive R-R intervals are plotted. The STA plot can be envisioned as values distributed over an area defined by four quadrants. Each quadrant indicates the direction of two consecutive changes in the interval length. Therefore, this plot demonstrate 1) the ratio of short to long term changes in the heart rate 2) the trends of variation in heart rate. The Δt values may be positive or negative so, the points in the STA plot are scattered around the point (0,0) i.e., origin. This plot has an advantage that it clearly segregates the moments that increase the RR interval and those which decreases it.

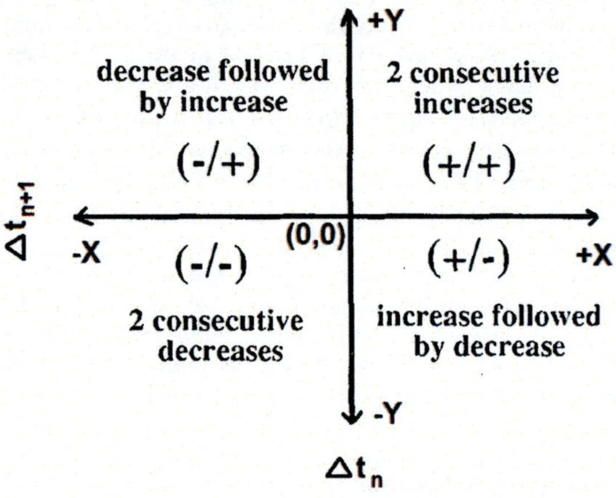

Fig. 1 Schematic diagram of a STA plot

Thus it separates the vagal activity from sympathetic activity and displays them separately in +/+ and -/- quadrants respectively. Whereas the other two quadrants -/+ and +/- show points of decrease in the R-R interval length followed by an increase and vice versa. The number of points in +/+ and -/- quadrant gives a measure of parasympathetic and sympathetic activity [8-9]. In this study, the STA plot is quantified by parameters which describe the sympathetic and parasympathetic tonic levels. Let us consider if there is no autonomous nervous system activity on the heart rate then there is no variation in the heart rate and STA plot show all the points at the origin i.e., at (0,0) position. Therefore the point (0,0) can be treated as a point of no activity. The distance from this point to any point in the four quadrants represents the tonic levels of the respective activities. Therefore, the mean of the distances in each quadrant is taken as a measure of STA plot and is given by the equation

$$\frac{1}{n-1}\sum_{k=1}^{n-1}\sqrt{{t_n}^2+{t_{n+1}}^2}$$

where ‘n’ is the total number of points in each quadrant.

The mean distances for four quadrants are calculated for each subject. They are designated as d_1, d_2, d_3 and d_4 for +/+, -/+, -/- and +/- quadrants respectively. Here the points on the axes are also counted into their respective quadrants depending on their directions (increase or decrease). The d_1, d_3 values of STA plot represents the parasympathetic, sympathetic tonic levels. Therefore we propose d_1 as a measure of parasympathetic nervous system activity (PNS measure) and d_3 as a measure of sympathetic nervous system activity (SNS measure). The mean distance from origin comprising of all the points in +/+ and -/- quadrants is a measure of total autonommic nervous system activity (ANS measure). The d_2, d_4 values of STA plot represents the change in R-R Interval while switching from parasympathetic to sympathetic activity and vice versa.

The parameters extracted from the STA plot are tested for null hypothesis using a significance test (T-test). T-test is the most commonly used method to evaluate the differences in means between the two groups. The significance level for rejection of null hypothesis is set to 0.001 in this study. The P value < 0.001 is considered to be statistically significant.

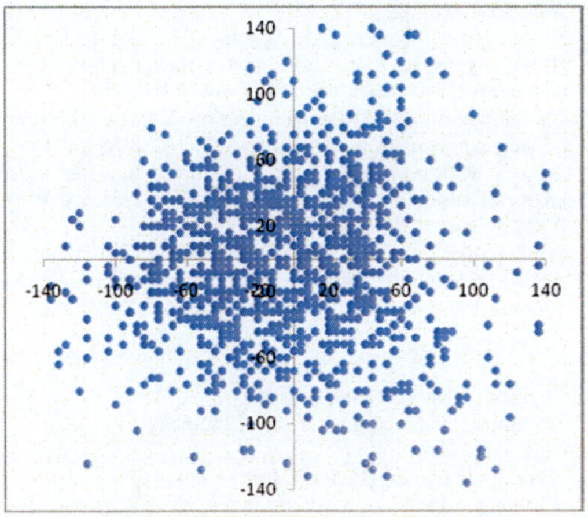

Fig. 2 STA plot of a young subject

III. Results

Figure 2 and Figure 3 represents the STA plots for young and elderly subjects respectively. These plots clearly show that there is a higher amount of variation in heart rate in young subject . Both the plots assume a circular shape, which reflects that they are healthy (where as for congestive heart failure subjects , the shape would be in the form of an ellipse) [10].

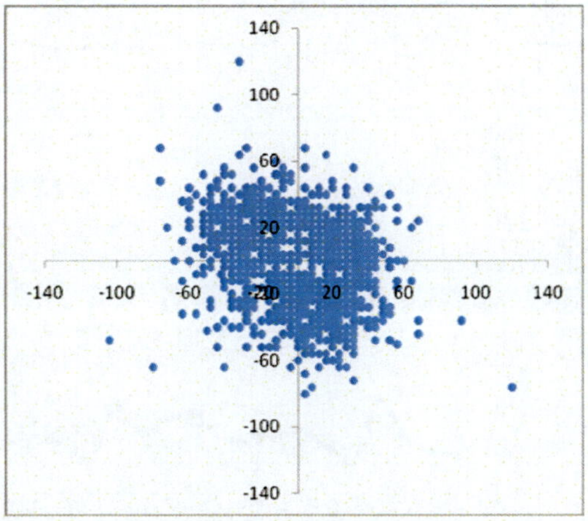

Fig. 3 STA plot of an old subject

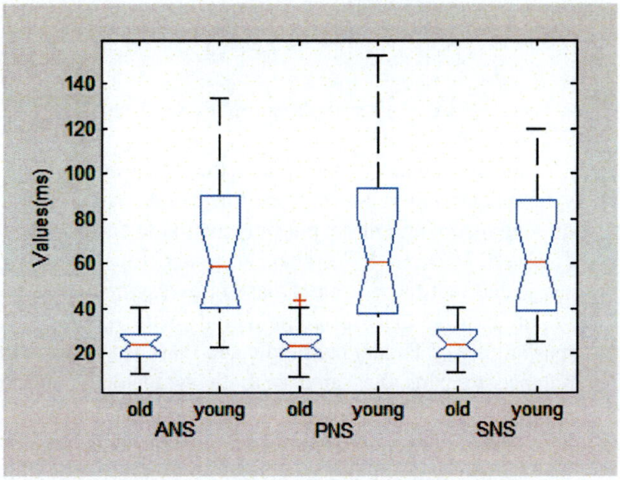

Fig. 4 Box – Whiskers plot for the parameters determined from STA plots for young and old groups

Box – Whiskers plot for the parameters determined from STA plots for young and old groups is drawn to compare the values (Fig. 4). Fig. 5 shows SNS measure and PNS measure values for old and young subjects. The figure shows a clear discrimination between the two groups (P value < 0.001). The SNS measure and PNS measure values for young subjects are greater than 40 (from fig. 5) with a average value of 63.82(±27.40) for SNS and 67.86(±34.22) for PNS (P value > 0.1). The SNS measure and PNS measure values for old subjects are less than 40 (from fig. 5) with an average value of 24.48(±7.44) for SNS and 24.673(±8.91) for PNS (P value > 0.1). The ANS measure for elder subjects is less than 40 with a average value of 24.51(±7.93) and for younger subjects, it is greater than 40 with a mean value of 65.82(±29.98) (P value < 0.001).

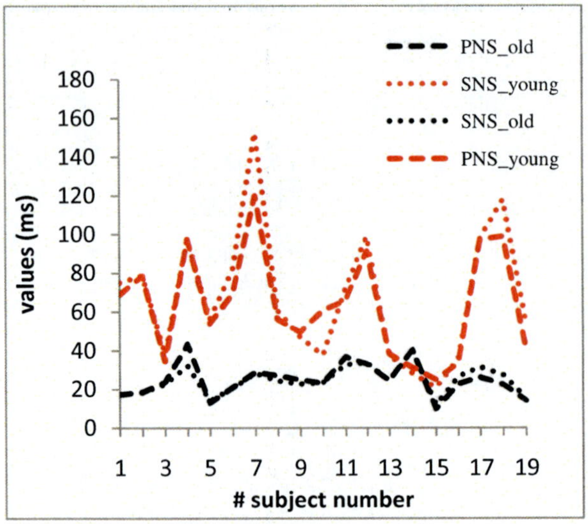

Fig. 5 STA plot measures for all the subjects

IV. Discussion and conclusion

This study mainly investigates the effect of aging on the sympathetic and vagal control of heart rate using HRV data obtained from an online, publicly available and widely used database. Vikram K Yeragani et al and Lipsitz L A et al suggested that there is a decrease in parasympathetic activity and increase in sympathetic activity with age [4-5]. Whereas studies of Pagani et al indicated that the sympathovagal balance remains the same as both sympathetic and parasympathetic components have decreased with aging [6]. In the present study also, it is observed that both the sympathetic and vagal components are decreased in elder subjects compared to younger subjects. Although, there is a decrease in sympathetic and vagal components the sympathovagal balance is maintained. Further, it can be seen from Fig. 5 above the sympathetic and vagal tonic levels are having the same values (p value > 0.1), representing the healthy condition of the subjects. The SNS and PNS measures for young subjects are having relatively large values of standard deviation, showing higher heart rate variability. The parameters extracted from the STA plots are analyzed using t-test. There is a significant difference between the ANS activity of younger and elder subjects (P value < 0.001). The method used in this study is very simple and easy to implement in real time hardware. This method is promising and can be used as a tool to determine the sympathovagal balance of heart function and their quantification.

References

1 Task Force of the European Society of Cardiology of the North American Society of Pacing Electrophysiology (1996) Heart rate variability standards of measurement, physiological interpretation, and clinical use. Circulation 93: 1043-1065.

2 Mathias Baumert, Gavin W Lambert et al. (2009) Short term heart rate variability and cardiac norepinephrine spill over in patients with depression and panic disorder, AJP-Heart 297(2):H674-H679.

3 R C Melo, M D B Santos et al. (2005) Effects of age and physical activity on the autonomic control of heart rate in healthy men, Braz J Med Biol Res 38(9):1331-1338.

4 Vikram K Yeragani, Edward Sobolewski et al. (1997) Effect of age on long term heart rate variability, cardiovascular Research 35:35-42.

5 Lipsitz LA, Mietus J, Moody GB & Goldberger AL (1990). Spectral characteristics of heart rate variability before and during postural tilt: Relations to aging and risk of syncope. Circulation 81: 1803-1810.

6 Pagani M, Lombardi F, Guzzetti S et al. (1986). Power spectral analysis of heart rate and arterial pressure variabilities as a marker of sympatho-vagal interaction in man and conscious dog. Circulation Research 59: 178-193.

7 N. Iyengar, C.-K. Peng, R. Morin, A.L. Goldberger, L.A. Lipsitz, (1996) Age-related alterations in the fractal scaling of cardiac inter beat interval dynamics, Am. J. Physiol. 271: 1078.

8 Carvalho et al. (2002) Development of a Matlab software for analysis of heart rate variability, ICSP proceedings 2:1488-1491.

9 Srinivas. K, Ram Gopal Reddy. L, Srinivas. R. (2007). Estimation of heart rate variability from peripheral pulse wave using PPG sensor, IFMBE proc. 15:325-328

10 Srinivas Kuntamalla, L. Ram Gopal Reddy (2010) Detecting congestive heart failure using heart rate sequential trend analysis plot, International Journal of Engineering Science and Technology 2(12):7329-7334.

Author: Dr. L. Ram Gopal Reddy
Institute: National Institute of Technology
Street: Kazipet
City: Warangal - 506004
Country: India
Email: lrgreddy@gmail.com

PENG Analysis for Evaluation of Telemedicine Projects

E. Rowe[1], S. Jonsson[1,2], and H. Teriö[1]

[1] Karolinska University Hospital, Dept. of Biomedical Engineering, Stockholm, Sweden
[2] Karolinska Institute, CLINTEC, Stockholm, Sweden

I. Abstract

The Swedish population is living longer than before due to advances in medical treatments, among other things, thus the demand on health care is increasing. Telemedicine may be a solution for increasing healthcare efficiency. However in order to justify an implementation of new telemedicine solutions proof of value is an important factor; economic evaluation plays an essential role in the healthcare sector for assessing costs and benefits of scarce resources. Traditional evaluation methods however disregard intangible benefits and costs though these are believed to be of significant importance for understanding the success of projects and organizations. The PENG model may serve as a tool for evaluating telemedicine projects and solutions.

Key words: PENG-model, Cost-Benefit Analysis, telemedicine.

II. Introduction

The objective of healthcare in Sweden is to provide good health and care on equal terms to the entire population. [1] This task might however be jeopardized since the Swedish population is living longer than before; the need for healthcare is increasing at the same time as healthcare costs are rising. [2][3] The increased demand has to however be solved with lesser resources; telemedicine solutions may be a tool for responding to the increased demand. [4] However, in order to justify the implementation of new telemedicine solutions proof of value is an important factor; economic evaluation plays an essential role in the health care sector for assessing costs and benefits of scarce resources. [5][6][7] The PENG model may be of interest for evaluating telemedicine projects.

PENG is a Swedish acronym which stands for *"Prioritering efter NyttoGrunder"*, this may be translated to "Prioritizing based on contribution of benefits". The PENG-model is a structural method of evaluating all the different types of benefits that Information and Communication Technology (ICT) generates within a project, both tangible and intangible benefits are evaluated in monetary units. [8][9] The model is based on Cost-Benefit Analysis (CBA) methodology: evaluation in monetary units. The PENG model consists of three phases which are composed of ten different steps and activities that ultimately lead to a benefit analysis. [9] The PENG model was created and evolved in response to the need to identify and evaluate the benefits of ICT projects, the model is however also applicable for evaluation of other types of projects. The PENG model could for example be a tool for identifying and valuing intangible benefits in telemedicine projects which could lead to a better understanding of the long term effects (benefits and costs) of the project. [9] It may also be of interest to utilize the PENG model as a framework for evaluating national information systems. [10] The interest of proposing the PENG model as a possible model for the evaluation of telemedicine projects lay in that there is a demand for information regarding economic and qualitative effects of telemedicine. There is however a lack of complete economic analysis of telemedicine and generally accepted methods for evaluating telemedicine. [11][12][13][14] There is a need for developing methodological evaluation methods which incorporate not only the evaluation of tangible benefits but also includes the evaluation of qualitative benefits regarding telemedicine projects and solutions. [11]

A project called WiPOX is being developed by Karolinska University Hospital, TeliaSonera and Aerotel Medical Systems, the aim of the project is to develop and evaluate a unique wireless wrist-wearable pulse oximeter and medical monitoring system, which is a telemedicine solution. Evidence of benefits and costs is an important factor for decision makers within the healthcare sector, in order to for example avoid implementation of medical devices based on inadequate information about its future implications, among other things. A PENG analysis of telemedicine projects, such as the WiPOX project at the department of Biomedical Engineering at Karolinska University Hospital, may be of interest in order to identify and illustrate all of the benefits and costs a telemedicine system could result in.

III. Method

The core of the PENG-model consists of the following three phases and ten steps:

K. Dremstrup, S. Rees, M.Ø. Jensen (Eds.): 15th NBC on Biomedical Engineering & Medical Physics, IFMBE Proceedings 34, pp. 249–252, 2011.
www.springerlink.com

Preparation phase	Step 1 Determine purpose.	
	Step 2 Create awareness and assemble analysis team.	
	Step 3 Determine area (processes/systems).	
	Step 4 Define and describe processes/systems.	
Analysis phase	Step 5 Identify effects of benefits.	
	Step 6 Clarify links in an objectives (benefit) structure.	
	Step 7 Evaluate benefit effects (gross benefits).	
	Step 8 Define and evaluate IT costs.	
Quality insurance phase	Step 9 Classify the benefit and conduct risk analysis.	
	Step 10 Calculate net benefits and determine the value receiver manager. [8][9]	

A PENG analysis of the WiPOX project has been planned but has yet not been conducted. In order to illustrate how the PENG model can be utilized for the evaluation of the WiPOX system, and therefore other telemedicine evaluations, the planned analysis of the WiPOX system will hereby be described.

It is planned that the PENG analysis of the WiPOX system should be conducted in four half days, spread over approximately one month. The first meeting will entail that a certified PENG leader and a task manager will address important aspects, related to the preparation phase, such defining the aim of the WiPOX project and solution. A possible aim of the WiPOX project and solution could be to reduce costs and to increase healthcare accessibility. The purpose of the PENG analysis regarding the WiPOX project will also be defined in order to set the ambition level of the overall evaluation and in order to motivate the choice of analysis team participants. The analysis team should consist of five to eight participants, who have adequate knowledge about the business at hand, of which two should have the authority to take responsibility for the monetary valuing of benefits. [9] In order to assure an evaluation from multiple perspectives the analysis team for the WiPOX system could consist of a financial controller, manager in medical technology, clinical doctor, engineer and a researcher. The time-span for obtaining the benefits, time of revision and follow up will also be addressed at this point.

The benefit evaluation itself is done in the analysis phase. This will be the first meeting with the entire analysis team; therefore the primary task for the PENG leader will be to inform the participants about the PENG model and the purpose of the analysis. Thereafter the identification of possible benefits will be done; this step should take no more than 15-20 minutes. The participants will individually identify possible benefits, which will be written down on post-it notes. The post-it notes will have different colors depending on who the beneficiary is; patient, organization, society at large or other stakeholders. The post-it notes will then be utilized to categorize the benefit effects in a so called benefit structure. The benefit structure will illustrate all of the possible benefit effects of an implementation of the WiPOX system, illustrate what will be required to obtain the benefit effects and illustrate the correlations between the benefit effects.

A second group meeting will be scheduled three to five days after the first group meeting; this is done in order to give the participants an opportunity to reflect over the identified benefits and in order to collect relevant data for the evaluation. Any comments or suggested changes to the benefit structure will be addressed initially. The main objective of the second meeting will however be to set monetary values on all benefits. In order to place a monetary value on a specific benefit all of the benefit effects correlated to the specific benefit must be identified, and in order to determine the degree of importance of the different benefit effects they should be placed in correlation to the overall aim of the WiPOX project. For example an overall aim of the WiPOX system could be to reduce healthcare costs and reduced hospitalization may be a way to achieve this. In order to place a monetary value on reduced hospitalization all effects leading to reduced hospitalization, such as early detection of exacerbations and improved quality of care, have to be identified and valued. In order to prioritize the importance of the benefit effects influence on obtaining reduced hospitalization the benefit effects must be weighed against the overall aim of the WiPOX project, in this case what significance does reduced hospitalization have on the overall aim of reducing healthcare costs.

The third group meeting will entail calculating the costs for obtaining the identified benefits and reviewing the quality of the benefit evaluation. The quality insurance phase will entail discussing if the benefit evaluation is satisfactory and classifying the benefits; the benefits will be classified into three levels depending on the degree of difficulty to implement them: green (direct benefits), yellow (indirect benefits) and red (intangible benefits), see figure 1. The importance here lies in that the costs should not surpass the amount of green benefits. A risk analysis will also be conducted, the participants will make notes of possible obstacles that may occur and thereafter the analysis team will determine what measures must be taken in order to eliminate or minimize the obstacles, the costs for addressing the obstacles should also be calculated. The final step of the analysis will be to determine the value receiver manager who will have the important responsibility of assuring that the identified benefits will be realized.

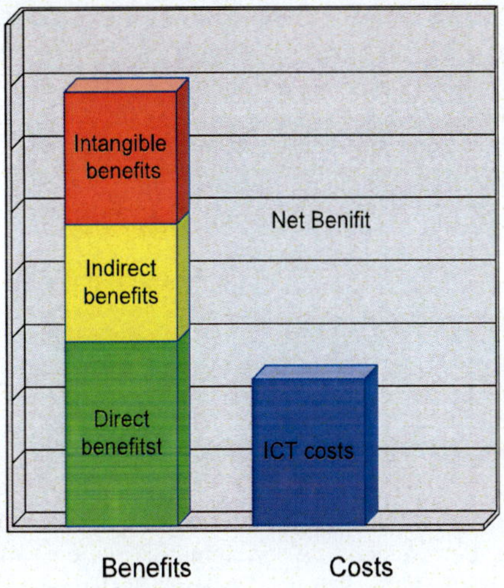

Fig. 1 PENG-model bar chart

The objective of the PENG model is not to calculate rigid financial terms, although the PENG model utilizes a method which measures benefits and costs in quantitative terms. The objective of the model is rather to evaluate the size of different types of benefits that may occur in an ICT project in order to achieve a multifaceted analysis. [10]

IV. DISCUSSION

The PENG model is a type of Cost-Benefit Analysis (CBA) method, hence all costs and benefits are measured in monetary units. There may lay an unwillingness to conduct CBA on projects within the healthcare sector as it would entail placing a monetary value on health and human life for example. The PENG model could however lead to a broader understanding of what can be expected by telemedicine, regarding costs and benefits, as the PENG model considers both tangible and intangible assets, which traditional assessment methods usually ignore. It is suggested that there are research gaps regarding the evaluation of telemedicine projects. Research of the economical outcomes of telemedicine solutions are lacking in aspects such as methodology, comparability, reliability, long-term benefit effects, scope of perspectives (patient, provider, other stakeholders) and benefit measures included in evaluations. [11][12][13][14] Telemedicine has the potential to decrease healthcare costs there is however a general lack of evidence proving the wide-range of possible benefits. Compared to other evaluation methods and measures, such as ROI (Return on investment) and Cost-Minimization Analysis, the PENG model provides a multi-facetted evaluation model.

A PENG analysis of the WiPOX system has been planned but not yet utilized. The interest of utilizing the PENG model for the evaluation of the WiPOX system lays in the ability to identify all possible benefit effects and costs in order to identify the possibilities and outcomes of an implementation of the system. It is also of interest to examine if the PENG model is an adequate evaluation model for the WiPOX project hence the results of a PENG analysis are dependent of the evaluator's judgments, which may affect issues regarding trustworthiness and whom should be included in the evaluation group. Another aspect of importance that is not addressed within the scope of the PENG model is how the assessment of health and human life should be conducted. Since the PENG model considers factors such as tangible and intangible benefits, follow-up and lead to clear and structured results, which may be of significance for decision makers within the healthcare sector as it could for example raise the awareness of the short and long term affects of a possible implementation of a innovative telemedicine system, it is of interest to examine how issues regarding reliability and evaluation of human health and life can be addressed within the scope of the model. It is also of interest to investigate if the PENG model is applicable for the economic evaluation of telemedicine hence there is a lack of methodological approaches which include the evaluation of qualitative benefits of telemedicine. [11]

The PENG model may contribute to added value in the evaluation process of not only the WiPOX project but also for further research within the area of economic evaluation of telemedicine. Due to factors such as scarce resources within the healthcare sector and a lack of holistic evaluation methods for telemedicine it is of interest to examine the possibilities of utilizing the PENG model for the evaluation of telemedicine projects, hence theoretically the PENG model is a multifaceted model which leads to important insight of which aspects create value within a project.

V. Conclusion

A PENG analysis may provide added value to the understanding of what creates value and what does not create value regarding telemedicine projects, a factor of importance hence the healthcare sector face challenges such as an aging population and scarce resources. No PENG analysis of the WiPOX system has yet been conducted there lays however a specific interest in investigating if the PENG model is suitable for economically evaluating telemedicine projects, such as the WiPOX system, hence there is a lack of evaluation methods which incorporate both tangible and intangible benefits and costs regarding telemedicine. Important results of the PENG analysis will be a report describing the evaluation of benefits, a benefit structure that illustrates the correlation between benefit and benefit value, a bar chart illustrating the net benefit and the annual sum of benefits compared to costs, a description of the investments and costs, a risk analysis and a description of the value receiving process.

References

[1] Svensk författningssamling (SFS) at http://www.riksdagen.se/webbnav/index.aspx?nid=3911&bet=1982:763

[2] Rae D (2005) Getting Better Value for Money from Sweden's Healthcare System OECD Economics Department Working Papers No. 443. OECD Publishing DOI 10.1787/082725005676

[3] Gustafsson A, Kilefors P, Andrén J et al. (2009) Huvudrapport, Framtidens hälso- och sjukvård, Långtidsutredning om sjukvården i Stockholms läns landsting 2008-2025 at http://www.sll.se/upload/Huvudrapport_slutlig.pdf

[4] Nyctelius H, Jostrup R, Lundh R et al. (2009) Delrapport 5, Framtidens hälso- och sjukvård, Läkemedel och medicinteknik- nuläge och prognos. Långtidsutredning om sjukvården i Stockholms läns landsting 2008-2025 at http://www.sll.se/upload/HSNf/Langtidsutredningen/Delrapport_5_Lakemedel-Medicinteknik.pdf

[5] Haley D, Jennet P (2004) The need for economic evaluation of telemedicine to evolve: The experience in Alberta, Canada. Telemed J e-Health 10(1):71-76.

[6] Reardon T (2005) Research findings and strategies for assessing telemedicine costs. Telemed e-health 11(3):348-369.

[7] Bashshur R, Shannon G, Sapci H (2005) Telemedicine Evaluation. Telemed e-health 11(3): 296-316

[8] Dahlgren L E, Lundgren G, Stigberg L (1997) MAKE IT PROFITABLE! PENG – a practical tool for financial evaluation of IT benefits. Ekerlids Förlag, Falun

[9] Dahlgren L E, Lundgren G, Stigberg L (2006) PENG-modellen. Ekerlids Förlag, Helsingborg

[10] Saluse J, Aaviksoo A, Ross P et al. (2010) Assessing the Economic Impact/Net Benefits of the Estonian Electronic Health Record System DIGIMPACT Final Report at http://www.praxis.ee/fileadmin/tarmo/Projektid/Tervishoid/Digimoju/Digimpact.pdf

[11] Johnsen E, Breivik E, Myrvang R et al. (2006) Benefits from telemedicine in Norway, An examination of available documentation. HØYKOM report 2006(1):1-22

[12] Bergmo S T (2009) Can economic evaluation in telemedicine be trusted? A systematic review of the literature. Cost Eff Resour Alloc 7(18):1-10 DOI 10.1186/1478-7547-7-18

[13] Dávalos M E, French M T, Burdick A E et al. (2009) Economic Evaluation of Telemedicine: Review of the Literature and Research Guidelines for Benefit-Cost Analysis. Telemed e-health 15(10):933-948 DOI 10.1089/tmj.2009.0067

[14] Polisena J, Coyle D, Coyle K et al. (2009) Home telehealth for chronic disease management: A systematic review and an analysis of economic evaluations. Intl J of Technology Assessment in Health Care 25(3):339-349 DOI 10.1017/S0266462309990201

Contact:
Eve-Marie Rowe/Sven Jonsson
Karolinska University hospital, Huddinge
Dept. of Biomedical Engineering
141 86 Stockholm
Sweden

Enhancing Control of Advanced Hand Prostheses Using a Tongue Control System

D. Johansen[1], D.B. Popović[1,2], F. Sebelius[3], S. Jensen[4], and L.N.S.A. Struijk[1]

[1] Department of Health Science and Technology, Aalborg University, Aalborg, Denmark
[2] School of Electrical Engineering, University of Belgrade, Belgrade, Serbia
[3] Department of Electrical Measurements, Lund University, Lund, Sweden
[4] Arm Center, Sahva A/S, Copenhagen, Denmark

Abstract— **This paper presents a novel control scheme for new advanced hand prostheses implementing multiple grasps and pinches. The control scheme implements myoelectric signals combined with signals generated through an inductive interface with a mouth piece. A pilot study using three healthy able-bodied subjects was performed to compare the new control scheme to a standard prosthesis control scheme based solely on myoelectric signals. Comparisons were made on the time used to activate a specific grasp or pinch pattern.**

From first to last (third) training session results showed a significant decrease in the time needed to activate a specific grasp or pinch. Also results suggest that amputees using prostheses with more than three functions could benefit from using the new control scheme.

This research is being continued in healthy subjects with the intention to translate the results to patients.

Keywords— **Hand Prosthesis, Prosthesis Control System, Inductive Tongue Control System.**

I. Introduction

Through surveys and questionnaires it has been found that a general demand among hand and arm prostheses users' is improved digit movement and grasp functions [1, 2]. New advanced hand prostheses are therefore implementing multiple grasps and pinches. Examples of such prostheses are the i-Limb, commercially available from Touch Bionics; the Michelangelo® Hand from Otto Bock®; the Vincethand, available from medical technics by Vincent Systems GmbH; and the SmartHand, which will be commercially available from Prensilia S.R.L.

When viewed in the context of Activities of Daily Living (ADL) these new advanced prostheses allows for the use of grasps and pinches covering up to 79% of ADL [3], and when compared to the 20% of ADL covered by the three-finger precision pinch [3], which is the only implemented pinch in most commercially available prostheses e.g. the SensorHand Speed® from Otto Bock®, Germany, it is evident that these new advanced prostheses implementing multiple grasps and pinches allow amputees a potentially enhanced functionality of their prosthetic device.

To fulfill this potential for enhanced functionality of advanced prostheses new control schemes are needed as well. These control schemes should allow for both easy and intuitive control of both the selection and the opening and closing of the different grasps and pinches implemented by the prosthetic device. Recent research on control signals and control schemes for hand and arm prostheses includes MyoElectric Signals (MES) as well as alternative control signals.

In terms of MES, pattern recognition algorithms are being used to extract more information from the MES recorded from the remaining muscles of the arm or forearm of the amputee [4, 5]. However the intuitiveness and the functionality of control schemes solely based on surface MES recordings are highly dependent on the level of amputation. The higher the level of amputation the fewer the number of muscles will be available for use in a MES control scheme, and at the same time the amount of functionality to be recovered through the use of a prosthesis increases with the level of amputation. Therefore studies have been conducted in relation to control signals that could be used as an alternative for, or an addition to the MES, e.g. foot switches, shoulder joysticks and throat microphones [6-8].

Recently, a control scheme combining myoelectric trigger control from the forearm with artificial vision for detection of the prehension and hand orientation has been tested in combination with one of the versions of SmartHand [9]. The tests suggested that adding another input to the myoelectric control is an important aspect in order to fulfill the potential of enhanced functionality of the multiactuated hand prostheses.

This paper presents a comparison of a simple MES control scheme with a novel control scheme for hand prostheses in which standard surface MES, recorded from the arm or forearm of the amputee are combined with control signals from an inductive tongue control system [10], to obtain an efficient control of multiple grasping and pinching modalities.

II. Methods

A. The Inductive Tongue Control System

The Inductive Tongue Control System (ITCS) is a com-

K. Dremstrup, S. Rees, M.Ø. Jensen (Eds.): 15th NBC on Biomedical Engineering & Medical Physics, IFMBE Proceedings 34, pp. 253–256, 2011.
www.springerlink.com

pletely integrated wireless inductive tongue interface, consisting of a mouth piece unit, an activation unit and a central unit.

The mouth piece unit incorporates 18 separate inductive sensors and it is placed in the upper palatal area, similar to a dental brace [11]. The activation unit is fixed to the tip of the tongue using bio-compatible tissue-glue. When the activation unit is placed on a sensor on the mouth piece, the sensor is activated. Every 33 ms the onboard electronics of the mouth piece unit scans the 18 inductive sensors of the mouth piece. The sensor output is amplified, rectified, low pass filtered and sent wirelessly to the central unit of the ITCS. The 18 different signal values are then processed to determine if and which sensor is activated [11]. Then, the central unit transmits the processed signal to the desired interface serial port or radio receiver (e.g. wheelchairs control BUS, computers USB-radio receiver, or prosthetics controller).

B. The ITCS Control Scheme

The ITCS control scheme for hand prostheses combines the use of an ITCS with MES. The control scheme enables the amputee to directly select a desired grasp, pinch or function of the prosthesis using the ITCS (Fig. 1A). MES recordings from the wrist flexor and extensor muscles of the forearm, are then used for opening and closing of a selected pinch or grasp or operating a function, e.g. wrist rotation. The MES control is implemented using thresholds set at 10% of Maximum Voluntary Contraction (MVC).

C. The MES Control Scheme

The MES control scheme used in this study is a simple control scheme using two surface electrodes placed on the forearm of the amputee. Co-contractions are used to switch to the next type of grasp or pinch pattern (Fig. 1B), and MES recorded from the wrist flexor and extensor muscles are then used respectively to close and open the selected grasp or pinch. The MES control is implemented using thresholds set at 10% of MVC.

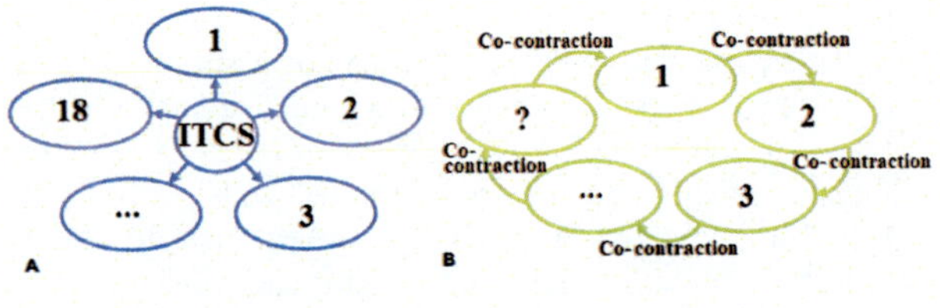

Fig. 1 A: The ITCS control scheme with the tongue control system allowing the user to directly select between or activate a desired function or grasp of hand prosthesis. With 18 sensors in the current ITCS up to 18 functions or grasps (dashed) could be implemented in the ITCS control scheme. B: The MES control scheme where co-contractions are used to switch to the next grasp, pinch or function of the hand prosthesis. The MES control scheme can implement whatever number of grasps, pinches and functions that is needed.

D. Experimental Setup

An ITCS and a MES control scheme based on a prototype of the SmartHand prosthesis was developed and implemented on an ITCS central unit. The control schemes both implements the use of five pinches and grasps; precision pinch, lateral pinch, diagonal volar grasp, transversal volar grasp and a tripod pinch (Fig. 2).

For the ITCS control scheme each of the five pinches and grasps were then allocated to two specific adjacent sensors of the mouth piece unit. This was done according to the individual representation of a pinch or grasp in ADL and the accessibility of the ITCS sensors (Fig. 3A) [12]. The layout of the five grasps and pinches is shown in (Fig. 3B).

MES for both control schemes were recorded using two 13E200 MYOBOCK electrodes, from Otto Bock®. The electrodes were connected to the central unit. The output of the 13E200 is an amplified, filtered, rectified and enveloped representation of the MES. Sampling of the electrode output was done every time the central unit received a transmission from the mouth piece unit performed. For the MES control scheme the mouth piece was left on the table next to the central unit to trigger the sampling of the electrode output. In both control schemes the amplification of the MES signals were adjusted to give a 0-3V range from none to maximum contraction of the flexor and extensor muscles.

The comparison of the ITCS control scheme and the MES control scheme was performed using a computer model resembling the movements of an actual SmartHand prosthesis.

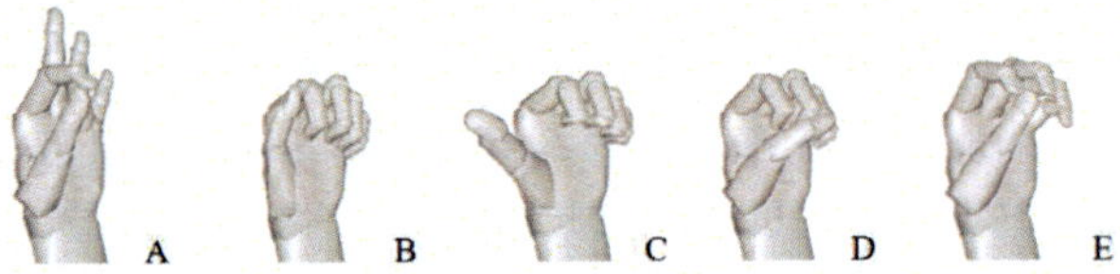

Fig. 2 The five implemented grasp patterns of the hand prosthesis computer model: A: Precision grasp, B: Lateral grasp, C: Transversal volar grasp, D: Diagonal volar grasp, E: Three-finger grasp.

The computer model used is based on a modified version of the 18 degree of freedom computer model VirtualHand from Virtual Technologies Inc., USA. The computer model was modified to take standard keyboard character inputs to allow for selection, closing and opening of the five implemented grasps and pinches. When a specific grasp or pinch was selected the matching preshape would be assumed by the computer model, and then the grasp or pinch could be closed or opened.

For both the ITCS and the MES control scheme the central unit would send commands to a wireless USB keyboard emulator, which then would generate the correct keyboard event and make the computer model respond accordingly.

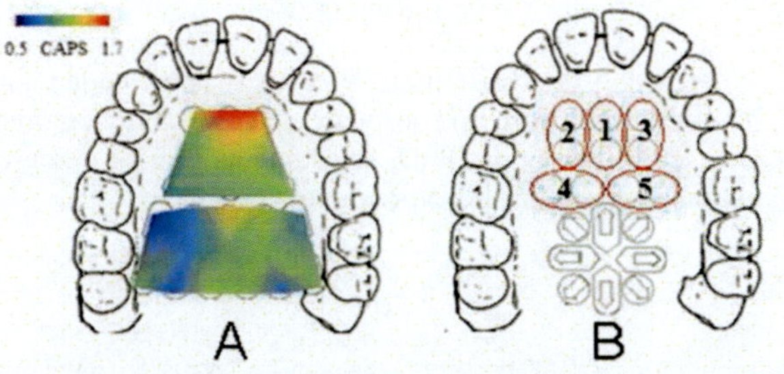

Fig. 3 A: The accessibility of the ITCS sensors measured as correct activations per second (CAPS) [11]. B: The layout of the five pinches and grasps of the ITCS control scheme; 1: precision pinch, 2: lateral pinch, 3: diagonal volar grasp, 4: transversal volar grasp, 5: tripod pinch.

With the ITCS control scheme the subject would select a specific pinch or grasp by activating the sensor of the ITCS mouth piece allocated to that desired pinch or grasp. With the MES control scheme the subject would switch to the next grasp or pinch by means of a co-contraction of the wrist extensor and flexor muscles. In both control schemes the closing and opening of the selected grasp or pinch would be done using MES recordings of the wrist flexor and extensor muscles respectively.

In this study the ITCS mouth pieces used are prototypes. With this prototype only the inductive sensor boards are fitted inside the mouth of the subject. The electronic circuit board of the ITCS mouth piece is placed externally and then connected to the sensor boards by means of copper wires insulated by a silicone tube. The silicone tube with the copper wires inside then exits the corner of the mouth.

E. Subjects

Three healthy able bodied subjects participated in this pilot study, two female subjects and one male. Mean age of the subjects is 27±2 years, and none of the subjects have had any prior training with the ITCS or the MES control schemes.

F. Experimental Protocol

The protocol used in this pilot study was approved by the local ethics committee. The subjects trained for three consecutive days with both the ITCS and the MES control scheme. During each training session the subjects completed 25 exercises in a randomized order. Each exercise was comprised by correctly activating a pinch or grasp, completely closing it and then completely opening it again. Timestamps were recorded for both activation of grasps and pinches as well as when a grasp or pinch was fully closed or opened.

Because of an ensuing study it was necessary to ensure that the training sessions resembled the normal daily use of a prosthesis. Therefore the number of times that each grasp or pinch was represented in the 25 exercises was based on their corresponding representation in ADL. Thus each training session included six precision pinch and lateral pinch exercises, five diagonal volar grasp and the transversal volar grasp exercises and three tripod pinch exercises.

Before starting each training session the subjects had recording electrodes placed on their left forearm. The electrode for closing pinches and grasps was placed on the anterior side and the electrode for opening was placed on posterior side. If the training session was on the ITCS control scheme the subject would then have the ITCS activation unit glued to the tip of the tongue using tissue glue Histoacryl®.

III. Results

Three subjects completed three training sessions of 25 grasping and pinching exercises, which corresponds to approximately 1½ hours of training with each of the two control schemes.

All of the recorded timestamps for both the ITCS and the MES control scheme were grouped by session number and type of control scheme used. The 95% confidence intervals for the time used to activate the correct grasp were then calculated across all three subjects. For sessions 1 and 3 these confidence intervals were then plotted (Fig. 4, Fig. 5).

For both the ITCS and the MES control schemes mean values as well as the standard deviations decreases, indicating that performance increases when comparing session 1 and 3.

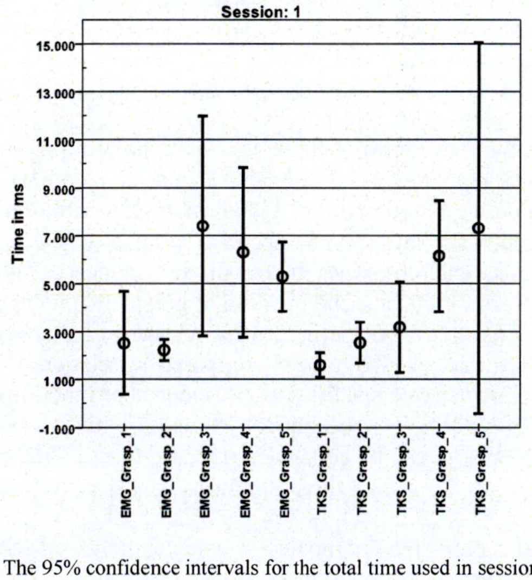

Fig. 4 The 95% confidence intervals for the total time used in session 1 to activate a specific preshape of the computer model using the MES or the ITCS control scheme.

From Fig. 5 it can be seen that when using the MES control scheme there is a linear dependency between the numbers of co-contractions needed to activate a desired grasp or pinch and the time used. Moreover it can be seen that when using the ITCS control scheme the time needed to activate a desired grasp or pinch converges towards the mean time it takes the subjects to simply activate the ITCS. Also it can be seen that if a control scheme implements more than three functions, then the mean time for activating a desired function will be lower when using an ITCS control scheme.

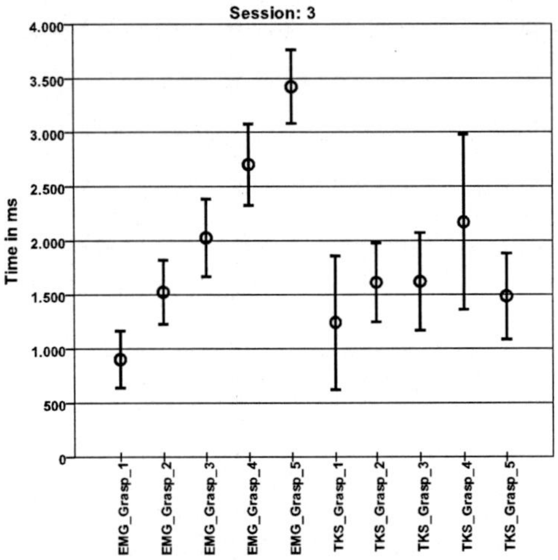

Fig. 5 The 95% confidence intervals for the total time used in session 3 to activate a specific preshape of the computer model using the MES or the ITCS control scheme.

IV. Conclusions

A novel control scheme for hand and arm prostheses has been proposed. The control scheme implements a novel inductive tongue control system that allows the amputee to directly activate specific grasps or pinches of the hand prosthesis, and combines this with the standard myoelectric signals for closing and opening of grasp and pinches.

A simple myoelectric control scheme and the novel ITCS control scheme have been compared in relation to the time used to activate specific grasps, pinches or functions.

After less than 1½ hours of training with the two control schemes, it can be concluded that amputees using a prosthesis with more than three functions would be able activate a desired grasp, pinch or function faster using the ITCS control scheme than compared to using a standard MES control scheme.

Acknowledgment

Henrik Vie Christensen, PhD. is acknowledged for his help with hardware and software design and debugging. Bo Bentsen, cand.odont., PhD, is acknowledged for being clinical responsible for the experimental protocol.

References

1. Kyberd P J, Wartenberg C, Sandsjö L, Jönsson S, Gow D, Frid J, Almström C and Sperling L, "Survey of Upper-Extremity Prosthesis Users in Sweden and the United Kingdom," JPO Journal of Prosthetics and Orthotics, vol. 19, pp. 55, 2007.
2. Pylatiuk C, Schulz S and Doderlein L, "Results of an Internet survey of myoelectric prosthetic hand users," Prosthet. Orthot. Int., vol. 31, pp. 362-370, Dec. 2007.
3. Sollerman C and Ejeskar A, "Sollerman hand function test. A standardised method and its use in tetraplegic patients," Scand. J. Plast. Reconstr. Surg. Hand Surg., vol. 29, pp. 167-176, Jun. 1995.
4. Guanglin L, Schultz A E, Kuiken T A, "Quantifying Pattern Recognition—Based Myoelectric Control of Multifunctional Transradial Prostheses," Neural Systems and Rehabilitation Engineering, IEEE Transactions on , vol.18, no.2, pp.185-192, 2010
5. Hargrove L, Scheme E, Englehart K, Hudgins B, "Multiple Binary Classifications via Linear Discriminant Analysis for Improved Controllability of a Powered Prosthesis," Neural Systems and Rehabilitation Engineering, IEEE Transactions on , vol.18, no.1, pp.49-57, 2010
6. Carrozza M C, Persichetti A, Laschi C, Vecchi F, Lazzarini R, Tamburrelli V, Vacalebri P and Dario P, "A novel wearable interface for robotic hand prostheses," in Proceedings of the 2005 IEEE 9th International Conference on Rehabilitation Robotics, 2005, pp. 109-112.
7. Losier Y, Englehart K and Hudgins B, "A control system for a powered prosthesis using positional and myoelectric inputs from the shoulder complex," Conf. Proc. IEEE Eng. Med. Biol. Soc., vol. 2007, pp. 6138-6141, 2007.
8. Mainardi E and Davalli A, "Controlling a prosthetic arm with a throat microphone," Conf. Proc. IEEE Eng. Med. Biol. Soc., vol. 2007, pp. 3035-3039, 2007.
9. Došen S, Cipriani C, Kostić M, Carrozza M C, Popović D B, "Cognitive vision system for the control of a dexterous prosthetic hand: An evaluation study" J NeuroEng Rehabil 2010, 7:42 doi:10.1186/1743-0003-7-42.
10. Struijk L N, "An inductive tongue computer interface for control of computers and assistive devices," *IEEE Trans. Biomed. Eng.,* vol. 53, pp. 2594-2597, Dec. 2006.
11. Struijk L N, Lontis E R, Bentsen B, Christensen H V, Caltenco H A and Lund M E, "Fully integrated wireless inductive tongue computer interface for disabled people," Conf. Proc. IEEE Eng. Med. Biol. Soc., vol. 2009, pp. 547-550, 2009.
12. Caltenco H A, Lontis E R, Boudreau S, Bentsen B, Struijk L N, "Tip of the tongue selectivity and motor learning around the palatal area" Submitted to: IEEE Transactions on Biomedical Engineering. (Provisionally accepted October 2010)

Author: Daniel Johansen
Institute: Department of Health Science and Technology
Street: Fredrik Bajers Vej 7D E1-111
City: 9000 Aalborg
Country: Denmark
Email: Djoh@hst.aau.dk

Model-Based Medical Decision Support – A Road to Improved Diagnosis and Treatment?

S. Andreassen[1], D. Karbing[1], U. Pielmeier[1], S. Rees[1], A. Zalounina[1], Line Sanden[1], M. Paul[2], and L. Leibovici[2]

[1] Center for Model-Based Medical Decision Support, Aalborg University, Aalborg, Denmark
[2] Rabin Medical Center, Petah Tiqva, Israel

Abstract— **The hypothesis is advanced that model-based medical decision support is a methodology, which may be appropriate for construction of medical decision support systems. The methodology, which is based on a combination of structural modeling and decision theory is outlined through three medical applications.**

All three applications have been brought to the point where they can be subjected to rigorous clinical testing. One of them, Treat for advising on antibiotic treatment, belongs to the small group of systems for which it has even been possible to show an improvement in patient outcome.

Keywords— **Medical decision support, decision theory, antibiotic treatment, mechanical ventilation, glycaemic control.**

I. Introduction

The development of IT for the health care sector in general and hospitals in particular is driven by the underlying assumption that IT eventually will improve the quality and safety of health care. Medical Decision Support (MDS) have attracted attention and funding as one of the possible ways to achieve this, but clinicians have argued that we are "Still waiting for Godot" [1] and that "systems that are in use in multiple locations, that have satisfied users, and that effectively and efficiently contribute to the quality and safety of care are few and far between" [1]. A review by Garg [2] identified 100 MDS systems, which had been subjected to randomized clinical trials. Out of those only 51 determined the effect of the systems on patient outcome and only 7 reported a positive effect. It was not stated how many of those 7 systems are in routine clinical use.

These considerations motivate speculations on the reasons for the meager results. Some of the reasons are outside the narrow field of MDS. For example, problems and delays in implementation the Electronic Health Record, which is the natural source of patient data for MDS, has made it more difficult to integrate MDS into the clinical workflow. However, this paper will offer speculations on how some of the popular methodologies for construction of MDS systems may contribute to the problems and will point to model-based methodologies as a promising alternative.

The paper will also review progress achieved using a model-based approach combined with decision theory to construct MDS systems in several important clinical areas, including control of hyperglycaemia by insulin administration to patients in the intensive care unit (ICU) (Glucosafe) [3-12], optimization of mechanical ventilation in ICU patients (INVENT) [13-17] and choice of antibiotic treatment in patients with severe infections (Treat) [18-21]. The model-based methodology will be explained using these three applications and results of clinical trials of these systems will be outlined.

II. The model-based methodology

A. A Motivation for a Model-Based Approach

A wide range of methodologies have been used to construct MDS systems. Several of these methodologies rely solely or mostly on data in the form of patient cases from which the system is constructed by supervised learning. This applies to methodologies like neural nets or automatic rule induction and for these methods to work, the number and representativeness of the cases must be sufficient to match the complexity of the decisions to be made. For the applications considered in this paper, the authors have felt that the procurement of a sufficient number of cases for the supervised learning to proceed satisfactorily would prove difficult. This is mainly due to the high dimensionality of the input data, which for example for the Treat system includes more than 150 data items.

An alternative could be to construct MDS systems manually, based on insight into the clinical decision process. This approach is used in the construction of rule-based systems and their cousins, fuzzy logic systems. These approaches have the advantage, that knowledge or insight into the problem at hand can be used in their construction, but the disadvantage that it is difficult to integrate knowledge with the

K. Dremstrup, S. Rees, M.Ø. Jensen (Eds.): 15th NBC on Biomedical Engineering & Medical Physics, IFMBE Proceedings 34, pp. 257–260, 2011.
www.springerlink.com

use of case data. In addition, reasoning with rule-based systems is brittle in the sense that exceptions to the rules are difficult to handle. Pearl [22] showed that this weakness is due to the lack of a representation of causality in rule-based systems.

Based on these observations the hypothesis was formed that a model-based approach is an appropriate methodology for construction of MDS systems. In this approach a model of the relevant physiology or pathophysiology is used to predict responses to treatment and decision theory is used to determine which treatment gives the best expected outcome.

B. *Structural Models*

In the proposed methodology only structural models will be considered, where we define a structural model as a model which attempts to capture the structure of the relevant (patho)physiology. For example, Glucosafe contains a compartment model of insulin pharmacokinetics and this we considered to be a structural model. In contrast, a scoring system or a logistic regression model is not considered a structural model.

Structural models have the advantage that their structure can be determined from insight, for example knowledge of relevant anatomy, physiology and pharmacology. In the Glucosafe system this insight is used to determine the structure of a set of ordinary coupled differential equations. These differential equations may contain many parameters, and in a clinical case with limited patient data, most of these will be unidentifiable. However, quite often most of these parameters can be estimated from experimental data published in the scientific literature, leaving only a small number of patient specific parameters, which then hopefully can be identified from typically available clinical data. Glucosafe only has one patient specific parameter, called insulin sensitivity, which can be readily estimated from measurements of blood glucose and data on past infusions of insulin and nutrition. This is an example of how structural models allow both knowledge and data to be integrated into the model.

The Glucosafe system is a deterministic model, which does not contain any explicit representation of uncertainty. In contrast the model used in the Treat system is a stochastic model. This is necessary, if the stochastic component plays a very strong role in the model and this is the case in Treat, since the ability of an antibiotic to eradicate an infection will always remain a question of probabilities. Treat is implemented as a causal probabilistic network, also called a Bayesian network. This is a technology, which allows structural stochastic models to be constructed in a manner which to the model builder does not feel much different from building deterministic models, except that variable have been replaced by stochastic variables and parameters with conditional probabilities.

C. *Decision Theory*

In principle it is a simple two stage process to use decision theory for example to choose between different therapeutic alternatives for a given patient. In the first step all relevant treatments are enumerated and a model is used to predict the outcome, i.e. the expected state of the patient for each of these treatments. In the second step utilities (positive) or penalties (negative) are assigned to the possible outcomes and the difference between expected utility and expected penalty is calculated for each treatment. Here the word expected is used to indicate that if the model is stochastic, then the expected utility should be calculated as a probability weighted sum of the utility. This corresponds to calculating the value of a lottery ticket as:

$$Value = \Sigma_i \, (Price_i \, {*}Prob_i) \qquad (1)$$

where $Price_i$ is the value of each of the i prices in the lottery and $Prob_i$ is the associated probability of winning each price. Usually the outcome space is not one-dimensional as in a lottery, but has several dimensions. For example, when ventilating a patient mechanically, it may be possible to increase the oxygenation of a patient's blood by increasing the oxygen concentration of the inspired air. Well oxygenated blood is beneficial and we have assigned a utility to oxygenation of blood, reflecting how well oxygenated the blood is. At the other hand, oxygen in high concentrations is toxic, and will lead to lung injury. Therefore we have assigned a penalty, which increases with the oxygen concentration in the inspired air. The treatment with the highest value will thus be the treatment which strikes the best compromise between the desire to achieve a high oxygenation of the patient's blood and the risk of inducing lung injury due to oxygen toxicity.

The proposed methodology can be summarized in the following steps: a) For a given clinical application a model of the relevant physiology is constructed. b) The values of different treatments are calculated by assigning utilities and penalties to outcome variables and according to decision theory, the treatment with the highest value is chosen. c) The resulting system is tested clinically, preferably in a randomized controlled trial.

III. Results obtained with the Methodology

The methodology has been applied to construct and test MDS systems in the three clinical areas mentioned in the introduction. A brief description of the systems will be given along with a description of the clinical trials of the systems.

A. Glucosafe

The clinical problem: Glucosafe is designed to control blood glucose in patients in the ICU. More than half of these patients, up to 71%, develop hyperglycaemia, even without previously diagnosed diabetes [3]. In several large landmark studies it was shown that tight control of blood glucose reduced relative ICU mortality by up to 45% and in-hospital mortality by up to 34% [4,5,6], although a less clear picture has emerged from later studies. In these studies hyperglycaemia is reduced by insulin therapy, but the improvement in blood glucose comes at the expense of an increased risk of hypo-glycaemia, which is an independent risk factor for death [7,8]. The purpose of the Glucosafe MDS system is to achieve a substantial reduction in blood glucose without increasing the risk of hypo-glycaemia.

A problem in some ICU patients is that their insulin sensitivity is so reduced, that blood glucose can only be normalized if the patient's intake of carbohydrates is reduced; however, a prolonged shortage in carbohydrates leads to a risk of underfeeding the patient.

The MDS system: The Glucosafe system is implemented as a set of ordinary, non-linear coupled differential equations, representing an underlying compartment model of insulin pharmacodynamics and glucose absorption and metabolism [9]. The system requires data on insulin infusions, enteral and parenteral nutrition and blood glucose measurements every one to two hours. Glucosafe includes penalties on hypo- and hyper-glycaemia, on excessive use of insulin, on underfeeding and on too low doses of enteral feed, which may lead to decay of the intestinal endothelium [10,11].

Clinical Trials: Glucosafe has been tested in a pilot trial with a before/after design with results indicating improved control of blood glucose [12]. Currently Glucosafe is undergoing a randomized clinical trial to show that reduced blood glucose can be achieved in ICU patients without increasing the risk of hypoglycaemia.

B. Invent

The clinical problem: The primary goal of mechanical ventilation is to secure gas exchange achieving sufficient oxygenation of arterial blood and sufficient removal of carbon dioxide. Gas exchange can be improved by increasing tidal volume, respiratory frequency and/or by increasing airway pressure. In addition, oxygenation can be improved by increasing the fraction of oxygen in inspired air (FiO_2). A multicenter randomized controlled clinical trial [13] has shown that careful setting of the ventilator, in particular low tidal volumes, can reduce mortality in the ICU, but there is still little clinical agreement on the choice of ventilator settings [14].

The MDS system: INVENT includes models of oxygen and carbon dioxide transport and storage, and a linear model of lung mechanics. The input to the system is current ventilator settings, measurements of airway pressures and expired gas concentrations and arterial blood gas measurements. The output from the system is advice on revised ventilator settings and a prediction of the state of the patient, when the ventilator settings have been changed.

The assessment of which ventilator setting is optimal for the patient is based on the predicted state of the patient and a set of penalty functions which quantify clinical preferences. Clinical preferences for securing oxygenation and removal of carbon dioxide are represented by penalty functions associated with blood oxygen contents and pH of blood, respectively. The risk of VILI is represented by a penalty function associated with high levels of FiO_2 to penalize oxygen toxicity, and a penalty function associated with high airway pressures and respiratory frequencies to penalize the risk of causing mechanical trauma to the lungs [15,16].

Clinical trials: A prototype of INVENT for managing inspired oxygen level has been prospectively evaluated in an ICU showing that INVENT suggested appropriate levels of FiO_2 in comparison to clinicians in attendance [17]. The MDS system has also been retrospectively evaluated for managing FiO_2, tidal volume and respiratory frequency in cardiac surgery patients [16] showing that INVENT advice on these three settings were in accordance with the goals of securing oxygenation, normalising pH of blood and avoiding high airway pressures.

C. Treat

The clinical problem: TREAT is a system for advising on antibiotic therapy for patients with severe infections. The clinical problem is that initially about 40% of these patients receive inappropriate treatment, i.e. antibiotic treatment, which can not eradicate the infection. This is a major clinical problem, because mortality is quite high, typically 10-15% and because inappropriate initial treatment approximately doubles the odds ratio for death. A secondary problem is that many patients receive treatment with broad-spectrum antibiotics, prone to promote bacterial resistance.

The MDS system: Treat is implemented as a Causal Probabilistic Network, which models symptoms from infected organs, sepsis and the effect of antibiotics on different types of bacteria. Input to the system is patient signs and symptoms, underlying non-infectious conditions, lab and imaging data. Output from the system is an assessment of the severity of the infection, of the organ(s) likely to be infected and the bacteria likely to cause the infection. A utility is associated with survival 30 days after the onset of the infection, and penalties are associated with the cost of purchasing antibiotics, with the side-effects caused by antibiotics and with the estimated ability of different antibiotics to promote bacterial resistance, the so-called ecological cost. Treat provides the user with a ranked list of the antibiotic treatments giving the overall largest difference between the expected utility and the expected costs.

Clinical trials: Treat has been tested on more than 5000 patients in 4 countries and Treat has been able to increase the fraction of patients receiving appropriate treatment, both in countries with low and with high levels of antibiotic resistance. The largest trial of Treat was a cluster randomized controlled clinical trial in 3 different countries with 2326 patients showing that Treat can both improve the fraction of patients receiving appropriate antibiotic treatment, and reduce the cost of purchasing the antibiotics and the ecological cost. An significant reduction in bed-day was shown and a reduction in mortality was also seen, although the study was not powered to give a statistically significant reduction in mortality. Treat has been installed in a Danish University hospital for regular clinical use and is currently being installed at another Danish hospital.

IV. Conclusions

A methodology for construction of MDS systems has been proposed and it has been applied to three different clinical problems. The hypothesis that model-based medical decision support may be an appropriate methodology has been confirmed in the sense that it has been possible to develop three systems and bring them to the point where rigorous clinical testing has been possible. One of the systems, Treat, has been subjected to a cluster randomized controlled clinical trial and results have shown that Treat can substantially improve antibiotic treatment and thereby patient outcome in countries with very different levels of bacterial resistance.

Given the modest number of MDS systems where improvements in patient outcome has been shown, this indicates that it may be worthwhile to pursue model-based medical decision support as a methodology for construction of MDS systems.

References

1. Wears RL, Berg. JAMA 2005;293(10):1261. Computer Technology and Clinical Work: Still Waiting for Godot
2. Garg AX, Adhikari NKJ, McDonald H, et al. JAMA (2005); 293(10):1223-1238; doi:10.1001/jama.293.10.1223)
3. Capes SE, Hunt D, Malmberg K, Gerstein HC (2000) Stress hyperglycaemia and increased risk of death after myocardial infarction in patients with and without diabetes: a systematic overview. Lancet 355:773-778
4. Krinsley JS (2004) Effect of an intensive glucose management protocol on the mortality of critically ill adult patients. Mayo Clin Proc 79:992-1000
5. Chase JG, Shaw G, LeCompte A et al (2008) Implementation and evaluation of the SPRINT protocol for tight glycaemic control in critically ill patients: a clinical practice change. Crit Care 12(2):R49
6. van den Berghe G, Wouters P, Weekers F et al (2001) Intensive insulin therapy in the critically ill patients. N Eng J Med 345:1359-1367
7. Griesdale DE, de Souza RJ, van Dam RM et al (2009) Intensive insulin therapy and mortality among critically ill patients: a meta-analysis including NICE-SUGAR study data. CMAJ 180:821-827
8. Preiser JC, Devos P, Ruiz-Santana S et al. (2009) A prospective randomised multi-centre controlled trial on tight glucose control by intensive insulin therapy in adult intensive care units: the Glucontrol study. Intensive Care Med 35:1738-1748
9. Pielmeier U, Andreassen S, Nielsen, BS et al (2010) A simulation model of insulin saturation and glucose balance for glycemic control in ICU patients. Comput Methods Progr Biomed 97:211-222
10. Lara TM, Jacobs, DO (1998) Effect of critical illness and nutritional support on mucosal mass and function. Clin Nutr 17(3):99-105
11. Pielmeier U, Boudreau S, Andreassen S (2010) A decision-theoretic approach to consistent tight glycemic control in critical care patients. UKACC Int. Conf. on Control, 7-10 September 2010, Coventry, UK, pp. 839-844.
12. Pielmeier U, Andreassen S, Juliussen B et al. (2010) The Glucosafe system for tight glycemic control in critical care: a pilot evaluation study. J Crit Care 25(1):97-104
13. The Acute Respiratory Distress Syndrome (ARDS) Network (2000) Ventilation with lower tidal volumes as compared with traditional tidal volumes for acute lung injury and the acute respiratory distress syndrome. N Engl J Med 342:1301–1308
14. Allerød C, Karbing DS, Thorgaard P et al. (2011) Variability of preference toward mechanical ventilator settings: A model-based behavioral analysis. J Crit Care "in press"
15. Rees SE, Allerød C, Murley D et al. (2006) Using physiological models and decision theory for selecting appropriate ventilator settings. J Clin Monit Comput 35:421-429
16. Allerød C, Rees SE, Rasmussen BS et al. (2008) A decision support system for suggesting ventilator settings: Retrospective evaluation in cardiac surgery patients ventilated in the ICU. Comput Methods Programs Biomed 92:205-212
17. Karbing DS, Allerød C, Thorgaard P et al. (2010) Prospective evaluation of a decision supports system for setting inspired oxygen in intensive care patients. J Crit Care 25:367-374
18. Andreassen S et al. 2005. A probabilistic network for fusion of data and knowledge in clinical microbiology. In: Probabilistic Modelling in Bioinformatics and Medical Informatics. (Ed.: Dybowski)
19. Paul M et al. 2006. Prediction of bacteremia using TREAT, a computerized decison support system. Clin Infect Dis 42:1274-82.
20. Paul M et al. 2006. Improving empirical antibiotic treatment using TREAT, a computerised decision support system: a cluster randomised trial. J Antimicrob Chemother 58:1238-45.
21. Leibovici L et al. 2007. The TREAT project: decision support and prediction using causal probabilistic networks. Int J Antimicrob Agents 30:93-102.
22. Pearl J (1988) Probabilistic reasoning in intelligent systems: networks of plausible inference, Morgan Kaufmann Publishers, San Mateo

Corresponding Author: Steen Andreassen
Institute: Center for Model-Based Medical Decision Support
Street: Fredrik Bajersvej 7
City: Aalborg
Country: Denmark
Email: sa@hst.aau.dk

Nerve Conduction Velocity Selective Recording Using a Multi-contact Cuff Electrode – A Case Study of In-Vitro Vagus Nerve Preparation

G.A.M. Kurstjens

Center for Sensory-Motor Interaction (SMI), Department of Health Science and Technology, Aalborg University, Aalborg, Denmark

Abstract— **In this study, a method for nerve conduction velocity selective recording using a multi-contact cuff electrode is evaluate using an in-vitro vagus nerve preparation. A delay adder only and a delay adder followed by a matched filter tuned to the different components of the elicited compound action potentials (CAP) were applied to extract the nerve conduction velocity (CV). Both filters were able to single out the conduction velocity of the fastest component of the elicited CAP. Selective tuning to the CV of the other components was only possible when considering corresponding sub-regions of interest for each particular component of the CAP.**

Keywords— **Multi-contact cuff electrode, vagus nerve, conduction velocity selective filter.**

I. Introduction

Nerve cuff electrodes are extraneural electrodes that can provide a stable neural interface for long-term recording of peripheral nerve activity [1]. They have therefore become a realistic alternative to artificial sensors for provide feedback or control signals in neuroprosthetic devices [2].

Although the traditional (tripolar) nerve cuff electrode recordings are considered to provide only a single channel of information, the amplitude and frequency spectrum of the recorded signals depend on the contact spacing as well as the conduction velocity of the nerve fibers recorded from. As such, the tripolar cuff configuration can be regarded as a signal velocity dependent bandpass filter [3].

This let to the introduction of a special multi-contact (MC) cuff electrode capable of perform conduction velocity selective recordings (VSR) [4]. This method consists of using an array of double differential amplifiers, artificial time delays for each selected conduction velocity, an adder and bandpass filters. Two later publications have described experimental evaluations of this method in frog peripheral nerve, were the conduction velocity of first one [5] and then two [6] fiber groups was estimated. More recently, Yoshida et al [7] evaluated the VSR technique using a new experimental model where a MC cuff electrode was places around an entire earthworm. In this way, a simplified peripheral nerve model containing only two fibers with distinctly different conduction velocities was obtained, allowing working with 'single fiber' action potentials rather that compound action potentials as before.

In the present study, the vagus nerve is for the first time explored as model for experimental evaluation of the velocity-selective recording technique. The advantage of this model is that the vagus nerve contains only four distinct groups of nerve fibers: large (≥ 10 um), medium (3-9 um) and small (< 3um) myelinated fibers and unmyelinated fibers [8], which will simplify the analysis of compound nerve activity.

II. Methods

A. *Experimental Set-Up*

An approximately 10 cm long section of a left cervical vagus nerve was excised from a sacrificed pig. The nerve was placed in a bath that was filled with oxygenated Krebs solution at room temperature (21 °C) for in-vitro experimentation [9]. Two cuff electrodes were then placed on the nerve: a tripolar stimulation cuff and a multicontact recording cuff electrode containing 11 contacts. The two cuff electrodes had both an inner diameter of 3 mm, contained 1 mm wide, circumferential platinum foil contacts with a pitch of 3mm, and were separated with a distance of 30 mm between their outer contacts. An electrode contact mounted on the outside of the MC cuff served as reference.

Compound nerve action potentials were evoked using a stimulator consisting of two synchronized current sources with a common cathode to apply single monophasic 400 μs long constant current pulses. Signals recorded from the MC cuff were amplified and filtered (total gain: 10,000x, band-pass: 10 Hz – 5 kHz) using a differential preamplifier and amplifier system (SMI, Aalborg University). The resulting ENG signals were sampled in sweeps of 50 ms, starting 1 ms before onset of stimulation, using a PC with data acquisition system (National Instruments) with a sample frequency of 50 kHz.

K. Dremstrup, S. Rees, M.Ø. Jensen (Eds.): 15th NBC on Biomedical Engineering & Medical Physics, IFMBE Proceedings 34, pp. 261–263, 2011.
www.springerlink.com

B. Protocol

First, the stimulation intensities for activation threshold and supra-maximal activation of the different fiber groups were established. Second, ENG responses elicited by 10 stimuli of each of the selected stimulation current amplitudes were recorded from each two subsequent MC electrode contacts in bipolar configuration. However, since a multi-channel amplifier with the necessary number of channels was not available, only one single differential amplifier was used to record from one contact pair at a time and multi-channel signal recording was simulated by offline post-processing.

C. Signal Processing and Analysis

The experimentally recorded bipolar signals were digitally post-processed to obtain tripolar signals. First, the conduction velocities of the different fiber groups was estimated considering the time delay of the corresponding negative peak in the signals from the most proximal and most distal tripole, and the distance between both tripoles.

The velocity selective filter was implemented by delaying each tripolar channel by a multiple of the channel number and a constant time delay dt, followed by a summation of the delayed outputs. After summation, the output was quantified by the peak amplitude of the rectified CAP response, Vp. So-called tuning curves were generated by doing this for a range of different time delays. To further enhance selectivity, the output of the delay adder was past through three different matched filters (MF1 to MF3), which were created using template regions-of-interest (ROI) corresponding to CV specific components of the elicited CAP [7]. Output of the filters was normalized to the CAP amplitude of the first tripole to quantify improvement in selectivity. Offline processing and analysis were performed in Matlab (The Mathworks, Inc).

III. RESULTS

The CAP's recorded from the MC cuff electrode consisted of up to three different components (labelled I-III in Fig 1), depending on the current level used for stimulation. The first two components had a low activation threshold (respectively 0.1 and 0.5 mA) and were identified as belonging to type Aα and Aβ fibers having conduction velocities of, respectively, 31 and 18 m/s. The third component had the highest threshold (2.0 mA), and was identified as belonging to type B fibers having a CV of 5 m/s. No component originating from type C fibers was observed.

Figure 2 shows tripolar signals constructed from the bipolar signals recorded during supra-maximal stimulation (I = 3.0 mA). The tripole number indicates its position along the cuff electrode. Separation between individual CAP components increases as during propagation through the cuff, resulting from the different fiber group CV's.

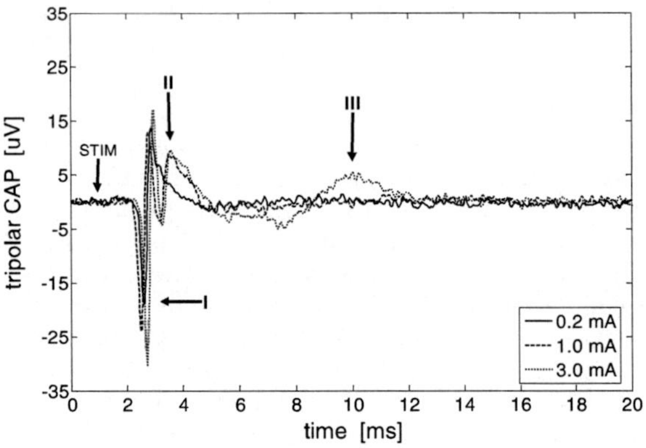

Fig. 1 Averaged CAP responses elicited by stimulation with different current levels, showing recruitment of up to three components (N=10). Arrow with label STIM indicates onset of stimulation at time = 1 ms

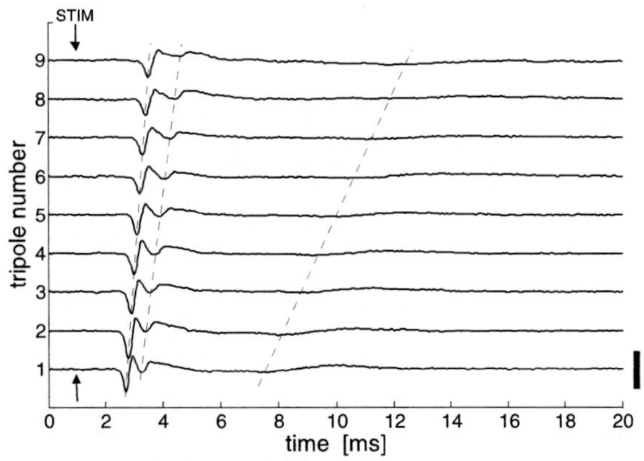

Fig. 2 Averaged tripolar signals recorded from the MC cuff electrode when a current amplitude of 3 mA (N=10). Onset of stimulation (STIM) indicated by arrows at time = 1 ms. The dashed lines mark the propagation of different CAP components through the cuff over time. Scale bar: 50 μV

The effect of the delay adder filter is shown in Fig 3. Tuning curves for the whole CAP as ROI and using three different current levels (Fig 3A) showed correctly the direction of propagation (increased output for positive delay times only). One distinct peak at dt = 0.1 ms was found. Considering the contact pitch of the cuff electrode was 3 mm, this translates to a CV of 30 m/s which corresponds to the fastest component of the recorded CAP (Fig. 1). Two smaller increases can also be seen, corresponding to the two longer latency CAP components. However, when considering smaller ROI's containing only the response of an individual CAP component, then single peaks at dt = 0.1, 0.18 and 0.64 ms were found, corresponding to the appropriate component CV's of respectively 30, 16.7 and 4.7 m/s.

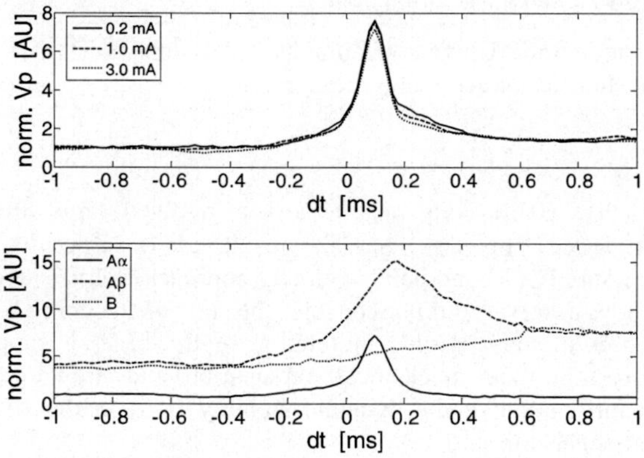

Fig. 3 Filter tuning curves for the delay adder, using full CAP ROI at different current levels (upper plot) and when using individual CAP component ROI's (lower plot, I = 3.0 mA).

Fig 4 shows that the application of a matched filter on the output of the delay adder resulted in a small improved in CV selectivity when using the whole CAP as ROI: the output of filters MF2 and MF3 was lower than MF1 and tuning curves peaks of MF1 and MF2 gave near correct CV estimates of 30 and 16.7 m/s, but MF3 gave a too large CV estimate of 18 m/s. Output of MF1 and MF2 did give largest tuning peaks for their respectively designed A fiber component, but the output by MF3 was largest for the Aβ fibers in stead of the B fiber component.

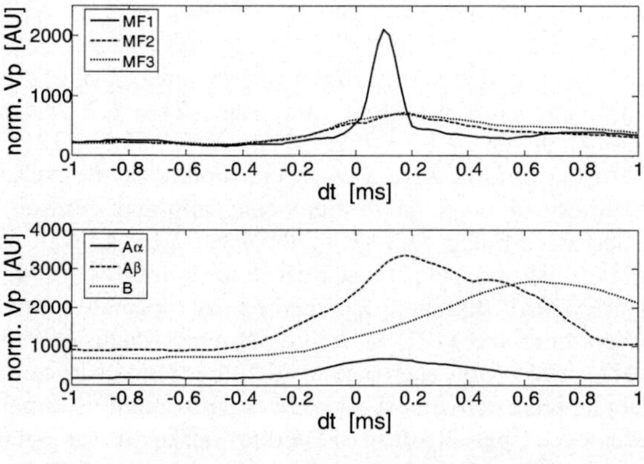

Fig. 4 Filter tuning curves for delay adder plus three different matched filters (MF1-3) each tuned to one of the three CAP components (upper plot), and output of MF3 for individual component ROI's (lower plot).

IV. Conclusions

Both conduction velocity selective filters were able to estimate the conduction velocity of the fastest component of elicited CAP's. Tuning selectively to one individual fiber group response was only possible when considering sub-ROI's in the CAP's, and estimated CV's were close to the measured CV's. Further in-vivo evaluation with natural nerve activity is needed because both filters may perform different as action potentials from different fibers appear randomly and not sequential as when elicited electrically.

Acknowledgment

The author would like to thank Aleksandra Vuckovic for use of the in-vitro preparation. This work was partially supported by the Danish National Research Foundation, the Danish Research Council, and the Danish National Advanced Technology Foundation.

References

1. Stein RB, Nichols TR, Jhamandas J et al. (1977) Stable longterm recordings from cat peripheral nerves. Brain Res 128:21–38
2. Sinkjaer T, Haugland M, Struijk JJ, and Riso R (1999) Long-term cuff electrode recordings from peripheral nerves in animals and humans. In: Windhorst U and Johansson H (eds) Modern techniques in neuroscience, Springer Verlag, p 787-802
3. Struijk, JJ (1997) On the spectrum of nerve cuff electrode recordings CDROM Conf Proc IEEE Eng Med Biol Soc, Chicago, USA.
4. Taylor J, Donaldson N, Winter J (2004) Multiple-electrode nerve cuffs for low-velocity and velocity-selective neural recording. Med Biol Eng Comput.42(5):634-643
5. Rieger R, Taylor J, Comi E, et al. (2004) Experimental determination of compound action potential direction and propagation velocity from multi-electrode nerve cuffs. Med Eng Phys. 26(6):531-534
6. Schuettler M, Seetohul V, Taylor J, et al. (2006) Velocity-selective recording from frog nerve using a multi-contact cuff electrode. Conf Proc IEEE Eng Med Biol Soc 1:2962-2965
7. Yoshida K, Kurstjens GA, Hennings K, (2009) Experimental validation of the nerve conduction velocity selective recording technique using a multi-contact cuff electrode, Med Eng Phys. 31(10):1261-70.
8. Schnitzlein HN, Rowe LC, Hoffman HH (1958) The myelinated component of the vagal nerve in human, Anat Rec/Am. Assoc. Anat. 131:649-667
9. Vuckovic A, Struijk JJ & Rijkhoff N (2004).Diameter selective nerve fiber stimulation in the vagal nerve using anodal block, depolarising prepulses and long exponentially rising pulses. Conf Proc 9th Ann. Conf. IFESS, Bournemouth, UK. Wood,D ; Taylor,P (eds.) , 330-332.

Author: Mathijs Kurstjens
Institute: Center for Sensory-Motor Interaction (SMI), Dept. Health Science and Technology, Aalborg University.
Street: Fredrik Bajers Vej 7-D3
City: 9200 Aalborg
Country: Denmark
Email: mku@hst.aau.dk

Evidence of Feedforward Postural Adjustments to Reduce Knee Joint Loading in ACL Deficient Patients at Cost of Dynamic Stability Control

K.D. Oberländer[1], K. Karamanidis[1], J. Höher[2], and G.-P. Brüggemann[1]

[1] Institute of Biomechanics and Orthopaedics, German Sport University Cologne, Germany;
[2] Clinic for Sports Traumatology at Merheim Medical Center, Cologne, Germany

Abstract— **The aim of this study was to examine the effect of anterior cruciate ligament (ACL) deficiency on joint kinetics and dynamic stability (DS) during landing after a single-leg hop test (SLHT). Twelve unilateral ACL deficient (ACLd) subjects performed a SLHT (both legs). Calculation of landing mechanics was done by means of a soft tissue artifact optimized rigid full body model. Margin of stability (MoS) was defined by the differences between the base of support and extrapolated centre of mass (X_{CoM}). During landing, the ACLd leg showed a lower external knee flexion and adduction moment but generated higher ankle dorsiflexion and hip flexion moments compared to the healthy leg. The kinetic changes at the joints were explained by a increased forward lean of the trunk resulting in a more anterior position of the centre of mass causing an anterior translation of the ground reaction force vector with respect to the joints of the lower extremity in the ACLd leg. The consequence of this ACLd-related control strategy was a greater X_{CoM} reducing the MoS during landing. Our results give evidence of a feedforward adaptive adjustment in the ACLd leg aimed to reduce mechanical loading at the knee joint at a cost of lesser DS control.**

Keywords— **Landing, anterior cruciate ligament, postural control, knee joint moment.**

I. Introduction

Ruptures of the anterior cruciate ligament (ACL) are common among athletes with approximately 80.000 injuries in the United States every year [1]. Living with an ACL deficient knee increasing the risk for suffering chronic joint disease e.g. articular cartilage damage [2]. Changes in joint kinetics are often proposed as a relevant issue for the initiation of degenerative joint changes in the ACL deficient (ACLd) knee. While walking the ACLd leg compared to the healthy leg (control) shows a lower magnitude of external knee flexion moment but higher ankle and hip joint moments providing evidence of a motor task reorganizing aimed to reduce knee joint mechanical loading [3, 4]. It is reported that ACLd patients showed a poorer ability to reach a stable body posture at the involved limb after a single-leg hop test (SLHT) quantified by a centre of pressure (CoP) and pelvis velocity approach [5]. The above study, however, did not consider the state of the entire body center of mass (CoM) in relation to the base of support (BoS) in their mechanical explanation and, therefore, a quantification of the dynamic stability (DS) in the ACLd patients is missing.

Based on these literature reports it seems reasonable to hypothesise that an ACLd patients may use at the involved limb a motion strategy aimed to reduce knee joint loading at a cost of DS control while performing functional motor tasks such as landing. This could result in a greater risk for further musculoskeletal injuries due to deficient DS control.

Therefore, the aim of the current study was to examine the effect of ACL deficiency on lower extremity joint kinetics and DS control during landing after a SLHT. In the current study we examined the landing phase of a SLHT because this task enables the analyses of each leg separately with an increased task demand.

II. Methods

A. Subjects

Twelve patients suffering from unilateral ACL rupture participated in the current study (age: 27 ± 6.2 yrs. mean and SD; height: 1.82 ± 0.1m; body mass: 83.5 ± 12.6 kg). Inclusion criteria were a complete rupture of the ACL and no injury of other major ligaments, minimal miniscii and articular cartilage damage of the knee joint as well as no other history of musculoskeletal injuries and no neurological disease (documented by specialist medical practitioner and MRI) as well as no pain. Inclusion criteria were proven for each subject by the same surgeon. All subject were active in sport were leg movement is important like soccer, basketball or ski. Ethics approval was obtained from the local institution and all participants provided written informed consent before enrolment.

K. Dremstrup, S. Rees, M.Ø. Jensen (Eds.): 15th NBC on Biomedical Engineering & Medical Physics, IFMBE Proceedings 34, pp. 264–267, 2011.
www.springerlink.com

B. Experimental design

All subjects performed a modified SLHT for distance keeping (putting) ones hand's on one's hip and wear sport shoes [6] This hop is performed with one leg over a given distance of 0.75*height. Landing has to be on the force plate (40cm *60cm, Kistler, Winterthur, Switzerland) within a target area corresponding to the jumping distance ± 5 cm. Subjects have to perform at least 5 to 10 SLHT's with both legs. The hop was deemed correct by the experimenter if the subject was able to reach the target area during initial contact while maintaining balance as long as possible after landing.

C. Kinematic and kinetic analysis

In order to determine landing mechanics we used a multi-segment system with 13 segments (head, torso, pelvis, upper arms, forearms, thighs, shanks and feet). Following anatomical landmarks were identified with respect to a corresponding cluster of 4 Markers (diameter 0.016m) by using a calibrated pointer: right and left femoral epicondyles, distal and medial apex of the malleolus. The hip joint centers were calculated by a functional calibration method SCoRE [7] during a StarArc movement [8] by minimizing the objective Function Eq. 1. This leads out a least square problem Eq. 2 and could be solved by using pseudo inverse approach [9].

$$f_{SCoRE}(c_1, c_2) = \sum_{i=1}^{n} \left\| R_i c_1 + p_i - (S_i c_2 + q_i) \right\|^2 \tag{1}$$

$$\begin{pmatrix} R_1 & S_1 \\ \vdots & \vdots \\ R_n & S_n \end{pmatrix} \begin{pmatrix} c_1 \\ c_2 \end{pmatrix} = \begin{pmatrix} q_1 - p_1 \\ \vdots \\ q_n - p_n \end{pmatrix} \tag{2}$$

With c_1, c_2 being the centers of rotation in the local coordinate system and (R_i, p_i), (S_i, q_i) the transformations from the local to the global systems of the adjoining segments. Remaining anatomical landmarks were identified by skin-mounted markers. To overcome soft tissue artifacts during the movement's lower extremity marker positions as well as the head and pelvis were optimized by the following least-squares problem:

$$\min_{R,d} \sum_{m=1}^{n} \left\| R x_m + d - y_m \right\|^2 \tag{3}$$

with x_m being the markers 3-D position in the neutral stance and y_m being the Marker position during the movement of the corresponding segment. A *Singular Value Decomposition* (SVD) Algorithm was used to solve Eq. (3) to get the rotation Matrix R and the translation vector d that maps the anatomical frames defined during neutral stance to the movement [10].

Lower extremity joint moments were calculated by inverse dynamics:

$$M_j = -M_F - (r_j \times F) - \sum_{i=1}^{n} (r_{ji} \times G_i) + \sum_{i=1}^{n} (r_{ij} \times \dot{p}_i) + \sum_{i=1}^{n} \dot{H}_i \tag{4}$$

where M_j is the jth joint moment, M_F the frictional moment, F the ground reaction force, r_j the position vector between j^{th} joint and the point of force application, r_{ij} the position vector between j^{th} joint and centre of mass of ith segment, G_i the force of gravity on the ith segment, $\dot{p}_i$ the first derivative of the ith segment impulse, $\dot{H}_i$ first derivative of the ith segment angular momentum, and n the number of segments.

DS during the landing phase was quantified for both legs using an inverted pendulum model by calculating the margin of stability (MoS) as the instantaneous differences between the anterior boundary of the base of support (BoS) and the extrapolated center of mass (X_{CoM}) defined as follows [11]:

$$X_{CoM} = P_{CoM} + \frac{V_{CoM}}{\sqrt{g/l}} \tag{5}$$

where P_{CoM} is the anteroposterior component of the projection of the CoM to the ground, V_{CoM} is the anteroposterior CoM velocity, g is the acceleration of gravity, and l is the distance between CoM and centre of the ankle joint in the sagittal plane.

For the statically analysis the mean values from the 5 to 10 trials were utilized for each participant and leg. In order to get relevant information about motion mechanics during different intervals of the landing phase the kinematic and kinetic variables were determined at means over 5 equal intervals. The landing phase was defined from the first foot contact after jumping (vertical ground reaction force > 10N) to the point where all landing energy was absorbed ($0 > F_x v_x + F_y v_y + F_z v_z$, where F_i is one component of the ground reaction force (GRF) an v_i the corresponding CoM velocity, $i \in \{x, y, x\}$). A t-test for dependent samples was used to identify possible differences on the kinetic data, on the components of the DS and on the MoS at the 5 intervals. A linear regression model was used in order to determine the relationship between DS control and knee joint kinetics.

III. RESULTS AND DISCUSSION

During the landing phase, the ACLd leg showed a lower joint moment output at the knee joint but generated higher ankle and hip moments in the sagittal plane compared to the control leg ($p<0.05$; Table 1). Our findings confirm the literature reports [3, 4] showing a redistribution of joint moment output at the lower extremity during dynamic motor task due to ACL deficiency [3].

The magnitude of the GRF showed no significant differences ($p>0.05$) between legs. Thus, the mechanical explanation for our observed joint moment changes in the ACLd leg was related to a more anterior position of the GRF vector relative to the joints of the lower extremity for the ACLd leg (i.e. greater lever arms at the ankle- and hip joint, shorter lever arms at the knee joint; $p<0.05$). We found that at the ACLd leg the subject landed with a increased forward lean of the trunk ($p<0.05$) caused by postural feed forward adaptations (i.e. increased forward lean of the trunk was present at initial portion of the ground contact) positioning the point of force application under the foot more anterior ($p<0.05$) during following ground contact phase and thereby causing the changes in lower extremity joint kinetics (strong positive correlation between the sagittal lever arm of the GRF at the knee joint and the forward lean of the trunk; $r^2=0.47$, $p<0.001$; Fig. 1). However, the consequence of the altered landing strategy in the ACLd leg (i.e. increased forward lean of the trunk) was a more anterior position of CoM resulting in a greater X_{CoM} and, thereby, reducing the MoS during landing ($p<0.05$) in the ACLd leg compared to the control leg (Fig. 2).

IV. Conclusions

Our results give evidence of a feedforward adaptive adjustment in the ACLd leg (i.e. increased forward lean of the trunk) aimed to reduce mechanical loading at the knee joint at a cost of lesser DS control. We concluded that ACL deficient knee is at greater risk for further musculoskeletal injuries due to diminished ability to control DS during landing maneuvers.

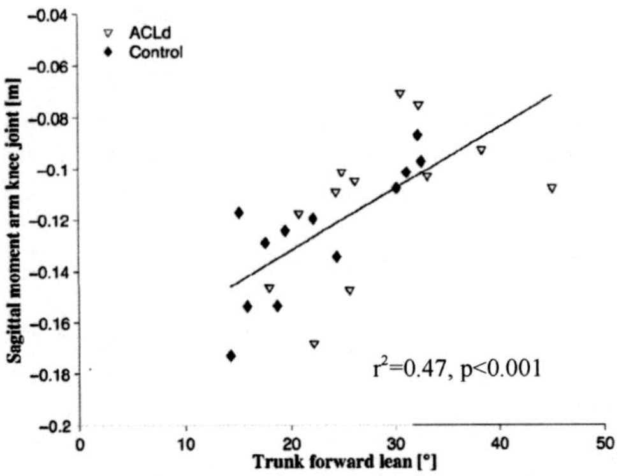

Figure 1: Relationship between trunk flexion angle at the initial landing phase (mean values over first 20% of absorption phase) and the mean sagittal moment arm at the knee joint at the terminal interval of the landing phase (81 to 100%).

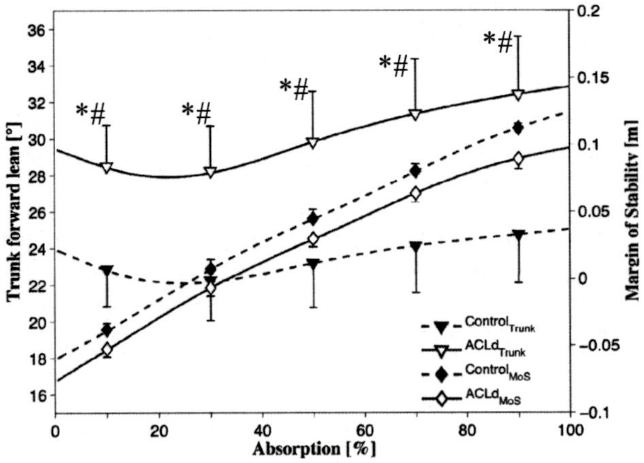

Figure 2: Forward lean of the trunk and MoS during the landing phase after a SLHT. Solid line represent the ACLd leg, dashed line the control leg. Statistically significant differences ($p<0.05$) between ACLd and control leg were marked with * for the MoS and # for the trunk flexion angle.

Table 1: Lower extremity joint moments [Nm/kg] during different intervals of the landing phase after a single leg hop for the anterior cruciate deficient leg (ACLd) as well for the contra lateral leg (Control).

Moment [Nm/kg]	Landing Phase									
	0% to 20%		21% to 40%		41% to 60%		61% to 80%		81% to 100%	
	ACLd	Control	ACLd	Control	ACLd	Control	ACLd	Control	ACLd	Control
$Ankle_{Flex}$	0.06 ±0.26	0.05 ±0.39	0.52 ±0.38	0.38 ±0.53	0.93 ±0.31	0.71 ±0.41	1.14 ±0.24	0.96 ±0.27	1.13 ±0.18	1.01 ±0.20
$Knee_{Flex}$	0.24 ±0.34	0.22 ±0.39	1.89 ±0.64	1.79 ±0.66	2.40 ±0.74	2.71 ±0.71	2.21 ±0.76	2.58 ±0.70	1.92 ±0.65	2.30 ±0.62
Hip_{Flex}	1.11 ±0.55	0.80 ±1.02	1.55 ±0.70	1.43 ±0.65	1.51 ±0.43	0.85 ±0.58	1.59 ±0.48	1.06 ±0.49	1.42 ±0.46	0.89 ±0.40
$Knee_{Add}$	0.40 ±0.14	0.41 ±0.27	0.80 ±0.19	0.83 ±0.41	1.10 ±0.28	1.30 ±0.31	1.09 ±0.28	1.30 ±0.25	1.06 ±0.30	1.19 ±0.20

REFERENCES

1. Griffin LY, Agel J, Albohm MJ et al. (2000) Noncontact anterior cruciate ligament injuries: risk factors and prevention strategies. J Am Acad Orthop Surg 8: 141:150
2. Fithian DC, Paxton L, Golizt D (2002) Fate of the anterior cruciate ligament-injured knee. Orthop Clin North Am. 33:621-636
3. Lewek M, Rudolph K, Axe M et al., (2002) The effect of insufficient quadriceps strength on gait after anterior cruciate ligament reconstruction. Clin Biomech. 17: 56-63,
4. Rudolph K, Axe M, Buchanan T et al. (2001) Dynamic stability in the anterior cruciate ligament deficient knee. Sports Traumatol. Arthrosc. 9:62-71
5. Phillips N, van Deursen R (2008) Landing stability in anterior cruciate ligament deficient versus healthy individuals: a motor control approach. Phys. Ther. Sport. 9:193-201
6. Daniel, DM., et al., Am J Knee Surg. **1**:212–214, 1988
7. Ehrig R, Taylor W, Duda G et al. (2006) A survey of formal methods for determining the centre of rotation of ball joints. J Biomech. 39: 2798-2809
8. Camomilla V, Cereatti A, Vannozzi G et al. (2006) An optimized protocol for hip joint centre determination using the functional method. J Biomech. 39:1096-1106
9. Deuflhard P, Hohmann A (2003). Numerical Analysis in Modern Scientific Computing: An Introduction. Springer, Berlin
10. Söderkvist I, Wedin P (1993) Determining the movements of the skeleton using well-configured markers. J Biomech. 26:1473-1477
11. Hof A (1982) An explicite expression for the moment in multibody systems. L Biomech. 25: 1209-1211
12. Hof A, Gazendam M, Sinke W (2005) The condition for dynamic stability. J Biomech. 38:1-8

Author:	Kai Daniel Oberländer
Institute:	Institute of Biomechanics and Orthopaedics
Street:	Am Sportpark Müngersdorf 6
City:	50933 Cologne
Country:	Germany
Email:	k.oberlaender@dshs-koeln.de

Reactive Response and Adaptive Modifications in Dynamic Stability to Changes in Lower Limb Dynamics in the Elderly while Walking

K. Karamanidis[1], F. Süptitz[1], M.M. Catalá[1], J. Piiroinen[2], K.D. Oberländer[1], J. Avela[2], and G.-P. Brüggemann[1]

[1] Institute of Biomechanics and Orthopaedics, German Sport University Cologne, Germany
[2] Department of Biology of Physical Activity, Neuromuscular Research Center, University of Jyväskylä, Finland

Abstract— **The aim of this study was to examine the reactive responses and adaptive modifications in dynamic stability resulting from a unilateral change in lower limb dynamics in older and younger adults while walking. Eleven older (62-76yrs) and eleven younger (22-30yrs) subjects walked on a treadmill and performed different gait conditions using an external resistance against lower limb movement. The margin of stability (MoS) at touchdown was calculated as the difference between base of support (BoS) and extrapolated centre of mass. After the resistance was turned on unexpectedly, older adults needed more steps to get back to the MoS baseline level due to a lower increase of the BoS. In the following protocol, a continuous resistance was applied over 11 consecutive steps. Adaptation level in MoS and BoS was lower in the early adaptation phase (trial 1-3) but not in the late adaptation phase (trial 9-11) for the older compared to the younger adults. After removing the resistance, both groups showed similar aftereffects (i.e. increased BoS). Our results indicate that elderly preserve their ability to recalibrate their feedforward motor commands to control dynamic stability during perturbed walking. However, the rate of adaptive improvements and feedback driven postural modifications is diminished in the elderly, increasing the risk of falling.**

Keywords— **Aging, falls, dynamic stability control, gait.**

I. Introduction

Human gait is a mechanically complex task and the patterns must be flexible enough to accommodate changing environmental demands and task constraints. As a consequence, effective postural modifications are required to produce successful and safe gait patterns without loss of stability. Such modifications in motor task execution take place on different time scales; while some are immediate reactions to a novel situation, others are slower adaptive changes that last longer. Reactive or feedback-driven movement corrections occur quickly, using ongoing afferent feedback information. Slower adaptive modifications require practice of the novel situation and result in storage of the new movement pattern. They result in new calibration of feedforward motor commands, seen as aftereffects that persist upon return to the original condition [1].

Previous work has documented that cognitive [2], and skeletal muscle-tendon [3] functions gradually decline with advancing age, and it would, therefore, not be surprising if the ability to use appropriate feedback corrections and the ability to adapt also decline. Accordingly, a number of studies have shown that, in general, older adults have a slower motor learning process and reduced adaptive improvement under a novel task constraint [4, 5]. However, most studies analyzing adaptation potential in the elderly are based on discrete upper-limb tasks and such findings do not necessarily apply to postural tasks such as gait. Falls during ongoing gait after a sudden postural perturbation are a major health threat in the elderly and, therefore, analyzing the motor learning processes during walking may be important for the implementation of early effective intervention to reduce the risk of falls in the elderly population. Based on upper-limb tasks, it is known that changes in the dynamics of the limb due to the application of an external resistance against limb movement will result in movement errors due to changes in the sensory input from the limb [1]. In the current study we used similar task constraints for the lower limbs, aimed at examining the reactive response and adaptive modifications in dynamic stability resulting from a unilateral change in lower limb dynamics in older and younger adults while walking. We hypothesized that older compared to younger adults will show (i) a less effective feedback-driven postural adjustment after an unexpected change in lower limb dynamics and (ii) a slower motor learning process and reduced magnitude of adaptive improvement in dynamic stability as a result of a sustained change in walking condition.

II. Methods

Eleven older women aged between 62 and 76 years and eleven younger women aged between 22 and 30 years with similar anthropometric data participated in the study. Women were chosen as it is known that they have a greater

K. Dremstrup, S. Rees, M.Ø. Jensen (Eds.): 15th NBC on Biomedical Engineering & Medical Physics, IFMBE Proceedings 34, pp. 268–270, 2011.
www.springerlink.com

risk for falls during daily activities than elderly men. All subjects had to walk on a motor driven treadmill with a belt speed of 1.4 m/s. A custom-built device including an electric driven brake-and-release system was used in order to unexpectedly apply and remove a resistance of 2.1 kg on the lower right limb during the swing phase while walking. For this reason a Teflon rope connected with the brake-and-release system was secured with Velcro straps around the right leg of the subjects just above the ankle joint. The gait protocol consisted of three Blocks starting with the baseline trials (Block 1: resistance was turned off while walking), followed by the reactive response trials (Block 2: resistance was turned on during the swing phase of the right leg for one step) and finally the adaptation trials (Block 3: resistance was turned on during the swing phase of the right leg for 11 consecutive steps followed by an additional step without resistance in order to examine aftereffects). Between Blocks 2 and 3, the resistance was always turned off and sufficient time (typically 2 to 4 minutes) was given for each subject to provide a "wash-out" effect of the postural adjustments experienced due to the applied resistance. The "wash-out" effect was controlled for each individual by an online expectation of the anteroposterior displacement of the toe markers while walking in relation to the base line level by using a motion capturing system (Vicon system, Oxford, UK). Subjects were not warned about the application or removal of the resistance.

Kinematic data while walking were recorded with a Vicon motion capture system using eight cameras operating at 120 Hz. Twenty-six reflective markers were fixed on anatomical landmarks on the skin to track a twelve segment full body model. The margin of stability (MoS) in anteroposterior direction while walking was determined using the "extrapolated centre of mass" concept provided by Hof et al. [5] and was calculated as the difference between the anterior boundary of the base of support (BoS, i.e. horizontal component of the projection of the toe from the corresponding limb to the ground) and the extrapolated center of mass (X_{CM}) in the anteroposterior direction. X_{CM} was defined as follows [5]:

$$X_{CM} = P_{XCM} + \frac{V_{XCM} + |V_{BoS}|}{\sqrt{g/l}}$$

where P_{XCM} is the horizontal (anteroposterior) component of the projection of the centre of mass (CM) to the ground, V_{XCM} is the horizontal (anteroposterior) CM velocity, V_{BoS} is the horizontal (anteroposterior) BoS velocity (approximately equal to the velocity of the treadmill), g is the acceleration of gravity, and l is the distance between CM and the centre of the ankle joint in the sagittal plane. Postural stability is maintained in circumstances where the position of the X_{CM} is within the BoS (positive values of MoS) while stability is lost in cases where the X_{CM} exceeds the anterior boundary of the BoS. All components of the dynamic stability were analyzed at the instants of touchdown (TD). TD was identified by the acceleration of the tibia determined by 2D accelerometers (1080 Hz). All events were further visually checked by the same examiner using a video camera (60 Hz). The mean values of eight consecutive baseline trials were used to determine the baseline level. For the reactive response condition in Block 2, the disturbed right leg and the following six consecutive steps (left and right) were analyzed in order to determine the reactive response in dynamic stability during ongoing gait. For the adaptation trials in Block 3, resistance trials 1-3 as well as resistance trials 9-11 were pooled together as representative of the early and late adaptive adjustments in dynamic stability, respectively. A two-way analysis of variance (with age and trial as factors) with repeated-measures was used in order to examine the age and trial-related differences in the analyzed dynamic stability parameters. Post-hoc testing (Duncan's Test) was applied for the pairwise comparison.

III. Results and Discussion

Compared to the baseline level, the MoS at TD of the right leg after the resistance was turned on unexpectedly decreased in both age groups significantly ($p<0.05$) with no significant differences ($p>0.05$) between older and younger adults (Figure: Block 2). This means that the consequence of the applied perturbation on dynamic stability was similar for both age groups, leading to a clear unstable body configuration at TD (i.e. negative values of MoS). However, the analysis of the following consecutive steps revealed that the older adults needed significantly more steps in order to get back to the baseline level (on average three more steps; Figure: Block 2), showing that the feedback corrections were less effective in the elderly. The main reason for this was the reduced ability of the elderly to increase their BoS following the unexpected change in lower limb dynamics ($p<0.05$).

During the following adaptation trials (Figure: Block 3), the MoS as well as the BoS at TD of the weighted right leg showed lower values in both the early and late adaptation phases compared to the baseline value ($p<0.05$) independent of the subject's age (no age-effect). However, a continuous increase in the MoS and BoS at TD from the first perturbation in Block 2 (reactive response trial), to the early until the late adaptation phase was present for the younger adults ($p<0.05$) but not for the older adults (i.e. no significant differences in MoS or BoS between the first unexpected perturbation and the early adaptation phase for the elderly;

p>0.05). Accordingly, there was a significant ($p<0.05$) age-effect in the MoS and BoS at TD for the early adaptation phase (lower values for the elderly) but not for the late adaptation phase ($p>0.05$). The above findings demonstrate that the older subjects achieved the same adaptation level as the younger ones after performing all 11 weighted gait trials; however, they adapted more slowly than the younger ones. After removing the resistance from the limb, both age groups showed clear aftereffects manifested in a significant increase ($p<0.05$) of the BoS at TD compared to the baseline level with a similar magnitude between age groups.

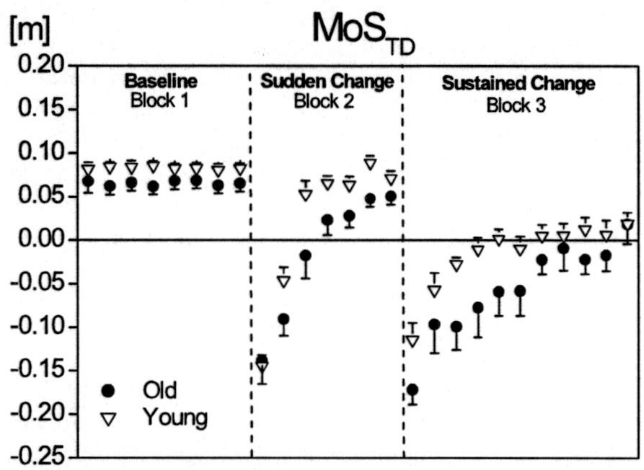

Figure: Margin of stability at touchdown (MoS_{TD}) for the older and younger adults for the three gait conditions (mean and standard error of mean). Block 1: resistance was turned off while walking; Block 2: resistance was turned on for one step of the right leg; Block 3: resistance was turned on for 11 consecutive steps of the right leg. Note that for Block 2, in addition to the single step with the disturbed right leg, the following six consecutive recovery steps (left and right) were analyzed.

IV. Conclusions

Our results provide evidence that older adults preserve their ability to recalibrate their feedforward motor commands to control dynamic stability during perturbed walking. However, the rate of adaptive improvements to a sustained change in lower limb dynamics and feedback-driven compensatory adjustments in dynamic stability are diminished in the elderly population, increasing the risk of falling while walking.

References

1. Sainburg RL, Ghez C, Kalakanis D (1999) Intersegmental dynamics are controlled by sequential anticipatory, error correction, and postural mechanisms. J Neurophysiol. 81:1045-56
2. Raz N, Williamson A, Gunning-Dixon F, Head D, Acker JD (2000) Neuroanatomical and cognitive correlates of adult age differences in acquisition of a perceptual-motor skill.. Microsc Res Tech. 51:85-93
3. Karamanidis K, Arampatzis A (2005) Mechanical and morphological properties of different muscle-tendon units in the lower extremity and running mechanics: effect of aging and physical activity. Exp Biol. 208 :3907-23
4. Fernández-Ruiz J, Hall C, Vergara P, Díiaz R (2000) Prism adaptation in normal aging: slower adaptation rate and larger aftereffect. Brain Res Cogn Brain Res. 9:223-6
5. Buch ER, Young S, Contreras-Vidal JL (2003) Visuomotor adaptation in normal aging. Learn Mem. 10:55-63
6. Hof A, Gazendam M, Sinke W (2005) The condition for dynamic stability. J Biomech. 38:1-8

Author: Kiros Karamanidis
Institute: Institute of Biomechanics and Orthopaedics
Street: Am Sportpark Müngersdorf 6
City: 50933 Cologne
Country: Germany
Email: karamanidisr@dshs-koeln.de

Gait Modulation for the Reactive Recovery of Balance

A.S. Oliveira[1,2], L. Gizzi[3,4], D. Farina[1,3], and U.G. Kersting[1]

[1] Center for Sensory-Motor Interaction, Aalborg University, Aalborg, Denmark
[2] The CAPES Foundation, Brasilia, Brazil
[3] Department of Neurorehabilitation Engineering, George-August University, Göttingen, Germany
[4] Dipartimento di Scienze del Movimento Umano e della Salute (DiSMUS) Università degli studi di Roma "Foro Italico"

Abstract— **Falls have a high impact worldwide with respect to health care and employer costs. In the last years there were advances in understanding mechanisms for balance recovery during walking but little is known about motor patterns during gait when experiencing slips. The aim of this study was to investigate gait pattern modulation during slips while walking. Eight healthy subjects walked along a walkway with a moveable force platform embedded in the center. Subjects stepped with the right foot on the platform which elicited perturbations at heel strike. Surface electromyography was collected from lower limbs, trunk and neck muscles, from which motor modules and activation coefficients were extracted by non-negative matrix factorization. Comparisons were made between normal, unperturbed gait and perturbed trials. Five modules were sufficient to account for more than 80% of the variation for both the normal and perturbed walking. Moreover, a robust inter-subject similarity for normal and perturbed walking (r=0.81±0.1 and r=0.79±0.1 respectively), as well as high similarity (r>0.75) between modules of normal and perturbed walking were found for four out of five modules. The second module for perturbed walking had lower similarity in comparison to modules of normal gait (r~0.70), which may represent the reactive responses from the perturbation during early stance phase.. It can be concluded that the recovery of balance during walking may not require higher complexity from CNS, which recruited mostly similar modules to perform the task.**

Keywords—**locomotion, balance, EMG activity, modularity.**

I. Introduction

Humans present specific strategies to regain balance after perturbations by activating appropriate muscles to relocate the body's center of mass with respect to the area of support, especially leg and trunk muscles [1, 2]. In addition, the exposure to a loss of balance may possibly change muscular activation [2, 3]. However, the way the CNS organizes these activation outputs is not well understood. The *muscle synergies* theory – i.e., a set of muscles that are activated simultaneously from coherent neural output - is a possible description of this organization in a modular way. Movements such as walking may usually be described by a set of four to five motor modules (synergies) [4], which remain mostly the same for walking at different speeds [5], and other combined tasks while walking [6].

These motor modules were successfully used to explain the recovery of balance during standing. Results were robust across subjects, describing a low-dimensional modularity controlling balance in different directions [7]. In addition, walking over slippery surfaces implies overall changes in muscular activation, by increased contributions of sensory inputs and changes in kinematics generating a unique but altered gait pattern [2]. This study was the first about motor modules and balance during walking. However, the loss of balance may imply a foot displacement such as during slips backwards, evoking a wide variety of responses from reflex to voluntary compensation to maintain the original walking task. Thus, the modular organization for regaining balance when experiencing slips during walking remains unknown.

The aim of the present study was to investigate whether the modular organization of gait while slipping can be described by a low-dimensional set of motor modules. We investigated this by having subjects walking on a level surface which had a moveable platform embedded in its center. We hypothesized that the modular organization when experiencing slips is similar to the organization verified during normal walking. Any alterations will be possible to describe by minor changes to existing modules or addition of a small number of additional motor modules related to regaining balance during the stance phase of walking.

II. Methods

A. Participants

Eight healthy men (age, 28±4 yrs; body mass, 71±10 kg; stature, 171±7 cm) volunteered for the experiment. All subjects provided written informed consent before participation and the procedures were approved by the ethical committee of Northern Jutland (N-20100042).

B. Experimental design

The experiment consisted of walking at each individuals preferred speed along a 7-m walkway. A force platform was positioned in the middle of the walkway. Subjects were asked to step with the right foot on the center of the force platform and continue walking. After familiarization, subjects performed 10 unperturbed trials at their preferred walking speed to establish a normal walking pattern. Following these trials, perturbations in the anterior-posterior directions (forward/backward) were elicited while walking. The force platform could be activated to perform translational movements (10-cm translation at 66.7 cm/s^{-1}). Subjects experienced catch trials and backward perturbations in random order. For this task, 10-12 trials were performed, totalizing 20-24 randomized trials.

C. Data recording

Kinematics. Retroreflective ball-shaped markers were affixed bilaterally to the skin over the heel, first and fifth metatarsal. Marker positions were tracked by a motion analysis system with eight infrared digital video cameras at 256Hz (Oqus 300, Qualisys, Gothenburg, Sweden).

Ground reaction forces. A three-dimensional force platform constructed over a hydraulic system (van Doornik and Sinkjær, 2007) provided perturbation stimuli and simultaneous measures of ground reaction forces and moments. Using a feedback electric circuit, the vertical force (*Fz*) forces served as trigger signals to initiate the force plate movement once the subject achieved a correct foot contact.

Surface EMG. Signals were recorded in bipolar derivations. The EMG signals were amplified with a gain of 2,000 (EMG-USB, LISiN; OT Bioelettronica, Turin, Italy), A/D converted at 12 bits per sample and band-pass filtered (second-order, zero lag Butterworth, bandwidth 10–450 Hz), sampled at 2048 Hz. A reference electrode was placed at the right wrist. The EMG signals were recorded unilaterally (right side) from the following muscles: peroneus longus (PER), gastrocnemius lateralis (GL), soleus (SOL), tibialis anterior (TA), vastus lateralis (VL), rectus femoris (RF), biceps femoris (BF), tensor fascia latae (TFL), gluteus maximus (GLU), rectus abdominis (RAB), external oblique (EOB), erector spinae at L1 (ESP), medial deltoideus (MD), upper trapezius (TRP), splenius capitis (SPL) and sternocleidomastoideus (SCM).

D. Data analysis

The gait cycle was defined as the time between two successive right heel strikes. EMG signals respective to gait cycles from 16 muscles were full-wave rectified, low pass filtered (10 Hz) and time interpolated in order to obtain 200 samples length for each trial [4]. Non negative matrix factorization algorithm (NMF) [8] was used in order to identify motor modules and activation coefficients.

Motor modules model. In the model we used, the EMG signals recorded from *M* muscles can be expressed as:

$$X(k) = [x_1(k), x_2(k), \cdots, x_M(k)]^T \quad (1)$$

where $x_m(k)$ is the activity of the *m*th muscle at the time instant *k*.

Muscle electrical activity depicts the activity of α-motor neurons from the spinal cord. The activation signals $P(k)$ were less than the number of muscles ($N<M$):

$$P(k) = [p_1(k), p_2(k), \cdots, p_N(k)]^T \quad (2)$$

The envelopes of muscle activities were obtained from the activation signals by linear transformation with gain factors s_{mn}. The matrix whose columns were the weights of each activation signal for each muscle was denoted as *S* and referred to as the motor module matrix (Lee et al. 2001). The relation between *X(k)* and *P(k)* was described as follows:

$$X(k) \approx X_r(k) = S \cdot P(k) \quad (3)$$

where $X_r(k)$ is the muscle activity vector reconstructed by the module matrix and the activation signals.

Dimensionality

The number of motor modules needed for accurate description of the movement was assessed by the dimensionality analysis proposed by d'Avella et al. [9]. According to this procedure, the quality of reconstruction of the muscle activation pattern is analyzed as a function of the number of modules and the minimum number of modules is identified as the point in which this curve changes its slope [9]. However, a minimum threshold for reconstruction quality was set at 80%. For quantifying the quality of reconstruction, the estimated muscular activation pattern was compared with the recorded pattern by means of the variation accounted for (VAF) value, defined as the variation that can be explained by the model: VAF = 1 – SSE/SST, where SSE (sum of squared errors) is the unexplained variation and SST (total sum of squares) is the total variation of the data.

The module matrices were compared by computing the scalar product of pairs of columns of the matrix *S* and normalizing by the product of the norms of each column. The degree of similarity between modules extracted for the unperturbed gait and perturbed gait was computed.

The activation signals were compared by cross-correlation at zero lag. Motor modules were first extracted from single trials for each condition and compared among trials for the same subject and condition. Similarity among individuals and conditions was calculated by the scalar product between pairs of columns of the matrix *S* and normalizing by the product of the norms of the (respective) columns. After computation of reconstruction quality and motor module similarity, motor modules were extracted from the concatenation of all trials in one condition and similarity between unperturbed and perturbed gait for the same individual and among different individuals in the same condition. Finally, the concatenation of all EMG envelopes for one condition was used to extract motor modules representative of each condition for the whole group of subjects. This way, all the variability in the dataset was taken into account. The presence of shared motor modules among conditions was investigated by computing specific similarities of modules. Each module of one condition was compared with all other modules of another condition until the best match was found.

III. Results

Figure 1 shows representative EMG traces for lower limbs, trunk and neck muscles for the perturbed and nonperturbed condition. Increased EMG activity is elicited during and after the perturbation event, returning to normal levels before the end of stance phase. Five motor modules were typically required to reconstruct unilateral muscular activation for both tasks. Reconstruction quality for normal gait was 0.85±0.02% and 0.73±0.03% using single trials and concatenated signal, respectively. Similarly, reconstruction quality for perturbed gait was 0.86±0.02% and 0.78±0.03% using single trials and concatenated signals. There was a high similarity among subjects, r=0.81±0.14 and r=0.79±0.14 for normal and perturbed gait trials.

For normal gait, the first module mainly consisted of TA, knee extensors and flexor activation during early stance and late swing phases of normal walking (Figure 2), likely to support the body during load acceptance. The second module consisted mainly of calf muscles (GL and SOL, PER), for body support and forward propulsion. The third module consisted mainly of trunk and neck muscles activation during early swing phase. The fourth module consisted mainly of hip/trunk muscles activated during stance phase and in the middle of swing phase and fifth module consists mainly of TA, RF and trunk/neck muscles, late in the swing phase.

For backward slips (Figure 3), the first module is also related to load acceptance The second module is related to the first reaction to the slip, by activating pairs of antagonist muscles in the ankle (TA, SOL) and knee (BF, VL), as well as hip and trunk muscles. The third module provided propulsion to initiate the swing phase. The fourth module consisted of trunk and neck muscle activations, but it maintains activation for some trunk muscles through the swing phase. Finally, the last module for the perturbed walking consists mainly of RA and upper trunk/neck muscles, slightly activating TA, RF and TFL in response to the perturbation and also during swing phase.

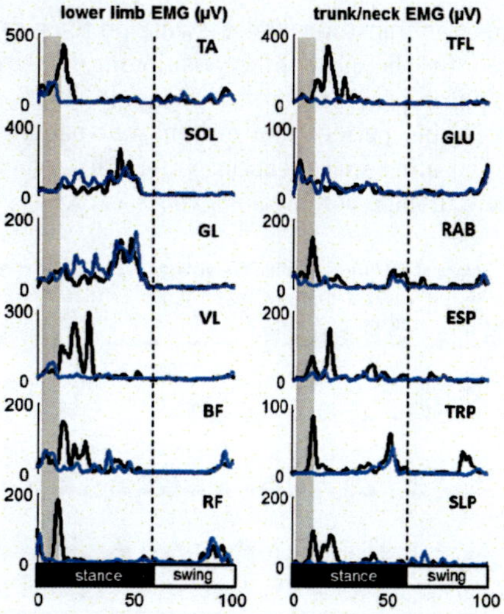

Figure 1. Representative EMG traces during normal walking (blue traces) and perturbed walking (black traces). Gray area represents the perturbation period of 150 ms. Refer to section IIC for muscles nomenclature.

Similarities between modules of normal gait and perturbed gait are shown in Table 1. A high similarity for all modules can be verified with the lowest similarity found between module 4 of normal gait and module 2 of perturbed gait.

IV. Discussion and Conclusion

We hypothesized that by eliciting a perturbation at heel strike additional modules to the original motor program of walking would to be required to describe muscle activations. However, our results indicate that by using the same number of modules, it is possible to reconstruct walking. Modules related to load acceptance, body support, propulsion and preparation for touch-down are not changed during slips. However, the module related to stance/swing transition is altered. During normal gait the third module consisted of trunk/neck muscles activation, in order to position the trunk in anticipation of the swing phase. Backward slips induce changes in trunk kinematics, positioning the center of mass backwards and it must be reversed [1, 7]. On the other hand, the contralateral side is still within the swing phase and its transition to the next step is changed [7].

Table 1. Mean (SD) similarity between modules of normal walking and perturbed walking (Back-slips).

Normal walking	Back-slips	Similarity (r)
Module 1	Module 1	0.78±0.20
Module 2	Module 3	0.85±0.11
Module 3	Module 4	0.83±0.22
Module 4	Module 2	0.70±0.11
Module 5	Module 5	0.83±0.10

Thus, the transition stance-swing on the perturbed side is affected by the need of repositioning the trunk and the swing phase is compromised, changing the modular organization in this period. The normal gait pattern described is similar to other investigations [4, 6, 10], which is essential to assure that the comparisons were correct.

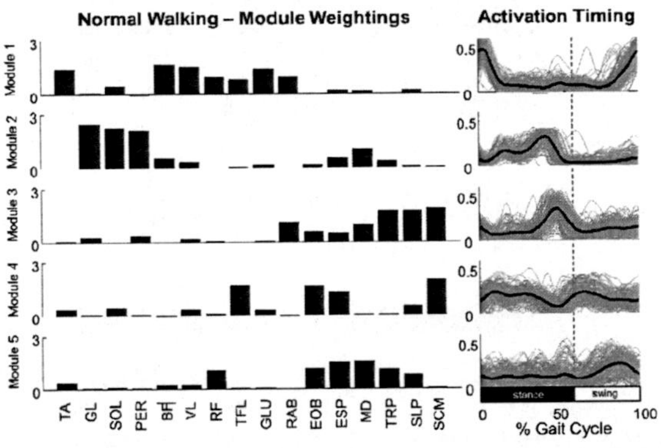

Figure 2. Concatenated module muscle weighting coefficients of individual muscles and averaged activation timing (black thick line) and single trials (gray thin lines) for each of the modules during normal walking.

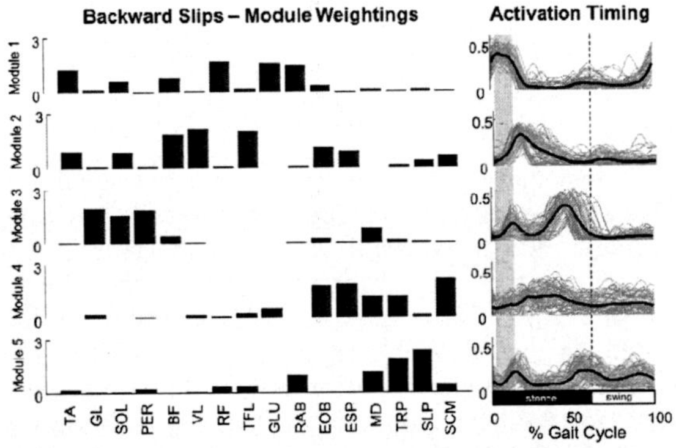

Figure 3. Concatenated module muscle weighting coefficients of individual muscles and averaged activation timing (black thick line) and and single trials (gray thin lines) for each of the modules during perturbed walking (light gray area indicates perturbation period).

Voluntary tasks may present similar modules to maintain walking, while one module may allow for the additional movement to maintain dynamic balance [6]. In case of perturbations the activation can be described by a different module. This different module immediately decreases the risk of falls by stiffening lower limbs and trunk. However, as a result of the trunk displacement and contralateral adjustments the pattern of the swing phase may change. Consequently, the modules are slightly altered. It is concluded that the reactive recovery of balance may be represented by a low-dimensional set of modules, which are similar across subjects and mostly similar to those from normal walking.

References

[1] S. L. Jones, S. M. Henry, C. C. Raasch, J. R. Hitt and J. Y. Bunn, "Responses to multi-directional surface translations involve redistribution of proximal versus distal strategies to maintain upright posture," *Experimental Brain Research,* vol. 187, pp. 407-417, 2008.

[2] G. Cappellini, Y. P. Ivanenko, N. Dominici, R. E. Poppele and F. Lacquaniti, "Motor patterns during walking on a slippery walkway," *J. Neurophysiol.,* vol. 103, pp. 746-760, 2010.

[3] P. F. Tang, M. H. Woollacott and R. K. Y. Chong, "Control of reactive balance adjustments in perturbed human walking: roles of proximal and distal postural muscle activity," *Experimental Brain Research,* vol. 119, pp. 141-152, 1998.

[4] Y. P. Ivanenko, R. E. Poppele and F. Lacquaniti, "Five basic muscle activation patterns account for muscle activity during human locomotion," *J. Physiol. (Lond.),* vol. 556, pp. 267-282, 2004.

[5] Y. P. Ivanenko, R. E. Poppele and F. Lacquaniti, "Motor control programs and walking," *Neuroscientist,* vol. 12, pp. 339-348, 2006.

[6] Y. P. Ivanenko, G. Cappellini, N. Dominici, R. E. Poppele and F. Lacquaniti, "Coordination of locomotion with voluntary movements in humans," *Journal of Neuroscience,* vol. 25, pp. 7238-7253, 2005.

[7] G. Torres-Oviedo and L. H. Ting, "Muscle synergies characterizing human postural responses," *J. Neurophysiol.,* vol. 98, pp. 2144-2156, 2007.

[8] S. Muceli, A. T. Boye, A. d'Avella and D. Farina, "Identifying Representative Synergy Matrices for Describing Muscular Activation Patterns During Multidirectional Reaching in the Horizontal Plane," *J. Neurophysiol.,* vol. 103, pp. 1532-1542, 2010.

[9] A. d'Avella, P. Saltiel and E. Bizzi, "Combinations of muscle synergies in the construction of a natural motor behavior," *Nat. Neurosci.,* vol. 6, pp. 300-308, 2003.

[10] D. J. Clark, L. H. Ting, F. E. Zajac, R. R. Neptune and S. A. Kautz, "Merging of healthy motor modules predicts reduced locomotor performance and muscle coordination complexity post-stroke," *J. Neurophysiol.,* vol. 103, pp. 844-857, 2010.

Author: Anderson de Souza Castelo Oliveira
Institute: Center for Sensory-Motor Interaction, AAU
Street:Fredrik Bajers vej 7E
City:Aalborg
Country: Denmark
Email: oliveira@hst.aau.dk

Author Index